The IMA Volumes
in Mathematics
and its Applications

Volume 89

Series Editors
Avner Friedman   Robert Gulliver

Springer Science+Business Media, LLC

Institute for Mathematics and its Applications
IMA

The **Institute for Mathematics and its Applications** was established by a grant from the National Science Foundation to the University of Minnesota in 1982. The IMA seeks to encourage the development and study of fresh mathematical concepts and questions of concern to the other sciences by bringing together mathematicians and scientists from diverse fields in an atmosphere that will stimulate discussion and collaboration.

The IMA Volumes are intended to involve the broader scientific community in this process.

Avner Friedman, Director

Robert Gulliver, Associate Director

* * * * * * * * * *

IMA ANNUAL PROGRAMS

Continued at the back

Donald G. Truhlar   Barry Simon
Editors

Multiparticle Quantum Scattering With Applications to Nuclear, Atomic and Molecular Physics

With 55 Illustrations

Springer

Donald G. Truhlar
Department of Chemistry and
  Supercomputer Institute
University of Minnesota
Minneapolis, MN 55455-0431, USA

Barry Simon
Department of Mathematics
California Institute of Technology
Pasadena, CA 91125-0001, USA

Series Editors:
Avner Friedman
Robert Gulliver
Institute for Mathematics and its
  Applications
University of Minnesota
Minneapolis, MN 55455, USA

Mathematics Subject Classifications (1991): 81-02, 81U10

Library of Congress Cataloging-in-Publication Data
Truhlar, Donald, G., 1944–
    Multiparticle quantum scattering with applications to nuclear, atomic,
and molecular physics / Donald G. Truhlar, Barry Simon.
      p. cm. – (The IMA volumes in mathematics and its
applications ; v. 89)
    This work is based on a workshop with the same title which was an
integral part of the 1994–1995 IMA program on 'Waves and
scattering'.
    ISBN 978-1-4612-7318-9   ISBN 978-1-4612-1870-8 (eBook)
    DOI 10.1007/978-1-4612-1870-8
    1. Multiple scattering (Physics) – Congresses.   2. Quantum theory –
Congresses.   3. Many-body problem – Congresses.   4. Mathematical
physics – Asymptotic theory – Congresses.   I. Simon, Barry, 1946–
II. Title.   III. Series.
QC173.4.M85T78   1997
539.7'58 – dc21   97-2459

Printed on acid-free paper.

Production managed by Karina Mikhli; manufacturing supervised by Jacqui Ashri.
Camera-ready copy prepared by the IMA.
9 8 7 6 5 4 3 2 1

ISBN 978-1-4612-7318-9

FOREWORD

This IMA Volume in Mathematics and its Applications

MULTIPARTICLE QUANTUM SCATTERING WITH APPLICATIONS TO NUCLEAR, ATOMIC AND MOLECULAR PHYSICS

is based on the proceedings of a workshop with the same title, which was an integral part of the 1994–1995 IMA program on "Waves and Scattering." We would like to thank Donald G. Truhlar and Barry Simon for their excellent work as organizers of this meeting and as editors of the proceedings. We also take this opportunity to thank the National Science Foundation (NSF), the Army Research Office (ARO), and the Office of Naval Research (ONR), whose financial support made the workshop possible.

Avner Friedman

Robert Gulliver

PREFACE

The workshop on Multiparticle Quantum Scattering with Applications to Nuclear, Atomic, and Molecular Physics was held June 12–16, 1995 at the Institute for Mathematics and Its Applications in the University of Minnesota Twin Cities campus as part of the 1994–95 Program on Waves and Scattering. There were about seventy participants including the plenary lecturers whose contributions are included in this volume. The workshop was preceded by a two-day tutorial featuring lectures by Donald J. Kouri and Gian Michele Graf, and we are pleased that both Professors Graf and Kouri were able to write up their tutorials as opening chapters of this volume.

Multiparticle scattering theory in quantum mechanics is technically complex because of the variety of scattering channels — for example, if one scatters two hydrogen atoms off each other the possible results include two free protons, two free electrons, or a proton and a hydride ion. A key issue in the mathematical foundations is that of asymptotic completeness which says that any state of the quantum system is a superposition of bound and scattering states.

Asymptotic completeness was proven by Kato and Birman in the late 1950's for two-body systems with short-range potentials and by Faddeev in the early 1980's for three-body problems. Going beyond three bodies turned out to be remarkably hard. There was important progress by Balslev and Combes, Enss, Mourre, and Perry, Sigal, and Simon, but it was only in 1985 that Sigal and Soffer solved the problem. Their proof is complex, but there has been a significantly simplified extension by Graf and Yafeev.

While mathematicians have been focused inwards on these foundational questions, users of quantum mechanical scattering — notably physicists studying atoms and nucleii and chemists studying atoms and molecules — have been developing computational techniques, and their calculations raise new mathematical issues concerned with the necessity to enforce explicit boundary conditions, the interaction of composite systems with electromagnetic fields, the properties of effective potentials for many-body systems, and the properties of the scattering matrix for resonances.

One of our goals in this workshop was to increase communication between the two sides, to allow the computational side to make contact with the new results of Sigal, Soffer, and Graf and to stimulate the mathematicians by greater contact with users of the theory and the new mathematical issues arising in their work.

The Workshop was a great scientific success, but there were some unfortunate events on the personal side. One of the original co-organizers, W. Hunziker of Zurich, became seriously ill and could not attend. We offer our hopes for his complete recovery. One of our invited speakers, Jens Bang of Copenhagen, broke six ribs the week before the workshop and could not attend. Fortunately, we are able to include his manuscript. Another of our invited participants, Per-Olov Löwdin of Uppsala, was unable to attend due to the death of his colleague Jean-Louis Calais. We offer our condolences to the family and colleagues of Professor Calais.

We would like to express our thanks and those of all the participants to Avner Friedman and Robert Gulliver of the IMA for their help in organizing and hosting the Workshop. It is also a pleasure to thank the IMA staff, especially Joy Paul Schwenke, for their very special assistance in the running of the Workhop and Patricia V. Brick for her usual expert assistance in the production of these proceedings.

Barry Simon

Donald G. Truhlar

September 1996

CONTENTS

N-BODY QUANTUM SYSTEMS: A TUTORIAL

GIAN MICHELE GRAF*

These notes correspond to a tutorial on N-body quantum systems given at the Institute for Mathematics and its Applications (IMA) of the University of Minnesota in June 95. We tried to keep the exposition elementary, at the expense of a certain lack of precision. Most results are given without proof. More complete reviews, on which the present one is partly based, are [1,2,6,15,24].

1. The geometry of N-body systems. A standard many-particle system consists of N non-relativistic particles moving in $\mathbb{R}^\nu$, $\nu \geq 1$, interacting by two-body potentials which vanish in the limit of large separation. More precisely, they are described by the Hamiltonian

$$(1) \qquad H = \sum_{i=1}^{N} \frac{p_i^2}{2m_i} + \sum_{i<j} V_{ij}(x^i - x^j) \,,$$

with $V_{ij}(y) \to 0$ as $|y| \to +\infty$. Here $m_i > 0$ are the masses, x^i the positions, and p_i the momenta of the particles. The configuration space is the real vector space

$$X := \left\{ x = (x^1, \ldots, x^N) \mid x^i \in \mathbb{R}^\nu \right\} = \mathbb{R}^{\nu N}$$

equipped with the inner product [25]

$$(2) \qquad x \cdot y := \sum_{i=1}^{N} m_i x^i \cdot y^i \,,$$

where $x^i \cdot y^i$ is the standard inner product on $\mathbb{R}^\nu$. We will also use the notation $x^2 = x \cdot x$. This allows to identify the dual X' of X through $x_i = m_i x^i$. An example is the identification of (classical) momenta and velocities, i.e., $p_i = m_i \dot{x}^i$. The Hilbert space of the quantum mechanical N-body system is $L^2(X)$, where the volume element of X is defined by the metric. The particle momenta p_i are the operators acting as $(p_i \psi)(x) = (1/i)\partial\psi/\partial x^i$ and are thus the covariant components of the vector operator p acting as $p\psi = (1/i)d\psi$, with d being the derivative. The kinetic energy then is $\sum_i p_i^2/2m_i = \sum_i p_i p^i/2 = p^2/2$. For the potentials we write $V(x) = \sum_{i<j} V_{ij}(x^i - x^j)$. With these notations (1) just reads

$$H = \frac{1}{2}p^2 + V(x) \,.$$

* Theoretische Physik, ETH-Hönggerberg, CH-8093 Zürich, Switzerland.

A cluster decomposition a is a partition of $\{1, \ldots, N\}$ into disjoint, nonempty subsets, called clusters. An example for $N = 4$ is $a = (13)(24)$, but also $a = (1234)$. For any cluster decomposition a we introduce the external configuration space

$$(3) \qquad X_a := \left\{ x \in X \mid x^i = x^j \text{ if } i, j \in C \text{ for some } C \in a \right\},$$

which is the configuration space for the centers of mass of the clusters of a, as well as the internal configuration space

$$X^a := \left\{ x \in X \mid \sum_{i \in C} m_i x^i = 0 \text{ for } C \in a \right\},$$

which is the configuration space for the particles with fixed centers of mass of the clusters. These spaces orthogonally complement each other: $X = X_a \oplus X^a$. Indeed, for any $x_a \in X_a$ and $x^a \in X^a$ we have $x_a \cdot x^a = \sum_{C \in a} \sum_{i \in C} m_i (x_a)^i (x^a)^i = 0$, since $(x_a)^i$ is independent of $i \in C$. On the other hand, any $x \in X$ has a decomposition $x = x_a + x^a$ with

$$(x_a)^i = \xi_C(x), \qquad (x^a)^i = x^i - \xi_C(x)$$

for $i \in C \in a$, where $\xi_C(x) = \left(\sum_{i \in C} m_i \right)^{-1} \left(\sum_{i \in C} m_i x^i \right)$ is the center of mass of the cluster C. In particular, for the trivial decomposition $0 = (1 \ldots N)$,

$$X_0 = \{ x = (x^1, \ldots, x^N) \mid x^1 = \ldots = x^N \},$$

$$X^0 = \left\{ x = (x^1, \ldots, x^N) \mid \sum_{i=1}^{N} m_i x^i = 0 \right\}$$

are the configuration spaces for center of mass of the N-body system, resp. for the system in its own center of mass frame.

Given a cluster decomposition a, the sum $X = X_a \oplus X^a$ induces the factorization $L^2(X) = L^2(X_a) \otimes L^2(X^a)$ w.r.t. which the kinetic energy decomposes as $p^2 = (p_a)^2 \otimes \mathbb{1} + \mathbb{1} \otimes (p^a)^2$. The potentials are split as to whether they link particles belonging to the same cluster of a, or not (intercluster interactions):

$$V(x) = V^a(x) + I_a(x),$$

with

$$V^a(x) = \sum_{\substack{i<j \\ i \underset{a}{\sim} j}} V_{ij}(x^i - x^j), \qquad I_a(x) = \sum_{\substack{i<j \\ i \underset{a}{\not\sim} j}} V_{ij}(x^i - x^j).$$

where $i \underset{a}{\sim} j$ means that i, j belong to the same cluster of a. We remark that

$$(4) \qquad V^a(x) = V^a(x^a), \qquad I_a(x) \xrightarrow[|x|_a \to +\infty]{} 0,$$

where $|x|_a = \min_{\substack{i \neq j \\ a}} |x^i - x^j|$ is the intercluster distance in the configuration x. Non-interacting clusters of a are the described by the Hamiltonian $H_a := H - I_a$. It decomposes into a part describing the center of mass motion of the clusters in a, and an internal part H^a:

$$(5) \qquad H_a = \frac{1}{2}(p_a)^2 \otimes \mathbb{1} + \mathbb{1} \otimes H^a , \qquad H^a = \frac{1}{2}(p^a)^2 + V^a .$$

In particular, for the decomposition $0 = (1 \ldots N)$, we have $I_0 = 0$, $H_0 = H$ and (5) is the separation of the center of mass motion. The definition (3) and the subsequent ones can be set in the center of mass frame, i.e., with X^0 instead of X, yielding spaces $X_a^0 \subset X^0$ and Hamiltonians H_a^0 on $L^2(X^0)$. We shall be henceforth concerned with these objects but will omit the superscript 0. With this notation we have e.g. $X_{(1\ldots N)} = \{0\}$ and $X_{(1)\ldots(N)} = X$. Some pictures may help to visualize the configuration space X and its subspaces X_a:

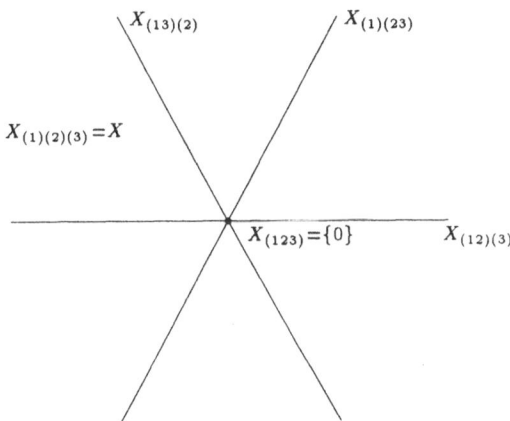

Figure 1. $N = 3$, $\nu = 1$: The configuration space X and its subspaces X_a.

Cluster decompositions are partially ordered by $b \geq a$, meaning that b is a refinement of a. For instance, $(12)(3)(4) \geq (123)(4)$. We have $b \geq a$ iff $X_b \supset X_a$. Clearly, by definition, in a configuration $x \in X_a$ particles belonging to the same cluster of a have coinciding positions. This, however, does not mean that particles belonging to different clusters of a are separated. The latter is true only for configurations $x \in X_a$ in the smaller but non-empty set

$$Z_a := X_a \setminus \bigcup_{b < a} X_b .$$

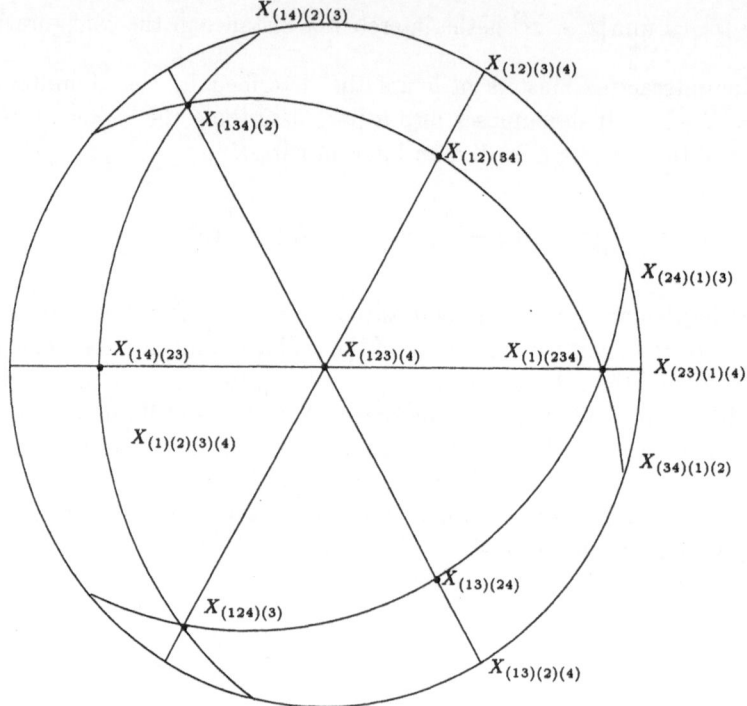

Figure 2. $N = 4$, $\nu = 1$: The subspaces X_a intersecting the unit sphere.

Then (see Fig. 2)

$$(6) \qquad x \in Z_a \iff x \in X_a, \ |x|_a > 0 \ .$$

Finally, as a point of rigor, we should make sure H is self-adjoint. To this end we assume that V is infinitesimally small w.r.t. p^2: For any $\varepsilon > 0$ there is b such that

$$(7) \qquad \|V\psi\| \le \varepsilon\|p^2\psi\| + b\|\psi\|$$

for all ψ in a core for p^2, e.g., $\psi \in C_0^\infty(X)$. Equivalently, one may assume the corresponding statement for all V_{ij} as operators on $L^2(\mathbb{R}^\nu)$. Then H is self-adjoint with the same domain as p^2, and is bounded below.

2. The essential spectrum. The spectrum of H is denoted by $\sigma(H)$. The discrete spectrum $\sigma_{\text{disc}}(H)$ consists of the eigenvalues of H which are isolated in $\sigma(H)$ and of finite multiplicity. The essential spectrum is then defined as

$$\sigma_{\text{ess}}(H) = \sigma(H) \setminus \sigma_{\text{disc}}(H) \ .$$

In principle, $\sigma_{\text{ess}}(H)$ may contain eigenvalues which are isolated but of infinite multiplicity. We shall see that there are none for the Hamiltonian (1). The following lemma proves useful in locating $\sigma_{\text{ess}}(H)$:

Lemma. (Weyl) *Let A, B be self-adjoint operators. If $A - B$ is relatively compact w.r.t. A (or B), then*

$$\sigma_{\text{ess}}(A) = \sigma_{\text{ess}}(B) . \tag{8}$$

We recall that, by definition, C is relatively compact w.r.t. A if $C(A - z)^{-1}$ is compact for some $z \notin \sigma(A)$. For instance, a bounded function on X vanishing at infinity is, taken as a multiplication operator, relatively compact w.r.t. H. Hence $\sigma_{\text{ess}}(H)$ is affected only by the potential $V(x)$ arbitrarily far out. As Fig. 2 suggests, out there H reduces to some H_a with $a > 0$.

Theorem. (HVZ)

$$\sigma_{\text{ess}}(H) = \bigcup_{a>0} \sigma(H_a) . \tag{9}$$

We remark that $\sigma(H_a) = \sigma(p_a^2/2) + \sigma(H^a)$ due to (5). If $a > 0$, then $\sigma(p_a^2/2) = [0, +\infty)$, so that

$$\sigma(H_a) = [\Sigma_a, +\infty) , \qquad \Sigma_a := \inf \sigma(H^a) .$$

The r.h.s. of (9) thus is

$$\bigcup_{a>0} \sigma(H_a) = [\Sigma, +\infty) , \qquad \Sigma := \min_{a>0} \Sigma_a . \tag{10}$$

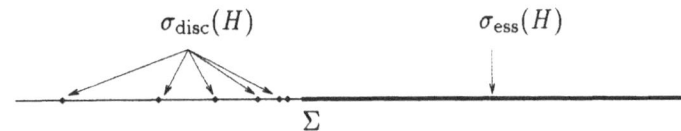

Figure 3. The spectrum of H.

Proof. We will show that the each side of (9) is contained in the other one.

(i) '$\supset$': We claim that $\sigma(H_a) \subset \sigma(H)$. Then, by (10), $[\Sigma, +\infty) \subset \sigma(H)$ and, actually, $[\Sigma, +\infty) \subset \sigma_{\text{ess}}(H)$ since a half-line has no isolated points. To prove the claim, pick $y \in Z_a$ and let $(T_s\psi)(x) = \psi(x - sy)$, $(s \in \mathbb{R})$ be the translation operators along y. Since T_s commutes with H_a we have

$$\|(H - \lambda)T_s\psi\| \leq \|(H_a - \lambda)\psi\| + \|I_a T_s\psi\| .$$

For $\lambda \in \sigma(H_a)$, the first term one r.h.s. can be made arbitrarily small for some suitable $\psi \in C_0^\infty(X)$, $\|\psi\| = 1$. The second one is bounded by $\sup_{x \in \text{supp } \psi} |I_a(x + sy)|$. This vanishes in the limit $s \to \infty$ because of (4) and since, due to (6), $|x + sy|_a \to +\infty$ uniformly for x in a compact set. This proves that the l.h.s. is arbitrarily small for some $T_s \psi$ and thus that $\lambda \in \sigma(H)$.

(ii) '⊂': [7] We will need a partition of unity on X,

$$(11) \qquad \sum_{a>0} j_a(x)^2 = 1 \, ,$$

where the j_a are smooth functions satisfying

$$(12) \qquad j_a(\lambda x) = j_a(x) \qquad (|x| \geq 1, \lambda \geq 1) \, ,$$

$$(13) \qquad |x|_a \geq \varepsilon |x| \qquad (|x| \geq 1, x \in \text{supp } j_a)$$

for some $\varepsilon > 0$. We first construct a partition of unity on the unit sphere $S = \{x \in X \mid |x| = 1\}$. An open cover of S is given by the sets $S_a = \{x \in S \mid |x|_a > 0\}, (a > 0)$. Indeed, if $|x|_a = 0$ for all $a > 0$, then $x = 0$, which is not in S. Thus there exists a partition of unity with $\text{supp } j_a \subset S_a$. Since $\text{supp } j_a$ is compact, we there have $|x|_a \geq \varepsilon$ for some $\varepsilon > 0$. The functions of this partition are then extended to $|x| > 1$ following (12), and to $|x| < 1$ in any smooth way preserving (11). We use the partition to localize H:

$$(14) \qquad \begin{aligned} H &= \frac{1}{2} \sum_{a>0} (j_a^2 H + H j_a^2) = \sum_{a>0} \left(j_a H j_a + \frac{1}{2} [j_a, [j_a, H]] \right) \\ &= \sum_{a>0} j_a H_a j_a + \sum_{a>0} j_a^2 I_a - \sum_{a>0} (\nabla j_a)^2 \, . \end{aligned}$$

Here, $\nabla j_a(x) = O(|x|^{-1})$ as $|x| \to \infty$ by (12), and $j_a(x)^2 I_a(x) \to 0$ as $|x| \to +\infty$ because of (4,13). Hence, both the last and the middle term in (14) are relatively compact w.r.t. H. Then (8) implies

$$\sigma_{\text{ess}}(H) = \sigma_{\text{ess}} \left(\sum_{a>0} j_a H_a j_a \right) \subset [\Sigma, +\infty) \, ,$$

where, in the second step, we used $H_a \geq \Sigma_a \geq \Sigma$ and hence $\sum_{a>0} j_a H_a j_a \geq \Sigma \sum_{a>0} j_a^2 = \Sigma$. $\qquad \square$

3. The Mourre estimate. The Mourre estimate is an important step in understanding the dynamical properties of states from the continuous spectral subspace $\mathcal{H}$. Heuristically, we expect that trajectories $\psi_t = e^{-itH} \psi \in \mathcal{H}$ correspond to clusters separating at positive speed, i.e.,

$$\langle x^2 \rangle_t = (\psi_t, x^2 \psi_t) \approx \theta t^2 \quad (t \to \infty)$$

for some $\theta > 0$. For the sake of definiteness we shall assume that ψ_t has energy distribution around E and describes freely moving bound clusters of $a > 0$. Actually, let

$$\psi_t \approx e^{-itp_a^2/2}\psi_a \otimes e^{-itH^a}\psi^a$$

with $H^a\psi^a = \lambda^a\psi^a$, and $\|((p_a^2/2) - (E - \lambda^a))\psi_a\|$ small. This describes bound clusters of a with total internal energy λ^a and kinetic energy about $E - \lambda^a \geq 0$. Then

$$\langle x^2 \rangle_t = \langle x_a^2 + (x^a)^2 \rangle_t \approx \langle x_a^2 \rangle_t \approx \langle p_a^2 \rangle_t t^2 \approx 2(E - \lambda^a)t^2 \ ,$$

$$\frac{d^2}{dt^2}\langle x^2/2 \rangle_t \approx 2(E - \lambda^a) \ .$$

Noting that

$$\frac{d}{dt}\langle x^2/2 \rangle_t = \langle A \rangle_t \ , \qquad A := i\left[H, \frac{x^2}{2}\right] = \frac{1}{2}(p \cdot x + x \cdot p) \ ,$$

$$\frac{d^2}{dt^2}\langle x^2/2 \rangle_t = \langle i[H, A] \rangle_t \ ,$$

we expect that for continuum states with $\|(H - E)\psi\|$ small

(15)
$$\langle i[H, A] \rangle \geq \inf 2(E - \lambda^a) \ ,$$

where the infimum is over $a > 0$ and over eigenvalues $\lambda^a \leq E$ of H^a. On the other hand for bound states $H\psi = E\psi$ we have the *virial theorem*

$$\langle i[H, A] \rangle = i(\psi, (EA - AE)\psi) = 0 \ .$$

In order to state the Mourre estimate (15) properly we introduce the spectral projection $E_\Omega(H)$ of H for $\Omega \subset \mathbb{R}$. Let furthermore $\mathcal{E}(H) := \{$ eigenvalues of $H \}$ and let $\mathcal{T}(H) := \cup_{a>0}\mathcal{E}(H^a)$ be the set of all thresholds. Then we set for $E \in \sigma(H)$

$$\theta(E) = \inf\{2(E - \lambda) \mid \lambda \in \mathcal{T}(H) \cup \{c\}, \lambda \leq E\} \ ,$$

where we added a fictitious threshold $c \leq H$ so as to make θ well defined on all of $\sigma(H)$.

Theorem. (Perry, Sigal, Simon) *Let both V and $x \cdot \nabla V$ satisfy (4,7).*

(i) Given $E \in \sigma(H)$ and $\varepsilon > 0$, there exists an open interval $\Delta \ni E$ and a compact operator K such that

(16)
$$E_\Delta(H)i[H, A]E_\Delta(H) \geq (\theta(E) - \varepsilon)E_\Delta(H) + K \ ;$$

(ii) Eigenvalues can only accumulate at thresholds: $\lambda_n \in \mathcal{E}(H)$, $E \neq \lambda_n \to E$ *implies* $E \in T(H)$.

We remark that the compact term K is needed for consistency with the virial theorem.

Corollary. *If* $E \notin \mathcal{E}(H)$, *then (16) holds true with K dropped.*

Proof. Upon multiplying both sides by $E_{\Delta'}(H)$ with $\Delta' \subset \Delta$, the term K is replaced by $E_{\Delta'}(H)KE_{\Delta'}(H)$. We then let $\Delta' \to \{E\}$. Since E is not an eigenvalue of H we have $E_{\Delta'}(H) \to 0$ strongly and $\|E_{\Delta'}(H)KE_{\Delta'}(H)\| \to 0$. This term can thus be absorbed in the first one on the r.h.s. of (16) at expense of doubling ε.   □

The proof of the Mourre estimate [13,15], which we omit, is by induction in subsystems H^a, $(a > 0)$. For those, the induction hypothesis consists of (i) and (ii) jointly. In particular, since thresholds are eigenvalues of subsystems, the set $\mathcal{E}(H) \cup T(H)$ is closed and countable.

4. Scattering theory. The first task of quantum scattering theory is to give a classification of the possible large time behaviors of Schrödinger orbits $e^{-itH}\psi$. In the intuitive picture of the scattering process, this system is well described at large times by a number of independently moving, bound clusters, i.e., by free composite particles. This statement is called asymptotic completeness. We discuss this issue for N particles interacting through pair potentials of short range, i.e., decaying at infinity faster than the Coulomb potential:

$$(17) \qquad\qquad (1 + |y|)^\mu V_{ij}(y)$$

is infinitesimally small w.r.t. Δ_y on $L^2(\mathbb{R}^\nu)$ for some $\mu > 1$. We denote by P^a the bound state projection of H^a. States in the range of P^a represent, physically speaking, a set of composite particles, each of them consisting of the 'elementary' particles of some cluster of a.

We say that the N-body short-range system above is *asymptotically complete,* if (a) for any state $\psi \in L^2(X)$ and any $\varepsilon > 0$ there are states $\psi^a = P^a\psi^a \in L^2(X)$ such that

$$(18) \qquad\qquad \left\| e^{-itH}\psi - \sum_a e^{-itH_a}\psi^a \right\| \leq \varepsilon$$

for $t > 0$ large enough, where the sum ranges over all cluster decompositions. This property is equivalent to the apparently stronger one, where one requires that (b) the l.h.s. of (18) tends to 0 as $t \to +\infty$. Given the existence of the wave operators $\Omega_a := s - \lim_{t \to \infty} e^{itH}e^{-itH_a}P^a$, asymptotic completeness is also equivalent to the statement (c)

$$\bigoplus_a \operatorname{Ran}\Omega_a = L^2(X).$$

Indeed, (c)⇒(b) and (b)⇒(a) are clear. Assuming the existence of Ω_a, (a) implies that $\|\psi - \sum_a \Omega_a \psi^a\| \leq \varepsilon$. The operators Ω_a are partial isometries and thus have closed range. Hence $\psi \in \bigoplus_a \operatorname{Ran} \Omega_a$, i.e., (a) ⇒(c).

Theorem. (Sigal, Soffer) *The quantum N-body system (1) satisfying (17) is asymptotically complete.*

The main tools for proving this result will be developed in the next section. Some of them are of independent interest. Moreover we should mention that the statement of asymptotic completeness, properly modified, also holds for some long-ranged interactions [5,6,28].

5. Propagation estimates. To each configuration of the particles one can associate a cluster decomposition in such a way that particles in the same cluster are close together, and that clusters are very distant from each other. If at some large time t one groups the particles as explained, then the motion of the centers of mass of these clusters should be almost free, i.e., $p_a \approx x_a/t$. In other words, the quantum dynamics is concentrating on states where these two quantities coincide. This is the main intermediate result of the proof. It will be cast into precise mathematical terms by means of a *propagation estimate*. Several sorts of propagation estimates have been used in scattering theory. The type we are going to describe goes back to [16-18], except for the use of time-dependent observables, which is due to [27].

Let H be a self-adjoint operator on a Hilbert space $\mathcal{H}$. Let $P(t)$ be a bounded, time-dependent positive operator, i.e.,

$$P : [1,\infty) \rightarrow \mathcal{L}(\mathcal{H}), \ t \mapsto P(t) ,$$

$$P(t) \geq 0 .$$

A propagation estimates is an estimate of the form

$$\int_1^\infty dt \, (\psi_t, P(t)\psi_t) \leq \operatorname{const} \|\psi\|^2 ,$$

where $\psi_t := e^{-itH}\psi$. It tells us that the trajectory $e^{-itH}\psi$ concentrates where $P(t)$ is small.

Propagation estimates can be generated by a *propagation observable*

$$\Phi : [1,\infty) \rightarrow \mathcal{L}(\mathcal{H}), \ t \mapsto \Phi(t),$$

where $\Phi(t)$ is uniformly bounded in t. We assume the existence (as a bounded operator) of the Heisenberg derivative $D\Phi(t)$, i.e.,

$$\frac{d}{dt}(\psi_t, \Phi(t)\psi_t) = (\psi_t, D\Phi(t)\psi_t) ,$$

which is formally given by $D\Phi(t) = i[H, \Phi(t)] + \partial\Phi/\partial t$. Assume it is positive in the sense that

$$D\Phi(t) \geq P(t) + R(t)$$

in $\mathcal{L}(\mathcal{H})$, where

$$P(t) \geq 0$$

and $R(t)$ satisfies the *remainder estimate*

$$\int\limits_1^\infty dt \, |(\psi_t, R(t)\psi_t)| \leq \text{const} \, \|\psi\|^2 \, .$$

Then

$$(\psi_t, \Phi(t)\psi_t)\big|_1^T \geq \int\limits_1^T dt \, (\psi_t, P(t)\psi_t) + \int\limits_1^T dt \, (\psi_t, R(t)\psi_t) \, ,$$

where both the l.h.s. and the last term on the r.h.s. are bounded by const $\|\psi\|^2$. Since the first integrand is nonnegative, we conclude that $P(t)$ satisfies a propagation estimate.

   We now mention two ways to prove a remainder estimate. We assume one of the following

1)
$$\int\limits_1^\infty dt \, \|R(t)\| < \infty \, ;$$

2)
$$R(t) = R_1(t)^* B(t) R_2(t) + R_2(t)^* B(t)^* R_1(t) \, ,$$

where $R_i(t)^* R_i(t)$, $i = 1, 2$ satisfy a propagation estimate and $B(t)$ is uniformly bounded. In fact

$$|(\psi_t, R(t)\psi_t)| \leq \text{const} \, (\psi_t, R_1(t)^* R_1(t)\psi_t)^{1/2}(\psi_t, R_2(t)^* R_2(t)\psi_t)^{1/2}$$

from which the remainder estimate follows by the Cauchy inequality. Coming back to (1), we will give a number of examples of propagation estimates, propagation observables, and remainder estimates.

Example 1. (Maximal velocity estimate) [27] Let $\Omega \subset \mathbb{R}$ be a bounded interval, and $\lambda > 0$ large enough. Then

$$\int\limits_1^\infty \frac{dt}{t} \, (\psi_t, E_\Omega(H)F(\lambda \leq |x/t| \leq 2\lambda)E_\Omega(H)\psi_t) \leq \text{const} \, \|\psi\|^2$$

for all $\psi \in L^2(X)$, where $E_\Omega(H)$ is the spectral projection for H associated with Ω and the constant depends on Ω, λ.

The idea is that a particle with finite energy and hence speed cannot propagate into regions moving away still faster. This propagation estimate is generated by the propagation observable

$$\Phi(t) = -E_\Omega(H)h(|x/\lambda t|)E_\Omega(H),$$

where $h' \in C_0^\infty(0,\infty)$, $h' \geq 0$ and $h'(y) \geq 1$ for $1 \leq y \leq 2$. One checks in fact that

$$D\Phi \geq \frac{\text{const}}{t} E_\Omega(H)F(\lambda \leq |x/t| \leq 2\lambda)E_\Omega(H) + \mathrm{O}(t^{-2}) .$$

Example 2. This example deals with a free particle, $H = p^2/2$, and is therefore not directly applicable to N-body systems. The basic observation is that a free particle has the classical trajectory $p = p_0$, $x = p_0 t + x_0$, where p_0, x_0 are the initial data. It follows that $(p - x/t)^2 = x_0^2/t^2$ is decreasing. This suggests that one can get a positive Heisenberg derivative by considering the propagation observable

$$-K_0(t) := -\frac{1}{2}\left(p - \frac{x}{t}\right)^2$$

(we ignore that $K_0(t)$ is not uniformly bounded for a while). In fact

$$(19) \qquad -DK_0(t) = \frac{1}{t}\left(p - \frac{x}{t}\right)^2 \geq 0 ,$$

which 'proves'

$$(\text{wrong}) \qquad \int_1^\infty \frac{dt}{t}\left(\psi_t, \left(p - \frac{x}{t}\right)^2 \psi_t\right) \leq \text{const } \|\psi\|^2 .$$

This means that the time evolved quantum state is concentrating on the region in phase-space where $(p - x/t)^2$ is small, i.e., around the classical trajectories. But the equation above is wrong as it stands, since $K_0(t)$ is not even a bounded operator. The next example will deal with this problem.

Example 2 (revisited). We replace $K_0(t)$ by a bounded operator by introducing some cutoffs, namely an energy cutoff $E_\Omega(H)$, where $\Omega \subset \mathbb{R}$ is some bounded interval, and a volume cutoff $f(x/\lambda t)$, where $f \in C_0^\infty(X)$, $f(x) = 1$ for $|x| \leq 1$, $f(x) = 0$ for $|x| \geq 2$, and $\lambda > 0$ is large. $-E_\Omega f K_0 f E_\Omega$ is now a suitable propagation observable, since it is uniformly bounded in t. Then

$$D(E_\Omega f K_0 f E_\Omega) = E_\Omega \left(f(DK_0)f + f K_0(Df) + (Df)K_0 f\right) E_\Omega .$$

Since Df is supported in $\lambda \leq |x/t| \leq 2\lambda$, it is easily checked that the terms above arising from Df satisfy a remainder estimate by Example 1. Hence

$$\int\limits_{1}^{\infty} \frac{dt}{t} \left(\psi_t, E_\Omega f \left(p - \frac{x}{t}\right)^2 f E_\Omega \psi_t\right) \leq \text{const} \, \|\psi\|^2 \, .$$

Example 3. We extend the previous result to the two-body case. Clearly $K_0(t)$ as defined above is not going to be non-increasing if the system is in a bound state or close to the scatterer. On the other hand the Hamiltonian H, being conserved, is trivially non-increasing (but, as a propagation observable, it does not provide any dynamical information). Let $v(x,t)$ be a vector field with $v = x/t$ for $|x| \gtrsim R$, resp. $v = 0$ for $|x| \lesssim R$ and with a smooth transition in between. Then

$$K(t) := \frac{1}{2} \left(p - v(x,t)\right)^2 + V(x)$$

reduces to $K_0(t)$, resp. H in those two regions and should thus be non-increasing. One computes

$$
\begin{aligned}
(20) \qquad -DK(t) ={}& (p-v)\frac{v_* + v_*^\dagger}{2}(p-v) \\
&+ \frac{1}{2}[(p-v) \cdot \mathcal{D}_v v + \mathcal{D}_v v \cdot (p-v)] \\
&- \frac{1}{4}\Delta(\nabla \cdot v) - v \cdot \nabla V \, ,
\end{aligned}
$$

where $v_*(x,t) : X \to X$ is the x-derivative of v, and $v_*^\dagger$ its adjoint with respect to the inner product (2), and $\mathcal{D}_v = v \cdot \nabla + \partial/\partial t$ is the material derivative for the vector field v. If v is chosen appropriately, the first term on the r.h.s. of (20) is nonnegative and reduces to (19) far from the the origin. For $R := t^{1-\delta}$, $\delta > 0$, one gets $\mathcal{D}_v v = O(t^{-(1+\delta)})$, $\Delta(\nabla \cdot v) = O(t^{-(3-2\delta)})$ and $v \cdot \nabla V = O(t^{-(1-\delta)\mu})$. Thus all terms on the r.h.s. of (20), except for the first one, satisfy a remainder estimate if δ is taken small enough. After adding the same cutoffs to $K(t)$ as in the free case, its Heisenberg derivative then satisfies

$$-D(E_\Omega f K f E_\Omega) \geq \frac{1}{t} E_\Omega f j \left(p - \frac{x}{t}\right)^2 j f E_\Omega + R(t)$$

for $j = j(x/t)$ bounded and supported away from the origin, with $R(t)$ satisfying a remainder estimate. Hence

$$(21) \qquad \int\limits_{1}^{\infty} \frac{dt}{t} \left(\psi_t, E_\Omega f j \left(p - \frac{x}{t}\right)^2 j f E_\Omega \psi_t\right) \leq \text{const} \, \|\psi\|^2 \, ,$$

i.e., the conclusion of the previous example still holds except for a region close to the origin.

Example 4. Next we extend the previous example to the N-body case. Consider a configuration x. Group the particles into clusters in such a way that particles in the same cluster are close together, and that clusters are far from each other (the scale being $R = t^{1-\delta}$ as before). Call a the resulting cluster decomposition. We express this by saying that the particles are a-clustered at x. The configuration space X then breaks up into regions corresponding to a-clustered particles, with a ranging over all cluster decompositions. In the context of Fig. 1 these regions are shown in the following picture.

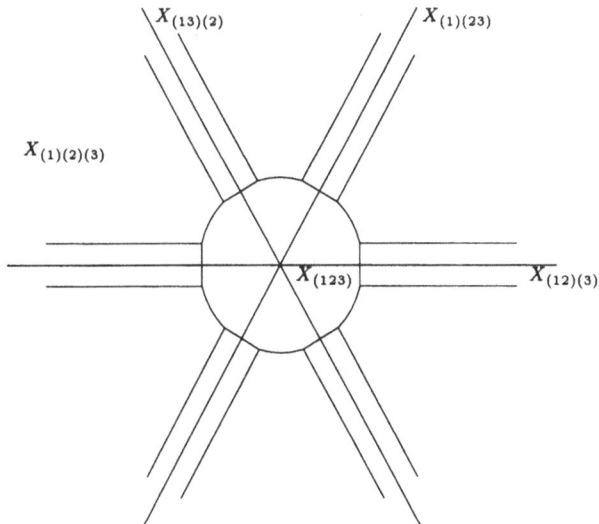

Figure 4. $N = 3$, $\nu = 1$: The partition of X.

Let $v = x_a/t$ if x is a-clustered. Then $v_*(x)$ reduces to $\mathbb{1}_a$ and the first term on the r.h.s. of (20) to

$$(p - v)\frac{v_* + v_*^\dagger}{2}(p - v) = \frac{1}{t}(p - v)_a^2 = \frac{1}{t}\left(p - \frac{x}{t}\right)_a^2 .$$

Thus (21) goes over to

$$(22) \qquad \int\limits_1^\infty \frac{dt}{t}\,(\psi_t, E_\Omega f j_a \left(p - \frac{x}{t}\right)_a^2 j_a f E_\Omega \psi_t) \leq \text{const } \|\psi\|^2$$

for $j_a = j_a(x/t)$ bounded and supported on a-clustered configurations. For large times, this estimate asserts that whenever the particles are a-clustered, the velocities of the centers of mass of the clusters of a are approximately given by their positions, divided by t. This is the result mentioned at the beginning of this section.

Example 5. (Minimal velocity estimate) As mentioned in connection with the Mourre estimate, at energies away from eigenvalues and thresholds the separation between clusters should occur at strictly positive speed. Let $E \notin \mathcal{E}(H) \cup \mathcal{T}(H)$. Then there is an open interval $\Delta \ni E$ such that

$$(23) \qquad \int_1^\infty \frac{dt}{t} \, (\psi_t, E_\Delta j_0^2(x/vt) E_\Delta \psi_t) \leq \text{const} \, \|\psi\|^2$$

for given j_0 supported on 0-clustered configuration and $v > 0$ small enough. This propagation estimate states that the region $|x| \lesssim vt$ is avoided by the trajectory of states in the energy range Δ. The propagation observable generating it is

$$\Phi(t) = E_\Delta j_0 \frac{A}{t} j_0 E_\Delta \, .$$

Then

$$D\Phi(t) = \frac{1}{t} E_\Delta \left[j_0 \left(i[H, A] - \frac{A}{t} \right) j_0 + (Dj_0) \frac{A}{t} j_0 + j_0 \frac{A}{t} (Dj_0) \right] E_\Delta \, ,$$

where the last two terms can be shown to be remainders by Example 4. Up to integrable remainders the first term is $\geq (\theta(E) - \varepsilon - \text{const} \, v) t^{-1} E_\Delta j_0^2 E_\Delta$ because of the Corollary to the Mourre estimate. This expression is positive for $v > 0$ small enough.

6. Asymptotic completeness. The propagation estimates of the preceding section can be used to prove asymptotic completeness. Let us first investigate its consequences heuristically. Suppose that at some large time t the particles are a-clustered. By (22) the velocities of the clusters of a are $p_a \approx x_a/t$. Hence the motion still to come will separate them even more from each other, making the intercluster interactions I_a negligible. In other words, H_a will account for this motion almost as well as H does. Mathematically, if j_a is supported on a-clustered configurations, given ψ there should be a state ψ^a such that

$$(24) \qquad \lim_{t\to\infty} \|j_a(x/t) e^{-itH} \psi - e^{-itH_a} \psi^a\| = 0 \, .$$

In particular, let the $\{j_a\}_a$ form a partition of unity on X indexed by all cluster decompositions, i.e., $\sum_a j_a = 1$. Then (24) implies

Lemma. *The Deift-Simon [3] wave operators*

$$(25) \qquad W_a(v) := s - \lim_{t\to\infty} e^{itH_a} j_a(x/vt) e^{-itH}$$

exist for any $v > 0$.

Thus

$$(26) \qquad e^{-itH}\psi = \sum_a j_a e^{-itH}\psi = \sum_a e^{-itH_a} W_a(v)\psi + o(1)$$

as $t \to \infty$, showing that the system is well described by independently moving clusters. Yet, for asymptotic completeness each cluster should also be bound, i.e., $\psi^a = P^a \psi^a$ in (18). Consider for simplicity the trivial cluster decomposition $0 = (1 \ldots N)$. Then $(\psi, W_0(v)\psi)$ is the asymptotic probability for a cluster where all particles are close together on the scale vt. The probability for being in a bound state is included in $(\psi, W_0(v)\psi)$ but does not exhaust it: If the energy of ψ is sufficiently close to a threshold, then the particles can break apart (thus not forming a bound state) with speed less than v. However no other possibilities for $W_0(v)\psi$ exist:

Lemma. *Let $\Omega \subset \mathbb{R}$ be a compact set, which does not contain any eigenvalues or thresholds. Then*

$$(27) \qquad\qquad W_0(v) E_\Omega(H) = 0 ,$$

provided $v > 0$ is small enough.

Proof. By compactness, it suffices to show that for any $E \notin \mathcal{E}(H) \cup T(H)$ there is an open interval $\Delta \ni E$ such that $W_0(v) E_\Delta(H) = 0$. By the convergence of (23), for given ψ there is a sequence of times t_n such that $\|j_0(x/vt_n) E_\Delta(H) e^{-it_n H}\psi\| \to 0$. Then (27) follows from (25). $\qquad\square$

These lemmas easily imply asymptotic completeness: Since (18) is trivial if ψ is a bound state, we may assume $\psi \perp \mathrm{Ran}\, P^{(1 \ldots N)}$. Given $\varepsilon > 0$, we can find, by part (ii) of Mourre's theorem, a compact set $\Omega \subset \mathbb{R}$ which does not contain any eigenvalues or thresholds, such that

$$\|(1 - E_\Omega(H))\psi\| \le \varepsilon$$

and then $v > 0$ such that (27) holds. Thus, by (26)

$$e^{-itH} E_\Omega(H)\psi = \sum_{\#(a) \ge 2} e^{-itH_a} W_a(v) E_\Omega(H)\psi + o(1) ,$$

where the sum now ranges only over decompositions with at least two clusters. Since each of these clusters consists of less than N particles, by induction we can apply (18) to the terms in the last sum above. We end up with asymptotic completeness.

Acknowledgments. This work was done in part at the Institute for Mathematics and its Applications (IMA) of the University of Minnesota. The author's research at the IMA was supported in part with funds provided by the U.S. National Science Foundation. These notes have been written in final form at the IHES, Bures-sur-Yvette.

REFERENCES

[1] H.L. Cycon, R.G. Froese, W. Kirsch, B. Simon, *Schrödinger Operators*, Springer-Verlag (1987).

[2] W.O. Amrein, A. Boutet de Monvel-Berthier, V. Georgescu, *Co-groups, commutator methods and spectral theory of N-body Hamiltonians*, Birkhäuser, (1996).

[3] P. Deift, B. Simon, A time-dependent approach to the completeness of multiparticle quantum systems. Commun. Pure Appl. Math. **30** (1977), pp. 573–583.

[4] J. Derezinski, Algebraic approach to the *N*-body quantum long range scattering, Rev. Math. Phys. **3** (1990), pp. 1–62.

[5] J. Derezinski, Asymptotic completeness for *N*-particle long-range quantum systems, Ann. Math. **138** (1993), pp. 427–476.

[6] J. Derezinski, C. Gérard, *Scattering theory of classical and quantum N-particle systems, tracts and monographs in physics*, Springer-Verlag, to appear in 1997.

[7] V. Enss, A note on Hunziker's theorem. Commun. Math. Phys. **52** (1977), pp. 233–238.

[8] V. Enss, Asymptotic completeness for quantum-mechanical potential scattering, I. Short-range potentials. Commun. Math. Phys. **61** (1978), pp. 285–291.

[9] V. Enss, "Geometric methods in spectral and scattering theory of Schrödinger operators", in *Rigorous atomic and molecular physics*, G. Velo, A.S. Wightman eds., Plenum (1981).

[10] V. Enss, "Completeness of Three-Body Quantum Scattering", in *Dynamics and Processes* ed. by P. Blanchard, L. Streit, Lecture Notes in Mathematics, Springer-Verlag, **1031** (1983), pp. 62–88.

[11] V. Enss, "Introduction to asymptotic observables for multiparticle quantum scattering", in *Schrödinger operators, Aarhus 1985*, E. Balslev ed., Lecture Notes in Mathematics, Springer-Verlag, **1218** (1986).

[12] L. Faddeev, *Mathematical Aspects of the Three Body Problem in Quantum Scattering Theory* (Steklov Institute 1963), Israel Program for Scientific Translation (1965).

[13] R. Froese, I. Herbst, A new proof of the Mourre estimate. Duke Math. J. **49** (1982), pp. 1075–1085.

[14] G.M. Graf, Asymptotic completeness for *N*-body short-range quantum systems: A new proof. Commun. Math. Phys. **132** (1990), pp. 73–101.

[15] W. Hunziker, I.M. Sigal, "The general theory of *N*-body quantum systems", to appear. Preliminary version in *Mathematical quantum theory: II. Schrödinger operators*, J. Feldman et al. eds., AMS-publ. (1994).

[16] T. Kato, Wave operators and similarity for some non-self-adjoint operators. Math. Ann. **162** (1966), pp. 258–279.

[17] T. Kato, Smooth operators and commutators. Stud. Math. **31** (1968), pp. 535–546.

[18] R. Lavine, Absolute continuity of Hamiltonian operators with repulsive potentials. Proc. Am. Math. Soc. **22** (1969), pp. 55–60.

[19] E. Mourre, Link between the geometrical and the spectral transformation approaches in scattering theory. Commun. Math. Phys. **68** (1979), pp. 91–94.

[20] E. Mourre, Absence of singular continuous spectrum for certain selfadjoint operators. Commun. Math. Phys. **78** (1981), pp. 391–408.

[21] E. Mourre, Opérateurs conjugués et propriétés de propagation. Commun. Math. Phys. **91** (1983), pp. 279–300.

[22] P. Perry, I.M. Sigal, B. Simon, Spectral analysis of N-body Schrödinger operators. Ann. Math. **114** (1981), pp. 519–567.

[23] A.J. Povzner, On eigenfunction expansions in terms of scattering solutions. Dokl. Akad. Nauk SSSR **104** (1955).

[24] M. Reed, B. Simon, *Methods of modern mathematical physics, III. Scattering theory*, Academic Press, (1979).

[25] I.M. Sigal, A.G. Sigalov, Description of the spectrum of the energy operator of quantum mechanical system that is invariant with respect to permutations of identical particles. Theor. Math. Phys. **5** (1970), pp. 990–1005.

[26] I.M. Sigal, A. Soffer, The N-particle scattering problem: asymptotic completeness for short-range systems. Ann. Math. **126** (1987), pp. 35–108.

[27] I.M. Sigal, A. Soffer, Local decay and propagation estimates for time-dependent and time-independent Hamiltonians. Princeton University preprint (1988).

[28] I.M. Sigal, A. Soffer, Asymptotic completeness for N-particle long-range scattering. J. AMS **7** (1994), pp. 307–334.

A TUTORIAL ON COMPUTATIONAL APPROACHES TO QUANTUM SCATTERING

DONALD J. KOURI* AND DAVID K. HOFFMAN†

1. Introduction. The following is a summary of a series of tutorial lectures given at the Institute for Mathematics and its Applications during a Workshop on Waves and Scattering during the Spring, 1995. The purpose was to provide a little background in computational methods that have been developed over the past several decades by quantum dynamicists. Most of these techniques have been developed without worrying about mathematical details, with the primary objective being to apply them to specific systems. It is not possible to give an exhaustive account of all the methods that have been used, but hopefully a sufficiently wide range is given to enable one to obtain a general sense of the types of methods that have been brought to bear. Both time-independent and time-dependent approaches are discussed, and representative references to the literature given to facilitate the interested reader to obtain more details. For the most part, we have restricted ourselves to simple one-dimensional (1D) problems in order to illustrate the strategies, but the methods are more general and have been applied to real problems of varied complexity. In general, the systems of interest are conservative ones, although at least one computational method is discussed which can be applied to systems with an explicitly time dependent Hamiltonian. Finally, in addition to discussing computational techniques, we also give a brief outline of how one extracts the physically meaningful or measureable quantities from solutions of the Schrödinger equation (since this also may not be familiar to the non-specialist in quantum scattering).

2. Time-dependent Schrödinger equation and wavepackets. Although the time-dependent Schrödinger equation (TDSE) has been known since the inception of quantum mechanics, it has not been used much as the basis of large scale calculations until fairly recently [1]. The reason is simply, why treat a partial differential equation (PDE) with one additional variable in it if it can be avoided? If the potential is independent of time (conservative forces), then the time variable is separable, and generally one always separates variables whenever possible. There are several

* Department of Chemistry and Department of Physics, University of Houston, Houston, TX 77204-5641.
Supported under National Science Foundation Grant CHE-9403416 and R.A. Welch Foundation Grant E-0608.
† Department of Chemistry and Ames Laboratory, Iowa State University, Ames, IA 50011.
The Ames Laboratory is operated for the Department of Energy by Iowa State University under Contract No. 2-7405-ENG82.

answers:

(1) The TDSE, given by

$$(2.1) \qquad i\hbar\frac{\partial\chi}{\partial t} = H\chi,$$

with the time-independent Hamiltonian H satisfying

$$(2.2) \qquad H = H_0 + V,$$

V being the perturbation causing scattering among the states of H_0, is an *initial value problem*, and these are generally simpler to solve than boundary value problems.

(2) Provided the Hamiltonian is self-adjoint (which is usually equated to Hermitian by the quantum scattering theorists), solutions to the TDSE are $\mathcal{L}^2$, provided the initial wavepacket is $\mathcal{L}^2$. By contrast, the solutions (for scattering) of the time-independent Schrödinger equation (TISE) are *not* $\mathcal{L}^2$, the Hamiltonian H has a continuous spectrum and improper eigenstates which reflect the fact that there is a nonzero (relative) probability of finding the projectile infinitely far from the target.

The simplest procedure for obtaining the time-independent Schrödinger equation (TISE) for the case of a time-independent Hamiltonian, is to separate variables, writing the solutions of (2.1) as a product:

$$(2.3) \qquad \chi = \exp(-iEt/t)\psi(E).$$

Plugging (2.3) into (2.1), and using the time-independence of H gives us

$$(2.4) \qquad (E - H)\psi(E) = 0$$

as the equation determining $\psi(E)$. Equation (2.4) is recognized as the TISE. The perturbation V causing scattering will be defined as

$$(2.5) \qquad V = H - H_0,$$

where

$$(2.6) \qquad H_0 = \lim_{\substack{\text{large separation} \\ \text{of projectile and target}}} H.$$

It then follows that

$$(2.7) \qquad \lim_{\substack{\text{large separation} \\ \text{of projectile and target}}} V \equiv 0.$$

The scattering solution of the TISE is obtained by re-arranging (2.4) into the form

$$(2.8) \qquad (E - H_0)\psi(E) = V\psi(E),$$

and solving this using the so-called "causal Green's operator" (or resolvent),

$$(2.9) \qquad \psi_p^+(E) = \frac{1}{(E - H_0 + i\epsilon)} V \psi_p^+(E).$$

The "+" superscript denotes the causal boundary condition of outgoing scattered waves. Clearly, $\psi_p(E)$ is a particular solution of (2.8), viewed as the inhomogeneous equation

$$(2.10) \qquad (E - H_0)\psi(E) = \mathcal{I}(E).$$

As usual, one may add to the particular solution a solution of the homogeneous equation,

$$(2.11) \qquad (E - H_0)\phi(E) = 0,$$

and the complete "solution" is then

$$(2.12) \qquad \psi^+(E) = \phi(E) + \frac{1}{E - H_0 + i\epsilon} V \psi^+(E).$$

The $+i\epsilon$ selects a unique inverse of $(E - H_0)$, so that $(E - H_0 + i\epsilon)^{-1}$ generates purely outgoing scattering waves at large separations. In fact, $\phi(E)$ for a structureless projectile scattering off an infinitely massive center of force satisfies

$$(2.13) \qquad \left(E + \frac{\hbar^2}{2m} \nabla^2 \right) \phi(E) = 0,$$

having plane wave solutions,

$$(2.14) \qquad \phi(E|\vec{r}) = e^{i\vec{k}\cdot\vec{r}}$$

$$(2.15) \qquad \frac{\hbar^2 k^2}{2m} = E.$$

The quantity $(E - H_0 + i\epsilon)^{-1}$ satisfies outgoing radiation boundary conditions and solves the inhomogeneous TISE,

$$(2.16) \qquad (E - H_0)G = I,$$

where here, I is the unit operator. The quantity $G^+(E) (\equiv (E - H_0 + i\epsilon)^{-1})$ is well known and in the coordinate representation is given (in Dirac notation) by [2]

$$(2.17) \qquad \langle \vec{r} | G^+(E) | \vec{r}' \rangle = \left\langle \vec{r} \left| \frac{1}{(E - H_0 + i\epsilon)} \right| \vec{r}' \right\rangle,$$

or

$$(2.18) \qquad G^+(E|\vec{r}, \vec{r}') = -\frac{m}{2\pi\hbar^2} \frac{e^{ik|\vec{r} - \vec{r}'|}}{|\vec{r} - \vec{r}'|}.$$

Equation (2.16), in the coordinate representation, reads

$$(2.19) \qquad \left(E + \frac{\hbar^2}{2m} \nabla^2 \right) G^+(E|\vec{r}, \vec{r}') = \delta(\vec{r} - \vec{r}').$$

It is important to label ϕ by a complete set of "good quantum numbers", and not just the energy. Thus, one also uses the momentum $\hbar \vec{k}$, or just $\vec{k}$, so that

$$(2.20) \qquad \phi_{\vec{k}}(E|\vec{r}) = e^{i\vec{k} \cdot \vec{r}}.$$

Similarly, we label the scattering state $\psi_{\vec{k}}^+(E|\vec{r})$, satisfying the celebrated Lippmann-Schwinger equation

$$(2.21) \qquad \psi_{\vec{k}}^+(E|\vec{r}) = e^{i\vec{k} \cdot \vec{r}} - \frac{m}{2\pi\hbar^2} \int d\vec{r}' \frac{e^{ik|\vec{r} - \vec{r}'|}}{|\vec{r} - \vec{r}'|} V(\vec{r}') \psi_{\vec{k}}^+(E(\vec{r}')$$

with causal boundary conditions.

One can also write solutions of the Schrödinger equation labelled by different sets of eigenvalues for different commuting operators. The above uses the eigenvalues for linear momentum and the asymptotic Hamiltonian; one could also use the angular momentum, its projection onto the z-axis, the magnitude of the linear momentum, and the asymptotic Hamiltonian. This leads to the "partial wave expansion", based on the fact that

$$(2.22) \qquad \phi(E) = Y_{\ell m}(\hat{r}) j_\ell(kr)$$

also satisfies (2.13). Here, $Y_{\ell m}(\hat{r})$ is the spherical harmonic (equal to the product of a normalized Legendre polynomial, $\mathcal{P}_{\ell m}(\cos\theta)$, and the periodic function $\exp(im\varphi)/\sqrt{2\pi}$), $\hat{r}$ denotes the polar angle θ and azimuthal angle φ, and $j_\ell(kr)$ is the regular spherical Bessel function satisfying the condition

$$(2.23) \qquad \lim_{r \to \infty} j_\ell(kr) = \frac{\sin(kr - \ell\pi/2)}{kr}.$$

It is easily (but tediously) shown that the corresponding partial wave expanded Green's function is

$$(2.24) \quad G^+(E|\vec{r}, \vec{r}') = -\frac{2mk}{\hbar} \sum_{\ell=0}^{\infty} \sum_{\mu=-\ell}^{\ell} Y_{\ell\mu}(\hat{r}) Y_{\ell\mu}^*(\hat{r}') h_\ell^+(kr_>) j_\ell(kr_<).$$

In this expression, $h_\ell^+(kr)$ is the spherical outgoing wave Hankel function satisfying

$$(2.25) \qquad \lim_{r \to \infty} h_\ell^+(kr) = \frac{e^{i(ir - \ell\pi/2)}}{kr},$$

and $r_>$ ($r_<$) is the greater (lesser) of r and r'.

An analogous partial wave expansion of $\psi_{\vec{k}}^+(E|\vec{r})$ is

$$(2.26) \qquad \psi_{\vec{k}}^+(E|\vec{r}) = 4\pi \sum_{\ell,\ell'} \sum_{\mu,\mu'} i^{\ell'} Y_{\ell'\mu'}(\hat{r}) Y_{\ell\mu}^*(\hat{k}) g^+(\ell'\mu'\ell\mu|r),$$

where the radial function $g^+(\ell'\mu'\ell\mu|r)$ satisfies the integral equation

$$(2.27) \qquad \begin{aligned} g^+(\ell'\mu'\ell\mu|r) &= \delta_{\ell\ell'}\delta_{\mu\mu'} j_\ell(kr) - \\ &- \frac{2mk}{\hbar^2} \sum_{\ell''} \sum_{\mu''} i^{\ell''-\ell'} \int d\vec{r}\, Y_{\ell'\mu'}^*(\hat{r}) h_{\ell'}^+(kr_>) j_{\ell'}(kr_<) \\ &\quad V(\vec{r}) Y_{\ell''\mu''}(\hat{r}') g^+(\ell''\mu''\ell\mu|r'). \end{aligned}$$

If the potential is spherically symmetric (a "central potential"), then

$$(2.28) \qquad g^+(\ell'\mu'\ell\mu|r) = \delta_{\ell\ell'}\delta_{\mu\mu'} g_\ell^+(r),$$

where $g_\ell^+(r)$ satisfies

$$(2.29) \quad g_\ell^+(r) = j_\ell(kr) - \frac{2mk}{\hbar^2} \int_0^\infty dr'\, r'^2 h_\ell^+(kr_>) j_\ell(kr_<) V(r') g_\ell^+(r').$$

In developing computational strategies, it is useful to know exactly what quantities contain the physically significant information. To determine this, we must consider both the TDSE, (2.1), and its complex conjugate,

$$(2.30) \qquad \left[-\frac{\hbar^2}{2m}\nabla^2 + V\right]\chi^* = -i\hbar\frac{\partial\chi^*}{\partial t},$$

where we have assumed a real potential. Multiplying (2.1) by χ^* and (2.30) by χ, and subtracting one from the other, we obtain

$$(2.31) \qquad -\frac{\hbar^2}{2m}[\chi^*\nabla^2\chi - \chi\nabla^2\chi^*] = i\hbar\left[\chi^*\frac{\partial\chi}{\partial t} + \chi\frac{\partial\chi^*}{\partial t}\right].$$

But we have the identities

$$(2.32) \qquad \chi^*\nabla^2\chi - \chi\nabla^2\chi^* = \vec{\nabla}\cdot[\chi^*\vec{\nabla}\chi - \chi\vec{\nabla}\chi^*]$$

and

$$(2.33) \qquad \chi^*\frac{\partial\chi}{\partial t} + \chi\frac{\partial}{\partial t}\chi^* = \frac{\partial}{\partial t}[\chi^*\chi].$$

We then write (2.31) as

$$(2.34) \qquad \vec{\nabla}\cdot\vec{j} = -\frac{\partial\rho}{\partial t},$$

where

(2.35)
$$\rho = \chi^*\chi,$$

and the "flux vector", $\vec{j}$, is defined to be

(2.36)
$$\vec{j} = \frac{-i\hbar}{2m}[\chi^*\vec{\nabla}\chi - \chi\vec{\nabla}\chi^*].$$

Equation (2.34) is recognized as a "continuity equation" for incompressible fluid flow, but $\vec{j}$ is the flux vector for the probability density [3]. (This identification can be made sharper by integrating (2.34) over an arbitrary volume.) Now we note that for solutions to the TDSE of the form (2.3), ρ is independent of time and $\vec{j}$ becomes a zero divergence vector. We can calculate the flux associated with the plane wave unperturbed state, $\phi_{\vec{k}}(E|\vec{r})$, as follows:

(2.37)
$$\vec{j} = -\frac{i\hbar}{2m}\left[e^{-i\vec{k}\cdot\vec{r}}(i\vec{k})e^{i\vec{k}\cdot\vec{r}} - e^{i\vec{k}\cdot\vec{r}}(-i\vec{k})e^{i\vec{k}\cdot\vec{r}}\right]$$

(2.38)
$$= \hbar\vec{k}/m,$$

(2.39)
$$= \vec{p}/m,$$

or $\vec{j}$ is equal to the velocity of the particle. Note that the plane wave has been normalized in a way that leads to a constant probability density of 1. If we had used the so-called particle-in-a-box normalization, then

(2.40)
$$\phi_{\vec{k}}(E|\vec{r}) = \frac{1}{L^{3/2}}e^{i\vec{k}\cdot\vec{r}},$$

where L^3 is the volume of the box. Calculating the flux for this state yields

(2.41)
$$\vec{j} = \frac{\vec{p}}{L^3 m}$$

(2.42)
$$= \text{density} \times \text{velocity},$$

which is the familiar form for the flux of an incompressible fluid. In all our discussions, we shall, however, use the normalization leading to (2.39) for the flux vector.

In an experiment, one arranges to have some incident flux, and one measures the probability that the projectile is scattered into a detector. The detector is characterized by a solid angle of acceptance, and one observes the rate at which projectiles are scattered into a given solid angle, $d\Omega$, subtended at the detector. Thus, one measures the probability per second of a projectile being scattered into the solid angle $\sin\theta d\theta d\varphi = d\Omega$. We accordingly define a scattered flux, $\vec{j}_{\text{scatt}}$, so that if $\hat{n}$ is a unit vector pointing in the direction of our detector,

(2.43) $\hat{n} \cdot \vec{j}_{\text{scatt}} = $ probability flux of scattering into $d\Omega$.

Both the incident and scattered flux have units of probability/unit area-sec. The area subtended at the detector by the solid angle $d\Omega$ is $r^2 d\Omega$, and the unit vector pointing in the direction of observation is $\hat{r}$. Therefore, the probability per sec of the projectile being scattered into the detector equals $\vec{j}_{\text{scatt}} \cdot \hat{r} r^2 d\Omega$. This should be proportional to the magnitude of the incident flux (provided only single collisions occur). Therefore,

$$(2.44) \qquad j_{\text{incid}}\, d\sigma = \frac{\text{probability of scattering into the detector}}{\text{sec}}$$

and the constant of proportionality, $d\sigma$, has units of area. It follows that

$$(2.45) \qquad j_{\text{incid}}\, d\sigma = (\hat{r} \cdot \vec{j}_{\text{scatt}}) r^2 \sin\theta d\theta d\varphi$$

or

$$(2.46) \qquad d\sigma = \frac{(\hat{r} \cdot \vec{j}_{\text{scatt}}) r^2 \sin\theta d\theta d\varphi}{j_{\text{incid}}}.$$

We have already calculated the incident flux, (2.38), for the incident plane wave portion of the Lippmann-Schwinger solution of the Schrödinger equation, and the scattered wavefunction is given by noting that (2.21) can be written as

$$(2.47) \qquad \psi_{\vec{k}}^{+}(E|\vec{r}) = \psi_{\text{incid}} + \psi_{\text{scatt}}$$

so

$$(2.48) \qquad \psi_{\vec{k},\text{scatt}}^{+}(E|\vec{r}) = -\frac{m}{2\pi\hbar^2} \int d\vec{r}'\, \frac{e^{ik|\vec{r}-\vec{r}'|}}{|\vec{r}-\vec{r}'|} V(\vec{r}') \psi_{\vec{k}}^{+}(E|\vec{r}').$$

Now the detector is typically located far from the target, so we really only need $\psi_{\vec{k},\text{scatt}}^{+}(E|\vec{r})$ at large $\vec{r}$ (the so-called "far field scattered wave"). The standard procedure is to perform Taylor expansions of $|\vec{r} - \vec{r}'|$ in the exponent and denominator in the integral in (2.48). It is important to notice, however, that $d\sigma$ contains a factor r^2, so that any contribution to $\vec{j}_{\text{scatt}}$ that decays faster than $1/r^2$ can be made arbitrarily small by positioning the detector a large enough r. (In fact, this argument requires a sufficiently fast decay of the potential as r becomes large.) In calculating the scattered flux, we must apply $\vec{\nabla}$, and it has the general structure

$$(2.49) \qquad \vec{\nabla} = \hat{r}\frac{\partial}{\partial r} + \frac{1}{r}\mathcal{O}(\theta,\varphi),$$

where $\mathcal{O}(\theta,\varphi)$ is an operator in the angular variables. The leading term of $\psi_{\vec{k},\text{scatt}}^{+}(E|\vec{r})$ decays as $1/r$ due to the $1/|\vec{r} - \vec{r}'|$ factor,

$$(2.50) \qquad \frac{1}{|\vec{r}-\vec{r}'|} = \frac{1}{r|\hat{r} - \vec{r}'/r|}.$$

The role of the potential is to bound the largest r' contributing significantly to the integral over $d\vec{r}'$ in (2.48); then large r means large with respect to the largest r' contributing to the integral.

The Taylor expansion of the phase of the Green's function, $ik|\vec{r} - \vec{r}'|$, must include one higher term than the expansion of $1/|\vec{r} - \vec{r}'|$, and the result for the far field scattered wave is

$$(2.51) \qquad \psi^+_{\vec{k},\text{scatt}}(E|\vec{r}) \rightarrow \frac{e^{ikr}}{r} f(\theta, \varphi),$$

where θ and φ are the angles of observation (which determine the unit vector, $\hat{r}$), and

$$(2.52) \qquad f(\theta, \varphi) = -\frac{m}{2\pi\hbar^2} \int d\vec{r}' e^{-ik\hat{r}\cdot\vec{r}'} V(\vec{r}') \psi^+_{\vec{k}}(E|\vec{r}').$$

A moment's consideration shows that $\chi^* \vec{\nabla} \chi$ and $\chi \vec{\nabla} \chi^*$ will fall off at least as fast as $1/r^3$ so far as contributions from the $\Theta(\theta, \varphi)/r$ portion of $\vec{\nabla}$ is concerned. It follows that these terms will be negligible at the detector, and the *only* term which survives the large r-limit comes from the $\hat{r}\frac{\partial}{\partial r}$ piece of $\vec{\nabla}$. It is a simple calculation to show that

$$(2.53) \qquad \vec{j}_{\text{scatt}} = -\frac{i\hbar}{2m}\hat{r}$$
$$\left[\frac{e^{-ikr}}{r} f^*(\hat{r})\frac{ik}{r}e^{ikr} f(\hat{r}) - \frac{e^{ikr}}{r} f(\hat{r})\frac{(-ik)}{r}e^{ikr} f^*(\hat{r})\right]$$

$$(2.54) \qquad = \hat{r}\frac{\hbar k}{mr^2}|f(\hat{r})|^2.$$

Thus, $\hat{r} \cdot \vec{j}_{\text{scatt}}$ is trivially calculated, and

$$(2.55) \qquad d\sigma = \frac{(\hbar k/mr^2)}{(\hbar k/m)}|f(\hat{r})|^2 r^2 \sin\theta d\theta d\varphi$$

or

$$(2.56) \qquad \frac{d\sigma}{d\Omega} = |f(\theta, \varphi)|^2 = |f(\hat{r})|^2.$$

The quantity $d\sigma/d\Omega$ is called the "differential scattering cross section", and it is totally determined by $f(\theta, \varphi)$, given by (2.52). It is the size that the target presents to the projectile which causes it to be scattered into the solid angle $d\Omega$. If one "sums" over all solid angles, the result is the total size, σ, presented by the target that causes scattering of the projectile:

$$(2.57) \qquad \sigma = \int d\Omega \frac{d\sigma}{d\Omega}.$$

An extremely important point is that although the analysis of the cross sections (and standard experimental geometries) involves the far field behavior of $\psi^+_{\vec{k}}(E|\vec{r})$, one does *not* have to carry out calculations all the way to the far field region in order to obtain $f(\hat{r})$. Rather, one needs only know $\psi^+_{\vec{k}}(E|\vec{r})$ in the region where the potential is significantly different from zero (since $f(\hat{r})$ is determined by the integral (2.52)).

We shall complete this Section by considering the derivation of the Lippmann-Schwinger equation and Schrödinger equation by a more general wavepacket approach [4]. We begin again with (2.1), but we now solve it formally, exactly, by the standard method of integrating factors. Multiplying (2.1) by $\exp(iHt/\hbar)$, we have

$$(2.58) \qquad i\hbar e^{iHt/\hbar}\frac{\partial\chi}{\partial t} = e^{iHt/\hbar}H\chi,$$

and since $\exp(iHt/\hbar)$ commutes with H, we find

$$(2.59) \qquad \frac{\partial}{\partial t}\left\{e^{iHt/\hbar}\chi(t)\right\} = 0,$$

implying that $\exp(iHt/\hbar)\chi(t)$ is constant in time (but not necessarily as a function of position). Therefore,

$$(2.60) \qquad \chi(t) = e^{-iHt/\hbar}C,$$

and substituting $t = 0$ yields

$$(2.61) \qquad C = \chi(0).$$

Therefore, our formal solution of (2.1) is

$$(2.62) \qquad \chi(t) = e^{-iHt/\hbar}\chi(0),$$

and knowledge of χ at time $t = 0$ determines $\chi(t)$ at *all* other times. Rather than seek solutions of the form of (2.3), we now *define* the definite energy time independent state $\xi^+(E)$ by the "half-Fourier transform",

$$(2.63) \qquad \xi^+(E) \equiv \frac{1}{2\pi}\int\limits_0^\infty \frac{dt}{\hbar}e^{iEt/\hbar}\chi(t).$$

The superscript "+" denotes the fact that the time integral runs from 0 to $+\infty$ (the causal sense). Thus, in contrast to the ansatz $\chi(t) = \exp(-iEt/\hbar)\psi(E)$, which yielded the TISE for $\psi(E)$ when substituted into the TDSE, the TI state vector $\xi^+(E)$ does *not* correspond to assuming $\chi(t)$ to contain a single energy.

It rather allows $\chi(t)$ to be a superposition of states of a range of energy, and therefore $\chi(t)$ so constructed is a *wavepacket*. We consider the projection of this state onto an eigenstate of position, $\langle x|\xi^+(E)\rangle \equiv \xi^+(E|x)$, associated with the position x. For pedagogic purposes, we treat a one dimensional (1D) system, but the analysis is general. By (2.63), $\xi^+(E|x)$ is given by

$$(2.64) \qquad \xi^+(E|x) = \frac{1}{2\pi} \int_0^\infty \frac{dt}{\hbar} e^{iEt/\hbar} \chi(t|x),$$

and although this integral does not converge absolutely (in either 1D or 2D; it does converge absolutely in $3 - D$), it does converge due to oscillations. For convenience we take $\chi(0|x)$ to have compact support on $[x_1, x_2]$. Thus, for x outside this region, $\chi(t|x)$ initially vanishes. As time progresses and the packet moves and spreads, $\chi(t|x)$ becomes nonzero. Eventually, $\chi(t|x)$ decays (as the packet moves and spreads elsewhere) except for highly oscillatory portions whose time integral is arbitrarily small. Thus, $\xi^+(E|x)$, for *any* particular x, has a well defined value. If (2.63) is viewed as an *abstract* equation, the time integral does not converge in norm because there is always a value of x beyond where the packet has moved and spread up to any time T. Using (2.62), we rewrite (2.64) as

$$(2.65) \qquad \xi^+(E|x) = \frac{1}{2\pi} \int_0^\infty \frac{dt}{\hbar} e^{iEt/\hbar} e^{-iHt/\hbar} \chi(0|x),$$

where H is understood to be in the coordinate representation,

$$(2.66) \qquad H = -\frac{\hbar}{2m}\frac{d^2}{dx^2} + V(x).$$

Now because the integral over t has a well defined value, we can formally integrate (2.65) by inserting a factor which only affects the integrand at times t which are longer than the time T at which the integral has attained its correct value. So long as these modified terms are sufficiently small, they will not contribute anything to the integral, and the value of the integral will be unaffected. We do this by inserting the factor $\exp(-\epsilon t/\hbar)$, where ϵ is so small that for $t < T$, $\exp(-\epsilon t/\hbar) \approx 1$, while for $t > T$, $\exp(-\epsilon t/\hbar) \approx 0$, and the value of the integral is then unchanged. Then

$$(2.67) \qquad \xi^+(E|x) = \frac{1}{2\pi} \int_0^\infty \frac{dt}{\hbar} e^{i(E-H+i\epsilon)/\hbar} \chi(0|x),$$

$$(2.68) \qquad = \frac{i}{2\pi} \left\langle x \left| \frac{1}{(E - H + i\epsilon)} \right| \chi(0) \right\rangle,$$

This equation looks similar to the formal solution of the Lippmann-Schwinger equation. To see this, note that

$$(2.69) \qquad E - H + i\epsilon = E - H_0 - V + i\epsilon,$$

apply $(E - H + i\epsilon)^{-1}$ from the left and $(E - H_0 + i\epsilon)^{-1}$ from the right, to obtain

$$(2.70) \quad \frac{1}{(E - H_0 + i\epsilon)} = \frac{1}{(E - H + i\epsilon)} - \frac{1}{(E - H + i\epsilon)} V \frac{1}{(E - H_0 + i\epsilon)}.$$

Similarly, one can show that

$$(2.71) \quad \frac{1}{(E - H_0 + i\epsilon)} = \frac{1}{(E - H + i\epsilon)} - \frac{1}{(E - H_0 + i\epsilon)} V \frac{1}{(E - H + i\epsilon)}.$$

Using (2.70) in (2.12) gives

$$\psi^+(E) = \phi(E) + \frac{1}{(E - H + i\epsilon)} V \psi^+(E)$$

$$(2.72) \qquad - \frac{1}{(E - H + i\epsilon)} V \frac{1}{(E - H_0 + i\epsilon)} V \psi^+(E),$$

and using (2.12) to substitute for $(E - H_0 + i\epsilon)^{-1} V \psi^+(E)$ on the right hand side (RHS) of (2.72) gives

$$(2.73) \qquad \psi^+(E) = \phi(E) + \frac{1}{(E - H + i\epsilon)} V \phi(E).$$

But we can consider the abstract form of (2.68):

$$(2.74) \qquad \xi^+(E) = \frac{i}{2\pi} \frac{1}{(E - H + i\epsilon)} \chi(0),$$

and use (2.70) to write this as

$$\xi^+(E) = \frac{i}{2\pi} \frac{1}{(E - H_0 + i\epsilon)} \chi(0) + \frac{1}{(E - H + i\epsilon)} V \frac{i}{2\pi} \frac{1}{(E - H_0 + i\epsilon)} \chi(0).$$
$$(2.75)$$

Clearly, by defining

$$(2.76) \qquad \xi_0^+(E) = \frac{i}{2\pi} \frac{1}{(E - H_0 + i\epsilon)} \chi(0),$$

we obtain

$$(2.77) \qquad \xi^+(E) = \xi_0^+(E) + \frac{1}{(E - H + i\epsilon)} V \xi_0^+(E),$$

which corresponds to (2.73). We refer to both (2.74) and (2.77) as the time-independent wavepacket Lippmann-Schwinger equation (TIWLSE).

There is also a form of the TIWLSE which corresponds to the LS equation, (2.12). We substitute for $(E - H + i\epsilon)^{-1}$ in (2.77) using (2.71):

$$\xi^+(E) = \xi_0^+(E) + \frac{1}{(E - H_0 + i\epsilon)} V \xi_0^+(E)$$
$$(2.78) \qquad + \frac{1}{(E - H_0 + i\epsilon)} V \frac{1}{(E - H + i\epsilon)} V \xi_0^+(E).$$

Then replace $(E - H + i\epsilon)^{-1}V\xi_0^+(E)$ on the RHS of (2.78) by (2.77), with the result being

$$(2.79) \qquad \xi^+(E) = \xi_0^+(E) + \frac{1}{(E - H_0 + i\epsilon)}V\xi^+(E).$$

In order to proceed further in comparing the TIWLSE with the standard TILSE, we must analyze $\xi_0^+(E)$ in more detail. To do this, we must be more specific about our choice of $t = 0$ wavepacket, $\chi(0|x)$, and the potential . We shall assume $\chi(0|x)$ to be nonzero on the closed interval $[x_1, x_2]$, the potential $V(x)$ to be nonzero on the closed interval $[x_3, x_4]$, and $x_2 < x_3$ (so $\chi(0)$ and V have compact support, are non-overlapping, and $\chi(0)$ lies to the left of V). Let us first examine what $\xi_0^+(E)$ looks like on the "potential side" of $\chi(0|x)$ (so $x > x_2$). Then

$$(2.80) \qquad \langle x|\xi_0^+(E)\rangle = \frac{i}{2\pi}\langle x|G_0^+(E)|\chi(0)\rangle,$$

and using the resolution of the identity,

$$(2.81) \qquad 1 = \int\limits_{-\infty}^{\infty} dx'|x'\rangle\langle x'|,$$

$$(2.82) \qquad \xi_0^+(E|x) = \frac{i}{2\pi}\int\limits_{-\infty}^{\infty} dx'\langle x|G_0^+(E)|x'\rangle\chi(0|x'),$$

with

$$(2.83) \qquad \langle x|G_0^+(E)|x'\rangle = -\frac{mi}{\hbar^2 k}e^{ik|x-x'|},$$

$$(2.84) \qquad\qquad\qquad\qquad k > 0.$$

It follows that for $x > x_2$,

$$(2.85) \qquad \xi_0^+(E|x) = \frac{me^{ikx}}{2\pi\hbar^2 k}\int\limits_{x_1}^{x_2} dx'e^{-ikx'}\chi(0|x'),$$

where we have used the fact that $\chi(0|x')$ is nonzero only on $[x_1, x_2]$. But defining $A(k)$ by

$$(2.86) \qquad A(k) = \frac{1}{2\pi}\int\limits_{x_1}^{x_2} dx'e^{-ikx'}\chi(0|x'),$$

and noting that due to the compact support of $\chi(0|x')$,

$$(2.87) \qquad A(k) = \frac{1}{2\pi} \int_{-\infty}^{\infty} dx' e^{-ikx'} \chi(0|x'),$$

we find

$$(2.88) \qquad \xi_0^+(E|x) = \frac{mA(k)}{\hbar^2 k} e^{ikx}, x > x_2, k > 0,$$

and $A(k)$ is the Fourier component of $\chi(0|x)$ with momentum $\hbar k$. This is precisely the piece of $\chi(0)$ which generates flux with the correct energy, $E = \hbar^2 k^2 / 2m$, and which moves toward the potential (target).

Next, we consider the region $x < x_1$ (on the side of $\chi(0)$ which is away from the region of the potential). It is easily seen from (2.82)–(2.83), that for $x < x_1$,

$$(2.89) \qquad \xi_0^+(E|x) = \frac{me^{-ikx}}{2\pi\hbar^2 k} \int_{x_1}^{x_2} dx' e^{ikx'} \chi(0|x')$$

$$(2.90) \qquad = \frac{me^{-ikx}}{\hbar^2 k} \frac{1}{2\pi} \int_{x_1}^{x_2} dx' (e^{i(-k)x'})^* \chi(0|x'),$$

from which we deduce that

$$(2.91) \qquad \xi_0^+(E|x) = \frac{mA(-k)}{\hbar^2 k} e^{-ikx}, x < x_1.$$

This is the piece of $\chi(0)$ which has momentum $(-\hbar k)$, consistent with the energy E, and it generates flux which moves *away* from the region of the potential. Using these results, we can now obtain $\xi^+(E|x)$ in the two regions of interest. For $x > x_2$ (the potential side of $\chi(0|x)$), we obtain

$$(2.92) \qquad \xi^+(E|x) = \frac{mA(k)}{\hbar^2 k} e^{ikx} + \int_{-\infty}^{\infty} dx' G^+(E|x, x') V(x') \xi_0^+(E|x'),$$

and since $V(x')$ is nonzero *only* for $x' > x_2, \xi_0^+(E|x')$ can be replaced by (2.88):

$$(2.93) \qquad \xi^+(E|x) = \frac{mA(k)}{\hbar^2 k} \left[e^{ikx} + \int_{x_3}^{x_4} dx' G^+(E|x, x') V(x') e^{ikx'} \right],$$
$$x > x_2.$$

The quantity in brackets is recognized as the coordinate representation of (2.73) for $\psi_{\underline{k}}^+(E)$, so

$$(2.94) \qquad \xi^+(E|x) = \frac{mA(k)}{\hbar^2 k} \psi_{\underline{k}}^+(E|x), x > x_2.$$

In the region $x < x_1$, we note that the first term on the RHS of (2.77) will give rise to $[mA(-k)/\hbar^2 k]\exp(-ikx)$, but the second term will be identical to the second term on the RHS of (2.93). This is the consequence of the presence of the potential in the integral, which has compact support in the region $x > x_2$. Thus, one obtains

$$\xi^+(E|x) = \frac{mA(-k)}{\hbar^2 k}e^{-ikx} + \frac{mA(k)}{\hbar^2 k}\int_{x_3}^{x_4} dx' G^+(E|x,x')V(x')e^{ikx'}, x < x_1,$$

(2.95)

so that

(2.96) $$\xi^+(E|x) = \frac{mA(-k)}{\hbar^2 k}e^{-ikx} + \frac{mA(k)}{\hbar^2 k}\psi_{k,SW}^+(E|x).$$

If we remove the incident packet so far from the target that one *cannot* probe it (so x_2 is moved infinitely far to the left of x_3), then in all accessible regions, $x > x_2$ and then

(2.97) $$\xi^+(E|x) = \frac{mA(k)}{\hbar^2 k}\psi_k^+(E|x).$$

The above can be viewed as a rigorous (from the *physical* point of view) derivation of the TILSE.

Finally, the abstract form of the TIWLSE can be used to derive the corresponding time-independent wavepacket Schrödinger equation (TIWSE). We simply apply $(E - H + i\epsilon)$ to (2.74), and then take the $\epsilon \to 0_+$ limit with impunity, obtaining thereby

(2.98) $$(E - H)\xi(E) = \frac{i}{2\pi}\chi(0).$$

It is distinguished from the standard TISE by the presence of the $t = 0$ wavepacket; thus, the TIWSE retains a memory of the experimental setup associated with the quantum dynamics. Again, in the limit that $\chi(0)$ is removed infinitely far from the region of the potential, (2.98) is replaced by the standard, homogeneous TISE. Equation (2.98) is a more general form, and it gives rise to very significant changes in one's computational strategy. The use of the TIWLSE and TIWSE as the bases of computational schemes will be the topic of a separate contribution to these Workshop proceedings. We now turn to a consideration of a variety of computational schemes for solving quantum scattering problems.

3. Time-independent quantum scattering computational approaches.

3.1. Basis set expansion solution of the TILSE. We begin by discussing one very robust method for converting the TI quantum scattering problem into one which deals with quadratically integrable functions (so one works in a Hilbert space). We define the "amplitude density", $\zeta_{\vec{k}}^+(E)$, to be [5]

$$(3.1) \qquad \zeta_{\vec{k}}^+(E) \equiv V\psi_{\vec{k}}^+(E).$$

Provided V tends to zero sufficiently rapidly, and is not too singular at the origin, the (state) vector $\zeta_{\vec{k}}^+(E)$ will be $\mathcal{L}^2$, and can be expanded in terms of a discrete, $\mathcal{L}^2$-basis [6]. By (2.21), the equation satisfied by $\zeta_{\vec{k}}^+(E|\vec{r})$ can be written as

$$(3.2) \qquad \zeta_{\vec{k}}^+(E|\vec{r}) = V(\vec{r})e^{i\vec{k}\cdot\vec{r}} - \frac{m}{2\pi\hbar^2}\int d\vec{r}' V(\vec{r})\frac{e^{ik(\vec{r}-\vec{r}')}}{|\vec{r}-\vec{r}'|}\zeta_{\vec{k}}^+(E|\vec{r}'),$$

and $\zeta_{\vec{k}}^+(E|\vec{r})$ is expanded as

$$(3.3) \qquad \zeta_{\vec{k}}^+(E|\vec{r}) = \sum_j z_j\Omega_j(\vec{r}),$$

and if the $\Omega_j(\vec{r})$ are orthonormal,

$$(3.4) \qquad z_{j'} = \langle\Omega_{j'}|V|\vec{k}\rangle - \frac{m}{2\pi\hbar^2}\sum_j \langle\Omega_{j'}|VG_0^+(E)|\Omega_j\rangle z_j.$$

The reader should note that the amplitude density is intimately connected with the differential scattering amplitude, $f(\hat{r})$, since by (2.52) and (3.1),

$$(3.5) \qquad f(\hat{r}) = -\frac{m}{2\pi\hbar^2}\int d\vec{r}' e^{-ik\hat{r}\cdot\vec{r}'}\zeta_{\vec{k}}^+(E|\vec{r}').$$

The linear, inhomogeneous algebraic equations for the z_j, (3.4), are truncated and solved. The resulting z_j, along with quantities $C_j(\hat{r})$ defined by

$$(3.6) \qquad C_j(\hat{r}) = \int d\vec{r}' e^{-ik\hat{r}\cdot\vec{r}'}\Omega_j(\vec{r}'),$$

are then used to construct $f(\hat{r})$ according to

$$(3.7) \qquad f(\hat{r}) = -\frac{m}{2\pi\hbar^2}\sum_j C_j(\hat{r})z_j.$$

Clearly, by (3.1), the amplitude density has the *same* compact support as the potential, $V(\vec{r})$.

Analogous $\mathcal{L}^2$-radial amplitude densities can be defined in connection with the partial wave decomposition of the TILSE, (2.27). In this case, a matrix notation is convenient, and we define the matrix radial amplitude density, $\overleftrightarrow{\zeta}^+(r)$, by

$$(3.8) \qquad [\overleftrightarrow{\zeta}(r)]_{\ell''\mu'',\ell\mu} = \sum_{\ell'}\sum_{\mu'} V(\ell''\mu''\ell'\mu'|r)g^+(\ell'\mu'\ell\mu|r)$$

$$(3.9) \qquad\qquad\qquad = \zeta(\ell''\mu''\ell\mu|r).$$

Then the matrix radial amplitude density equation becomes

$$\overset{\leftrightarrow}{\zeta}(r) = \overset{\leftrightarrow}{V}(r) \cdot \overset{\leftrightarrow}{j}(r) - \frac{2m}{\hbar^2} \overset{\leftrightarrow}{V}(r) \cdot \overset{\leftrightarrow}{k} \cdot \int_0^\infty dr' r'^2 \, \overset{\leftrightarrow}{h}{}^+ (r_>) \cdot \overset{\leftrightarrow}{j}(r_<) \cdot \overset{\leftrightarrow}{\zeta}(r'),$$

(3.10)
where

(3.11) $$[\overset{\leftrightarrow}{V}(r)]_{\ell'\mu'\ell\mu} = V(\ell'\mu'\ell\mu|r),$$

(3.12) $$[\overset{\leftrightarrow}{j}(r)]_{\ell'\mu'\ell\mu} = \delta_{\ell\ell'}\delta_{\mu\mu'} j_\ell(kr),$$

(3.13) $$[\overset{\leftrightarrow}{h}{}^+(r)]_{\ell'\mu'\ell\mu} = \delta_{\ell\ell'}\delta_{\mu\mu'} h_\ell^+(kr),$$

(3.14) $$[\overset{\leftrightarrow}{k}]_{\ell'\mu'\ell\mu} = \delta_{\ell\ell'}\delta_{\mu\mu}k,$$

and

(3.15) $$\frac{\hbar^2 k^2}{2m} = E.$$

For reasonable potentials (i.e., that die off sufficiently rapidly at large r and are not too singular at $r = 0$), each element of $\overset{\leftrightarrow}{\zeta}(r)$ is $\mathcal{L}^2$ and can be expanded in a discrete, $\mathcal{L}^2$-basis. This again generates an infinite linear system of algebraic equations which must be truncated and solved. The scattering amplitude is constructed from the solution by using (3.5), combined with (2.26), and the partial wave expansion of the plane wave,

(3.16) $$e^{-ik\hat{r}\cdot\vec{r}'} = 4\pi \sum_{\ell''} \sum_{\mu''} i^{-\ell''} Y_{\ell''\mu''}(\hat{r}) Y_{\ell''\mu''}^*(\hat{r}') j_{\ell''}(kr').$$

The result is

$$f(\hat{r}) = -\frac{8\pi m}{\hbar^2} \sum_{\ell''\mu''} \sum_{\ell\ell'} \sum_{\mu\mu'} i^{\ell-\ell''}$$

$$\left[\int_0^\infty dr' r'^2 j_{\ell''}(kr') \int d\hat{r}' Y_{\ell''\mu''}^*(\hat{r}') Y_{\ell\mu}(\hat{r}') V(\vec{r}') g^+(\ell\mu\ell'\mu'|r') \right]$$

(3.17) $$Y_{\ell''\mu''}(\hat{r}) Y_{\ell'\mu'}^*(\hat{k}),$$

and noting the definitions

(3.18) $$V(\ell''\mu''\ell\mu|r) = \int d\hat{r}' Y_{\ell''\mu''}^*(\hat{r}') Y_{\ell\mu}(\hat{r}') V(\vec{r}'),$$

$$(3.19) \qquad \zeta^+(\ell''\mu''\ell'\mu'|r') = \sum_\ell \sum_\mu V(\ell''\mu''\ell\mu|r')g^+(\ell\mu\ell'\mu'|r'),$$

one obtains

$$f(\hat{r}) = -\frac{8\pi m}{\hbar^2} \sum_{\ell''\mu''} \sum_{\ell'\mu'} i^{\ell'-\ell''} Y_{\ell''\mu''}(\hat{r}) Y^*_{\ell'\mu'}(\hat{k}) \int_0^\infty dr' r'^2 j_{\ell''}(kr') \zeta^+(\ell''\mu''\ell'\mu'|r').$$

(3.20)

Obviously, knowing the expansion coefficients for expanding the $\zeta^+(\ell''\mu''\ell'$ $\mu'|r')$ in terms of an $\mathcal{L}^2$-basis provides the information to calculate $f(\hat{r})$. Such a computational approach does work (and has been applied to inelastic atom-diatom scattering and to reactive atom-diatom scattering, below the three-body threshold). We comment in closing our discussion of this approach that it has some similarity to a well known procedure in the analysis of the Lippmann-Schwinger equation. Thus, a standard method for deriving an integral equation with a compact kernel is to consider $V^{1/2}\psi^+_{\vec{k}}(E)$, rather than $\psi^+_{\vec{k}}(E)$ [2]. The kernel then becomes $V^{1/2}G_0^+(E)V^{1/2}$, and provided V behaves appropriately, is compact. Thus, defining $\tilde{\zeta}^+_{\vec{k}}(E)$ by

$$(3.21) \qquad\qquad \tilde{\zeta}^+_{\vec{k}}(E) = V^{1/2}\psi^+_{\vec{k}}(E),$$

and using (2.11), we have

$$(3.22) \qquad \tilde{\zeta}^+_{\vec{k}}(E) = V^{1/2}\phi_{\vec{k}}(E) + V^{1/2}G_0^+(E)V^{1/2}\tilde{\zeta}^+_{\vec{k}}(E)$$

as the "modified LSE". Provided the potential is $\mathcal{L}^1$, the resulting $\tilde{\zeta}^+_{\vec{k}}(E)$ will be $\mathcal{L}^2$. So far as we are aware, (3.22) has never been used as the basis of a computational algorithm. The success of the amplitude density as a computational tool suggests that it might be worthwhile to pursue similar methods with equations like (3.22).

3.2. Solution by conversion of the TILSE to Volterra equation form.

An alternative, non-iterative method for solving the radial TILSE is based on the fact that rather than treating it is an inhomogeneous Fredholm integral equation of the second kind, it can be converted into an inhomogeneous Volterra integral equation of the second kind. This equation has the extremely attractive property of possessing a triangular kernel, implying that under very weak conditions on the potential, *iterative* solutions converge absolutely irrespective of the strength of the interaction [2]. (By contrast, the radial TILSE Fredholm integral equation converges under iteration only under extremely restrictive conditions on the potential.) We do not advocate iterative methods, however, because the non-iterative approach is extremely simple and robust. To illustrate, it is sufficient to consider scattering by a spherically symmetric (central) potential [7]. In that case, we deal with (2.29), which we write in terms of Ricatti-Bessel functions by defining

$$(3.23) \qquad\qquad U_\ell^+(r) = r g_\ell^+(r),$$
$$(3.24) \qquad\qquad \tilde{j}_\ell(kr) = kr j_\ell(kr),$$
$$(3.25) \qquad\qquad \tilde{h}_\ell^+(kr) = kr h_\ell^+(kr),$$

so that

$$(3.26) \qquad\qquad \tilde{j}_\ell(kr) \xrightarrow[\text{large } r]{} \sin(kr - \ell\pi/2),$$
$$(3.27) \qquad\qquad \tilde{h}_\ell(kr) \xrightarrow[\text{large } r]{} e^{i(kr - \ell\pi/2)}.$$

It will also be convenient to introduce the Ricatti-Neumann function $\tilde{n}_\ell(kr)$, such that

$$(3.28) \qquad\qquad \tilde{h}_\ell^+(kr) = \tilde{n}_\ell(kr) + i\tilde{j}_\ell(kr).$$

It follows that

$$(3.29) \qquad\qquad \tilde{n}_\ell(kr) \xrightarrow[\text{large } r]{} \cos(kr - \ell\pi/2).$$

Using (3.23)–(3.25), we find that $U_\ell^+(r)$ satisfies

$$(3.30) \quad U_\ell^+(r) = \frac{\tilde{j}_\ell(kr)}{k} - \frac{2m}{\hbar^2 k} \int_0^\infty dr' \tilde{h}_\ell^+(kr_>) \tilde{j}_\ell(kr_<) V(r') U_\ell^+(r').$$

Substituting (3.28) into (3.30) above results in the equation

$$U_\ell^+(r) = \tilde{j}_\ell(kr) \left[\frac{1}{k} - \frac{2mi}{\hbar^2 k} \int_0^\infty dr' \tilde{j}_\ell(kr') V(r') U_\ell^+(r') \right]$$

$$(3.31) \qquad\qquad - \frac{2m}{\hbar^2 k} \int_0^\infty dr' \tilde{n}_\ell(kr_>) \tilde{j}_\ell(kr_<) V(r') U_\ell^+(r').$$

Now although we do not know the value of the factor $[1/k - (2mi/\hbar^2 k) \int_0^\infty dr' \tilde{j}_\ell(kr') V(r') U_\ell^+(r')]$, we *do* know that it is a *constant*, C_ℓ, independent of r:

$$(3.32) \qquad\qquad C_\ell = \left[\frac{1}{k} - \frac{2mi}{\hbar^2 k} \int_0^\infty dr' \tilde{j}_\ell(kr') V(r') U_\ell^+(r') \right].$$

Therefore, (3.30) can be written as

$$(3.33) \quad U_\ell^+(r) = \tilde{j}_\ell(kr)C_\ell - \frac{2m}{\hbar^2 k} \int_0^\infty dr' \tilde{n}_\ell(kr_>)\tilde{j}_\ell(kr_<)V(r')U_\ell^+(r').$$

This suggests that we define an auxiliary function, $\mathcal{V}_\ell(r)$, by

$$(3.34) \qquad\qquad \mathcal{V}_\ell(r) \equiv U_\ell^+(r)/C_\ell,$$

and one easily verifies that $\mathcal{V}_\ell(r)$ satisfies the integral equation

$$(3.35) \quad \mathcal{V}_\ell(r) = \tilde{j}_\ell(kr) - \frac{2m}{\hbar^2 k} \int_0^\infty dr' \tilde{n}_\ell(kr_>)\tilde{j}_\ell(kr_<)V(r')\mathcal{V}_\ell(r'),$$

where $\mathcal{V}_\ell(r)$ is *real*. (This is because the real and imaginary parts of $\mathcal{V}_\ell(r)$ satisfy uncoupled equations, with the imaginary part being a *homogenous* equation. At positive or scattering energies, the homogenous equation has only the trivial solution, $Im\mathcal{V}_\ell(r) = 0$, except for possible bound states embedded in the continuum [2].) The equation for $\mathcal{V}_\ell(r)$ is, of course, also an inhomogeneous Fredholm equation of the second kind, with the same structure as the equation for $U_\ell(r)$. However, we show that it can be converted into an inhomogeneous Volterra integral equation of the second kind by the following procedure. We eliminate the $r_>$ and $r_<$ variables in (3.35) by splitting the r'-integral up into an integral from 0 to r, plus an integral from r to ∞:

$$\mathcal{V}_\ell(r) = \tilde{j}_\ell(kr) - \frac{2m}{\hbar^2 k}\tilde{n}_\ell(kr) \int_0^r dr'\tilde{j}_\ell(kr')V(r')\mathcal{V}_\ell(r')$$

$$(3.36) \qquad\qquad - \frac{2m}{\hbar^2 k}\tilde{j}_\ell(kr) \int_r^\infty dr'\tilde{n}_\ell(kr')V(r')\mathcal{V}_\ell(r').$$

We next add and subtract

$$(3.37) \qquad\qquad - \frac{2m}{\hbar^2 k}\tilde{j}_\ell(kr) \int_0^r dr'\tilde{n}_\ell(kr')V(r')\mathcal{V}_\ell(r'),$$

so (3.35) becomes

$$\mathcal{V}_\ell(r) = \tilde{j}_\ell(kr) \left[1 - \frac{2m}{\hbar^2 k} \int_0^\infty dr'\tilde{n}_\ell(kr')V(r')\mathcal{V}_\ell(r') \right]$$

$$- \frac{2m}{\hbar^2 k}\tilde{n}_\ell(kr) \int_0^r dr'\tilde{j}_\ell(kr')V(r')\mathcal{V}_\ell(r')$$

$$(3.38) \qquad\qquad + \frac{2m}{\hbar^2 k} \tilde{j}_\ell(kr) \int_0^r dr' \tilde{n}_\ell(kr') V(r') \mathcal{V}_\ell(r').$$

In analogy to the conversion from an equation for $U_\ell(r)$ to one for $\mathcal{V}_\ell(r)$, we note that the quantity, D_ℓ, defined as

$$(3.39) \qquad\qquad D_\ell \equiv 1 - \frac{2m}{\hbar^2 k} \int_0^\infty dr' \tilde{n}_\ell(kr') V(r') \mathcal{V}_\ell(r'),$$

although unknown, is nevertheless a *constant* independent of r. Thus, (3.38) is of the form

$$\mathcal{V}_\ell(r) = \tilde{j}_\ell(kr) D_\ell - \frac{2m}{\hbar^2 k} \tilde{n}_\ell(kr) \int_0^r dr' \tilde{j}_\ell(kr') V(r') \mathcal{V}_\ell(r')$$

$$(3.40) \qquad\qquad + \frac{2m}{\hbar^2 k} \tilde{j}_\ell(kr) \int_0^r dr' \tilde{n}_\ell(kr') V(r') \mathcal{V}_\ell(r'),$$

and we factor out D_ℓ in a manner similar to our factoring out C_ℓ. We define $W_\ell(r)$ by

$$(3.41) \qquad\qquad \mathcal{V}_\ell(r) = W_\ell(r) D_\ell,$$

and easily find that

$$W_\ell(r) = \tilde{j}_\ell(kr) - \frac{2m}{\hbar^2 k} \tilde{n}_\ell(kr) \int_0^r dr' \tilde{j}_\ell(kr') V(r') W_\ell(r')$$

$$(3.42) \qquad\qquad + \frac{2m}{\hbar^2 k} \tilde{j}_\ell(kr) \int_0^r dr' \tilde{n}_\ell(kr') V(r') W_\ell(r'),$$

or by an obvious definition of the kernel $K_\ell(r, r')$,

$$(3.43) \qquad W_\ell(r) = \tilde{j}_\ell(kr) - \frac{2m}{\hbar^2 k} \int_0^r dr' K_\ell(r, r') V(r') W_\ell(r').$$

Equations (3.42)–(3.43) are recognized as inhomogeneous Volterra integral equations of the second kind. They are used extensively (although derived by totally different methods) in the analytical theory of the scattering matrix, as well as in the mathematical theory of the Lippmann-Schwinger equation.

We now demonstrate a noniterative computational algorithm based on (3.42) [7]. We approximate the integrals on the RHS by Newton-Cotes quadrature. For a uniform grid, we evaluate r at the point r_p, so by (3.43), this gives

$$(3.44) \quad W_\ell(r_p) = \tilde{j}_\ell(kr_p) - \frac{2m}{\hbar^2 k} \sum_{p'=0}^{p} q_{p'} \dot{K}_\ell(r_p, r_{p'}) V(r_{p'}) W_\ell(r_{p'});$$

for the trapezoidal rule, $q_0 = \Delta/2$, $q_{p'} = \Delta$ for $0 < p' < p$, and $q_p = \Delta/2$. Now consider $K_\ell(r_p, r_{p.})$:

$$(3.45) \qquad K_\ell(r_p, r_{p'}) = \tilde{n}_\ell(kr_p)\tilde{j}_\ell(kr_{p'}) - \tilde{n}_\ell(kr_{p'})\tilde{j}_\ell(kr_p),$$

so

$$(3.46) \qquad\qquad\qquad K_\ell(r_p, r_p) \equiv 0.$$

It follows then that

$$(3.47) \quad W_\ell(r_p) = \tilde{j}_\ell(kr_p) - \frac{2m}{\hbar^2 k} \sum_{p'=0}^{p-1} q_{p'} K_\ell(r_p, r_{p'}) V(r_{p'}) W_\ell(r_{p'}),$$

which we recognize as a recursion relation expressing $W_\ell(r_p)$ solely in terms of its values at preceding grid points. It follows that if we know $W_\ell(r_{p'})$ at *one point* (an initial point, usually the origin where $W_\ell(0) = 0$ or a known constant value), we can use (3.47) to generate $W_\ell(r_p)$ at all succeeding points. This really only must be done out to the point where the potential is negligible (and can be set equal to zero). It is convenient in the computational algorithm to use the form (3.42), whereby

$$W_\ell(r_p) = \tilde{j}_\ell(kr_p) - \frac{2m}{\hbar^2 k}\tilde{n}_\ell(kr_p) \sum_{p'=0}^{p-1} q_{p'}\tilde{j}_\ell(kr_{p'}) V(r_{p'}) W_\ell(r_{p'})$$

$$(3.48) \qquad\qquad + \frac{2m}{\hbar^2 k}\tilde{j}_\ell(kr_p) \sum_{p'=0}^{p-1} q_{p'}\tilde{n}_\ell(kr_{p'}) V(r_{p'}) W_\ell(r_{p'}),$$

or

$$W_\ell(r_p) = \tilde{j}_\ell(kr_p) - \frac{2m}{\hbar^2 k}\tilde{n}_\ell(kr_p) I_{1\ell}(p-1)$$

$$(3.49) \qquad\qquad + \frac{2m}{\hbar^2 k}\tilde{j}_\ell(kr_p) I_{2\ell}(p-1),$$

with

$$(3.50) \qquad I_{1\ell}(p-1) \equiv \sum_{p'=0}^{p-1} q_{p'}\tilde{j}_\ell(kr_{p'}) V(r_{p'}) W_\ell(r_{p'})$$

and

$$(3.51) \qquad I_{2\ell}(p-1) \equiv \sum_{p'=0}^{p-1} q_{p'} \tilde{n}_\ell(kr_{p'}) V(r_{p'}) W_\ell(r_{p'}).$$

This form is readily adapted to digital computers. Notice that as soon as $V(r_p) = 0$, I_1 and I_2 are converged, and in fact, they are quadrature approximations to the integrals

$$(3.52) \qquad I_{1\ell} \to \int_0^{r_{max}} dr' \tilde{j}_\ell(kr') V(r') W_\ell(r'),$$

$$(3.53) \qquad I_{2\ell} \to \int_0^{r_{max}} dr' \tilde{n}_\ell(kr') V(r') W_\ell(r').$$

Next consider D_ℓ, given by (3.39), along with (3.41). This leads to the equation

$$(3.54) \qquad D_\ell = 1 - \frac{2m}{\hbar^2 k} \int_0^\infty dr' \tilde{n}_\ell(kr') V(r') W_\ell(r') D_\ell,$$

or

$$(3.55) \qquad D_\ell = 1/[1 + \frac{2m}{\hbar^2 k} I_2].$$

We next note that by (3.32) and (3.34),

$$(3.56) \qquad C_\ell = \frac{1}{k} \left[1 - \frac{2mi}{\hbar^2} \int_0^\infty dr' \tilde{j}_\ell(kr') V(r') \mathcal{V}_\ell(r') C_\ell \right],$$

so

$$(3.57) \qquad C_\ell = \frac{(1/k)}{[1 + \frac{2mi}{\hbar^2 k} \int_0^\infty dr' \tilde{j}_\ell(kr') V(r') \mathcal{V}_\ell(r')]};$$

then use (3.41) to obtain

$$(3.58) \qquad C_\ell = \frac{(1/k)}{[1 + \frac{2mi}{\hbar^2 k} I_{1\ell} D_\ell]}.$$

Finally, using (3.55), C_ℓ is given by

$$(3.59) \qquad C_\ell = \frac{(1/k)[1 + \frac{2m}{\hbar^2 k} I_{2\ell}]}{[1 + \frac{2m}{\hbar^2 k} I_{2\ell} + \frac{2mi}{\hbar^2 k} I_{1\ell}]}.$$

Thus, we can compute *both* D_ℓ and C_ℓ once $I_{1\ell}$ and $I_{2\ell}$ are known. Furthermore, knowing D_ℓ and C_ℓ, we can compute $U_\ell^+(r)$ (on the grid), and from it, $g_\ell^+(r)$. Finally, knowing $g_\ell^+(r)$, we can compute $\psi_{\vec{k}}^+(E|\vec{r})$ by the appropriate sum over partial waves:

$$(3.60) \qquad \psi_{\vec{k}}^+(E|\vec{r}) = 4\pi \sum_\ell \sum_\mu i^\ell Y_{\ell\mu}(\hat{r}) Y_{\ell\mu}^*(\hat{k}) g_\ell^+(r).$$

The differential scattering amplitude is constructed as discussed earlier, and for the spherically symmetric potential, one can show that (taking a z-axis along the incident relative momentum, $\hbar\vec{k}$)

$$(3.61) \qquad f(\theta) = -\frac{2m}{\hbar^2 k} \sum_\ell (2\ell + 1) P_\ell(\cos\theta) C_\ell D_\ell I_{1\ell}.$$

Clearly, knowing $I_{1\ell}$ and $I_{2\ell}$ gives all the information needed to calculate the differential scattering amplitude and the related cross sections.

It is perhaps worthwhile to comment on the mechanism that leads to convergence of the partial wave expansion. The quantum number ℓ determines the magnitude of the orbital angular momentum of the projectile. Classically the magnitude of the orbital angular momentum is determined by the speed of the projectile, s, and its "impact parameter", d (the distance of closest approach in the absence of the potential). Thus, the correspondence principle leads to the association $msd \cong \hbar(\ell + 1/2)$. As ℓ increases (for fixed speed, s, determined by the total energy according to $ms^2/2 = E$), d also increases, leading to more grazing collisions. Eventually, ℓ is large enough so that d is larger than the range of the potential, and so for larger ℓ-values, there is no scattering. Quantum mechanically, due to tunneling, things are not so precise but essentially the tunneling is insufficient to permit significant sampling of the potential, and the sum over ℓ converges. Classically, the size of the target is infinite if the potential does not cut off at some finite r. Quantum mechanically, the target has finite size provided the potential decays faster than $1/r$.

When there is structure in the target or projectile, the radial wavefunction becomes a matrix, as in (3.8)–(3.20). In addition, one can apply the method to the amplitude density, as well as the radial wavefunction. We can illustrate the "multi-channel" generalization of the non-iterative Volterra algorithm using (3.10)–(3.14) [7]. One still can introduce the spherical Neumann functions (note that one does *not* have to use Ricatti-functions, although it is convenient), so that

$$(3.62) \qquad \overset{\leftrightarrow}{h}^+(r) = \overset{\leftrightarrow}{n}(r) + i\,\overset{\leftrightarrow}{j}(r),$$

and

$$(3.63) \qquad [\overset{\leftrightarrow}{n}(r)]_{\ell'\mu'\ell\mu} = \delta_{\ell\ell'}\delta_{\mu\mu'} n_\ell(kr).$$

Then (3.10) becomes

$$\overset{\leftrightarrow}{\zeta}(r) = \overset{\leftrightarrow}{V}(r) \cdot \overset{\leftrightarrow}{j}(r) - \frac{2m}{\hbar^2} \overset{\leftrightarrow}{V}(r) \cdot \overset{\leftrightarrow}{k} \cdot \int_0^\infty dr' r'^2 \{\overset{\leftrightarrow}{n}(r_>) + i\,\overset{\leftrightarrow}{j}(r_>)\}$$

$$(3.64) \qquad\qquad \cdot \overset{\leftrightarrow}{j}(r_<) \cdot \overset{\leftrightarrow}{\zeta}(r')$$

However, the $r_>$ and $r_<$ variables are irrelevant in the $\overset{\leftrightarrow}{j}(r_>) \cdot \overset{\leftrightarrow}{j}(r_<)$ terms, so we obtain a structure analogous to (3.31):

$$\overset{\leftrightarrow}{\zeta}(r) = \overset{\leftrightarrow}{V}(r) \cdot \overset{\leftrightarrow}{j}(r) \cdot \left[\overset{\leftrightarrow}{1} - \frac{2mi}{\hbar^2} \overset{\leftrightarrow}{k} \cdot \int_0^\infty dr' r'^2 \overset{\leftrightarrow}{j}(r') \cdot \overset{\leftrightarrow}{\zeta}(r') \right]$$

$$(3.65) \qquad\qquad - \frac{2m}{\hbar^2} \overset{\leftrightarrow}{V}(r) \cdot \overset{\leftrightarrow}{k} \cdot \int_0^\infty dr' r'^2 \overset{\leftrightarrow}{n}(r_>) \cdot \overset{\leftrightarrow}{j}(r_<) \cdot \overset{\leftrightarrow}{\zeta}(r').$$

(We have used the fact that the diagonal matrices $\overset{\leftrightarrow}{k}$ and $\overset{\leftrightarrow}{j}(r)$ commute.)

The matrix $\overset{\leftrightarrow}{C}$ defined by

$$(3.66) \qquad\qquad \overset{\leftrightarrow}{C} \equiv \overset{\leftrightarrow}{1} - \frac{2mi}{\hbar^2} \overset{\leftrightarrow}{k} \cdot \int_0^\infty dr' r'^2 \overset{\leftrightarrow}{j}(r') \cdot \overset{\leftrightarrow}{\zeta}(r')$$

is unknown, but independent of r, so it may be factored out by introducing $\overset{\leftrightarrow}{U}(r)$,

$$(3.67) \qquad\qquad \overset{\leftrightarrow}{\zeta}(r) = \overset{\leftrightarrow}{U}(r) \cdot \overset{\leftrightarrow}{C},$$

and thus

$$(3.68) \qquad \overset{\leftrightarrow}{C} = \left[\overset{\leftrightarrow}{1} + \frac{2mi}{\hbar^2} \overset{\leftrightarrow}{k} \cdot \int_0^\infty dr' r'^2 \overset{\leftrightarrow}{j}(r') \cdot \overset{\leftrightarrow}{U}(r') \right]^{-1},$$

and

$$\overset{\leftrightarrow}{U}(r) = \overset{\leftrightarrow}{V}(r) \cdot \overset{\leftrightarrow}{j}(r) - \frac{2m}{\hbar^2} \overset{\leftrightarrow}{V}(r) \cdot \overset{\leftrightarrow}{k} \cdot \int_0^\infty dr' r'^2 \overset{\leftrightarrow}{n}(r_>) \cdot \overset{\leftrightarrow}{j}(r_<) \cdot \overset{\leftrightarrow}{U}(r').$$

(3.69)

The $r_>$ and $r_<$ are eliminated by dividing the integral over r' into 0 to r and r to ∞ segments, and we add and subtract

$$(3.70) \qquad \frac{2m}{\hbar^2} \overset{\leftrightarrow}{V}(r) \cdot \overset{\leftrightarrow}{k} \cdot \overset{\leftrightarrow}{j}(r) \cdot \int_0^r dr' r'^2 \overset{\leftrightarrow}{n}(r') \cdot \overset{\leftrightarrow}{U}(r'),$$

to obtain

$$\vec{U}(r) = \vec{V}(r) \cdot \vec{j}(r) \cdot \left[\vec{1} - \frac{2m}{\hbar^2} \vec{k} \cdot \int_0^\infty dr' r'^2 \vec{n}(r') \cdot \vec{U}(r') \right]$$

$$- \frac{2m}{\hbar^2} \vec{V}(r) \cdot \vec{k} \cdot \vec{n}(r) \cdot \int_0^r dr' r'^2 \vec{j}(r') \cdot \vec{U}(r')$$

$$(3.71) \qquad + \frac{2m}{\hbar^2} \vec{V}(r) \cdot \vec{k} \cdot \vec{j}(r) \cdot \int_0^r dr' r'^2 \vec{n}(r') \cdot \vec{U}(r').$$

As in the central potential case, we introduce the constant matrix $\vec{D}$, defined by

$$(3.72) \qquad \vec{D} \equiv \vec{1} - \frac{2m}{\hbar^2} \vec{k} \cdot \int_0^\infty dr' r'^2 r'^2 \vec{n}(r') \cdot \vec{U}(r'),$$

and the auxiliary matrix function $\vec{V}(r)$,

$$(3.73) \qquad \vec{U}(r) = \vec{V}(r) \cdot \vec{D},$$

whereby

$$(3.74) \qquad \vec{D} = \left[\vec{1} - \frac{2m}{\hbar^2} \vec{k} \cdot \int_0^\infty dr' r'^2 \vec{n}(r') \cdot \vec{V}(r') \right]^{-1},$$

and

$$\vec{V}(r) = \vec{V}(r) \cdot \vec{j}(r) - \frac{2m}{\hbar^2} \vec{V}(r) \cdot \vec{k} \cdot \vec{n}(r) \cdot \int_0^r dr' r'^2 \vec{j}(r') \cdot \vec{V}(r')$$

$$(3.75) \qquad + \frac{2m}{\hbar^2} \vec{V}(r) \cdot \vec{k} \cdot \vec{j}(r) \cdot \int_0^r dr' r'^2 \vec{n}(r') \cdot \vec{V}(r').$$

Now $\vec{n}$ and $\vec{j}$ commute (they are diagonal matrices), so that when a quadrature approximation is introduced, it remains true that

$$(3.76) \qquad \vec{K}(r_p, r_p) = \vec{k} \cdot \{ \vec{n}(r_p) \cdot \vec{j}(r_p) - \vec{j}(r_p) \cdot \vec{n}(r_p) \}$$
$$(3.77) \qquad \equiv 0,$$

and we again obtain the recursion

$$\overleftrightarrow{V}(r_p) = \overleftrightarrow{V}(r_p) \cdot \overleftrightarrow{j}(r_p) - \frac{2m}{\hbar^2} \overleftrightarrow{V}(r_p) \cdot \overleftrightarrow{k} \cdot \overleftrightarrow{n}(r_p) \cdot \overleftrightarrow{I}_1(p-1)$$

(3.78)
$$+ \frac{2m}{\hbar^2} \overleftrightarrow{V}(r_p) \cdot \overleftrightarrow{k} \cdot \overleftrightarrow{j}(r_p) \cdot \overleftrightarrow{I}_2(p-1).$$

Obviously,

(3.79)
$$\overleftrightarrow{I}_1(p-1) = \sum_{p'=0}^{p-1} q_{p'} r_{p'}^2 \ \overleftrightarrow{j}(r_{p'}) \cdot \overleftrightarrow{V}(r_{p'})$$

and

(3.80)
$$\overleftrightarrow{I}_2(p-1) = \sum_{p'=0}^{p-1} q_{p'} r_{p'}^2 \ \overleftrightarrow{n}(r_{p'}) \cdot \overleftrightarrow{V}(r_{p'}).$$

Construction of the physical solution and scattering information requires that one compute the inverses (3.74) and (3.68) (or equivalently, solve two systems of linear algebraic equations). It is here that the multichannel case can be more involved. The "channel potentials", $V(\ell'\mu'\ell\mu|r)$ can have quite different behavior as a function of r, and in particular, at the total energy E, they can produce widely differing classical turning points (points where $V(\ell\mu\ell\mu|r) - E$ equals zero). Inside the classical turning points, the diagonal components of $\overleftrightarrow{V}$ possess exponentially growing contributions. The most rapidly growing of these leads to loss of linear independence of the columns of $\overleftrightarrow{V}$, and this causes the matrices in (3.66) and (3.72) to become singular. To avoid this, it is necessary to periodically "stabilize" the calculation. This essentially requires that the matrix $[\overleftrightarrow{1} - \frac{2m}{\hbar^2} \overleftrightarrow{k} \cdot \int_0^r dr' r'^2 \ \overleftrightarrow{n}(r') \cdot \overleftrightarrow{V}(r')]$ periodically be put into upper triangular form. This transformation is also applied to $\overleftrightarrow{V}(r)$ and $-\frac{2m}{\hbar^2} \overleftrightarrow{k} \cdot \int_0^r dr' r'^2 \ \overleftrightarrow{j}(r') \cdot \overleftrightarrow{V}(r')$; as a result, one need not save the transformation in order to extract the scattering information, once one is outside the range of the potential, because the transformations all cancel. (This is *not* true if one desires to know the wavefunction at interior points, where the potential is nonzero. Then one must have saved the transformations.) The necessity to stabilize a set of solutions in order to maintain linear independence is an essential feature of computational methods based on the homogeneous time-independent Schrödinger equation. It may be disguised in some methods, but typically shows up in the necessity to carry out matrix inversions as the solution matrix is propagated from one region to another. The present approach avoids computing a full matrix inverse, necessitating instead transforming a matrix to triangular form. Time-dependent based approaches behave in a fundamentally different way, as we shall see later. Essentially this is because the time-independent approaches typically must generate a set of linearly independent solutions in order to impose boundary conditions, while time-dependent wavepacket methods are *initial value* approaches [8].

3.3. The transition operator equations. We now must develop some additional formalism in order to provide the framework for discussing another class of computational approaches. Consider the two forms of the TILSE:

$$(3.81) \qquad |\psi_k^+\rangle = |\phi_k\rangle + G_0^+(E)V|\psi_k^+\rangle$$
$$(3.82) \qquad \qquad = |\phi_k\rangle + G^+(E)V|\phi_k\rangle.$$

This implies that for *every* state $|\phi_k\rangle$,

$$(3.83) \qquad G_0^+(E)V|\psi_k^+\rangle \equiv G^+(E)V|\phi_k\rangle.$$

We can define the "transition" or "T"-operator as that operator which, when acting on $|\phi_k\rangle$, produces the same vector as V acting on $|\psi_k^+\rangle$:

$$(3.84) \qquad T|\phi_k\rangle \equiv V|\psi_k^+\rangle.$$

It follows that

$$(3.85) \qquad G_0^+(E)V|\psi_k^+\rangle = G_0^+(E)T|\phi_k\rangle$$
$$(3.86) \qquad \qquad = G^+(E)V|\phi_k\rangle,$$

and since this holds for *all* $|\phi_k\rangle$ in the complete set, we have the *operator* relation

$$(3.87) \qquad G_0^+(E)T = G^+(E)V.$$

Recall we also showed earlier, that

$$(3.88) \qquad G^+(E) = G_0^+(E) + G^+(E)VG_0^+(E)$$

so

$$(3.89) \qquad G^+(E) = G_0^+(E) + G_0^+(E)TG_0^+(E).$$

But we also showed that

$$(3.90) \qquad G^+(E) = G_0^+(E) + G_0^+(E)VG^+(E)$$

so that

$$(3.91) \qquad TG_0^+(E) = VG^+(E).$$

Now we substitute for $|\psi_k^+\rangle$ in (3.84), to obtain

$$(3.92) \qquad T|\phi_k\rangle = V\{|\phi_k\rangle + G_0^+(E)V|\psi_k^+\rangle\}$$
$$(3.93) \qquad = V\{|\phi_k\rangle + G_0^+(E)T|\phi_k\rangle\},$$

and since this holds for all $|\phi_k\rangle$ in a complete set, we have the operator equation

$$(3.94) \qquad T = V + VG_0^+(E)T.$$

But by (3.87), this yields

$$(3.95) \qquad T = V + VG^+(E)V,$$

and by (3.91), this gives

$$(3.96) \qquad T = V + TG_0^+(E)V.$$

Finally, we use (3.94) to write

$$(3.97) \qquad V = T - VG_0^+(E)T,$$

and combining (3.96) and (3.97) leads to

$$(3.98) \qquad T = V + TG_0^+(E)T - TG_0^+(E)VG_0^+(E)T.$$

Equations (3.94), (3.96) and (3.98) form the basis for several approaches to scattering calculations. In fact, (3.94) is equivalent to the amplitude density equation discussed earlier. To see this, we apply (3.94) to an initial state $|\phi_k\rangle$, where $E = \hbar^2 k^2/2m$:

$$(3.99) \qquad T|\phi_k\rangle = V|\phi_k\rangle + VG_0^+(E)T|\phi_k\rangle.$$

Defining

$$(3.100) \qquad |\zeta_k^+\rangle = T|\phi_k\rangle,$$

(3.98) becomes

$$(3.101) \qquad |\zeta_k^+\rangle = V|\phi_k\rangle + VG_0^+(E)|\zeta_k^+\rangle.$$

Projecting this onto the position eigenstate $\langle x|$, and inserting the resolution of the identity,

$$(3.102) \qquad 1 = \int\limits_{-\infty}^{\infty} dx'|x'\rangle\langle x'|,$$

we obtain

$$(3.103) \qquad \langle x|\zeta_k^+\rangle = \zeta_k^+(x),$$

so

$$(3.104) \quad \zeta_k^+(x) = V(x)\phi_k(x) + V(x) \int\limits_{-\infty}^{\infty} dx' \langle x|G_0^+(E)|x'\rangle \zeta_k^+(x').$$

In the simple ID case,

$$(3.105) \qquad \langle x|G_0^+(E)|x'\rangle = -\frac{mi}{\hbar^2 k} e^{ik|x-x'|},$$

and

$$(3.106) \quad \zeta_k^+(x) = V(x)e^{ikx} - V(x)\frac{mi}{\hbar^2 k} \int\limits_{-\infty}^{\infty} dx' e^{ik|x-x'|}\zeta_k^+(x'),$$

which is readily seen to be identical to the amplitude density equation resulting from multiplying $\psi_k^+(x)$ by $V(x)$:

$$(3.107) \qquad \psi_k^+(k) = e^{ikx} - \frac{mi}{\hbar^2 k} \int\limits_{-\infty}^{\infty} dx' e^{ik|x-x'|} V(x')\psi_k^+(x'),$$

$$(3.108) \qquad V(x)\psi_k^+(x) \equiv \zeta_k^+(x),$$

so

$$(3.109) \quad \zeta_k^+(x) = V(x)e^{ikx} - V(x)\frac{mi}{\hbar^2 k} \int\limits_{-\infty}^{\infty} dx' e^{ik|x-x'|}\zeta_k^+(x').$$

It is seen that (3.106) and (3.109) are identical. Thus, one can use (3.94) or (3.96) as the starting point for deriving Fredholm- or Volterra-based approaches.

However, of interest to us here is the derivation of variational principles. In general, these are "stationary principles", and provide no bounds. They result in quantities which possess second order errors. In general, the derivation of variational principles which deliver the basic dynamical equations (e.g., the TISE or TILSE) as their "Euler equations" requires deducing bilinear or quadratic forms, subject to the requirement that the value of the functional equals the dynamical quantity of interest when exact solutions to the dynamical equations are used to compute the functional. The simplest route to such bilinear forms of which we are aware employs the trivial device of deriving three alternative expressions for the dynamical quantity of interest, one of which is *quadratic* in the dynamical quantity (e.g., (3.94), (3.96), and (3.98)). Then one adds two of the expressions and subtracts the third or multiplies two and divides by the third. We will illustrate with the expression

$$(3.110) \qquad\qquad [T] = T + T - T,$$

which trivially equals the T-operator if one uses the exact solutions of (3.94), (3.96) and (3.98). Thus, the functional $[T]$ becomes

$$[T] = V + VG_0^+(E)T + TG_0^+(E)V - TG_0^+T$$

(3.111)
$$+ TG_0^+(E)VG_0^+(E)T$$

$$= V + VG_0^+(E)T + TG_0^+(E)V$$

(3.112)
$$- T[G_0^+ - G_0^+VG_0^+(E)]T.$$

This variational functional was given by Newton [2]. It is easy to show that it is stationary with respect to variations δT about the exact solutions of (3.94) and (3.96), since

$$\delta[T] = [VG_0^+(E) - TG_0^+(E) + TG_0^+(E)VG_0^+(E)]\delta T$$

(3.113)
$$+ \delta T[G_0^+(E)V - G_0^+(E)T + G_0^+(E)VG_0^+(E)T],$$

or

$$\delta[T] = [V - T + TG_0^+(E)V]G_0^+(E)\delta T$$

(3.114)
$$+ \delta TG_0^+(E)[V - T + VG_0^+(E)T].$$

It is evident that $\delta[T]$ vanishes identically if T satisfies (3.94) and (3.96), and these are the Euler equations for this functional.

Another way to view the functional is to calculate the error produced by [T] when *incorrect* values of T are substituted. Thus, let

(3.115)
$$T = T_{\text{ex}} + T_{\text{er}},$$

where T_{ex} is the exact solution of (3.94), (3.96), and (3.98), and T_{er} is the error. Then

$$[T] = V + VG_0^+(E)T_{\text{ex}} + VG_0^+(E)T_{\text{er}}$$
$$+ V + T_{\text{ex}}G_0^+(E)V + T_{\text{er}}G_0^+(E)V$$

(3.116)
$$- V - (T_{\text{ex}} + T_{\text{er}})[G_0^+(E) - G_0^+(E)VG_0^+(E)](T_{\text{ex}} + T_{\text{er}}),$$

and using the exact equations, this yields

(3.117)
$$[T] = T_{\text{ex}} - T_{\text{er}}[G_0^+(E) - G_0^+(E)VG_0^+(E)]T_{\text{er}}.$$

This says that when one evaluates [T] by substituting an approximation to T, the result is accurate up to second order in the error; i.e., the first order error is identically zero because the functional is stationary. This is computationally extremely useful because it means that substituting an approximation to T back into the functional delivers a new estimate of the transition operator accurate to second order in the error of the original approximation.

In actual applications, one looks at matrix elements of the operator. The Newton variational function for the scattering amplitude for the $\vec{k} \to \vec{k}'$ transition, e.g., is given by

$$\begin{aligned}
[T(\vec{k}'|\vec{k})] &= \langle \phi_{\vec{k}'}|V|\phi_{\vec{k}}\rangle + \langle \phi_{\vec{k}'}|VG_0^+(E)T|\phi_{\vec{k}}\rangle \\
&\quad + \langle \phi_{\vec{k}'}|TG_0^+(E)V|\phi_{\vec{k}}\rangle \\
&\quad - \langle \phi_{\vec{k}'}|T[G_0^+(E) - G_0^+VG_0^+]T|\phi_{\vec{k}}\rangle.
\end{aligned}$$

(3.118)

Now recall that

(3.119)
$$G_0^+(E)T|\phi_{\vec{k}}\rangle = |\psi^+_{\vec{k},SW}\rangle,$$

and since the anti-causal state satisfies

(3.120)
$$|\psi^-_{\vec{k}'}\rangle = |\phi_{\vec{k}'}\rangle + \frac{1}{E - H_0 - i\epsilon}V|\psi^-_{\vec{k}'}\rangle,$$

then

(3.121)
$$\langle\psi^-_{\vec{k}'}| = \langle\phi_{\vec{k}'}| + \langle\psi^-_{\vec{k}'}|VG_0^+(E),$$

and therefore

(3.122)
$$\langle\phi_{\vec{k}'}|TG_0^+(E) = \langle\psi^-_{\vec{k}',SW}|.$$

It follows that

$$\begin{aligned}
[T(\vec{k}'|\vec{k})] &= \langle\phi_{\vec{k}'}|V|\phi_{\vec{k}}\rangle + \langle\phi_{\vec{k}'}|V|\psi^+_{\vec{k},SW}\rangle \\
&\quad + \langle\psi^-_{\vec{k}',SW}|V|\phi_{\vec{k}}\rangle \\
&\quad - \langle\phi_{\vec{k}'}|T\{G_0^+(E)[G_0^+(E)]^{-1}G_0^+(E) \\
&\quad - G_0^+(E)VG_0^+(E)\}T|\phi_{\vec{k}}\rangle,
\end{aligned}$$

(3.123)

or

(3.124)
$$\begin{aligned}
[T(\vec{k}'|\vec{k})] &= \langle\phi_{\vec{k}'}|V|\phi_{\vec{k}}\rangle + \langle\phi_{\vec{k}'}|V|\psi^+_{\vec{k},SW}\rangle \\
&\quad + \langle\psi^-_{\vec{k}',SW}|V|\phi_{\vec{k}}\rangle - \langle\psi^-_{\vec{k}',SW}|\{[G_0^+(E)]^{-1} - V\}|\psi^+_{\vec{k},SW}\rangle.
\end{aligned}$$

Next we observe that

(3.125)
$$[G_0^+(E)]^{-1} = E - H_0,$$

where ϵ can be set equal to zero. Then we note that

(3.126)
$$[G_0^+(E)]^{-1} - V = E - H_0 - V = E - H,$$

with the final form of the variational functional being

$$[T(\vec{k}'|\vec{k})] = \langle \phi_{\vec{k}'}|V|\phi_{\vec{k}} \rangle + \langle \phi_{\vec{k}'}|V|\psi^+_{\vec{k},SW} \rangle$$

$$(3.127) \qquad + \langle \psi^-_{\vec{k}',SW}|V|\phi_{\vec{k}} \rangle - \langle \psi^-_{\vec{k}',SW}|(E-H)|\psi^+_{\vec{k},SW} \rangle,$$

and this form is known as the "scattered wave-Kohn variational principle" [9].

Yet another form for the same functional is obtained by expressing it in terms of the amplitude density. Using (3.100), we can re-write (3.118) as

$$[T(\vec{k}',\vec{k})] = \langle \phi_{\vec{k}'}|V|\phi_{\vec{k}} \rangle + \langle \phi_{\vec{k}'}|VG^+_0(E)|\zeta^+_{\vec{k}} \rangle$$

$$+ \langle \zeta^-_{\vec{k}'}|G^+_0(E)V|\phi_{\vec{k}} \rangle$$

$$(3.128) \qquad - \langle \zeta^-_{\vec{k}'}|[G^+_0(E) - G^+_0(E)VG^+_0(E)]|\zeta^+_{\vec{k}} \rangle,$$

which is known as the Newton variational principle for the amplitude density [10].

These equations are used for computations in the following way. We can expand $|\psi^+_{\vec{k},SW} \rangle$ and $\langle \psi^-_{\vec{k}',SW}|$ in a basis and evaluate the coefficients in the expansion by requiring that $[T(\vec{k}',|\vec{k})]$ be stationary with respect to variations in the coefficients, or we can expand the amplitude density in an $\mathcal{L}^2$ - basis and evaluate those expansion coefficients by requiring that $[T(\vec{k}'|\vec{k})]$ be stationary with respect to variations in those expansion coefficients. In the amplitude density case, we write

$$(3.129) \qquad |\zeta^+_{\vec{k}} \rangle = \sum_s a_s|\beta_s \rangle,$$

$$(3.130) \qquad \langle \zeta^-_{\vec{k}'}| = \sum_r b_r \langle \beta_r|,$$

and then require that $\delta[T(\vec{k}'|\vec{k})]$, given by

$$\delta[T(\vec{k}'|\vec{k})] = \sum_s \delta a_s \langle \phi_{\vec{k}'}|VG^+_0(E)|\beta_s \rangle$$

$$+ \sum_r \delta b_r \langle \beta_r|G^+_0(E)V|\phi_{\vec{k}} \rangle$$

$$(3.131) \qquad + \sum_s \sum_r \{b_r \delta a_s + a_s \delta b_r\} \langle \beta_r|G^+_0(E)[E-H]G^+_0(E)|\beta_s \rangle,$$

vanish for arbitrary δa_s and δb_r. We emphasize that $\langle \phi_{\vec{k}'}|VG^+_0(E)|\beta_s \rangle$, $\langle \beta_r|G^+_0(E)V|\phi_{\vec{k}} \rangle$, and $\langle \beta_r|G^+_0(E)[E-H]G^+_0(E)|\beta_s \rangle$ are *numbers*. We must choose a_s and b_r so that the coefficients of each δa_s and δb_r separately vanish. This implies that we solve the linear algebraic equations

$$(3.132) \qquad \sum_r b_r C_{rs} = B_s,$$

$$(3.133) \qquad \sum_s C_{rs} a_s = \tilde{B}_r,$$

where

$$(3.134) \qquad B_s \equiv \langle \phi_{\vec{k}'} | V G_0^+(E) | \beta_s \rangle,$$

$$(3.135) \qquad \tilde{B}_r \equiv \langle \beta_r | G_0^+(E) V | \phi_{\vec{k}} \rangle,$$

and

$$(3.136) \qquad C_{rs} \equiv \langle \beta_r | G_0^+(E)[E - H] G_0^+(E) | \beta_s \rangle.$$

In matrix notation, this yields

$$(3.137) \qquad \vec{b} = \vec{B} \cdot \overleftrightarrow{C}^{-1}$$

and

$$(3.138) \qquad \vec{a} = \overleftrightarrow{C}^{-1} \cdot \vec{\tilde{B}}$$

In addition, it is easy to write $[T(\vec{k}'|\vec{k})]$ as

$$(3.139) \qquad [T(\vec{k}'|\vec{k})] = \langle \phi_{\vec{k}'} | V | \phi_{\vec{k}} \rangle + \vec{B} \cdot \vec{a} + \vec{b} \cdot \vec{\tilde{B}} - \vec{b} \cdot \overleftrightarrow{C} \cdot \vec{a},$$

and combining this with (3.137) – (3.138) yields

$$(3.140) \qquad [T(\vec{k}'|\vec{k})] = \langle \phi_{\vec{k}'} | V | \phi_{\vec{k}} \rangle + \vec{B} \cdot \vec{a},$$

or

$$(3.141) \qquad [T(\vec{k}'|\vec{k})] = \langle \phi_{\vec{k}'} | V | \phi_{\vec{k}} \rangle + \vec{b} \cdot \vec{\tilde{B}}.$$

Thus, it is only necessary to calculate $\vec{a}$ or $\vec{b}$, but not both. Additionally, one does not compute the inverse $\overleftrightarrow{C}^{-1}$, but rather solves the linear algebraic equations either for the a_s or the b_r. The value of $[T(\vec{k}'|\vec{k})]$ generated by either (3.140) or (3.141) is accurate to second order in the errors.

One important practical note concerns the calculation of the B_s, $\tilde{B}_r$, and C_{rs} matrix elements. It is quite costly to generate matrix elements of $G_0^+(E|\vec{r}, \vec{r}')$, and then have to integrate over both $\vec{r}$ and $\vec{r}'$. Instead, we defined the "half-integrated Green's functions" (HIGF) [11]

$$(3.142) \qquad |h_s^+\rangle \equiv G_0^+(E) | \beta_s \rangle$$

and

$$(3.143) \qquad \langle h_r^-| \equiv \langle \beta_r | G_0^+(E).$$

It follows that $|h_s^+\rangle$ and $\langle h_r^-|$ satisfy inhomogeneous, unperturbed Schrödinger equations:

$$(3.144) \qquad (E - H_0)|h_s^+\rangle = |\beta_r\rangle$$

and

$$(3.145) \qquad \langle h_r^-|(E - H_0) = \langle \beta_r|.$$

Solution of these equations for the $|h_s^+\rangle$ and $\langle h_r^-|$ is much more efficient than direct calculation of the Green's function matrix elements. Having the HIFGs, we calculate C_{rs}, B_s, $\tilde{B}_r$ as

$$(3.146) \qquad C_{rs} = \langle h_r^-|(E - H)|h_s^+\rangle$$
$$(3.147) \qquad B_s = \langle \phi_{\bar{k}'}|V|h_s^+\rangle,$$

and

$$(3.148) \qquad \tilde{B}_r = \langle h_r^-|V|\phi_{\bar{k}}\rangle.$$

The expression (3.146) can also be written in the form

$$(3.149) \qquad C_{rs} = \langle h_r^-|\beta_s\rangle - \langle h_r^-|V|h_s^+\rangle.$$

The introduction of the HIGF makes the amplitude density approach appear very similar to the scattered wave Kohn approach. As basis functions for expanding $|\psi_{\bar{k},SW}^+\rangle$ and $\langle\psi_{\bar{k}',SW}^-|$, one can, e.g., use a small number of HIGF (say one for $|\psi_{\bar{k},SW}^+\rangle$ and one for $\langle\psi_{\bar{k}',SW}^-|$) and use $\mathcal{L}^2$-basis functions for the rest. The HIGF incorporate part of the dynamics (they are a dynamically determined basis) and can significantly improve the rate of convergence. Finally, if one includes some energy independent $\mathcal{L}^2$ basis functions but views them as HIGFs, then the fact that they satisfy

$$(3.150) \qquad (E - H_0)|h_s^+\rangle = |\beta_s\rangle$$

implies that the $|\beta_s\rangle$ must have an energy dependence which eliminates any energy dependence from the HIGF basis-function, $|h_s^+\rangle$. We now turn to a description of how such variational approaches can be implemented for state-of-the-art collision problems, specifically, atom-diatom collisions including rearrangements.

4. Algebraic variational treatment of rearrangement scattering. The above approaches have been utilized for some of the most demanding calculations ever done for quantum scattering. A fundamentally important area in the theory of chemical dynamics is the quantum theory of atom-diatom collisions, in which exchanges and total dissociation can occur. In fact, successful calculations in the full three physical dimensional space have *only* been carried out for collisions well below the threshold for producing three free atoms. In this Section, we describe in more detail how the algebraic variational approach can be implemented for such challenging problems. We consider a triatomic system in which the atoms interact via a *single* "Born-Oppenheimer potential surface." Such a potential is obtained by invoking the Born-Oppenheimer (BO) separation of the electronic and nuclear degrees of freedom (which is based on the large time scale difference between slow nuclear motions and fast electronic motions). The electronic energy eigenvalue problem is solved for fixed nuclear configuration, and the resulting lowest electronic energy (which is parametrically a function of nuclear relative positions) is added to the Coulombic nuclear-nuclear repulsion energy. The result is the energy of the system for that nuclear configuration, and it therefore serves as the potential energy for the nuclear dynamics. In general, the potential can include three pairwise interactions, as well as a three body interaction. The system of three nuclei is described by a total of 9 degrees of freedom, and because no forces act on the center-of-mass of the nuclei one introduces two relative vectors plus the position of the center-of-mass. Out of the infinite number of possible choices for the two relative vectors, there are several "natural" choices. If we label the nuclei 1,2,3, then in general, there are four arrangements of interest (a given system may not include all four, depending on the nature of the potential). The arrangements are conveniently indexed by the free atom label; thus, in one arrangement, atom 1 is unbound and atoms 2 and 3 form a cluster (molecule). We label this as arrangement 1. In the arrangement α, atom α is free and β and γ are bound, as $\alpha(\beta\gamma)$. The totally free arrangement will be indexed by zero, and the arrangement with all three atoms bound as a triatomic molecule is labeled 4. In fact, arrangement 4 is dynamically uncoupled from the others and need not be considered further. We use the standard "Jacobi coordinates", where for arrangement α, $\vec{R}_\alpha$ is the vector from the center of mass of the (β, γ) pair to atom α, and $\vec{r}_\alpha$ is the relative vector from atom β to atom γ. The breakup (three free atom) arrangement is most easily described using the hyperspherical radius, R_0, given by

$$(4.1) \qquad R_0^2 = R_\alpha^2 + r_\alpha^2,$$

which holds for $\alpha = 1, 2, 3$, and five internal coordinates (usually taken to be angle variables). Obviously, any point in the 6-dimensional space of the relative configuration of the three atoms can be described using *any* of the sets of relative coordinates.

The Schrödinger equation is

$$(4.2) \qquad (E - H)|\psi_{\alpha_0 n_0}\rangle = 0,$$

where α_0 labels the initial arrangement and n_0 denotes a complete set of quantum numbers describing the initial state in the α_0 arrangement. Generally, n_0 represents the quantum numbers $JM\nu_0 j_0 \ell_0$, where J is the total angular momentum quantum number, $M\hbar$ is the z-component of total angular momentum measured along a laboratory oriented, center-of mass z-axis, ν_0 is the initial vibrational state quantum number, j_0 is the initial rotational quantum number for the diatom in arrangement α_0, and ℓ_0 is the initial orbital angular momentum quantum number for rotation of atom α_0 relative to the center-of-mass of the diatom in arrangement α_0. If we consider the limit of the total Hamiltonian as $R_\alpha \to \infty$, we can define the unperturbed Hamiltonian for arrangement α by

$$(4.3) \qquad\qquad H_\alpha = \lim_{R_\alpha \to \infty} H.$$

Then the perturbation, V_α, causing scattering in arrangement α is defined to be

$$(4.4) \qquad\qquad V_\alpha = H - H_\alpha;$$

note that

$$(4.5) \qquad\qquad \lim_{R_\alpha \to \infty} V_\alpha = 0.$$

For breakup,

$$(4.6) \qquad\qquad \lim_{R_0 \to \infty} H = K,$$

where K is the relative kinetic energy for the three atoms. Then

$$(4.7) \qquad\qquad H_0 = K,$$
$$(4.8) \qquad\qquad V_0 = V,$$

where V is the full BO potential for the system and K is the kinetic energy operator of relative motion. For arrangements $1 - 3$, we construct projection operators, J_α, onto the (discrete) partially bound states of the α-arrangement molecular cluster:

$$(4.9) \qquad\qquad J_\alpha = \sum_{n=N_\alpha^i}^{N_\alpha^f} |\phi_n\rangle\langle\phi_n|,$$

where N_α^i is the lowest energy bound state of the arrangement α-diatom, and N_α^f is the highest energy bound state of the arrangement α-diatom. The orthogonal complement of J_α is

$$(4.10) \qquad\qquad Q_\alpha = 1 - J_\alpha,$$

and it contains the breakup for the α-cluster. We also introduce J_0, defined by

$$(4.11) \qquad\qquad J_0 = 1 - \sum_{\alpha=1}^{3} J_\alpha.$$

We now wish to decompose $|\psi_{\alpha_0 n_0}\rangle$ into pieces which contain the information about scattering into the various possible arrangements. This is achieved by introducing operators $\mathcal{J}$, defined by

$$(4.12) \qquad \mathcal{J} = \begin{pmatrix} J_1 \\ J_2 \\ J_3 \\ 1 - \sum_{\alpha=1}^{3} J_\alpha \end{pmatrix}$$

and $\tilde{\mathcal{J}}$, defined by

$$(4.13) \qquad \tilde{\mathcal{J}} = \overbrace{\begin{pmatrix} 1 & 1 & 1 & 1 \end{pmatrix}}.$$

It is trivial to show that

$$(4.14) \qquad \tilde{\mathcal{J}}\mathcal{J} = 1,$$

and

$$(4.15) \qquad \mathcal{J}\tilde{\mathcal{J}} = \begin{pmatrix} J_1 & J_1 & J_1 & J_1 \\ J_2 & J_2 & J_2 & J_2 \\ J_3 & J_3 & J_3 & J_3 \\ J_0 & J_0 & J_0 & J_0 \end{pmatrix}.$$

It should be noted that while $J_\alpha, \alpha = 1, 2, 3$ are projection operators,

$$(4.16) \qquad J_\alpha^2 = J_\alpha,$$

J_0 is *not*, and

$$(4.17) \qquad J_0^2 \neq J_0.$$

We now apply $\mathcal{J}$ to (4.2) from the left, and insert the identity using (4.14), to re-write the Schrödinger equation as

$$(4.18) \qquad [\mathcal{J}(E - H)\tilde{\mathcal{J}}]\mathcal{J}|\psi_{\alpha_0 n_0}\rangle = 0;$$

explicitly, this is the matrix equation

$$\begin{pmatrix} J_1(E-H) & J_1(E-H) & J_1(E-H) & J_1(E-H) \\ J_2(E-H) & J_2(E-H) & J_2(E-H) & J_2(E-H) \\ J_3(E-H) & J_3(E-H) & J_3(E-H) & J_3(E-H) \\ J_0(E-H) & J_0(E-H) & J_0(E-H) & J_0(E-H) \end{pmatrix} \begin{pmatrix} |\psi_1(\alpha_0 n_0)\rangle \\ |\psi_2(\alpha_0 n_0)\rangle \\ |\psi_3(\alpha_0 n_0)\rangle \\ |\psi_0(\alpha_0 n_0)\rangle \end{pmatrix} = 0.$$

(4.19)

Here, $|\psi_\lambda(\alpha_0 n_0)\rangle$ is defined as

$$(4.20) \qquad |\psi_\lambda(\alpha_0 n_0|\rangle = J_\lambda|\psi(\alpha_0 n_0)\rangle, \lambda = 0, 1, 2, 3.$$

In many chemical systems, the breakup arrangement is neglected. If the energy is sufficiently far below the threshold for dissociation, its effect *can* be taken into account by using large enough bases in the other three arrangements to converge the result. However, if the expansion is restricted to internal states of the three partially bound arrangements, and the energy is not far from the breakup threshold, this may not provide a sufficient basis to converge the results. Then either one must augment the partially bound arrangement bases or include the breakup explicitly. We can carry out a simple analysis of the coupling of partially bound arrangement α to the breakup by noting that this coupling is of the form

$$J_\alpha(E - H)|\psi_0(\alpha_0 n_0)\rangle = J_\alpha(E - H)J_0|\psi(\alpha_0 n_0)\rangle,$$

(4.21) $$\alpha = 1, 2, 3.$$

Using (4.11), this gives

$$J_\alpha(E - H)|\psi_0(\alpha_0 n_0)\rangle = J_\alpha(E - H)|\psi(\alpha_0 n_0)\rangle$$

(4.22) $$- J_\alpha(E - H)\sum_{\alpha'=1}^{3} J_{\alpha'}|\psi(\alpha_0 n_0)\rangle.$$

The first term vanishes due to (4.2), leaving the term

(4.23) $$- J_\alpha(E - H)\sum_{\alpha'=1}^{3} J_{\alpha'}|\psi(\alpha_0 n_0)\rangle.$$

Obviously, if H commutes with $\sum_{\alpha'=1}^{3} J_\alpha$, this term also vanishes due to (4.2). Of course,

(4.24) $$[H, \sum_{\alpha'=1}^{3} J_{\alpha'}] \neq 0.$$

However,

(4.25) $$[H_\alpha, J_\alpha] = 0$$

does hold, so that (4.24) is equivalent to

(4.26) $$\sum_{\alpha'=1}^{3} [V_{\alpha'}, J_{\alpha'}].$$

Now $V_{\alpha'}$ is equal to the sum of the pairwise interactions of atom α' with each the other two atoms plus the three body interaction, V_{123}. We denote these pairwise interactions by $V^{\alpha'\alpha''}$, $\alpha' \neq \alpha''$, and note that $V^{\alpha'\alpha''}$ is significant *only* when atom α' is close to atom α''. The projector $J_{\alpha'}$ is composed of internal states of the $\alpha''\alpha'''$, $\alpha'' \neq \alpha'''$, diatom. These are nonzero *only* when atoms α'' and α''' are close. We conclude that the commutators $[V_{\alpha'}, J_{\alpha'}]$ are significant only when all three atoms are close together. This requires a simultaneous overlap of the various wavefunctions, and the less localized these wavefunctions are, the larger the region of overlap becomes. This suggests that when the $J_{\alpha'}$ projectors must include the higher excited internal molecular states, the larger will be the commutators. This occurs at higher collision energies, and the conclusion is that the dissociation component of the wavefunction can be completely neglected so long as the energy is well below the dissociation threshold.

To carry out the traditional sort of formulation of scattering solutions to (4.18) or (4.19), it is necessary to decide how to split the Hamiltonian into an unperturbed piece and a perturbation. The principal difficulty in reactive scattering is that there are multiple asymptotic limits that must be incorporated, each of which defines its own unperturbed Hamiltonian and corresponding perturbation. The simplest way to split (4.18) and (4.19) into an "H_0" and "V" is to write

$$(4.27) \qquad (E\,\overset{\leftrightarrow}{1} - \overset{\leftrightarrow}{H}_0)|\vec{\psi}\rangle = \overset{\leftrightarrow}{V}\,|\vec{\psi}\rangle,$$

where

$$(4.28) \qquad (\overset{\leftrightarrow}{1})_{\lambda\lambda'} = \delta_{\lambda\lambda'},$$

$$(4.29) \qquad (\overset{\leftrightarrow}{H}_0)_{\lambda\lambda'} = \delta_{\lambda\lambda'}J_\lambda H_\lambda, \lambda = 1,2,3$$

$$(4.30) \qquad\qquad\qquad = \delta_{\lambda\lambda'}H_0, \lambda = 0,$$

$$(4.31) \qquad (|\vec{\psi}\rangle)_\lambda = |\psi_\lambda(\alpha_0 n_0)\rangle = J_\lambda|\psi(\alpha_0 n_0)\rangle,$$

and

$$(4.32) \qquad \overset{\leftrightarrow}{V} = \begin{pmatrix} J_1 V_1 & J_1(H-E) & J_1(H-E) & J_1(H-E) \\ J_2(H-E) & J_2 V_2 & J_2(H-E) & J_2(H-E) \\ J_3(H-E) & J_3(H-E) & J_3 V_3 & J_3(H-E) \\ J_0(H-E) & J_0(H-E) & J_0(H-E) & \{V - \sum_{\alpha=1}^{3} J_\alpha(H-E)\} \end{pmatrix}.$$

We note that

$$(4.33) \qquad \sum_{\alpha=0}^{3} J_\alpha = 1,$$

so it follows that

$$(4.34) \qquad \sum_{\alpha=0}^{3} |\psi_\alpha n_0)\rangle = \sum_{\alpha=0}^{3} J_\alpha|\psi(\alpha_0 n_0)\rangle \equiv |\psi(\alpha_0 n_0)\rangle$$

and

$$(4.35) \qquad \sum_{\alpha=0}^{3} J_\alpha (H - E) = (H - E).$$

We can formally solve each of the 4 equations represented by (4.27) to obtain states of the form

$$
\begin{aligned}
(4.36) \qquad |\psi_\lambda^+(\alpha_0 n_0)\rangle &= \delta_{\lambda\alpha_0}|\phi(\alpha_0 n_0)\rangle \\
&+ \sum_{\lambda'=0}^{3}(E + i\epsilon - H_\lambda)^{-1}\mathcal{V}_{\lambda\lambda'}|\psi_{\lambda'}^+(\alpha_0 n_0)\rangle,
\end{aligned}
$$

where we note that for $\lambda = 1, 2, 3$, the $\mathcal{V}_{\lambda\lambda'}$ contains as its first factor the projector operator, J_λ. Thus, for the Green's operators associated with the partially bound arrangements, only the projection onto the bound molecular states included in the J_λ will contribute. *In particular, no contribution arises from the dissociative continuum states of arrangements* 1, 2, 3. Thus, $G_\lambda^+(E)\mathcal{V}_{\lambda\lambda}$, generates $J_\lambda G_\lambda^+(E)$, $\lambda = 1, 2, 3$ and this produces the correct outgoing waves in arrangement λ. In the breakup limit, $J_\lambda G_\lambda^+(E)$ will tend to zero. The equation for $|\psi_0^+(\alpha_0 n_0)\rangle$ involves the free Green's function for the relative kinetic energy, and it acts on $\mathcal{V}_{0\lambda'}$. But the elements $\mathcal{V}_{0\lambda'}$, do *not* contain a projection operator, but rather $J_0 \equiv (1 - \sum_{\alpha=1}^{3} J_\alpha)$. In the breakup asymptotic regions, the J_α tend to zero rapidly and J_0 behaves like the identity in this region. This ensures that the totally free Green's function generates the correct outgoing free waves. In calculations, one works with the amplitude density form of these equations, and with the variational approach (this is characterized by much more rapid convergence). Furthermore, *all* applications of the formalism have neglected the explicit coupling to the breakup, and have dealt with collisions at energies far below the breakup threshold.

Then

$$(4.37) \qquad \sum_{\lambda=1}^{3} J_\alpha \cong 1,$$

and we solve [11]

$$
\begin{aligned}
(4.38) \qquad & \begin{pmatrix} J_1(E - H_1) & 0 & 0 \\ 0 & J_2(E - H_2) & 0 \\ 0 & 0 & J_3(E - H_3) \end{pmatrix} \begin{pmatrix} |\psi_1(\alpha_0 n_0)\rangle \\ |\psi_2(\alpha_0 n_0)\rangle \\ |\psi_3(\alpha_0 n_0)\rangle \end{pmatrix} \\
&= \begin{pmatrix} J_1 V_1 & J_1(H - E) & J_1(H - E) \\ J_2(H - E) & J_2 V_2 & J_2(H - E) \\ J_3(H - E) & J_3(H - E) & J_3 V_3 \end{pmatrix} \begin{pmatrix} |\psi_1(\alpha_0 n_0)\rangle \\ |\psi_2(\alpha_0 n_0)\rangle \\ |\psi_3(\alpha_0 n_0)\rangle \end{pmatrix}
\end{aligned}
$$

or with obvious definitions,

$$(4.39) \qquad (E\overset{\leftrightarrow}{1} - \overset{\leftrightarrow}{H}_0) \cdot |\vec{\psi}(\alpha_0 n_0)\rangle = \overset{\leftrightarrow}{V} \cdot |\vec{\psi}(\alpha_0 n_0)\rangle.$$

Then

$$(4.40) \qquad |\vec{\psi}^{+}(\alpha_0 n_0)\rangle = |\vec{\psi}(\alpha_0 n_0)\rangle + \overset{\leftrightarrow}{G_0}^{+}(E) \cdot \overset{\leftrightarrow}{V} \cdot |\vec{\psi}^{+}(\alpha_0 n_0)\rangle,$$

with

$$(4.41) \qquad (|\vec{\phi}(\alpha_0 n_0)\rangle)_\lambda = \delta_{\lambda \alpha_0} \phi(\alpha_0 n_0),$$

$$(4.42) \qquad (E \overset{\leftrightarrow}{1} - \overset{\leftrightarrow}{H_0}) \cdot |\vec{\phi}(\alpha_0 n_0)\rangle = 0,$$

$$(4.43) \qquad (\overset{\leftrightarrow}{G_0}^{+}(E))_{\lambda\lambda'} = \delta_{\lambda\lambda'} J_\lambda \frac{1}{(E - H_\lambda + i\epsilon)}.$$

The generalized amplitude density, $|\vec{\zeta}^{+}(\alpha_0 n_0)\rangle$, is defined as

$$(4.44) \qquad |\vec{\zeta}^{+}(\alpha_0 n_0)\rangle = \overset{\leftrightarrow}{V} \cdot |\vec{\psi}^{+}(\alpha_0 n)\rangle,$$

and one can prove that

$$(4.45) \qquad T(\alpha n | \alpha_0 n_0) \equiv \langle \vec{\phi}(\alpha n) | \vec{\zeta}^{+}(\alpha_0 n_0)\rangle$$

is the correct, physical transition matrix element for the transition $\alpha_0 n_0 \rightarrow \alpha n$. We derive the amplitude density equation by substituting (4.40) into (4.44), and the result is

$$(4.46) \qquad |\vec{\zeta}^{+}(\alpha_0 n_0)\rangle = \overset{\leftrightarrow}{V} \cdot |\vec{\phi}(\alpha_0 n_0)\rangle + \overset{\leftrightarrow}{V} \cdot \overset{\leftrightarrow}{G_0}^{+}(E) \cdot |\vec{\zeta}^{+}(\alpha_0 n_0)\rangle.$$

Just as in Section 3.3, we can define the arrangement channel generalization of the T-operator according to

$$(4.47) \qquad |\vec{\zeta}^{+}(\alpha_0 n_0)\rangle = \overset{\leftrightarrow}{V} \cdot |\vec{\psi}^{+}(\alpha_0 n_0)\rangle \equiv \overset{\leftrightarrow}{T} \cdot |\vec{\phi}(\alpha_0 n_0)\rangle,$$

and it easily follows that

$$(4.48) \qquad \overset{\leftrightarrow}{T} = \overset{\leftrightarrow}{V} + \overset{\leftrightarrow}{V} \cdot \overset{\leftrightarrow}{G_0}^{+}(E) \cdot \overset{\leftrightarrow}{T}.$$

We can iterate this equation, according to

$$(4.49) \qquad \overset{\leftrightarrow}{T} = \overset{\leftrightarrow}{V} + \overset{\leftrightarrow}{V} \cdot \overset{\leftrightarrow}{G_0}^{+}(E) \cdot \overset{\leftrightarrow}{V} + \overset{\leftrightarrow}{V} \cdot \overset{\leftrightarrow}{G_0}^{+}(E) \cdot \overset{\leftrightarrow}{V} \cdot \overset{\leftrightarrow}{G_0}^{+}(E) \cdot \overset{\leftrightarrow}{V} + ...,$$

and factor out $\overset{\leftrightarrow}{G_0}(E) \cdot \overset{\leftrightarrow}{V}$ to the right,

$$(4.50) \qquad \overset{\leftrightarrow}{T} = \overset{\leftrightarrow}{V} + [\overset{\leftrightarrow}{V} + \overset{\leftrightarrow}{V} \cdot \overset{\leftrightarrow}{G_0}^{+}(E) \cdot \overset{\leftrightarrow}{V} + ...] \cdot \overset{\leftrightarrow}{G_0}^{+}(E) \cdot \overset{\leftrightarrow}{V},$$

and identify the quantity in brackets as $\overleftrightarrow{T}$:

$$(4.51) \qquad \overleftrightarrow{T} = \overleftrightarrow{V} + \overleftrightarrow{T} \cdot \overleftrightarrow{G}_0^{+} (E) \cdot \overleftrightarrow{V}.$$

The equations (4.48) and (4.51) are immediately recognized as reactive scattering generalizations of (3.94) and (3.96). Now rearrange (4.51) to express $\overleftrightarrow{V}$ as

$$(4.52) \qquad \overleftrightarrow{V} = \overleftrightarrow{T} - \overleftrightarrow{T} \cdot \overleftrightarrow{G}_0^{+} (E) \cdot \overleftrightarrow{V},$$

which when substituted into (4.48) yields the generalization of (3.98):

$$(4.53) \qquad \overleftrightarrow{T} = \overleftrightarrow{V} + \overleftrightarrow{T} \cdot \overleftrightarrow{G}_0^{+} (E) \cdot \overleftrightarrow{T} - \overleftrightarrow{T} \cdot \overleftrightarrow{G}_0^{+} (E) \cdot \overleftrightarrow{V} \cdot \overleftrightarrow{G}_0^{+} (E) \cdot \overleftrightarrow{T}.$$

This gives us once again three expressions for the transition operator, and the variational functional is easily constructed as [11]

$$(4.54) \qquad \begin{aligned} [\overleftrightarrow{T}] &= \overleftrightarrow{V} + \overleftrightarrow{V} \cdot \overleftrightarrow{G}_0^{+} (E) \cdot \overleftrightarrow{T} + \overleftrightarrow{T} \cdot \overleftrightarrow{G}_0^{+} (E) \cdot \overleftrightarrow{V} \\ &\quad - \overleftrightarrow{T} \cdot \{ \overleftrightarrow{G}_0^{+} (E) - \overleftrightarrow{G}_0^{+} (E) \cdot \overleftrightarrow{V} \cdot \overleftrightarrow{G}_0^{+} (E) \} \cdot \overleftrightarrow{T} \end{aligned}$$

$$(4.55) \qquad \begin{aligned} &= \overleftrightarrow{V} + \overleftrightarrow{V} \cdot \overleftrightarrow{G}_0^{+} (E) \cdot \overleftrightarrow{T} + \overleftrightarrow{T} \cdot \overleftrightarrow{G}_0^{+} (E) \cdot \overleftrightarrow{V} \\ &\quad - \overleftrightarrow{T} \cdot \overleftrightarrow{G}_0^{+} (E) \cdot [E \overleftrightarrow{1} - \overleftrightarrow{H}] \cdot \overleftrightarrow{G}_0^{+} (E) \cdot \overleftrightarrow{T}. \end{aligned}$$

Just as before, in practice we work with the matrix elements of $\overleftrightarrow{T}$, rather than with the operator. By (4.45), we have

$$(4.56) \qquad \begin{aligned} [T(\alpha_n | \alpha_0 n_0)] &= \langle \vec{\phi}(\alpha n)| \cdot \overleftrightarrow{V} \cdot |\vec{\phi}(\alpha_0 n_0)\rangle + \langle \vec{\phi}(\alpha n)| \cdot \overleftrightarrow{V} \cdot \overleftrightarrow{G}_0^{+} (E) \cdot |\vec{\zeta}^{+}(\alpha_0 n_0)\rangle \\ &\quad + \langle \vec{\zeta}^{-}(\alpha n)| \cdot \overleftrightarrow{G}_0^{+} (E) \cdot \overleftrightarrow{V} \cdot |\vec{\phi}(\alpha_0 n_0)|\rangle \\ &\quad - \langle \vec{\zeta}^{-}(\alpha n)| \cdot \overleftrightarrow{G}_0^{+} (E) \cdot [E \overleftrightarrow{1} - \overleftrightarrow{H}] \cdot \overleftrightarrow{G}_0^{+} (E) \cdot |\vec{\zeta}^{+}(\alpha_0 n_0)\rangle. \end{aligned}$$

The components of the amplitude density vectors can be expanded in $\mathcal{L}^2$-bases just as before. One can also write a generalized scattered wave Kohn variational principle by noting

$$(4.57) \qquad |\vec{\psi}_{SW}^{+}(\alpha_0 n_0)\rangle = \overleftrightarrow{G}_0^{+} (E) \cdot |\vec{\zeta}^{+}(\alpha_0 n_0)\rangle$$

and

$$(4.58) \qquad \langle \vec{\psi}_{SW}^{-}(\alpha n)| = \langle \vec{\zeta}^{-}(\alpha n)| \cdot \overleftrightarrow{G}_0^{+} (E).$$

Then (4.56) becomes [9]

$$
\begin{aligned}
[T(\alpha n|\alpha_0 n_0)] = \; & \langle\vec{\phi}(\alpha n)|\cdot\overset{\leftrightarrow}{V}\cdot|\vec{\phi}(\alpha_0 n_0)\rangle \\
& +\langle\vec{\phi}(\alpha n)|\cdot\overset{\leftrightarrow}{V}\cdot|\vec{\psi}_{SW}^{+}(\alpha_0 n_0)\rangle \\
& +\langle\vec{\psi}_{SW}^{-}(\alpha n)|\cdot\overset{\leftrightarrow}{V}\cdot|\vec{\phi}(\alpha_0 n_0)\rangle \\
& -\langle\vec{\psi}_{SW}^{-}(\alpha n)|\cdot[E\,\overset{\leftrightarrow}{1}-\overset{\leftrightarrow}{H}]\cdot|\vec{\psi}_{SW}^{+}(\alpha_0 n_0)\rangle.
\end{aligned}
$$

(4.59)

The solution of these by expansion in appropriate bases, solving the linear algebraic equations for the expansion coefficients, and substituting the results back into $[T(\alpha n|\alpha_0 n_0)]$ to obtain second order correct T-matrix elements proceeds as before.

5. Computational methods based on differential equation solvers. As the final example of solutions of the TISE for scattering, we comment on methods that solve the equations as coupled ordinary differential equations [8]. This approach can treat completely general scattering problems, including those involving rearrangements and breakup. However, to do so in a reasonably straight forward way, one must use coordinates that are able, with a single distance variable, to describe all possible arrangements. This is most easily achieved by use of hyperspherical coordinates (of which there are many possible choices). Our interest here is simply to indicate how solutions are generated that satisfy the boundary conditions, and what the principal complication is. It suffices to consider again the scattering of a projectile in $3D$ by a nonspherical scatterer. Thus, the potential, $V(\vec{r})$, depends on r, θ, and φ, and the TISE is

$$
-\frac{\hbar^2}{2m}\nabla^2\psi + V(\vec{r})\psi = E\psi.
$$

(5.1)

We expand ψ in a truncated basis of spherical harmonics:

$$
\psi(\vec{r}) = \sum_{\ell'}^{\ell_{max}}\sum_{m'}Y_{\ell'm'}(\hat{r})\mathcal{U}_{\ell'm'}(r)/r.
$$

(5.2)

We substitute (5.2) into (5.1) and project with a particular basis function, $Y_{\ell m}^{*}(\hat{r})$, to obtain the coupled equations

$$
\left(E+\frac{\hbar^2}{2m}\frac{d^2}{dr^2}-\frac{\hbar^2\ell(\ell+1)}{2mr^2}\right)\mathcal{U}_{\ell m}(r) = \sum_{\ell'}\sum_{m'}V(\ell m\ell'm'|r)\mathcal{U}_{\ell'm'}(r).
$$

(5.3)

In fact, if one includes a total of N-basis functions in the expansion, (5.3) will have $2N$ linearly independent solutions.

However, only N of these can be made to be regular at $r = 0$, and these are the physically relevant ones. It is therefore convenient to add a label to the $\mathcal{U}_{\ell m}$ to signify the linearly independent solution it represents:

$$\left(E + \frac{\hbar^2}{2m}\frac{d^2}{dr^2} - \frac{\hbar^2\ell(\ell+1)}{2mr^2}\right)\mathcal{U}(\ell m\ell_0 m_0|r)$$

$$(5.4) \qquad = \sum_{\ell'}\sum_{m'} V(\ell m\ell' m'|r)\mathcal{U}(\ell' m'\ell_0 m_0|r),$$

or in obvious matrix notation,

$$(5.5) \qquad (E\overleftrightarrow{1} - \overleftrightarrow{K})\cdot\vec{\mathcal{U}}(r) = \overleftrightarrow{V}(r)\cdot\vec{\mathcal{U}}(r).$$

Note that $\overleftrightarrow{1}$ and $\overleftrightarrow{K}$ are diagonal matrices, and we interpret the ℓm indices of $\mathcal{U}(\ell m\ell_0 m_0|r)$ as "final state" indices and $\ell_0 m_0$ as "initial state" indices. In addition to the regularity condition at $r = 0$ (which for typical potentials in atom-diatom scattering becomes the condition that $\mathcal{U}(\ell m\ell_0 m_0|0) = 0$), we want to impose the asymptotic condition

$$\lim_{r\to\infty}\mathcal{U}(\ell m\ell_0 m_0|r) = \delta_{\ell\ell_0}\delta_{mm_0}j_\ell(kr)$$

$$(5.6) \qquad\qquad\qquad\qquad +h_\ell^+(kr)T(\ell m\ell_0 m_0).$$

This means that we must deal with a "two point" boundary condition problem, and we do not know the value of $T(\ell m\ell_0 m_0)$. This is equivalent to saying that although we know that $\mathcal{U}(\ell m\ell_0 m_0|0) = 0$, we do *not* know what the correct value is for the derivatives, $d\mathcal{U}(\ell m\ell_0 m_0|r)/dr$ at $r = 0$. The usual way to circumvent this difficulty is to convert the problem into a *pseudo-initial value problem*. The initial conditions typically are taken to be

$$(5.7) \qquad\qquad (a) \qquad \mathcal{U}(\ell m\ell_0 m_0|0) = 0$$

and

$$(5.8) \qquad\qquad (b) \qquad \frac{d}{dr}\mathcal{U}(\ell m\ell_0 m_0|r)|_{r=0} = \delta_{\ell\ell_0}\delta_{mm_0}.$$

These initial conditions will *not*, in general produce the correct large-r behavior. However, knowing the N-linearly independent solutions allows one to take linear combinations that possess the correct asymptotic behavior. This procedure generally requires solving a system of linear algebraic equations that yields the scattering amplitudes.

In principle, the above procedure works. However, there is a practical difficulty that makes additional computational effort necessary. This has already been mentioned in Section 3.2 above, and is the fact that different ℓ-values lead to different classical turning points at a given total energy E. As a result, the pieces of the solution matrix grow at different rates in their nonclassical regions. The most rapidly growing component can become dominant in several (or even all) of the solution vectors, causing them to be multiples of one another; i.e., they become linearly dependent. If this occurs, it is no longer possible to construct the physical solution. This is manifested by the relevant algebraic equations becoming ill-conditioned. Various techniques are employed to avoid this difficulty, but they all essentially amount to periodically scrambling the column solutions (*before* they become linearly dependent) so as to maintain explicitly linearly independent solutions. Many methods are used to solve the coupled equations, (5.4), including varieties of the Numerov method, the so-called log-derivative method (which essentially calculates $\overleftrightarrow{U}^{-1} \cdot d\overleftrightarrow{U} /dr$), etc.

Finally, we should comment on how the computational effort scales with problem size, since some methods may be very efficient for small problems but rapidly become impractical as the complexity increases. Similarly, some methods may not be so efficient for smaller problems, but have slow scaling with problem size. Generally, all of the methods discussed involve either matrix-matrix multiplications or solutions of linear algebraic equations. Both procedures scale as the cube of the matrix dimensions. However, the number of matrix multiplications, the number of times the linear equations must be solved, and the dimensions of the relevant matrices can be quite different for the various methods. The non-iterative Volterra equation propagation method, and the propagation methods which solve coupled differential equations both must generate N-linearly independent solutions. The size of the matrices is the same for both these types of approaches (it equals the number of internal states, including those associated with any angular momentum in the system). Thus, these types of approaches involve of the order of N^3 multiplications at each step in propagating from $r = 0$ out beyond the range of the potential. If P steps are required, on the order of $N^3 P$ multiplications must be performed. This work must be repeated at every desired energy. (There are methods in which the number of internal states changes as the propagation moves further out and the potential becomes weaker. These scale more slowly.)

The algebraic variational methods involve expansion in f_n-basis functions for each internal quantum state, n (f_n can differ for different quantum states). The total number of unknowns, U, is then the sum over n of the f_n,

$$(5.9) \qquad U = \sum_{n=1}^{N} f_n.$$

The computational effort then scales as U^3. Some of the work is independent of the energy of interest, but in general, a substantial amount of work must be repeated at each new energy.

We can introduce an "average" number of basis functions, $\langle f_n \rangle$, such that

$$(5.10) \qquad U = N \langle f_n \rangle,$$

and the computational effort scales as $N^3 \langle f_n \rangle^3$. Clearly, if $\langle f_n \rangle^3$ is larger than P, the variational method will not be as efficient as the propagation methods, for a given energy. It is also possible to solve the variational equations for a restricted set of initial states (rather than for all N initial states). If one can construct reasonable starting "guesses" for the solution vectors, then iterative methods can be used to solve the algebraic equations. In this case, the effort scales more slowly than U^3, and this can lead to a very efficient computational approach.

This completes our discussion of illustrative time-independent methods. We now turn to consider direct solutions of the TDSE.

6. Computational approaches based on the TDSE.

6.1. Path integral approaches using a Fourier filter. We return to a consideration of the TDSE,

$$(6.1) \qquad i\hbar \frac{\partial \chi}{\partial t} = H\chi,$$

and note that the development of the standard scattering theory required the separation of H into an unperturbed Hamiltonian, H_0 (defined by the limit of H as a scattering distance became large), and the concommitant perturbation, $V(= H - H_0)$. However, one can actually divide H into *any* convenient "reference Hamiltonian", H_r, and the concommitant "disturbance Hamiltonian", $H_d(= H - H_r)$. Thus, we write

$$(6.2) \qquad H = H_r + H_d,$$

and we shall suppose that in the most general case, H_d may even be time-dependent. One no longer can write the solution simply as $\exp(-iHt/\hbar)\chi(0)$ (except by time-ordering or other more elegant procedures). We can write (6.1) as

$$(6.3) \qquad i\hbar \frac{\partial \chi}{\partial t} - H_r\chi = H_d\chi,$$

and the *exact* solution of this equation (even for a time-dependent H_d) is

$$\chi(t) = -\frac{i}{\hbar} \int_{t-\Delta t}^{t} dt' e^{-iH_r(t-t')/\hbar} H_d(t')\chi(t')$$

$$(6.4) \qquad\qquad\qquad + e^{-iH_r\Delta t/\hbar}\chi(t - \Delta t).$$

Here we explicitly include the possible dependence of H_d on time. We can use the exact solution (6.4) as the basis of a computational algorithm as follows. We approximate the integral over dt' by, e.g., the trapezoidal rule, obtaining

$$\chi(t) = \frac{-i\Delta t}{2\hbar} \left[H_d(t)\chi(t) + e^{-iH_r\Delta t/\hbar} H_d(t - \Delta t)\chi(t - \Delta t) \right]$$

$$(6.5) \qquad\qquad\qquad + e^{-iH_r\Delta t/\hbar}\chi(t - \Delta t).$$

We rearrange and solve for $\chi(t)$, obtaining [12]

$$(6.6) \qquad \chi(t) = [1 + \frac{i\Delta t}{2\hbar} H_d(t)]^{-1} e^{-iH_r\Delta/\hbar}$$
$$[1 - \frac{i\Delta t}{2\hbar} H_d(t - \Delta t)]\chi(t - \Delta t),$$

which expression is called the "Modified Cayley" propagation formula. We note that when H_d is time-dependent, this approach automatically yields a time-ordered propagator, $\mathcal{P}(t, \Delta t)$,

$$(6.7) \quad \mathcal{P}(t, \Delta t) = [1 + \frac{i\Delta t}{2\hbar} H_d(t)]^{-1} e^{-iH_r\Delta t/\hbar} [1 - \frac{i\Delta t}{2\hbar} H_d(t - \Delta t)].$$

We shall restrict ourselves to time-independent Hamiltonians. Then the modified Cayley short time propagation depends solely on the time step, Δt, and is given by

$$(6.8) \qquad \mathcal{P}(\Delta t) = [1 + \frac{i\Delta t}{2\hbar} H_d]^{-1} e^{-iH_r\Delta t/\hbar} \left[1 - \frac{i\Delta t}{2\hbar} H_d\right].$$

The reader should note that at no point in the derivation of either (6.7) or (6.8) did we make arguments based on neglect of commutators above a certain level. Rather, the approximation governing the accuracy of these expressions is determined by the trapezoidal rule. Thus, Δt must be small enough so that one obtains an adequate approximation to the integral over dt' in (6.4).

It is worthwhile to compare (6.8) with another extremely popular short time approximation to the time evolution operator, $\exp(-iH\tau/\hbar)$. For short enough times τ, one may neglect commutators of H_d and H_r above a certain level, and approximate $\exp(-iH\tau/\hbar)$ by the "symmetric split operator" expression [13,14]

$$(6.9) \qquad e^{-iH\tau/\hbar} \cong e^{-iH_d\tau/2\hbar} e^{-iH_r\tau/\hbar} e^{-iH_d\tau/2\hbar}.$$

Obviously, for small Δt, one can write

$$(6.10) \qquad [1 + \frac{i\Delta t}{2\hbar} H_d]^{-1} \cong 1/e^{i\Delta t H_d/2\hbar}$$

$$(6.11) \qquad\qquad\qquad = e^{-i\Delta t H_d/2\hbar}$$

and

$$(6.12) \qquad [1 - \frac{i\Delta t}{2\hbar} H_d] \cong e^{-i\Delta t H_d/2\hbar}.$$

It is *not* correct to say that (6.8) is a Taylor expansion approximation to the symmetric split operator approximate propagator. They are derived under different approximation conditions. However, computationally they will require similar amounts of effort.

To proceed further, we must specify the representation to be used in calculating the action of $P(\Delta t)$ on $\chi(t - \Delta t)$. The two obvious representations are the one in which H_r is diagonal, and the one in which H_d is diagonal. Clearly, these cannot be the same (otherwise, H_r and H_d would commute, one would have the exact solution, and there would be no scattering). We illustrate with the simple example of 1D scattering, with

$$(6.13) \qquad H = K + V,$$

K being the kinetic energy and V being the potential. Normally we take $H_r = K$ and $H_d = V$. Then $\chi(t)$ is given by

$$(6.14) \qquad \chi(t) = \left[1 + \frac{i\Delta t}{2\hbar} V\right]^{-1} e^{-iK\Delta t/\hbar} \left[1 - \frac{i\Delta t}{2\hbar} V\right] \chi(t - \Delta t).$$

Obviously, the factors involving the potential are diagonal in the coordinate representation (assuming a "local" potential), while $\exp(-iK\Delta t/\hbar)$ is diagonal in the momentum representation. Since the leading operator is diagonal in the coordinate representation, as is also the operator acting directly on $\chi(t - \Delta t)$, the natural representation for $\chi(t)$ is also the coordinate representation. We therefore form $\langle x|\chi(t)\rangle$ ($\equiv \chi(t|x)$) according to

$$\chi(t|x) = \left[1 + \frac{i\Delta t}{2\hbar} V(x)\right]^{-1} \langle x|e^{-iK\Delta t/\hbar} \int_{-\infty}^{\infty} dx'|x'\rangle \left[1 - \frac{i\Delta t}{2\hbar} V(x')\right] \chi(t - \Delta t|x'),$$

(6.15)

where use has been made of the resolution of the identity as

$$(6.16) \qquad 1 = \int_{-\infty}^{\infty} dx'|x'\rangle\langle x'|,$$

and the result

$$\langle x'| \left[1 - \frac{i\Delta t}{2\hbar} V\right] |\chi(t - \Delta t)\rangle$$

$$(6.17) \qquad = \left[1 - \frac{i\Delta t}{2\hbar} V(x')\right] \chi(t - \Delta t|x').$$

We note that the inverse $[1 + \frac{1\Delta t}{2\hbar} V(x)]^{-1}$ is trivial to calculate, and, we arrive at the equation

$$\chi(t|x) = \left[1 + \frac{i\Delta t}{2\hbar} V(x)\right]^{-1} \int_{-\infty}^{\infty} dx' \langle x|e^{-iK\Delta t/\hbar}|x'\rangle \left[1 - \frac{i\Delta t}{2\hbar} V(x')\right] \chi(t - \Delta t|x'),$$

(6.18)

where $\langle x|e^{-iK\Delta t/\hbar}|x'\rangle$ is the short time free particle propagator in the co-ordinate representation. One may formally write a long time propagation in terms of a sequence of such short time propagations, by dividing the full time t into N short time segments of duration Δt,

$$(6.19) \qquad t = N\Delta t,$$

and the result is [13,15]

$$\chi(t|x_N) = \int_{-\infty}^{\infty} dx_{N-1} \cdots \int_{-\infty}^{\infty} dx_0 \left[1 + \frac{i\Delta t}{2\hbar}V(x_N)\right]^{-1} \langle x_N|e^{-iK\Delta t/\hbar}|x_{N-1}\rangle$$

$$\left[1 - \frac{i\Delta t}{2\hbar}V(x_{N-1})\right]\left[1 + \frac{i\Delta t}{2\hbar}V(x_{N-1})\right]^{-1} \langle x_{N-1}|e^{-iK\Delta t/\hbar}|x_{N-2}\rangle$$

$$(6.20) \qquad \cdots \langle x_1|e^{-iK\Delta t/\hbar}|x_0\rangle \left[1 + \frac{i\Delta t}{2\hbar}V(x_0)\right]\chi(0|x_0).$$

Here, x_j is the set of coordinates at the jth time interval, and we recognize (6.20) as a Feynman path integral, but with the Modified Cayley short time propagator in place of the more commonly used "symmetrized Trotter" expression (which is identical to the symmetric split operator expression (6.9)). The $\langle x_j|\exp(-iK\Delta t/\hbar)|x_{j-1}\rangle$ matrix element can be evaluated analytically exactly, yielding [13]

$$(6.21) \qquad \langle x_j|e^{-iK\Delta t/\hbar}|x_{j-1}\rangle = \left(\frac{m}{2\pi i\hbar\Delta t}\right)^{1/2} e^{im(x_j-x_{j-1})^2/2\Delta t\hbar},$$

and this is interpreted as the amplitude for traveling from the point x_{j-1} to the point x_j in time interval Δt, and in the absence of any potential. The connection to Feynman's approach is obtained by noting that

$$(6.22) \qquad \frac{1}{2}m(x_j - x_{j-1})^2/\Delta t = \frac{1}{2}m\frac{(x_j - x_{j-1})^2}{(\Delta t)^2}\Delta t$$

$$(6.23) \qquad = \int_{(j-1)\Delta t}^{j\Delta t} dt\, K,$$

which is the *action* accumulated during the time interval Δt due to the kinetic energy. Note that since this is free motion, $(x_j - x_{j-1})/\Delta t$ is *exactly* the velocity, and not a finite difference approximation. The problem with (6.21) is that although it is an *exact* expression, it never directly corresponds to a situation which is realized experimentally. To see this, we note that the magnitude of the amplitude $\langle x_j|\exp(-iK\Delta t/\hbar)|x_{j-1}\rangle$ is a *constant*, independent of the distance traveled:

$$(6.24) \qquad |\langle x_j|e^{-iK\Delta t/\hbar}|x_{j-1}\rangle|^2 = \frac{m}{2\pi\hbar\Delta t}.$$

It also *increases* as Δt becomes smaller, and the amplitude itself oscillates more and more rapidly as either $|x_j - x_{j-1}|$ increases or $\Delta t \to 0$. As a result, (6.20) is not readily amenable to the usual approximations employed in developing quantum mechanical algorithms *if the exact free propagator is used.* (In fact, these troublesome oscillations are the feature used in the extraction of the classical limit from a path integral approach.) The fundamental source of this difficulty is readily apparent from considering how the exact value of $\langle x_j | \exp(-iK\Delta t/\hbar) | x_{j-1} \rangle$ is calculated. One inserts (judiciously) the resolution of the identity in the momentum representation,

$$(6.25) \qquad 1 = \int\limits_{-\infty}^{\infty} dk |k\rangle \langle k|,$$

where

$$(6.26) \qquad \langle x|k \rangle = \frac{1}{\sqrt{2\pi}} e^{ikx},$$

so that

$$(6.27) \quad \langle x_j | e^{-iK\Delta t/\hbar} | x_{j-1} \rangle = \frac{1}{2\pi} \int\limits_{-\infty}^{\infty} dk e^{ik(x_j - x_{j-1})} e^{-i\hbar k^2 \Delta t/2m}.$$

Note that (a) the integrand does *not* tend to zero as $k \to \pm\infty$, (and in fact, has *constant* magnitude) and (b) the integral converges due to the rapid oscillations of $\exp(-i\hbar k^2 \Delta t/2m)$. These features underly the properties of the exact $\langle x_j | \exp(-iK\Delta t/\hbar) | x_{j-1} \rangle$, and essentially reflect the fact that the exact short time propagator must be capable of propagating *any* wavepacket.

Thus, even though any particular physical wavepacket will have negligible contributions from infinitely large momenta, in order to be able to propagate all possible wavepackets, the exact propagator must include all momenta, with equal weighting. It follows that one really should *filter* the free propagator so as to remove the momenta larger than some maximum value. The maximum value is determined by the momentum composition of the incident packet, $\chi(0|x_0)$, and the maximum boosts that can occur due to the potential, during the course of the collision. The simplest way to do this is to use a "square filter" in momentum space, so that [16]

$$(6.28) \quad \langle x_j | e^{-iK\Delta t/\hbar} | x_{j-1} \rangle \cong \frac{1}{2\pi} \int\limits_{-k_{max}}^{k_{max}} dx e^{ik(x_j - x_{j-1})} e^{-i\hbar k^2 \Delta t/2m}$$

and then to approximate the integrals in (6.20) by the simple trapezoidal rule. The square wave filter produces a large $|x_j - x_{j-1}|$ behavior of the form

$$(6.29) \qquad \frac{\sin(|x_j - x_{j-1}|k_{\max})}{|x_j - x_{j-1}|},$$

so the magnitude of the filtered free propagator is no longer independent of $|x_j - x_{j-1}|$, but rather decays as $1/|x_j - x_{j-1}|$. The quadrature approximation then produces a discrete, matrix representation of $\langle x_j | \exp(-iK\Delta t/\hbar)|x_{j-1}\rangle$ which decays as one moves away from the diagonal. The discretized matrices $[1 - i\Delta t V(x_{j,\ell})/2\hbar]$ and $[1 + i\Delta t V(x_{j,\ell})/2\hbar]^{-1}$ are *diagonal*, $\chi(0|x_{0,\ell})$ is a column vector, and the propagation according to the discretized path integral becomes a matter of successive matrix-vector multiplications. If the number of grid points is P, the free propagator matrix is $P \times P$ and each time step involves on the order of P^2 multiplications. In fact, for Cartesian-type variables, the structure of the free propagator allows one to do the calculation by (a) applying the potential dependent matrix in the coordinate representation, then Fast Fourier transforming to momentum space to apply the filtered free propagator, then Fast Fourier transforming back to coordinate space to apply the next potential dependent matrix, etc. This leads to $P \log_2 P$ scaling. For multidimensional problems, the dimensionality for standard product quadratures grows like P^d, where d is the number of degrees of freedom. This places limits on the kinds of problems that can be addressed in this way (basically, one can treat up to three degrees of freedom by such quadrature methods; more degrees of freedom typically are dealt with by mixtures of quadratures and more traditional basis sets). It should be obvious that the structure of the calculation is identical whether one uses the Modified Cayley or symmetrized Trotter short time propagator, since they differ only in the dependence on the potential.

Before discussing an alternative to the square filter in momentum space, it is of interest to point out that one can make other choices for the reference Hamiltonian. Thus, one can include part of the potential with the kinetic energy, thereby reducing the size of the commutator between H_r and H_d. This makes it possible to use longer time steps, Δt, but may make evaluation of $\exp(-iH_r\Delta t/\hbar)$ more time consuming. As long as one need do this only once, this may well be worth the effort. However, it is also possible to make very unorthodox choices of H_r. For example, one may choose [17]

$$(6.30) \qquad\qquad H_r = V$$

and then

$$(6.31) \qquad\qquad H_d = K,$$

so that the reference Hamiltonian is the potential and the disturbance is the kinetic energy. In this case, (6.11) is replaced by

$$(6.32) \quad \chi(t) = \left[1 + \frac{i\Delta t}{2\hbar}K\right]^{-1} e^{-iV\Delta t/\hbar} \left[1 - \frac{i\Delta t}{2\hbar}K\right] \chi(t - \Delta t),$$

and the natural representation for $\chi(t)$ is the momentum representation. This leads to the equation

$$\chi(t|k) = \left[1 + \frac{i\Delta t\hbar k^2}{4m}\right]^{-1} \int_{-\infty}^{\infty} dk' \langle k|e^{-iV\Delta t/\hbar}|k'\rangle \left[1 - \frac{i\Delta t\hbar k^2}{4m}\right] \chi(t - \Delta t|k')$$

(6.33)

Now the non-diagonal matrix is $\langle k| \exp(-iV\Delta t/\hbar)|k'\rangle$, which can be interpreted as the actions of the potential that cause changes in the momentum. A Taylor expansion gives

$$(6.34) \quad \langle k|e^{-iV\Delta t/\hbar}|k'\rangle = \sum_n \frac{(-i\Delta t/\hbar)^n}{n!} \langle k|V^n|k'\rangle,$$

so $n = 1$ is the Born amplitude, $\langle k|V|k'\rangle$, $n = 2$ is the amplitude $\langle k|V^2|k'\rangle$, etc. It will again be reasonable to filter the momentum, so as to include only $-k_{max} \leq k \leq k_{max}$. One may also evaluate the action of $\exp(-iV\Delta t/\hbar)$ in the momentum space by Fourier transforming from $k' \to x$, applying $\exp(-iV(x)\Delta t/\hbar)$, and transforming back to k. Since both the x and k spaces will be truncated and discretized, this again leads to evaluating (6.33) by diagonal matrix multiplications alternating with Fast Fourier transformations between the x and k-spaces. The scaling will again be $P\log_2 P$. In fact, the choice of $H_r = V$ and $H_d = K$ has been used in calculations for atom-diatom rotational excitation and has been found to be very efficient. The analysis of the scattering information essentially follows the discussion in Section II.

There is, of course, an analogous symmetrized Trotter form of the short time propagator, given by

$$(6.35) \quad e^{-iH\Delta t/\hbar} \cong e^{-iK\Delta t/2\hbar} e^{-iV\Delta t/\hbar} e^{-iK\Delta t/2\hbar},$$

and one can derive the analogous path integral and discretized path integral expressions based on it. We now turn to a discussion of a different way to filter the short time free propagator, $\exp(-iK\Delta t/\hbar)$, which we believe has some very nice features for computations.

6.2. Path integral approach using the distributed approximating functional filter. In our discussion of path integrals, we found that the exact coordinate representation matrix element of $\exp(-iK\Delta t/\hbar)$ involved the integral expression

$$(6.36) \quad \langle x'|e^{-iK\Delta t/\hbar}|x\rangle = \frac{1}{2\pi} \int_{-\infty}^{\infty} dk e^{ik(x-x')} e^{-i\hbar k^2 \Delta t/2m}.$$

This can also be written as

$$(6.37) \qquad \langle x'|e^{-iK\Delta t/\hbar}|x\rangle = e^{-iK\Delta t/\hbar}\delta(x - x'),$$

where we turn a blind eye to the mathematical issues of interchanging integration and the action of $\exp(-iK\Delta t/\hbar)$. Thus, using the standard representation of $\delta(x - x')$ as

$$(6.38) \qquad \delta(x - x') = \frac{1}{2\pi} \int\limits_{-\infty}^{\infty} dx' e^{ik(x-x')},$$

we obtain (6.36) from (6.37), since

$$(6.39) \qquad e^{-iK\Delta t/\hbar}e^{ikx} = e^{-i\hbar k^2 \Delta t/2m}e^{ikx}.$$

Thus, we can generate the x', x matrix element of the operator by applying it to $\delta(x - x')$ (in fact, this is true for any operator function of the momentum or kinetic energy). If we Fourier transform $\delta(x - x')$ to momentum space, we see that (a) the transform is diagonal, with eigenvalue δ_k, since $\delta(x-x')$ depends only on $x - x'$ (b) its momentum space eigenvalue is equal to 1, independent of k (see (6.38)). The square filter used in the preceding discussion was simply a step function,

$$(6.40) \qquad \delta_{k,k_{max}} = 1, |k| \leq k_{max} > 0$$
$$(6.41) \qquad \qquad\quad = 0, |k| > k_{max} > 0.$$

The main drawback of such a filter is that, due to the sharp edges at $\pm k_{max}$, the coordinate representation of $\delta_{k,k_{max}}$, $\delta_F(x - x')$, decays slowly away from $x = x'$, and it has substantial oscillations. Here, our notation is intended to indicate that $\delta_F(x - x')$ is the filtered delta function based on a Fourier representation. However, there are many other representations of $\delta(x - x')$ which could be used, and one could, e.g., have employed Hermite polynomials, weighted by their generating function. Thus, one can write [18]

$$(6.42) \qquad \delta(x - x') = \frac{e^{-(x-x')^2/2\sigma^2}}{\sqrt{2\pi}\sigma} \sum_{n=0}^{\infty} \left(-\frac{1}{4}\right)^n \frac{1}{n!} H_{2n}\left(\frac{x - x'}{\sqrt{2}\sigma}\right),$$

where only even Hermite polynomials appear due to the fact that $\delta(x - x')$ is an even function of $(x - x')$. The Fourier transform of (6.42) is

$$(6.43) \qquad \delta_k = e^{-\sigma^2 k^2/2} \sum_{n=0}^{\infty} \left(\frac{\sigma k}{\sqrt{2}}\right)^{2n} \frac{1}{n!},$$

which in fact, equals 1 because

$$(6.44) \qquad \sum_{n=0}^{\infty} \left(\frac{\sigma k}{\sqrt{2}}\right)^{2n} \frac{1}{n!} = e^{+\sigma^2 k^2/2}.$$

A filter can be constructed by truncating the infinite sum, so that

$$(6.45) \quad \delta_M(x - x'|\sigma) = \frac{e^{-(x-x')^2/2\sigma^2}}{\sqrt{2\pi}\sigma} \sum_{n=0}^{M/2} \left(-\frac{1}{4}\right)^n \frac{1}{n!} H_{2n}\left(\frac{x-x'}{\sqrt{2}\sigma}\right)$$

in the coordinate representation, and

$$(6.46) \qquad \delta_{k,M}(\sigma) = e^{-\sigma^2 k^2/2} \sum_{n=0}^{M/2} \left(\frac{\sigma k}{\sqrt{2}}\right)^{2n} \frac{1}{n!}.$$

These are referred to as "distributed approximating functionals" (DAFs) in the literature, and $\delta_F(x - x')$ also can be viewed as a DAF (although this has not been the view normally taken). The DAF is a *functional*, rather than a function, and it can be used for a wide variety of purposes. For example, the Dirac delta function has the property

$$(6.47) \qquad f(x) = \int_{-\infty}^{\infty} dx' \delta(x - x') f(x'),$$

which in Fourier space translates to

$$(6.48) \qquad \overline{f}(k) = \delta_k \overline{f}(k),$$

and since δ_k is equal to 1 for all k, (6.48) is seen to be equivalent to (6.47). In the case of the DAF, we must consider the Fourier space - DAF in order to deduce what replaces (6.47). For example, in the case of the Fourier DAF, (6.48) is replaced by

$$(6.49) \qquad \overline{f}_F(k) = \delta_{k,k_{max}} \overline{f}(k),$$

where $\overline{f}_F(k)$ is the Fourier transform of the result of applying $\delta_F(x - x')$ to $f(x')$:

$$(6.50) \qquad f_F(x) = \int_{-\infty}^{\infty} dx' \delta_F(x - x') f(x').$$

Now if $\overline{f}(k)$ is zero for $|k| > k_{max}$, then $f_F(x)$ will be *identical* to $f(x)$, since $\delta_{k,k_{max}}$ equals 1 for $|k| \leq k_{max} > 0$. To the degree that $\overline{f}(k)$ extends beyond k_{max}, the truncated Fourier transform will be in error.

Similarly, we must consider how the Fourier transformed Hermite- DAF behaves to understand the DAF approximation to $f(x)$ that is delivered by

$$(6.51) \qquad f_{M,\sigma}(x) = \int_{-\infty}^{\infty} dx' \delta_M(x - x'|\sigma) f(x'),$$

or in momentum space,

$$\overline{f}_{M,\sigma}(k) = \delta_{k,M}(\sigma)\overline{f}(k).$$ (6.52)

By rewriting (6.46) as

$$\delta_{kM}(\sigma) = \sum_{n=0}^{M/2} \left(\frac{\sigma k}{\sqrt{2}}\right)^{2n} \frac{1}{n!}/e^{\sigma^2 k^2/2}$$ (6.53)

and noting that every term in the sum is positive, and the sum converges to $\exp(\sigma^2 k^2/2)$, we conclude that (a) when $k = 0$, $\delta_{k,M}(\sigma)$ equals 1 (b) $\delta_{k,M}(\sigma)$ decays *very slowly* away from 1 as $|k|$ increases, until $|k| \cong \sqrt{2}/\sigma$ (c) beyond $|k| \cong \sqrt{2}/\sigma$, $\delta_{k,M}(\sigma)$ decays exponentially rapidly to zero. Thus, $\delta_{kM}(\sigma)$ is a smoothed, step filter that retains almost all of the momentum content for $|k| < \sqrt{2}/\sigma$, and it will accurately reproduce functions whose momentum transform, $\overline{f}(k)$, is nonzero primarily in this range. In addition, the DAF representation of such functions will also yield accurate representations of their derivatives, so long as the Fourier transform of the derivative also lies mainly in this range of k. Such functions are said to belong to the "DAF-class".

The calculation of derivatives using the DAFs is quite simple, and we illustrate it for the Hermite DAF. With the exact Dirac delta function, we know that

$$\frac{df}{dx} = \int_{-\infty}^{\infty} dx' \frac{d}{dx}\delta(x-x')f(x')$$ (6.54)

since use of $\frac{d}{dx}\delta(x-x') = -\frac{d}{dx'}\delta(x-x')$, along with parts integration gives

$$\int_{-\infty}^{\infty} dx' \frac{d}{dx}\delta(x-x')f(x') = \int_{-\infty}^{\infty} dx'\delta(x-x')\frac{df}{dx'}.$$ (6.55)

Now

$$\frac{d}{dx}f_{M,\sigma}(x) = \int_{-\infty}^{\infty} dx' \frac{d}{dx}\delta_M(x-x'|\sigma)f(x')$$ (6.56)

$$= -\int_{-\infty}^{\infty} dx' \frac{d}{dx'}\delta_M(x-x'|\sigma)f(x'),$$ (6.57)

and integration by parts (noting that surface terms vanish) yields

$$\frac{d}{dx}f_{M,\sigma}(x) = \int_{-\infty}^{\infty} dx'\delta_M(x-x'|\sigma)\frac{df}{dx'}.$$ (6.58)

This says that one can generate the derivative of the DAF-approximated function *either* by applying the DAF to df/dx or by differentiating the DAF, and applying the result to the function. (All the above assumes that the function is of DAF-class.) Using the fact that in (6.45), the Hermite polynomial can be expressed as the appropriate derivative of the Gaussian generating function, it is easy to show that the qth derivative is given by

$$(6.59) \qquad \frac{d^q}{dx^q} \delta_M(x - x')\sigma) \equiv \delta_M^{(q)}(x - x'|\sigma)$$

$$(6.60) \qquad = \frac{e^{-(x-x')^2/2\sigma^2}}{(\sqrt{2}\sigma)^{q+1}\sqrt{\pi}} \sum_{n=0}^{M/2} \left(-\frac{1}{4}\right)^n \frac{1}{n!} H_{2n+q}\left(\frac{x - x'}{\sqrt{2}\sigma}\right).$$

The most useful of these in quantum dynamics are for $q = 1$ and 2. It should be noted that $\delta_M^{(q)}(x - x'|\sigma)$ is the kernel of an integral operator representation of d^q/dx^q, where $q = 0$ is the DAF-identity. A 1D Hamiltonian is therefore given by

$$H(x, x') = -\frac{\hbar^2 e^{-(x-x')^2/2\sigma^2}}{2m\sqrt{\pi}(\sqrt{2}\sigma)^3} \sum_{n=0}^{M/2} \left(-\frac{1}{4}\right)^n \frac{1}{n!} H_{2n+2}\left(\frac{x - x'}{\sqrt{2}\sigma}\right)$$

$$(6.61) \qquad\qquad + \delta(x - x')V(x).$$

(Note that we do *not* need to approximate $\delta(x - x')$ in the potential term.) In using (6.61) to apply the Hamiltonian, one must evaluate the integral over x' from the kinetic energy term; this is generally done by quadrature or by some other discrete sampling procedure. The most useful feature of the DAF-kinetic energy kernel is the fact that the presence of the Gaussian causes an exponential decay away from the $x = x'$ line. When discretized, this implies that the *DAF-kinetic energy is a highly banded matrix.*

Finally, we wish to evaluate the DAF-free propagator by applying $\exp(-iK\Delta t/\hbar)$ to $\delta_M(x - x'|\sigma)$. This is easily done either by Fourier transforming the product in momentum space,

$$(6.62) \qquad \frac{1}{2\pi} \int_{-\infty}^{\infty} e^{-i\hbar k^2 \Delta t/2m} \delta_{kM}(\sigma) e^{ik(x-x')},$$

or by directly calculating

$$(6.63) \qquad e^{-iK\Delta t/\hbar} \delta_M(x - x'|\sigma).$$

Both can be carried out analytically exactly, with the latter being the result of (a) writing the Hermite polynomials as derivatives of the Gaussian generating function (b) noting that $\exp(-iK\Delta t/\hbar)$ commutes with the derivatives of the generator and (c) noting that the free propagation of a Gaussian can be evaluated analytically, producing another Gaussian, but with a complex width parameter that depends on the time interval Δt. Thus,

$$(6.64) \qquad e^{-iK\Delta t/\hbar}\delta_M(x - x'|\sigma) \equiv \delta_M(x - x'|\sigma(\Delta t))$$

where

$$(6.65) \qquad \begin{aligned} \delta_M(x - x'|\sigma(\Delta t)) &= \frac{e^{-(x-x')^2/2\sigma^2(\Delta t)}}{\sqrt{2\pi}\sigma(\Delta t)} \sum_{n=0}^{M/2} \left(-\frac{1}{4}\right)^n \frac{1}{n!} \\ &\quad \left[\frac{\sigma}{\sigma(\Delta t)}\right]^{2n} H_{2n}\left(\frac{x - x'}{\sqrt{2}\sigma(\Delta t)}\right), \end{aligned}$$

$$(6.66) \qquad \sigma^2(\Delta t) = \sigma^2 + i\hbar\Delta t/m.$$

It is extremely interesting to note that the DAF-filtered free propagator is exponentially damped by the Gaussian factor. This can have a substantial impact on the computational usefulness of path integrals when expressed using the DAF free propagator.

It should be noted that when $\delta_M^{(2)}(x - x'|\sigma)$ or $\delta_M(x - x'|\sigma(\Delta t))$ is applied to some function,

$$(6.67) \qquad \frac{d^2 f}{dx^2} = \int_{-\infty}^{\infty} dx' \delta_M^{(2)}(x - x'|\sigma)f(x'),$$

$$(6.68) \qquad f(t + \Delta t|x) = \int_{-\infty}^{\infty} dx' \delta_M(x - x'|\sigma(\Delta t))f(t|x'),$$

and one approximates the integrals by the trapezoidal rule,

$$(6.69) \qquad \frac{d^2 f}{dx^2} \cong \Delta \sum_p \delta_M^{(2)}(x - x_p|\sigma)f(x_p),$$

$$(6.70) \qquad f(t + \Delta t|x) \cong \Delta \sum_p \delta_M(x - x_p|\sigma(\Delta t))f(t|x_p),$$

it is customary in the literature to define the fully discretized DAF-matrices as

$$(6.71) \qquad \begin{aligned} \delta_M^{(2)}(x_{p'} - x_p|\sigma) &= \frac{\Delta e^{-(x_{p'}-x_p)^2/2\sigma^2}}{\sqrt{\pi}(\sqrt{2}\sigma)^3} \\ &\quad \sum_{n=0}^{M/2} \left(-\frac{1}{4}\right)^n \frac{1}{n!} H_{2+n}\left(\frac{x_{p'} - x_p}{\sqrt{2}\sigma}\right), \end{aligned}$$

and

$$(6.72) \qquad \delta_M(x_{p'} - x_p | \sigma(\Delta t)) = \frac{\Delta e^{-(x_{p'} - x_p)^2 / 2\sigma}}{\sqrt{2\pi}\sigma(\Delta t)}$$

$$\sum_{n=0}^{M/2} \left(-\frac{1}{4}\right)^n \frac{1}{n!} \left[\frac{\sigma}{\sigma(\Delta t)}\right]^{2n} H_{2n}\left(\frac{x_{p'} - x_p}{\sqrt{2}\sigma(\Delta t)}\right).$$

Thus, the discretized DAFs include the trapezoidal rule weight, Δ (the uniform spacing between quadrature points).

Having obtained the DAF-free propagator, (6.65), one can then write real-time path integral expressions in which there naturally occurs a Gaussian factor. The Modified Cayley path integral, (6.10), then becomes

$$\chi(t|x_N) = \int_{-\infty}^{\infty} dx_{N-1} \cdots \int_{-\infty}^{\infty} dx_0 \left[1 + \frac{i\Delta t}{2\hbar}V(x_N)\right]^{-1} \delta_M(x_N - x_{N-1}|\sigma(\Delta t))$$

$$\left[1 - \frac{i\Delta t}{2\hbar}V(x_{N-1})\right]\left[1 + \frac{i\Delta t}{2\hbar}V(x_{N-1})\right]^{-1} \delta_M(x_{N-1} - x_{N-2}|\sigma(\Delta t))$$

$$(6.73) \qquad \ldots \delta_M(x_1 - x_0|\sigma(\Delta t))\left[1 + \frac{i\Delta t}{2\hbar}V(x_0)\right]\chi(0|x_0)$$

and analogously for the symmetrized Trotter formula in which the factors containing the potential are replaced by factors $\exp(-iV(x_j)\Delta t/2\hbar)$. Some success has been achieved evaluating the multiple integral over the x_j's by Monte Carlo sampling [15]. More recently, quasi-Monte Carlo sampling has been employed to integrate each time step *separately*. Thus at time step j, one calculates [19]

$$\chi(t|x_{j+1}) = \int_{-\infty}^{\infty} dx_j \left[1 + \frac{i\Delta t}{2\hbar}V(x_{j+1})\right]^{-1} \delta_M(x_{j+1} - x_j|\sigma(\Delta t))$$

$$(6.74) \qquad \left[1 - \frac{i\Delta t}{2\hbar}V(x_j)\right]\chi(t|x_j)$$

by sampling x_j at specific points determined by number theoretic methods. The values of x_{j+1} are *also* set equal to the same numerical values, so that they can be used in integrating the coordinates at time step $(j+1)\Delta t$ by the same sampling. This has been applied successfully to collinear atom-diatom reactive scattering, including the possibility of rearrangement. The approach is extremely promising as a means for treating systems in which the number of intrinsic degrees of freedom (d) is larger than 3. Although it has not been tried, the same technique also should work with the more usual Monte Carlo sampling. A crucial aspect of the DAF-free propagator that makes the method feasible is the highly banded nature of the DAF-free propagator. This decreases enormously the number of $\delta_M(x_{j+1,\ell} - $

$x_{j,\ell'}|\sigma(\Delta t))$ that contribute (where $x_{j+1,\ell}$ is the ℓth sampling point of the $(j+1)\Delta t$ time-step coordinates and $x_{j,\ell'}$ is the ℓ'th sampling point of the $j\Delta t$ time-step coordinates).

Another fine point of detail on making such calculations feasible is the use of negative imaginary absorbing potentials, placed in boundary regions outside the range of the physical interaction, V [20]. This causes the pieces of the wavepacket that reach these regions to be "absorbed" (exponentially damped), thereby eliminating having to continue to propagate them. This decreases the region in which sampling must be carried out. One also eliminates regions where the potential is so large that, at the energies significantly contained in the system wavepacket, it corresponds to a highly non-classical region. The wavepacket cannot penetrate into these regions and they are eliminated from the sampling.

Finally, we note that rather than use Monte Carlo-type sampling, we can discretize the coordinates and evaluate each time step expression, (6.74), by matrix-vector multiplication. In addition, because of the convolution form of (6.74), it is possible to carry out the discretized matrix-vector product using the "Fast DAF Transform" analogue of the Fourier approach [21]. The scaling is of the order $P \log_2 \omega$, where P is the number of grid points and ω is the band width of the discretized DAF-free propagator matrix. Since ω is independent of P, this results in linear scaling of the effort with the number of grid points.

6.3. Polynomial expansions of the evolution operator.

The final approach to direct solution of the TDSE which we discuss is the use of expansions of $\exp(-iHt/\hbar)$ in terms of polynomials. The most general such development is that using the so-called "Faber" or "regional polynomials" [22], and it includes as a special case the popular expansion in Chebychev polynomials [23]. Our purpose is to give the overall flavor of the approach, and so we refer the reader to the literature for the general discussion using Faber polynomials. (They are particularly important if one wants to use negative imaginary absorbing potentials to eliminate problems with reflections off grid boundaries and in order to reduce the grid size as much as possible.) We shall give the details for the choice of Chebychev polynomials. The simplest way to derive such an expansion is to recall the Jacobi-Anger formula [24]

$$(6.75) \qquad e^{-ikr\cos\theta} = \sum_{n=-\infty}^{\infty} (-i)^n J_n(kr) e^{in\theta},$$

where the J_n are cylinder Bessel functions of integer order. This can be written as a sum over positive integers:

$$(6.76) \qquad e^{-ikr\cos\theta} = J_0(kr) + \sum_{n=1}^{\infty} [J_n(kr)(-i)^n e^{in\theta} +$$
$$+ J_{-n}(kr)(-i)^{-n} e^{-in\theta}].$$

We use the relations

(6.77) $$J_{-n}(kr) = (-1)^n J_n(kr)$$

and

(6.78) $$\begin{aligned}(-i)^{-n} &= (-i)^n (-i)^{-n} (-i)^{-n} \\ &= (-1)^n (-i)^n,\end{aligned}$$

and write (6.76) as

(6.79) $$e^{-ikr\cos\theta} = J_0(kr) + \sum_{n=1}^{\infty} J_n(kr)(-i)^n \{e^{in\theta} + e^{-in\theta}\}$$

(6.80) $$= \sum_{n=0}^{\infty} [2 - \delta_{n0}](-i)^n J_n(kr)\cos n\theta.$$

Finally, $\cos n\theta$ is a well known expression for the Chebychev polynomials, T_n, with argument $\cos\theta$. Thus,

(6.81) $$e^{-ikr\cos\theta} = \sum_{n=0}^{\infty} [2 - \delta_{n0}](-i)^n J_n(kr) T_n(\cos\theta).$$

The next step is put $\exp(-iHt/\hbar)$ into the form $\exp(-ikr\cos\theta)$. To do this, we make use of the fact that, one way or another, the Hamiltonian H will be approximated. Generally this is done by obtaining some representation of H, which is then truncated, resulting in a finite matrix approximation to H. Then there exists a maximum and minimum eigenvalue, H_{max} and H_{min}, and we can define

(6.82) $$H_{norm} = (H - \overline{H})/\Delta H,$$

where

(6.83) $$\overline{H} = (H_{max} + H_{min})/2$$

and

(6.84) $$\Delta H = (H_{max} - H_{min})/2.$$

It is easily seen that H_{norm} has eigenvalues bounded by ± 1, and it therefore can be used as the argument of the Chebychev polynomial. We then write [23]

(6.85) $$e^{-iHt/\hbar} = e^{-i(H-\overline{H})t/\hbar} e^{-i\overline{H}t/\hbar}$$

(6.86) $$= e^{-iH_{norm}t\Delta H/\hbar} e^{-i\overline{H}t/\hbar},$$

and can make the associations $kr \leftrightarrow t\Delta H/\hbar$ and $\cos\theta \leftrightarrow H_{\mathrm{norm}}$ in (6.81):

$$(6.87) \quad e^{-iHt/\hbar} = e^{-i\overline{H}t/\hbar} \sum_{n=0}^{\infty} [2 - \delta_{n0}](-i)^n J_n\left(\frac{t\Delta H}{\hbar}\right) T_n(H_{\mathrm{norm}}).$$

The source of convergence for a given time of propagation, t, comes from the factor $J_n(t\Delta H/\hbar)$, since when $n > t\Delta H/\hbar$, the cylindrical Bessel function decays rapidly to zero. (Note that the convergence is *not* uniform in t.) The *only* place that the Hamiltonian occurs is in the Chebychev polynomials and the only place the time occurs is in the scalar factors $\exp(-i\overline{H}t/\hbar)J_n(t\Delta H/\hbar)$. Thus, there is an extremely convenient separation of these quantities. Application of (6.87) to the $t = 0$ wavepacket generates the wavepacket at time t:

$$(6.88) \quad \chi(t) = e^{-i\overline{H}t/\hbar} \sum_{n} [2 - \delta_{n0}](-i)^n J_n\left(\frac{t\Delta H}{\hbar}\right) T_n(H_{\mathrm{norm}})\chi(0).$$

The initial packet $\chi(0)$ is under our control, and therefore to generate $\chi(t)$, we must generate the η_n, defined by

$$(6.89) \qquad\qquad \eta_n \equiv T_n(H_{\mathrm{norm}})\chi(0).$$

This is done using the standard Chebychev recursion,

$$(6.90) \qquad\qquad T_0(H_{\mathrm{norm}}) = 1,$$

$$(6.91) \qquad\qquad T_1(H_{\mathrm{norm}}) = H_{\mathrm{norm}},$$

$$(6.92) \quad T_n(H_{\mathrm{norm}}) = 2H_{\mathrm{norm}}T_{n-1}(H_{norm}) - T_{n-2}(H_{\mathrm{norm}}),$$

so that

$$(6.93) \qquad\qquad \eta_0 \equiv \chi(0),$$
$$(6.94) \qquad\qquad \eta_1 \equiv H_{\mathrm{norm}}\chi(0),$$

and

$$(6.95) \qquad\qquad \eta_n = 2H_{\mathrm{norm}}\eta_{n-1} - \eta_{n-2}.$$

Discretization of the Hamiltonian also implies a corresponding discretization of $\chi(0)$, and the η_n's become discrete vectors. The action of the matrix H_{norm} on the η_n's causes succeeding ones to be shifted to different regions of the grid; when summed according to

$$(6.96) \qquad \chi(t) = e^{-i\overline{H}t/\hbar} \sum_{n} [2 - \delta_{n0}](-i)^n J_n\left(\frac{t\Delta H}{\hbar}\right) \eta_n,$$

they interfere in just the right way to describe the propagation of the packet through space. In fact, the η_n are a sort of "Krylov basis" which is dynamically shaped to describe the spatial evolution of the packet. As has been shown, this can be within the time-dependent wavepacket approach, or the newly developed time-independent wavepacket approach. The latter comes about from the time-to-energy half Fourier transform, discussed earlier, that gives rise to the $\xi^+(E|x)$. In that case, one gets a polynomial expansion of $G^+(E)$, *but the same vectors η_n appear in it as the* $\exp(-iHt/\hbar)$ *expansion* [21,24].

The major attraction of such approaches is the fact that one obtains results for as many energies as desired. That this is possible is again a consequence of the fact that the *same η_n* vectors determine both $\chi(t)$ *and* $\xi^+(E)$. If one wishes to include a negative imaginary absorbing potential to reduce the grid size and eliminate boundary reflections, care must be taken in normalizing H. The Faber polynomial generalization of the Chebychev expansion shows how this is done. The main complication is that the spectral range of H increases, and this requires more terms in the polynomial expansion. Again, we refer the reader to recent discussions in the literature [22].

The last point of discussion regards how one chooses to discretize the Hamiltonian. We favor the use of the Hermite DAF-Hamiltonian for many problems [22,25,26]. This is because it is capable of arbitrarily high accuracy and it leads to a highly banded kinetic energy matrix. The potential is diagonal (if it is local). The continuous Hamiltonian in 1D is given in (6.61), and discretizing it yields

$$\overset{\leftrightarrow}{(H)}_{pp'} = -\frac{\hbar^2 e^{-(x_p - x_{p'})^2/2\sigma^2}\Delta x}{2m\sqrt{\pi}(\sqrt{2}\sigma)^3}$$

$$(6.97) \quad \sum_{n=0}^{M/2}\left(-\frac{1}{4}\right)^n\frac{1}{n!}H_{2n+2}\left(\frac{x_p - x_{p'}}{\sqrt{2}\sigma}\right) + \delta_{pp'}V(x_p).$$

The Gaussian ensures the bandedness. Similarly, $\chi(0|x)$ would become the vector $\chi_{p'} \equiv \chi(0|x_{p'})$. The values of H_{max} can be estimated using the Fourier relationship between the grid spacing Δx and the maximum momentum, to obtain the maximum kinetic energy, to which is added the maximum value of the potential matrix. The value of H_{min} can be estimated using a minimum value of zero for the kinetic energy, plus the minimum value of the potential matrix. Otherwise, one can use an iterative method like that of Lanczos to obtain accurate values of H_{max} and H_{min}. These are used to normalize $H_{pp'}$, and the $\eta_n(x_p)$ are generated from (6.93)–(6.95). Once these are known, the $\chi(t|x_p)$ for any time, up to $t_{max} \cong n_{max}\hbar/\Delta H$, can be computed (where n_{max} is the largest order computed in the Chebychev recursion).

Other techniques can be employed for representing the matrix H. One can use standard basis set methods, including the "discrete variable representation" or DVR approach [27]. This makes use of the relation between Gaussian quadrature and orthogonal polynomials and is very popular among computational scattering theorists. It corresponds to a particular choice of basis which has the convenience of not requiring the calculation of numerical integrals in order to obtain the DVR Hamiltonian matrix elements. (The DAF-Hamiltonian approach shares this property as well.) This concludes our overview of some computational methods which have been used in the past.

REFERENCES

[1] Extensive references and many recent papers describing computational methods for the TDSE can be found in *Time-Dependent Quantum Molecular Dynamics*, eds. J. Broeckhove and L. Lathouwers, NATO ASI Series B: Physics, Vol. 299 (Plenum, New York, 1992) and in the thematic volume of *Computer Physics Communications*, Ed. K. Kulander, Vol. 63 (North-Holland, Amsterdam, 1991).

[2] See, e.g., any text on quantum scattering theory: R.G. Newton, *Scattering Theory of Waves and Particles* (Springer-Verlag, New York, 1982).

[3] See, e.g., J.L. Powell and B. Crasemann, *Quantum Mechanics* (Addison-Wesley, Reading, MA, 1961) pp. 96–98.

[4] This analysis follows the discussion in D.J. Kouri and D.K. Hoffman, *Few-Body Systems* 18, 203 (1995) and W. Zhu, Y. Huang, D.J. Kouri, M. Arnold, and D.K. Hoffman, Phys. Rev. Lett. 72, 1310 (1994) and 73, 1733 (1994); see also D.K. Hoffman, Y. Huang, W. Zhu, and D.J. Kouri, J. Chem. Phys. 101, 1242 (1994).

[5] The amplitude density for non-reactive scattering was introduced by B.R. Johnson and D. Secrest, J. Math Phys. 7, 2187 (1966) and generalized to reactive scattering by D.J. Kouri, J. Chem. Phys. 51, 5204 (1969).

[6] This solution method for the amplitude density was introduced by M. Baer and D. J. Kouri; see, e.g., J. Chem. Phys. 56, 1758 (1972).

[7] This noniterative computational approach was introduced by W.N. Sams and D.J. Kouri, J. Chem. Phys. 51, 4809 and 4815 (1969).

[8] A discussion of stabilization issues can be found in the article by D. Secrest, in *Atom-Molecule Collision Theory*, Ed. R.B. Bernstein (Plenum, New York, 1979) Chp. 8.

[9] Y. Sun, D.J. Kouri, D.G. Truhlar, and D.W. Schwenke, Phys. Rev. A41, 4857 (1990); see also C.W. McCurdy, T.N. Rescigno, and B.I. Schneider, ibid. A 36, 2061 (1987); R.R. Lucchese, K. Takatsuka, and V.McKoy, Phys. Rep. 131, 147 (1986) W. Kohn, Phys. Rev. 74, 1763 (1948), and W.H. Miller and B.M.D.D. Janson op de Haar, J. Chem. Phys. 86, 2061 (1987).

[10] I.H. Sloan and T.J. Brady, Phys. Rev. C6, 701 (1972); D.J. Kouri and F.S. Levin, ibid. C11, 352 (1975); G. Staszewska and D.G. Truhlar, Chem. Phys. Lett. 130, 341 (1986).

[11] D.W. Schwenke, K. Haug, M. Zhao, D.G. Truhlar, Y. Sun, J.Z.H. Zhang, and D.J. Kouri, J. Phys. Chem. 92, 3202 (1988).

[12] R.S. Judson, D.B. McGarrah, O.A. Sharafeddin, D.J. Kouri, and D.K. Hoffman, J. Chem. Phys. 94, 3577 (1991).

[13] R.P. Feynman, Rev. Mod. Phys. 20, 367 (1948).

[14] M.D. Feit and J.A. Fleck, J. Chem. Phys. 78, 301 (1982); see also M.F. Trotter, Proc. Am. Math. Soc. 10, 545 (1959) and E. Nelson, J. Math. Phys. 5, 332 (1964).

[15] X. Ma, D.J. Kouri, and D.K. Hoffman, Chem. Phys. Lett. 208, 207 (1993).

[16] N. Makri, Comp. Phys. Comm. 63, 389 (1991).

[17] O.A. Sharafeddin, R.S. Judson, D.J. Kouri, and D.K. Hoffman, J. Chem. Phys. 93, 5580 (1990); O.A. Sharafeddin, D.J. Kouri, and D.K. Hoffman, Can. J. Chem. 70, 686 (1992).

[18] A summary of the DAF theory, along with extensive references to the original literature may be found in D.K. Hoffman and D.J. Kouri, in Proc. Third Int. Conf. on Math. and Num. Aspects of Wave Prop., Ed. G. Cohen (SIAM, Philadelphia, 1995) pp. 56–83. The first paper introducing the discretized DAF is D.K. Hoffman, N. Nayar, O.A. Sharafeddin, and D.J. Kouri, J. Phys. Chem. 95, 8299 (1991); the first paper introducing the continuous DAF is D.J. Kouri, W. Zhu, X. Ma, B.M. Pettitt, and D.K. Hoffman, J. Phys. Chem. 96, 9622 (1992).

[19] D.J. Kouri, Y. Huang and D.K. Hoffman, Phys. Rev. Lett. 75, 49 (1995); see also Y. Huang, D.J. Kouri, and D.K. Hoffman, Chem. Phys. Lett. 238, 387 (1995).

[20] D. Neuhauser and M. Baer, J. Chem. Phys. 90, 4351 (1989); D. Neuhauser, M. Baer, R.S. Judson, and D.J. Kouri, Comp. Phys. Comm. 63, 460 (1991).

[21] Y. Huang, D.J. Kouri, M. Arnold, T.L. Marchioro, and D.K. Hoffman, Comp. Phys. Comm. 80, 1 (1994).

[22] Y. Huang, D.J. Kouri, and D.K. Hoffman, J. Chem. Phys. 101, 10493 (1994).

[23] H. Tal-Ezer and R. Kosloff, J. Chem. Phys. 81, 3967 (1984).

[24] This derivation was given by W. Zhu, Ph.D. Thesis, Dept. of Physics, University of Houston, 1995.

[25] Y. Huang, W. Zhu, D.J. Kouri, and D.K. Hoffman, Chem. Phys. Lett. 206, 96 (1993) and 213, 209(E) (1993).

[26] D.J. Kouri, M. Arnold, and D.K. Hoffman, Chem. Phys. Lett. 203, 166 (1993).

[27] J.V. Lill, G.A. Parker, and J.C. Light, Chem. Phys. Lett. 89, 483 (1982).

TIME-INDEPENDENT WAVEPACKET QUANTUM MECHANICS

DONALD J. KOURI*, YOUHONG HUANG†, AND DAVID K. HOFFMAN‡

1. Introduction. Quantum mechanics is a highly developed subject which has provided the foundation for describing physical and chemical phenomena since its inception in the 1920's [1]. For non-relativistic systems, the fundamental equation describing a well specified state of such a system is the time-dependent Schrödinger equation (TDSE). For systems which are in eigenstates of the energy, the state vector has the form

$$(1.1) \qquad |\chi(E|t)\rangle = e^{iEt/\hbar}|\psi(E)\rangle,$$

and substitution of this expression into the TDSE,

$$(1.2) \qquad i\hbar\frac{\partial}{\partial t}|\chi(E|t)\rangle = H|\chi(E|t)\rangle,$$

leads to the celebrated time-independent Schrödinger equation (TISE),

$$(1.3) \qquad (E - H)|\psi(E)\rangle = 0.$$

Here, H is the (time-independent) system Hamiltonian and we employ the Dirac notation. The form (1.1) is easily deduced from (1.2), for a time-independent H, by the standard technique of separation of variables.

It is well known that the TDSE possesses more general solutions than those given by (1.1), these solutions being constructed as superpositions of the definite energy states, $|\psi(E)\rangle$, and being called wavepackets [2]. It has long been appreciated that a rigorous theory of quantum scattering should be based on such states, and these have been used to derive the correct solutions of (1.3) which satisfy boundary conditions appropriate to collision processes [2]. The resulting solutions may be understood by rearranging (1.3) to the form

$$(1.4) \qquad (E - H_0)|\psi(E)\rangle = V|\psi(E)\rangle,$$

---

* Supported in part under National Science Foundation Grant CHE-9403416 and R.A. Welch Foundation Grant E-0608.
Department of Chemistry and Department of Physics, University of Houston, Houston, TX 77204-5641.
† Supported under National Science Foundation Grants ASC-9310235 and CHE-9403416.
Department of Chemistry and Department of Physics, University of Houston, Houston, TX 77204-5641.
‡ The Ames Laboratory is operated for the Department of Energy by Iowa State State University under Contract No. 2-7405-ENG82.
Department of Chemistry and Ames Laboratory, Iowa State University, Ames, IA 50011.

83

where H_0 is the limit of H when the collision partners are outside their range of interaction, and V is the perturbation responsible for the scattering:

$$(1.5) \qquad\qquad H = H_0 + V.$$

Since $H_0 = H$ in the limit that the collision partners are far apart, V tends to zero in this limit. If we view $V|\psi(E) >$ as an inhomogeneity, the homogeneous equation becomes

$$(1.6) \qquad\qquad (E - H_0)|\phi_k(E)\rangle = 0,$$

where k specifies the relative momentum of the collision partners. Then the wavepacket analysis results in the Lippmann-Schwinger solution

$$(1.7) \qquad |\psi_k^+(E)\rangle = |\phi_k(E)\rangle + \frac{1}{E - H + i\epsilon} V|\phi_k(E)\rangle,$$

where $(E - H + i\epsilon)^{-1} (\equiv G^+(E))$ is the Green's operator, or resolvent, for the Schrödinger equation, satisfying outgoing scattered wave (causal) boundary conditions. The limit $\epsilon \to 0_+$ is understood. The expression (1.7) conveniently divides the solution into the so-called "incident" wave, $|\phi_k(E)\rangle$, and the scattered wave, $G^+(E)V|\phi_k(E)\rangle$. The fact that there are substantial subtleties involved in obtaining this solution is suggested by the fact that for structureless particle scattering in one degree of freedom ($1D$), $|\phi_k(E)\rangle$ is a plane wave,

$$(1.8) \qquad\qquad \langle x|\phi_k(E)\rangle = e^{ikx},$$

which is nonzero over the *entire* x-axis. Therefore, referring to it as an "incident" wave is justified on the basis of the origin of this term in the solution, rather than on the basis of it representing a wave starting out in some local region, and subsequently impinging on a target.

The derivation of (1.7) has been given in the Tutorial Chapter, and it was shown there that (1.7) holds on the "target side" of an initial, localized wavepacket. Thus, implicit in the use of (1.7), or the equation it solves, (1.3), is the fact that one begins with the $t = 0$ wavepacket infinitely far from the target. Then (1.7) implies that one is totally ignorant of the composition of the initial wavepacket, and the $\hbar k$-momentum piece (which gives rise to the $|\phi_k(E)\rangle$ plane wave term in (1.7)) has a coefficient of 1, rather than its amplitude in the initial packet. All these remarks provide motivation for exploring in more detail the situation when the initial wavepacket is *not* taken to be infinitely far from the target. Indeed, there is considerable interest in studying the situation where an initial packet starts out *on top of the target*. Essentially what one finds is that the time-independent Schrödinger equation is modified from the form of (1.3), and becomes inhomogeneous. The inhomogeneity turns out to depend on the initial packet, $|\chi(0)\rangle$. But most importantly, the resulting equation, and its Lippmann-Schwinger-like solution have extremely convenient computational properties which are only now being exploited [3-10]. We turn now to describe this new approach to quantum dynamics.

2. Time-independent wavepacket quantum equations. The derivation of the time-independent wavepacket (TIW) quantum equations is based on considering solutions of the TDSE,

$$(2.1) \qquad i\hbar \frac{\partial}{\partial t}|\chi(t)\rangle = H|\chi(t)\rangle,$$

for a general wavepacket located sufficiently close to the target (and measuring apparatus) that its detailed composition is known. For the present, we restrict ourselves to packets which do not overlap the target, and to simplify the mathematics, we take a packet which has compact support. Thus,

$$(2.2) \qquad \langle x|\chi(0)\rangle = \chi(0|x),\ x_1 \le x \le x_2$$
$$(2.3) \qquad \qquad\qquad = 0,\ x < x_1,\quad x > x_2.$$

We shall also assume a potential with compact support, lying to the right of $\chi(0|x)$, so $V(x)$ is nonzero in a region $[x_3, x_4]$, with $x_3 > x_2$. To keep the notation simple, we consider a 1D system (no internal degrees of freedom). For a time-independent Hamiltonian,

$$(2.4) \qquad |\chi(t)\rangle = e^{-iHt/\hbar}|\chi(0)\rangle,$$

and as discussed in the Tutorial, we can transform to the energy domain state $|\xi^+(E)\rangle$ by means of a half-Fourier transform,

$$|\xi^+(E)\rangle = \frac{1}{2\pi}\int_0^\infty \frac{dt}{\hbar}e^{iEt/\hbar}|\chi(t)\rangle = \frac{1}{2\pi}\int_0^\infty \frac{dt}{\hbar}e^{iEt/\hbar}e^{-iHt/\hbar}|\chi(0)\rangle.$$
(2.5)

The projection of $|\xi^+(E)\rangle$ onto any position eigenstate, $\langle x|\xi^+(E)\rangle = \xi^+(E|x)$, has a unique, well defined value, which is preserved even when the integral over time is performed formally by introducing the factor $\exp(-\epsilon t/\hbar)$:

$$(2.6) \qquad |\xi^+(E)\rangle = \frac{i}{2\pi}\frac{1}{(E - H + i\epsilon)}|\chi(0)\rangle.$$

This is referred to as the TIW-Lippmann-Schwinger equation (TIWLSE), and the differential form is

$$(2.7) \qquad (E - H)|\xi(E)\rangle = \frac{i}{2\pi}|\chi(0)\rangle.$$

Obviously, in the differential form, one no longer retains the causal (or anti-causal, if the time-to-energy transform was carried out from $t = 0$ to $t =$

$-\infty$) designation since this form has infinitely many solutions. Similarly, the $(i/2\pi)$ factor can be absorbed into the wavepacket, so the general form of the equation is

$$(2.8) \qquad (E - H)|\xi(E)\rangle = |\chi(0)\rangle,$$

and we refer to these equations as the TIW-Schrödinger equation (TIWSE).

We now compare the TIW equations with the standard equations. Since the detailed analysis has been carried out in the Tutorial, we concentrate here on the qualitative aspects. First, the TIWSE is inhomogeneous, and the inhomogeneity is the $t = 0$ wavepacket. Physically, this is a natural consequence of the fact that the packet has been positioned a finite distance from the target, and within the region which can be scrutinized. This implies that, while the standard TISE is certainly correct, it is *not* the most general form of time-independent quantum wave equation since its form reflects rather special constraints on the initial packet (i.e., either it is totally delocalized, consisting of a *single energy*, or it is localized infinitely far from the region of interest). An interesting comparison can also be made between (2.6) (written in terms of the packet in (2.8)) and the equation determining the scattered wave portion of $|\psi_k^+(E)\rangle$;

$$(2.9) \qquad |\psi_{k,sw}^+(E)\rangle = \frac{1}{(E - H + i\epsilon)}V|\phi_k(E)\rangle$$

versus

$$(2.10) \qquad |\xi^+(E)\rangle = \frac{1}{(E - H + i\epsilon)}|\chi(0)\rangle.$$

These have extremely similar structures (but they determine fundamentally different state vectors); the essential difference is that $V|\phi_k(E)\rangle$ changes as the energy changes, while $|\chi(0)\rangle$ remains unchanged. As a result, if one can develop a procedure for calculating the action of $G^+(E)$ which separates the energy and H dependences, (2.10) will make it possible to calculate scattering as a function of energy much more efficiently [4,6,8,9]. We shall discuss this in detail later.

It is also instructive to write (1.7) and (2.10) in different forms, by using the identity

$$(2.11) \qquad \frac{1}{E - H + i\epsilon} = \frac{1}{E - H_0 + i\epsilon} + \frac{1}{E - H_0 + i\epsilon}V\frac{1}{E - H + i\epsilon},$$

leading to

$$(2.12) \qquad |\xi^+(E)\rangle = |\xi_0^+(E)\rangle + \frac{1}{E - H_0 + i\epsilon}V|\xi^+(E)\rangle$$

and

$$(2.13) \qquad |\psi_k^+(E)\rangle = |\phi_k(E)\rangle + \frac{1}{E - H_0 + \epsilon}V|\psi_k^+(E)\rangle,$$

and the identity

$$(2.14) \qquad \frac{1}{E - H + i\epsilon} = \frac{1}{E - H_0 + E} + \frac{1}{E - H + i\epsilon} V \frac{1}{E - H_0 + i\epsilon},$$

leading to [7,10]

$$(2.15) \qquad |\xi^+(E)\rangle = |\xi_0^+(E)\rangle + \frac{1}{E - H + i\epsilon} V |\xi_0^+(E)\rangle.$$

The state $|\xi_0^+(E)\rangle$ is given by

$$(2.16) \qquad |\xi_0^+(E)\rangle = \frac{1}{E - H_0 + iE} |\chi(0)\rangle.$$

Thus, (2.12) is the TIWLSE parallel to (2.13) for the standard LS-state, and (2.15) is the TIWLSE parallel to (1.7) for the standard LS-state. Now because of the compact support of $|\chi(0)\rangle$, we note that in the regions $x < x_1$, $x > x_2$, the TIWSE is *identical* to the TISE. This suggests that in these regions, there can be a relatively simple relation between the states $|\xi^+(E)\rangle$ and $|\psi_k^+(E)\rangle$ [4-7,10]. On physical grounds, we expect, in fact, that $|\xi^+(E)\rangle$ must be *proportional* to $|\psi_k^+(E)\rangle$ on the potential side of $|\chi(0)\rangle$, since in that region, there is nothing to distinguish the TIW-approach from that normally used to derive the TILSE. That is, the behavior of the TIW state in the region $x > x_2$ should not depend on the *value* of x_2. The precise value of the proportionality factor is worked out in the Tutorial. On the other hand, behind the initial packet ($x < x_1$), there can be no "incoming plane waves" (plane waves generating flux moving *toward* the potential). Instead, in this region there can only be waves moving towards decreasing x (away from the potential), and these can arise from two sources. One is the initial packet, which could have contained a component $|\phi_{-k}(E)\rangle$, $k > 0$; the other is the wave reflected back by the potential. Thus, for $x < x_1$, we expect $|\xi^+(E)\rangle$ to be the sum of a term coming from the initial packet, with momentum eigenvalue $-\hbar k, k > 0$, and the scattered wave, $|\psi_{k,sw}^+(E)\rangle$. The state $|\phi_{-k}(E)\rangle$ will be weighted by its amplitude in $|\chi(0)\rangle$, while $|\psi_{k,sw}^+(E)\rangle$ must be weighted by the amplitude in $|\chi(0)\rangle$ associated with the state $|\phi_k(E)\rangle$, with momentum $\hbar k, k > 0$. Essentially this says that for $x > x_2$, $|\xi^+(E)\rangle$ satisfies the homogeneous TISE,

$$(2.17) \qquad (E - H)|\xi^+(E)\rangle = 0, x > x_2,$$

with solution [3-7,10]

$$(2.18) \qquad |\xi^+(E)\rangle = \frac{2\pi m A(k)}{\hbar^2 k} |\psi_k^+(E)\rangle.$$

This result makes the calculation of the scattering information from knowledge of $|\xi^+(E)\rangle$ in this region trivial. In the coordinate representation,

(2.18) becomes

$$(2.19) \qquad \xi^+(E|x) = \frac{2\pi m A(k)}{\hbar^2 k} \psi_k^+(E|x),$$

and for $x_2 < x < x_3$,

$$(2.20) \qquad \psi_k^+(E|x) = c e^{ikx} + r e^{-ikx}.$$

Therefore, evaluating $\xi^+(E|x)$ at some point x_5 in this region allows r to be calculated as

$$(2.21) \qquad r = \left[\frac{\hbar^2 k}{2\pi m A(k)} \xi^+(E|x_5) c e^{ikx_5} - e^{2ikx_5} \right].$$

For $x_6 > x_4$, we have

$$(2.22) \qquad \psi_k^+(E|x_6) = t e^{ikx_6},$$

so that

$$(2.23) \qquad t = \frac{\hbar^2 k}{2\pi m A(k)} \xi^+(E|x_6) e^{-ikx_6}.$$

One can average out random errors by using "test functions" with compact support in either of the two regions [7]. Thus, let $f_5(x)$ have compact support within the region $x_2 < x < x_3$, or $[x_{5L}, x_{5R}]$, where x_{5L} is the left-hand boundary and x_{5R} is the right-hand boundary of the support for $f_5(x)$. Then

$$(2.24) \int_{x_{5L}}^{x_{5R}} dx f_5(x) \xi^+(E|x) = \frac{2\pi m A(k)}{\hbar^2 k} \int_{x_{5L}}^{x_{5R}} dx [e^{ikx} + r e^{-ikx}] f_5(x),$$

and the integral over $\xi^+(E|x)$ is done numerically, yielding a simple algebraic expression for r. A similar test function can be used in the $x > x_4$ region, so that (2.23) is replaced by

$$(2.25) \quad t = \frac{i\hbar^2 k^2}{2\pi m A(k)} \left[\int_{x_{6L}}^{x_{6R}} dx f_6(x) \xi^+(E|x) \right] \Big/ \int_{x_{6L}}^{x_{6R}} dx f_6(x) e^{ikx}.$$

For $x < x_1$, it satisfies the homogeneous TISE

$$(2.26) \qquad (E - H_0)|\xi^+(E)\rangle = 0, \quad x < x_1$$

(where everywhere in this region, $V \equiv 0$), with solution [10]

$$(2.27) \quad |\xi^+(E)\rangle = \frac{2\pi m A(-k)}{\hbar^2 k} |\phi_{-k}(E)\rangle + \frac{2\pi m A(k)}{\hbar^2 k} |\psi_{k,sw}^+(E)\rangle.$$

The $A(k)$ and $A(-k)$ are the Fourier amplitudes of $|\chi(0)\rangle$, and are given by

$$(2.28) \quad A(k) = \frac{1}{2\pi} \int_{-\infty}^{\infty} dx \; e^{-ikx} \chi(0|k) \equiv \frac{1}{2\pi} \int_{x_1}^{x_2} dx e^{-ikx} \chi(0|x),$$

where here, k can be positive or negative.

It should be clear that analogous expressions hold for an initial packet with compact support to the *right* of the target, so that $x_4 < x_1$ holds. If we evolve the packet from $t = 0$ to $t = +\infty$, as before, it is the negative momentum state components of $|\chi(0)\rangle$ which will generate scattering, and the $k > 0$ states will enter only to the *right* of this $|\chi(0)\rangle$, in the same manner as the $|\phi_{-k}(E)\rangle$ in (2.27). Specifically, we will find for $x < x_1$,

$$(2.29) \qquad |\xi^+(E)\rangle = \frac{2\pi m A(-k)}{\hbar^2 k} |\psi^+_{-k}(E)\rangle, \; x < x_1,$$

and

$$(2.30) \quad |\xi^+(E)\rangle = \frac{2\pi m A(k)}{\hbar^2 k} |\phi_k(E)\rangle + \frac{2\pi m A(-k)}{\hbar^2 k} |\psi^+_{-k,sw}(E)\rangle, \; x > x_2.$$

The expression (2.29) can also be used to obtain simple expressions for the scattering information. One can also evolve either initial packet from $t = 0$ to $t = -\infty$, thereby generating anti-causal scattering states. The first case is where $|\chi(0)\rangle$ is positioned to the left of the potential, so $x_2 < x_3$. Then in the region $x > x_2$, we would find

$$(2.31) \qquad |\xi^-(E)\rangle = \frac{2\pi m A(-k)}{\hbar^2 k} |\psi^-_{-k}(E)\rangle, \; x > x_2,$$

and for $x < x_1$,

$$(2.32) \quad |\xi^-(E)\rangle = \frac{2\pi m A(k)}{\hbar^2 k} |\phi_k(E)\rangle + \frac{2\pi m A(-k)}{\hbar^2 k} |\psi^-_{-k,sw}(E)\rangle,$$

Use of (2.31) to obtain the scattering information yields r^* and t^*, since $|\psi^-_{-k}(E)\rangle = |\psi^+_k(E)\rangle^*$. The second case results when $|\chi(0)\rangle$ is to the right of the interaction. In the region $x < x_1$, we obtain

$$(2.33) \qquad |\xi^-(E)\rangle = \frac{2\pi m A(k)}{\hbar^2 k} |\psi^-_k(E)\rangle, \; x < x_1,$$

and for $x > x_2$, the result is

$$(2.34) \quad |\xi^-(E)\rangle = \frac{2\pi m A(-k)}{\hbar^2 k} |\phi_{-k}(E)\rangle + \frac{2\pi m A(k)}{\hbar^2 k} |\psi^-_{k,sw}(E)\rangle.$$

Again, (2.33) provides alternative simple expressions for the scattering information. An interesting corollary of this analysis concerns the effect of the unperturbed resolvents, $G_0^{\pm}(E)$, on the initial packet [7]. Thus, $G_0^{+}(E)$ acting on the positive momentum states in a packet produces a wave proportional to $exp(ikx)$ to the right of the packet, and *zero* to the left of the packet, while $G_0^{+}(E)$ acting on the negative momentum states in the packet produces a wave proportional to $\exp(-ikx)$, $k > 0$, to the left of the packet and zero to the right of the packet. Similarly, $G_0^{-}(E)$ acting on the initial packet behaves in an opposite fashion: $G_0^{-}(E)$ acting on positive momentum states in a packet produces a wave proportional to $\exp(ikx)$ to the left of the packet and zero to the right of the packet, while $G_0^{-}(E)$ acting on negative momentum states in a packet produces a wave proportional to $\exp(-ikx)$, $k > 0$, to the right of the packet and zero to the left of the packet. The action of the full Green's operator on the various components of a packet can be deduced using the same ideas. The initial position of the packet is important in treating the effects of $G^{\pm}(E)$ because it influences whether the potential is present in the regions where waves are produced. In general, when $G^{+}(E)$ or $G^{-}(E)$ produces a nonzero wave, the presence of the potential in the region where the wave is produced causes the wave to be either the full LS-state or the scattered wave portion, plus waves moving away from the potential that were already present in the initial packet.

The last situation to be considered is where the initial packet sits over the target region [7]. From the above discussion, one expects $|\xi^{+}(E)\rangle$ or $|\xi^{-}(E)\rangle$ to be a complicated mixture of the scattering states. However, it is not true that the solution is purely a linear combination of the $|\psi_k^{+}(E)\rangle$ and $|\psi_{-k}^{+}(E)\rangle$; one also must include a particular solution of the inhomogeneous equation in which the potential is also present. However, when $|\chi(0)\rangle$ coincides with the target region, a much simpler approach can be implemented. It is based on the observation that one does not *care* about the particular solution of (2.8), because the homogeneous solution solves the full Schrodinger equation; and therefore should contain the scattering information desired. To see this, consider either the causal or anti-causal solutions of (2.8):

$$(2.35) \qquad |\xi^{\pm}(E)\rangle = G^{\pm}(E)|\chi(0)\rangle$$

$$(2.36) \qquad = \left[\frac{P}{E - H} \mp i\pi\delta(E - H)\right]|\chi(0)\rangle,$$

where $P/(E - H)$ is the principal value Green's operator.

Clearly, $\frac{P}{E-H}|\chi(0)\rangle$ is the particular solution, and $\mp i\pi\delta(E - H)|\chi(0)\rangle$ solves the homogeneous equation, since

$$(2.37) \qquad a\delta(a) = 0.$$

Thus [7],

$$(2.38) \qquad |\psi(E)\rangle = \delta(E - H)|\chi(0)\rangle$$

solves the homogeneous equation and must contain not only the scattering information, but for appropriate energy E, also the bound state information for the system. The quantity $\delta(E - H)$ is the so-called "spectral density" operator and is also the microcanonical ensemble equilibrium density operator. An important property of the operator $\delta(E - H)$ is the fact that it is real. Consequently, $|\psi(E)\rangle$ is real or complex if $|\chi(0)\rangle$ is real or complex. We also note that $|\psi(E)\rangle$ can be obtained from

$$(2.39) \qquad |\psi(E)\rangle \;=\; \frac{i}{2\pi}[|\xi^+(E)\rangle - |\xi^-(E)\rangle],$$

$$(2.40) \qquad\qquad =\; \frac{1}{2\pi}\int_{-\infty}^{\infty}\frac{dt}{\hbar}e^{i(E-H)t/\hbar}|\chi(0)\rangle.$$

The last expression is clearly consistent with the fact that

$$(2.41) \qquad 2\pi\delta(E - H) = \int_{-\infty}^{\infty}\frac{dt}{\hbar}e^{i(E-H)t/\hbar}$$

There are numerous ways to extract the scattering information from (2.38) [7]. The procedure we choose here makes use of the fact that $|\psi_k^+(E)\rangle$ is complex, and its real and imaginary parts are separately linearly independent solutions of the Schrodinger equation. For the simple $1D$ case, there are only two linearly independent solutions, so $|\psi_{k,R}^+(E)\rangle$ and $|\psi_{k,I}^+(E)\rangle$ are sufficient to express *any* solution of (1.3). We therefore write [7]

$$(2.42) \qquad \langle x|\psi(E)\rangle \;\equiv\; \psi(E|x)$$

$$(2.43) \qquad\qquad =\; A\psi_{k,R}^+(E|x) + B\psi_{k,I}^+(E|x).$$

Here, $\chi(0|x)$ is taken to be complex, so that $\psi(E|x)$ is complex, as are A and B. The functions ψ_{kR}^+ and ψ_{kI}^+ are real. If we choose x to the right of the potential and $\chi(0|x)$ to the left of the potential, then

$$(2.44) \qquad \psi_k^+(E|x) \;=\; te^{ikx}, \quad x > x_4$$
$$=\; [t_R\cos kx - t_I\sin kx]$$
$$(2.45) \qquad\qquad +i[t_R\sin kx + t_I\cos kx]$$

It then follows that

$$\psi_R(E|x) \;=\; A_R[t_R\cos kx - t_I\sin kx]$$
$$(2.46) \qquad\qquad +B_R[t_r\sin kx + t_I\cos kx], \, x > x_4,$$

and

$$\psi_I(E|x) \;=\; B_I[t_R\sin kx + t_I\cos kx],$$
$$(2.47) \qquad\qquad +A_I[t_r\cos kx - t_I\sin kx], \, x > x_4.$$

However, we also have to the left of the potential, $x < x_3$, the relation

$$(2.48) \qquad \psi_k^+(E|x) = e^{ikx} + re^{-ikx}, x < x_3$$

$$= [(1 + r_R)\cos kx + r_I \sin kx]$$

$$(2.49) \qquad\qquad\qquad + i[(1 - r_R)\sin kx + r_I \cos kx].$$

It follows that in this region

$$\psi_R(E|x) = A_R[(1 + r_R)\cos kx + r_I \sin kx]$$

$$(2.50) \qquad\qquad + B_R[(1 - r_R)\sin kx + r_I \cos kx], \quad x < x_3$$

and

$$\psi_I(E|x) = A_I[(1 + r_R)\cos kx + r_I \sin kx]$$

$$(2.51) \qquad\qquad + B_I[(1 - r_R)\sin kx + r_I \cos kx], x < x_3.$$

We evaluate $\psi_R(E|x)$ at 4 points, x^λ, $\lambda = 1, 2, 3, 4$, where $x^\lambda < x_3$. This yields, by (2.50), 4 equations in the unknowns $A_R(1 + r_R)$, $B_R(1 - r_R)$, $A_R r_I$, and $B_R r_I$, and these are easily solved:

$$(2.52) \qquad\qquad\qquad A_R(1 + r_R) = \alpha_1,$$

$$(2.53) \qquad\qquad\qquad B_R(1 - r_R) = \alpha_2,$$

$$(2.54) \qquad\qquad\qquad A_R r_I = \alpha_3,$$

and

$$(2.55) \qquad\qquad\qquad B_R r_I = \alpha_4.$$

But by (2.54)-(2.55),

$$(2.56) \qquad\qquad\qquad \frac{A_R}{B_R} = \frac{\alpha_3}{\alpha_4},$$

and, e.g., (2.52) can then be written as

$$(2.57) \qquad\qquad\qquad B_R \frac{\alpha_3}{\alpha_4}(1 + r_R) = \alpha_1.$$

We form the ratio of (2.57) and (2.53) to obtain an equation for r_R,

$$(2.58) \qquad\qquad\qquad \frac{\alpha_3(1 + r_R)}{\alpha_4(1 - r_R)} = \frac{\alpha_1}{\alpha_2},$$

which is solved for r_R. Having r_R, we solve (2.52) for A_R and either (2.53) or (2.57) for B_R, and (2.39) for r_I. This gives us the values of A_R, B_R, r_R and r_I. If we use the values of $\psi_I(E|x)$ at two of the points x^λ in (2.51), we will have two equations in the two unknowns, A_I and B_I, which can be solved. Once these are known, we can use either (2.46) or (2.47), with

$\psi_R(E|x)$ or $\psi_I(E|x)$ evaluated at two points x^λ, $\lambda = 5, 6$, and $x^\lambda > x_4$, to generate two equations in the two unknowns, t_R and t_I. This completes the calculation of the complex reflection and transmission amplitudes. Another approach can be based on expanding $\delta(E - H)$ in terms of complete set of eigenstates of H at energy E; either the pair $|\psi_k^+(E)\rangle$, $|\psi_{-k}^+(E)\rangle$ or $|\psi_k^-(E)\rangle$, $|\psi_{-k}^-(E)\rangle$ can be used (or for that matter, $|\psi_k^+(E)\rangle$ and $|\psi_{-k}^-(E)\rangle$), since both are eigenstates of H with energy E and since

$$(2.59) \qquad |\psi_k^+(E)\rangle = (|\psi_{-k}^-(E)\rangle)^*,$$

they are linearly independent. The reader is referred to the original literature for a discussion of more complicated systems involving internal states and the possibility of chemical reactions [7,11]. We now turn to a discussion of computational methods based on the TIW equations.

3. Computational approaches to the time-independent wavepacket quantum equations. The strategy we have pursued in developing computational schemes for solving the TIW equations has focussed on procedures for constructing the operators $G^\pm(E)$ and $\delta(E - H)$ so that the energy dependence is separated from the Hamiltonian dependence [4,6-9, 11-20]. This strategy is aimed at taking maximum advantage of the fact that $|\chi(0)\rangle$ does not depend explicitly on the energy E (except that it must have a nonvanishing Fourier component at that energy).

A secondary, although important, consideration is to evaluate $|\xi^\pm(E)\rangle$ or $|\psi(E)\rangle$ on the smallest possible grid or region. Connected with such "efficiency" aspects of the approach is the manner in which the Hamiltonian is represented, i.e., we shall advocate the use of the distributed approximating functional (DAF) form of H. However, the basic framework of the computational schemes *can* be used with any discretization or finite matrix representation of H. Background for the discussion in this section can be found in the Tutorial treatments of DAFs and polynomial expansions of operator functions.

We begin with a consideration of the solution

$$(3.1) \qquad |\xi^+(E)\rangle = \frac{i}{2\pi} G^+(E)|\chi(0)\rangle.$$

If the Hamiltonian is represented by a truncated matrix, we desire a polynomial expansion of $(E - H + i\epsilon)^{-1}$. Now the ϵ can be viewed either as an arbitrarily small, positive number or as a positive function with compact support only in the boundary region. Note that $-H + i\epsilon$ is equal to $-(H - i\epsilon)$, so $i\epsilon$ can be thought of as being subtracted from the Hamiltonian. It then is the result of a negative imaginary absorbing potential added to H. In either case, $(E - H')^{-1}$, $H' = H - i\epsilon$, is an analytic function both of E and the eigenvalues of H' in any region excluding points $E = H'$. If we enclose the region containing the eigenvalues of H' by the simply connected curve, L_γ, then for E outside L_γ, $(E - H')^{-1}$ is analytic as a function of E and of H'. G. Faber [21,22] showed that this could be used to expand $(E - H')^{-1}$ in a double, uniformly convergent series, involving the Faber or regional polynomials. If the curve L_γ is represented by the conformal mapping

$$(3.2) \qquad \Phi(s) = E,$$

the result is the domain exterior to L_γ maps to the exterior of a disk of radius γ, and the region enclosed by L_γ, including the curve L_γ maps to the disk itself. Faber showed that

$$(3.3) \qquad \frac{1}{E - H'} = \frac{1}{\Phi'(s)} \sum_{n=0}^{\infty} \frac{F_n(H')}{s^{n+1}},$$

where by (3.2),

$$(3.4) \qquad E = s + \alpha_0 + \frac{\alpha_1}{s} + \frac{\alpha_2}{s^2} + \dots.$$

We note also that the expansion (3.3) can be used to develop Faber polynomial expansions of operator functions of H', since

$$(3.5) \qquad f(H') = \frac{1}{2\pi i} \oint \frac{dE f(E)}{E - H'},$$

and by (3.2),

$$(3.6) \qquad dE = \Phi'(s)ds,$$

so

$$(3.7) \qquad f(H') = \sum_n \frac{1}{2\pi i} \left[\oint ds \frac{f(\Phi)}{s^{n+1}} \right] F_n(H'),$$

or

$$(3.8) \qquad f(H') = \sum_n \lambda_n F_n(H'),$$

$$(3.9) \qquad \lambda_n = \oint ds \frac{f(\Phi)}{s^{n+1}}.$$

In general, we can write $\Phi'(s)$ as

$$(3.10) \qquad \Phi'(s) = \sum_n \frac{\beta_n}{s^n},$$

and $\beta_n = -(n-1)\alpha_{n-1}$. Faber showed that the $F_n(H')$ are polynomials generated by the recursion

(3.11) $\qquad F_0(H') = \beta_0,$

(3.12) $\qquad F_1(H') = (H' - \alpha_0)F_0(H') + \beta_1,$

$$\vdots \qquad\qquad \vdots$$

$$F_n(H') = (H' - \alpha_0)F_{n-1}(H') - \sum_{k=1}^{n-1} \alpha_k F_{n-k-1}(H')$$

(3.13) $\qquad\qquad\qquad\qquad + \beta_n, n \geq 2.$

Furthermore, he proved that the more tightly L_γ encloses the spectrum of H', the faster the convergence. Some examples of mappings with a finite number of terms are:

(3.14) $\qquad\qquad \Phi(s) = E = s + \alpha,$

for which L_γ is a circle and the series generated is a McLaurin series;

(3.15) $\qquad\qquad \Phi(s) = E = s + \alpha_0 + \dfrac{\alpha_1}{s},$

for which L_γ is an ellipse. In this case, one obtains the Faber-Chebychev polynomials, and when ϵ is an infinitesimal number (rather than an absorbing potential), the tightest possible curve, L_γ, will be a straight line, or *degenerate ellipse*. It follows that in the case of a Hermitian Hamiltonian, the best choice of polynomial is the Faber-Chebychev [21-23]. In this situation, our conformal mapping can be written as

(3.16) $\qquad\qquad \Phi(s) = s + \overline{H} + (\dfrac{\Delta H}{2})^2/s,$

the inverse mapping is

(3.17) $\qquad\qquad s = \dfrac{(E - \overline{H}) + i\sqrt{(\Delta H)^2 - (E - \overline{H})^2}}{2},$

the α_n and β_n are given by

(3.18) $\qquad\qquad \alpha_0 = \overline{H},$

(3.19) $\qquad\qquad \alpha_1 = (\dfrac{\Delta H}{2})^2,$

(3.20) $\qquad\qquad \alpha_n = 0, n > 1,$

(3.21) $\qquad\qquad \beta_0 = 1,$

(3.22) $\qquad\qquad \beta_1 = 0,$

(3.23) $\qquad\qquad \beta_2 = -\left(\dfrac{\Delta H}{2}\right)^2,$

(3.24) $\qquad\qquad \beta_n = 0, n > 2,$

and the quantities $\overline{H}$ and $\Delta \overline{H}$ are

$$(3.25) \qquad \overline{H} = (H_{max} + H_{min})/2,$$
$$(3.26) \qquad \Delta H = (H_{max} - H_{min})/2,$$

with H_{max} being the largest and H_{min} the smallest of the eigenvalues of H. The sign of the imaginary part of s in (3.17) is fixed by the fact that the eigenvalues of $H' = H - i\epsilon$ lie in the lower half plane, since H itself is Hermitian. (The limit $\epsilon \to 0_+$ has, however, been taken.) Using (3.18)-(3.24), we obtain the recursion for the Faber-Chebychev polynomials:

$$(3.27) \qquad F_0(H') = 1,$$
$$(3.28) \qquad F_1(H') = (H' - \overline{H})F_0(H'),$$
$$(3.29) \qquad F_2(H') = (H' - \overline{H})F_1(H') - 2(\frac{\Delta H}{2})^2,$$
$$(3.30) \qquad F_n(H') = (H' - \overline{H})F_{n-1}(H') - (\frac{\Delta H}{2})^2 F_{n-2}(H'), \quad n > 2.$$

This can be transformed to the usual Chebychev recursion using the relation

$$(3.31) \qquad F_n(H') = (2 - \delta_{n0})(\frac{\Delta H}{2})^n T_n(H_{scaled}),$$

where

$$(3.32) \qquad H_{scaled} = (H - \overline{H})/\Delta H.$$

Then combining (3.31) with (3.27)-(3.30), we obtain

$$(3.33) \qquad T_0(H_{scaled}) = 1,$$
$$(3.34) \qquad T_1(H_{scaled}) = H_{scaled},$$
$$(3.35) \qquad T_h(H_{scaled}) = 2H_{scaled}T_{n-1}(H_{scaled}) - T_{n-2}(H_{scaled}),$$

which is the standard Chebychev recursion. However, when ϵ is a function producing an absorbing potential in the boundary region, $\overline{H}$ and ΔH are complex, and it is the *magnitude* of H_{scaled} which is bounded by +1. One must also take account of the effect of the imaginary part of the eigenvalues in scaling the Hamiltonian. In fact, the definition of the curve, L_γ, *automatically* determines the scaling, since it is the result of the values of the α_j in the mapping, $\Phi(s)$ [8,9].

When the limit $\epsilon \to 0_+$ is taken, H' is Hermitian and the Faber-Chebychev expansion of $(E - H + i\epsilon)^{-1}$ becomes

$$(3.36) \qquad G^+(E) = \sum_n a_n^+(E)T_n(H_{scaled}),$$

where

$$(3.37) \quad a_n^+(E) = \frac{-i(2 - \delta_{n0})[(E - \overline{H}) - i\sqrt{(\Delta H)^2 - (\overline{E} - \overline{H})^2}]^n}{(\Delta H)^n \sqrt{(\Delta H)^2 - (E - \overline{H})^2}}$$

ΔH is the spectral range and $\overline{H}$ is the average eigenvalue. The anti-causal Green's operator is obtained by simply complex conjugating $G^+(E)$. This can be written in several alternative forms. For example, defining the scaled energy,

$$(3.38) \qquad\qquad E_{scaled} = (E - \overline{H})/\Delta H,$$

we have

$$(3.39) \qquad a_n^+(E) = \frac{-i(2 - \delta_{n0})[E_{scaled} - i\sqrt{1 - E_{scaled}^2}]^n}{\Delta H \sqrt{1 - E_{scaled}^2}}$$

For energies satisfying $H_{min} \leq E \leq H_{max}$, we can define θ, so that

$$(3.40) \qquad\qquad \cos\theta = E_{scaled},$$

and it follows that

$$(3.41) \qquad\qquad a_n^+(E) = \frac{-i(2 - \delta_{n0})e^{-in\theta}}{\Delta H \sin\theta}.$$

This leads immediately to simple polynomial expansion expressions for the operator $\delta(E - H)$ and the principal value Green's operator, $G^p(E)$. The former is given by

$$(3.42) \qquad\qquad \delta(E - H) = -\frac{1}{\pi}ImG^+(E),$$

so that

$$(3.43) \quad \delta(E - H) = \frac{1}{\pi\Delta H \sin\theta}\sum_n (2 - \delta_{n0})\cos n\theta T_n(H_{scaled}),$$

and noting the well-known expression for Chebychev polynomials,

$$(3.44) \qquad\qquad T_n(\cos\theta) = \cos n\theta,$$

we obtain

$$(3.45) \quad \delta(E - H) = \frac{1}{\pi\Delta H \sin\theta}\sum_n (2 - \delta_{n0})T_n(\cos\theta)T_n(H_{scaled})$$

or

$$(3.46) \quad \delta(E-H) = \frac{1}{\pi\sqrt{1 - E_{\text{scaled}}^2}} \sum_n (2 - \delta_{n0}) T_n(E_{\text{scaled}}) T_n(H_{\text{scaled}}).$$

These expressions, (3.36)-(3.46), have proved to be very robust computationally.

The implementation in a computational algorithm essentially reduces now to evaluating $|\xi^\pm(E)\rangle$ by

$$(3.47) \quad |\xi^\pm(E)\rangle = \frac{\mp i}{\Delta H \sqrt{1 - E_{\text{scaled}}}} \sum_n (2 - \delta_{n0})$$

$$[E_{\text{scaled}} \mp i\sqrt{1 - E_{\text{scaled}}^2}]^n T_n(H_{\text{scaled}})|\chi(0)\rangle$$

or $|\psi(E)\rangle$ by

$$|\psi(E)\rangle = \frac{1}{\pi\sqrt{1 - E_{\text{scaled}}^2}} \sum_n (2 - \delta_{n0}) T_n(E_{\text{scaled}}) T_n(H_{\text{scaled}})|\chi(0)\rangle.$$

(3.48)

These are the expressions which have been applied most in computations [4-8,12,15,16]. In order to use them, one generates $T_n(H_{\text{scaled}})|\chi(0)\rangle$, which is done by the recursion. The reader should note that *both* $|\psi(E)\rangle$ and the $|\xi^\pm(E)\rangle$ are generated in terms of the same quantities (as well as the time-dependent wavepacket, $|\chi(t)\rangle$) [8,9],

$$(3.49) \qquad\qquad |\eta_n\rangle = T_n(H_{\text{scaled}})|\chi(0)\rangle.$$

By (3.33)-(3.35), these are given as

$$(3.50) \qquad\qquad |\eta_0\rangle = |\chi(0)\rangle,$$
$$(3.51) \qquad\qquad |\eta_1\rangle = H_{\text{scaled}}|\chi(0)\rangle,$$
$$(3.52) \qquad\qquad |\eta_n\rangle = 2H_{\text{scaled}}|\eta_{n-1}\rangle - |\eta_{n-2}\rangle, n \geq 2.$$

As discussed in the preceding Section, the procedure for extraction of the scattering information depends on where the initial packet is placed. It is simplest when $\chi(0|x)$ is located outside the potential. However, locating $\chi(0|x)$ on top of the potential can reduce the number of terms needed in the sums in order to "complete" the collision [7,11]. It is not difficult to convince oneself that the principal feature of the $|\eta_n\rangle$ vectors is that they are responsible for the "propagation" of $|\chi(0)\rangle$ from the initial region of compact support out into the rest of space. In the coordinate representation then, $\eta_n(x)$ decreases in the initial region $[x_1, x_2]$ of $\chi(0|x)$, and moves both to the right (toward the potential) and to the left (due to spreading associated with negative momentum components of $\chi(0|x)$). Eventually, as n increases, the $\eta_n(x)$ begins to overlap the target, and at some point,

n', the $\eta_{n'}(x)$ will have essentially "left" the target region. *Provided these* $\eta_{n'}(x)$ *'s do not yet overlap the boundary of the finite region of x used to discretize H*, the expressions (3.47)–(3.48) will generate the functions with the indicated boundary behavior. Thus, if one chooses not to use an absorbing potential (and uses the expressions in which $\epsilon \to 0_+$ has been imposed, (3.36)–(3.37) or (3.45)–(3.46)), then one must require a grid size that is sufficiently large that the highest $\eta_n(x)$ which contributes, while having negligible overlap with the target, does not significantly overlap the grid boundary.

If an absorbing potential [24,25] is used to eliminate reflection of the $\eta_n(x)$ by the grid boundary, the more general, complex Faber-Chebychev polynomials must be used, along with the corresponding $\overline{H}$ and ΔH for a complex Hamiltonian [8,9]. In this case, faster convergence can be achieved by using more sophisticated curves, L_γ, which lead to more than a three term recursion for the Faber polynomials [8,9]. However, even in this case, each stage of the recursion requires only a *single* application of H_{scaled} to the preceding Faber polynomial. All other terms in the recursion involve *scalar* coefficients. Of course, higher numbers of terms in the recursion do increase the storage by a factor of 2 times the number of grid points for every additional term. The gain is that *much* smaller grids can be used, thereby greatly reducing the computational effort.

The most important feature of the above techniques is the fact that the $\eta_n(x)$ vectors are independent of the energy E. As a consequence, (3.47)–(3.48) allow one to assemble solutions at *any energy*, E, in the range between H_{min} and H_{max} (assuming that the grid details are fine enough to ensure a converged answer at the highest desired energy). A consideration of the results in the preceeding Section shows that one actually only needs $\xi^+(E|x)$ at a *single* point to extract the transmission amplitude, t, and at another single point to extract the reflection amplitude, r. Thus, once enough $\eta_n(x)$ vectors have been generated, only the values at the small number of "analysis points" need be saved in order to generate r and t at any desired energy. In addition, all the dependence on energy is *analytical*, e.g., if one used a Gaussian initial packet (with negligible overlap with the target), so one can evaluate time delays and resonance information without any difficulty [12].

All these results generalize to more complex systems having internal degrees of freedom. In addition, although we have spoken in terms of the "target" as a structureless entity, the entire analysis can be carried through for situations where one has identical particles present, for which the total wavefunction must possess exchange symmetry. In fact, in such a situation, the Hamiltonian will be symmetric under exchange of identical particles. Then provided $|\chi(0)\rangle$ possesses the correct symmetry, it will be maintained in the above computational strategy. This is the consequence of the fact that the repeated application of the Hamiltonian cannot destroy the symmetry built into the initial wavepacket.

The last aspect of computations which we discuss concerns the use of the distributed approximating functional (DAF)-Hamiltonian as a means of obtaining *linear* scaling of the computational effort with a matrix dimension [26]. As discussed in the Tutorial on computational methods, the DAF-Hamiltonian is given (in $1D$) by [3]

$$(3.53) \qquad H_{\ell j} = K_{\ell j} + V_{\ell j},$$

where

$$(3.54) \quad K_{\ell j} = -\frac{\hbar^2 \Delta x}{4m\sqrt{2\pi}\sigma^3} e^{-(x_\ell - x_j)^2/2\sigma^2} \sum_{n=0}^{M/2} (-\frac{1}{4})^n \frac{1}{n!} H_{2n+2} \left(\frac{x_\ell - x_j}{\sqrt{2}\sigma} \right),$$

$$(3.55) \qquad V_{\ell j} = \delta_{\ell j} V(x_j),$$

and $x_\ell(x_j)$ is the ℓth (jth) grid point. In multi-dimensions, the structure (3.54) holds separately for each variable, leading to an extremely sparse kinetic energy matrix (both because of the Gaussian factor in each degree of freedom DAF-kinetic energy matrix, and the fact that the multidimensional kinetic energy is a sum of separable operators). The potential matrix remains diagonal for conservative systems (i.e., it remains a local operator in the coordinate representation). Now the primary computational step in computing the $\xi^+(E)$ or $\psi(E)$ is the generation of the η_n, and the main computational step in generating the η_n is carrying out the product

$$(3.56) \qquad (H_{\text{scaled}} \eta_{n-1})_\ell = \sum_j (H_{\text{scaled}})_{\ell j} \eta_{n-1}(x_j).$$

If the matrix were full, and of dimension D, this would involve D^2 multiplications. But the $H_{\ell j}$ matrix is extremely sparse. The application of the potential requires D-multiplications. Each piece of the kinetic energy matrix is banded; assuming equal band widths, ω, this results in $Dd\omega$ multiplications, where d is the number of degrees of freedom. The total number of multiplications is then $(1 + d\omega)D$, which is *linear* in D.

4. Conclusions. In this paper, we have provided a brief introduction to the treatment of quantum dynamics using the new, time-independent wavepacket formulation of the Schrödinger and Lippmann-Schwinger equations. There are a number of aspects of TIW-theory which we have not touched upon. For example, corresponding to the popular variational approaches to quantum scattering are TIW-variational principles [6]. These incorporate the essential feature of concentrating the difficult dependence on energy in the full Green's operator, $G^+(E)$, but they add the significant improvement in accuracy that variational principles can provide. An extremely powerful approach to bound state quantum calculations has been developed using the polynomial expansion of $\delta(E - H)\chi(0)$ [7]. The approach leads to diagonalization methods which scale linearly with the dimension of the Hamiltonian matrix. Finally, modifications of the Chebychev recursion have been used to introduce new polynomial expansions which allow one to damp the wavepacket in the boundary region (as when a negative imaginary absorbing potential is used [24,25], but retaining a Hermitian Hamiltonian [18,19,11]. As a result, grid boundaries can be greatly reduced, leading to *much* more rapid rates of convergence for the polynomial expansions. It is expected that the TIW-equation plus polynomial expansion methods will provide some of the most powerful computational tools available for quantum dynamics.

REFERENCES

[1] P.A.M. Dirac, *Quantum Mechanics*, 4th Ed. (Oxford, London, 1958).

[2] M.L. Goldberger and K.M. Watson, *Collision Theory*, (Wiley, New York, 1964).

[3] D.J. Kouri, M. Arnold, and D.K. Hoffman, *Chem. Phys. Lett*, 203, 166(1993).

[4] Y. Huang, W. Zhu, D.J. Kouri, and D.K. Hoffman, *Chem. Phys. Lett.*, 206, 96 (1993); 213, 209 E(1993).

[5] W. Zhu, Y. Huang, D.K. Kouri, M. Arnold, and D.K. Hoffman, *Phys. Rev. Lett.* 72, 1310 (1994); 73, 1733 E(1994).

[6] D.J. Kouri, Y. Huang, W. Zhu, and D.K. Hoffman, *J. Chem. Phys.* 100, 3662 (1994).

[7] D.K. Hoffman, Y. Huang, W. Zhu, and D.J. Kouri, *J. Chem. Phys.* 101, 1242 (1994).

[8] Y. Huang, D.J. Kouri, and D.K. Hoffman, *Chem. Phys. Lett.* 225, 37 (1994).

[9] Y. Huang, D.J. Kouri, and D.K. Hoffman, *J. Chem. Phys.* 101, 10493 (1994).

[10] D.J. Kouri and D.K. Hoffman, *Few Body Sys.* 18, 203 (1995).

[11] Y. Huang, S. Iyengar, D.J. Kouri, and D.K. Hoffman, *J. Chem. Phys.* 105, 927 (1996).

[12] D.J. Kouri, W. Zhu, Y. Huang, and D.K. Hoffman, *Chem. Phys. Lett.* 220, 312 (1994).

[13] W. Zhu, Y. Huang, D.J. Kouri, C. Chandler, and D.K. Hoffman, *Chem. Phys. Lett.* 217, 73 (1994).

[14] Y. Huang, D.J. Kouri, and D.K. Hoffman, *Chem. Phys. Lett* 238, 387 (1995).

[15] W. Zhu, Y. Huang, G.A. Parker, D.J. Kouri, and D.K. Hoffman, *J. Phys. Chem.* 98, 12516 (1994).

[16] Y. Huang, W. Zhu, D.J. Kouri, and D.K. Hoffman, *J. Phys. Chem.* 98, 1868 (1994).

[17] G.A. Parker, W. Zhu, Y. Huang, D.J. Kouri, and D.K. Hoffman, *Comp. Phys. Comm.* 96, 27 (1996).

[18] V.A. Mandelshtam and H.S. Taylor, *J. Chem. Phys.* 102, 7390 (1995).

[19] V.A. Mandelshtam and H.S. Taylor, *J. Chem. Phys.* 103, 2903 (1995).

[20] H.W. Jang and J.C. Light, *J. Chem. Phys.* 102, 3262 (1995).

[21] G. Faber, *Math. Ann.* 57, 398 (1903).

[22] G. Faber, *J. Reine Angew. Math.* 150, 79 (1920).

[23] H. Tal-Ezer and R. Kosloff, *J. Chem. Phys.*, 81, 3967 (1984).

[24] D. Neuhauser and M. Baer, *J. Chem. Phys.* 90, 4351 (1989).

[25] D. Neuhauser, M. Baer, R.S. Judson, and D.J. Kouri, *J. Chem. Phys.* 93, 312 (1990).

[26] D.K. Hoffman, N. Nayar, O.A. Sharafeddin, and D.J. Kouri, *J. Phys. Chem.* 95, 8299 (1991); D.J. Kouri, W. Zhu, X. Ma, B.M. Pettitt, and D.K. Hoffman, *ibid.* 96, 9622 (1992).

CLASSICAL ACTION AND QUANTUM N-BODY ASYMPTOTIC COMPLETENESS

GIAN MICHELE GRAF* AND DANIEL SCHENKER*

Abstract. The quantum propagation of N-body systems is asymptotically constrained to Lagrangian manifolds corresponding to particular solutions of the free Hamilton-Jacobi equation. This is used to give a proof of asymptotic completeness for short-range interactions.

1. Introduction. The issue of asymptotic completeness for N-body systems has risen some interest in the past decade [3, 6-8, 10, 12, 13, 19, 21-24]. Here we report on some of these contributions by freely combining them to yet another variant of proof of asymptotic completeness for short-ranged interactions. The meaning of the references occurring in the proof will hence range from compelling analogy to literal citation. For a comprehensive discussion of N-body systems we refer to [2, 4, 8, 13], and for a historical perspective to [18, 19].

A generalized N-body system [1] is characterized by some geometric and dynamical structures. We begin by explaining the geometric ones.
  – A configuration space X: a finite-dimensional Euclidean space.
  – A lattice L: a finite set of subspaces $a \subset X$, closed under intersections, with $\{0\}$, $X \in L$.
For each $a \in L$ there is an orthogonal decomposition

$$(1.1) \qquad X = a \oplus a^{\perp}.$$

Correspondingly, any $x \in X$ is decomposed as $x = x_a + x^a$ with $x_a = \mathbf{1}_a x \in a$, $x^a \in a^{\perp}$. We denote by $\leq$ the partial order on L given by inclusion. For any $a \in L$ the intercluster distance is defined as

$$(1.2) \qquad |x|_a = \min_{b \not\geq a} |x^b|.$$

In particular, $|x|_{\{0\}} = +\infty$.

The Hilbert space is $L^2(X)$. Let p be the vector operator $p = (1/i)d$, where d is the derivative. Then $p^2 = -\Delta$ is the Laplacian on X. The dynamical structure [13] is given by
  – A potential $V(x)$: a real-valued function on X, which for any $a \in L$ has a decomposition

$$(1.3) \qquad V(x) = V^a(x^a) + I_a(x),$$

such that for some $\mu > 1$

$$(1.4) \qquad (|x|_a + 1)^{\mu} I_a(x) << p^2.$$

---

* Theoretische Physik, ETH-Hönggerberg, CH–8093 Zürich, Switzerland.

In particular, $I_{\{0\}} = 0$ and $I_X = V(x) + \text{const}$. The Hamiltonian

$$H = \tfrac{1}{2}p^2 + V$$

is self-adjoint on $L^2(X)$ with domain $D(H) = D(p^2)$. It is to be compared with the operators

$$H_a = \tfrac{1}{2}p^2 + V^a = \tfrac{1}{2}p_a^2 \otimes \mathbb{1} + \mathbb{1} \otimes H^a,$$

where the second representation is w.r.t. the decomposition $L^2(X) = L^2(a) \otimes L^2(a^\perp)$ induced by (1.1). We denote by P^a the bound state projection of $H^a = \tfrac{1}{2}(p^a)^2 + V^a$. One verifies [13] that H^a is the Hamiltonian for the system with data (X^a, L^a, V^a), where $X^a = a^\perp$, $L^a = \{b \cap a^\perp \mid b \in L, b \geq a\}$.

For our purposes, asymptotic completeness is the following statement:

THEOREM 1.1. *[19] For any state $\psi \in L^2(X)$ and any $\varepsilon > 0$ there are states $\psi^a \in L^2(X)$ such that*

$$(1.5) \qquad \left\| e^{-itH}\psi - \sum_{a \in L} e^{-itH_a}(\mathbb{1} \otimes P^a)\psi^a \right\| \leq \varepsilon.$$

for $t > 0$ large enough.

A standard N-body system in its own center of mass frame corresponds to the following special choices. The configuration space

$$X = \left\{ x = (x^1, \ldots, x^N) \mid x^i \in \mathbb{R}^\nu, \ \sum_{i=1}^{N} m_i x^i = 0 \right\}$$

($\nu \geq 1$, $m_i > 0$) is equipped with the metric $x \cdot y := \sum_{i=1}^{N} m_i x^i y^i$, where $x^i y^i$ is the inner product on $\mathbb{R}^\nu$. A cluster decomposition a is a partition of $\{1, \ldots, N\}$ into disjoint, nonempty subsets, called clusters. It is to be identified with a subspace $X_a \in L$ given by

$$X_a = \left\{ x \in X \mid x^i = x^j \text{ if } i \underset{a}{\sim} j \right\},$$

where $i \underset{a}{\sim} j$ iff i, j belong to the same cluster of a. Then $a \leq b$ iff b, as a decomposition, is a refinement of a in the wide sense. Moreover,

$$|x|_a = \min_{i \underset{a}{\nsim} j} \left(\frac{m_i m_j}{m_i + m_j} \right)^{1/2} |x^i - x^j|$$

explains the origin of the name 'intercluster distance'. The potential $V(x)$ is

$$V(x) = \sum_{i<j} V_{ij}(x^i - x^j),$$

where $(|y| + 1)^\mu V_{ij}(y) << -\Delta_y$ with $\mu > 1$. Then (1.3, 4) hold with

$$I_a(x) = \sum_{\substack{i<j \\ i \not\sim j \\ a}} V_{ij}(x^i - x^j).$$

We remark that H^a describes the dynamics of the clusters of a in their own center of mass frame. States in the range of P^a represent bound clusters. Asymptotic completeness thus asserts that the N-body system is well described at large times by a number of independently moving, bound clusters, i.e., by free composite particles.

Let us dwell upon some of the heuristics involved in the proof presented in the following sections. It is centered around the classical action of free particles or, more generally, of free centers of mass of clusters of particles. More precisely, consider the solution

$$S(x, t) = \frac{x^2}{2t}$$

of the free Hamilton-Jacobi equation

$$\frac{(\nabla S)^2}{2} + \frac{\partial S}{\partial t} = 0.$$

It defines a time-dependent Lagrangian submanifold

$$p = \nabla S(x, t)$$

of phase space T^*X in terms of a graph over configuration space X. This manifold is absorbing [9] (or: is the propagation set [19]) in the sense that

(1.6) $$p - \nabla S = p - \frac{x}{t} \xrightarrow[t \to +\infty]{} 0$$

along any free trajectory. Moreover, $S(x, t)$ has two convexity properties, namely

(1.7)
$$\partial^2 S = \frac{1}{t} \geq 0,$$

$$\frac{d^2 S}{dt^2} = \frac{1}{t}\left(p - \frac{x}{t}\right) \geq 0,$$

where $\partial^2 S$ is the Hessian of S w.r.t. x and $d/dt = \{p^2/2, \cdot\} + \partial/\partial t$ is the time derivative along trajectories in T^*X generated by the free Hamiltonian $p^2/2$. We also note that

$$\frac{dS}{dt} = \frac{p^2}{2} - \frac{1}{2}\left(p - \frac{x}{t}\right)^2$$

asymptotically tends to the kinetic energy of the particles. Similar considerations can be made for the function $S(x_a, t) = x_a^2/2t$ on a in relation with the dynamics generated by H_a. In this case the propagation set is given by $p_a = \nabla S(x_a, t)$ with (x^a, p^a) left unspecified.

In the context of standard N-body systems, one can associate a cluster decomposition a to each configuration x of the particles, in such a way that particles in the same cluster are close together, but the clusters are far from each other. The length scale R underlying this definition should be taken so that $1 \ll R(t) \ll t$. For any two decompositions $b > a$, this choice allows to tell weakly bound a-clusters from slowly separating b-clusters. Asymptotic completeness entails the formation of stable clusters. Accordingly, each of the functions $S(x_a, t)$ should properly account for the part of the propagation set above a-clustered configurations. To establish this, we will construct a single function $S(x, t)$ which equals $x_a^2/2t$ on a-clustered configurations x, up to an error of relative order $(R/t)^n$ with $n = 2$. The same will be true for the respective derivatives, although n is lowered by 1 at each x-derivation. So, for instance, dS/dt should asymptotically tend to the kinetic energy associated with the centers of mass of the clusters formed by the scattering process. Quantum-mechanically, this quantity should

be positive at energies away from eigenvalues and thresholds. This will be confirmed by the Mourre estimate [4, 13, 16, 17].

2. Geometric constructions. Given $\sigma = \{\sigma_a\}_{a \in L}$ with $\sigma_a > 0$ we define the sets

$$(2.1) \qquad \Gamma_a(\sigma) = \{x \in X \,|\, a \text{ is minimal in } L \text{ with } x_a^2 + \sigma_a$$
$$\geq x_b^2 + \sigma_b \text{ for } b \in L\}$$

$$(2.2) \qquad = \{x \in X \,|\, a \text{ is minimal in } L \text{ with } (x^a)^2 - \sigma_a$$
$$\leq (x^b)^2 - \sigma_b \text{ for } b \in L\},$$

where the second equality follows from $x^2 = x_b^2 + (x^b)^2$. For suitable σ, the property $x \in \Gamma_a(\sigma)$ is a precise formulation of the configuration $x \in X$ being a-clustered on scale 1.

LEMMA 2.1. *[6, 12] There are $\sigma^\pm$ with $0 < \sigma_a^- < \sigma_a^+ < 1$ defining $\Sigma = \{\sigma \,|\, \sigma_a \in [\sigma_a^-, \sigma_a^+]\}$ such that for $\sigma, \sigma' \in \Sigma$:*

$$(2.3) \qquad i) \qquad \inf_{\substack{x \in \Gamma_a(\sigma) \\ \sigma \in \Sigma}} |x|_a > 0 \,.$$

$$(2.4) \qquad ii) \qquad \Gamma_a(\sigma) \cap \Gamma_b(\sigma) = \varnothing \qquad (a \neq b),$$

$$(2.5) \qquad \bigcup_{a \in L} \Gamma_a(\sigma) = X \,.$$

iii) Let $k \geq 2$. Then, for $kx \in \Gamma_b(\sigma')$,

$$(2.6) \qquad x \in \Gamma_a(\sigma) \Longleftrightarrow x_b \in \Gamma_a(\sigma)$$

and, if either side holds true, $a \leq b$.
Proof. The seminorm $((x^a)^2 + (x^b)^2)^{1/2}$ is a norm on $(a \cap b)^\perp$, hence

$$(2.7) \qquad \varepsilon(x^{a \cap b})^2 \leq (x^a)^2 + (x^b)^2$$

for some $0 < \varepsilon < 1$ and all $a, b \in L$. We then pick $\sigma^\pm$ such that $1 < \sigma_b^+/\sigma_b^- \leq 2$, as well as $\sigma_a^+/\sigma_b^- \leq \varepsilon/3$ whenever $b < a$.
  i) We recall (1.2). For $b \not\geq a$, i.e., $a \cap b < a$, and $x \in \Gamma_a(\sigma)$, we have from (2.7, 2)

$$(2.8) \qquad (x^b)^2 \geq \varepsilon[(x^{a \cap b})^2 - (x^a)^2] - (1 - \varepsilon)(x^a)^2 \geq \varepsilon(\sigma_{a \cap b} - \sigma_a)$$
$$-(1 - \varepsilon)\sigma_a = \varepsilon\sigma_{a \cap b} - \sigma_a \geq \tfrac{2}{3}\varepsilon\sigma_{a \cap b}.$$

  ii) Property (2.5) directly follows from (2.1). We claim that

$$\Gamma_a(\sigma) = \{x \in X \mid x_a^2 + \sigma_a \geq x_b^2 + \sigma_b \text{ for } b > a; \ x_a^2 + \sigma_a > x_b^2 + \sigma_b \text{ for } b < a\}.$$
$$(2.9)$$
Indeed, call $\tilde{\Gamma}_a(\sigma)$ the r.h.s. of (2.9). Then $\Gamma_a(\sigma) \subset \tilde{\Gamma}_a(\sigma)$, so that (2.4, 9) will follow from $\tilde{\Gamma}_a(\sigma) \cap \tilde{\Gamma}_b(\sigma) = \emptyset$ for $a \neq b$. The latter is immediate if $b > a$ or $b < a$. Otherwise, i.e., if $a \cap b < a, b$, then $x \in \tilde{\Gamma}_a(\sigma)$ implies, as in (2.8), $(x^b)^2 \geq \tfrac{2}{3}\varepsilon\sigma_{a \cap b} \geq 2\sigma_b$, whereas $x \in \tilde{\Gamma}_b(\sigma)$ implies $(x^b)^2 \leq \sigma_b$.
  iii) We note that (2.9) implies

$$(2.10) \qquad x \in \Gamma_a(\sigma) \Longrightarrow x_a \in \Gamma_a(\sigma).$$

Next we claim for $kx \in \Gamma_b(\sigma')$:
a) If $b \not\geq a$, then $x \notin \Gamma_a(\sigma)$;
b) If $b \geq a$, then

$$x \in \Gamma_a(\sigma) \Longleftrightarrow x_b \in \Gamma_{b;a}(\sigma),$$

where the sets $\Gamma_{b;a}(\sigma) \subset b$ with $a \leq b$ are defined as in (2.1) but w.r.t. the lattice $L_b = \{c \in L \mid c \leq b\}$. They enjoy properties analogous to (2.4, 5, 9).
  These facts imply (2.6): By (2.10) we also have $kx_b \in \Gamma_b(\sigma')$. So, if $b \not\geq a$, both sides of (2.6) are false. If $b \geq a$, both sides are equivalent to $x_b \in \Gamma_{b;a}(\sigma)$. To prove (a), assume the contrary and distinguish between $b < a$ and $b \not\leq a$. In the first case

$$k^2(x_a^2 - x_b^2) \geq k^2(\sigma_b - \sigma_a) \geq 4(\sigma_b^- - \sigma_a^+) \geq 2\sigma_b^- \geq \sigma_b^+ > \sigma_b' - \sigma_a';$$

in the second one (2.8) implies $k^2(x^b)^2 \geq (x^b)^2 \geq \tfrac{2}{3}\varepsilon\sigma_{a \cap b}^- > \sigma_b'$. Both inequalities contradict $kx \in \Gamma_b(\sigma')$. To prove (b), note that the implication to the right follows from (2.9). Conversely, if $x \notin \Gamma_a(\sigma)$, then $x \in \Gamma_c(\sigma)$

for some $c \neq a$. By (a), $b \geq c$, and, by the direction of (b) already proved, $x_b \notin \Gamma_{b;a}(\sigma)$.   □

A function related to the construction (2.1) is

$$S_0(\sigma; x) = \tfrac{1}{2} \max_{a \in L}(x_a^2 + \sigma_a) = \tfrac{1}{2} \sum_{a \in L} \chi_{\Gamma_a(\sigma)}(x)(x_a^2 + \sigma_a).$$

In the sense of distributions it satisfies

$$\nabla S_0(\sigma; x) = \sum_{a \in L} \chi_{\Gamma_a(\sigma)}(x) x_a, \qquad \partial^2 S_0(\sigma; x) \geq \sum_{a \in L} \chi_{\Gamma_a(\sigma)}(x) \mathbf{1}_a.$$

(2.11)

Indeed, $S_0(\sigma; x)$ is convex, being a maximum of convex functions. This and the fact that (2.11) holds true on the interior of each $\Gamma_a(\sigma)$ imply (2.11). The functions $\chi_{\Gamma_a(\sigma)}(x)$ and $S_0(\sigma; x)$ have smooth counterparts given by

$$(2.12) \quad j_a(x) = \int d\sigma\, m(\sigma) \chi_{\Gamma_a(\sigma)}(x), \qquad S_0(x) = \int d\sigma\, m(\sigma) S_0(\sigma; x)$$

where $0 \leq m \in C_0^\infty(\Sigma)$ with $\int d\sigma\, m(\sigma) = 1$. Property (ii) above implies

$$(2.13) \qquad\qquad \sum_{a \in L} j_a(x) = 1,$$

and $S_0(x)$ is convex. The function $S(x,t)$ mentioned in the introduction will be a scaled version of $S_0(x)$.

LEMMA 2.2. [6] The functions j_a, $(a \in L)$, and S_0 are smooth. They satisfy:

i) Let $\sigma \in \Sigma$ and $k \geq 2$. Then, for $kx \in \Gamma_b(\sigma)$,

$$(2.14) \qquad\qquad j_a(x) = j_a(x_b), \qquad S_0(x) = S_0(x_b).$$

ii) $\quad \|(x \cdot \nabla)^n \partial^\alpha j_a\|_\infty < +\infty, \qquad \|(x \cdot \nabla)^n \partial^\alpha (S_0 - \tfrac{x^2}{2})\|_\infty < +\infty$

(2.15)

for any $n \in \mathbb{N}$ and any multi-index α.

Proof. We prove only the statements concerning S_0, the others being similar. On $\mathbb{R}^{\#L}$ we consider the functions $s(\sigma) = \tfrac{1}{2} \max_{a \in L} \sigma_a$ and $m_-(\sigma) = m(-\sigma)$. Then $S_0(x) = (s * m_-)|_{\sigma_a = x_a^2,\, a \in L}$, which exhibits the smoothness of S_0. Part (i) follows from (2.6). To prove (ii), we note that for $\sigma \in \Sigma$ the maximum appearing in

$$S_0(x) - \frac{x^2}{2} = \tfrac{1}{2} \int d\sigma\, m(\sigma) \max_{a \in L}(\sigma_a - (x^a)^2)$$

is positive, due to $x^a = 0$ for $a = X$, and bounded by 1. Thus

$$S_0(x) - \frac{x^2}{2} = \tfrac{1}{2} \int d\sigma\, m(\sigma) \max_{a \in L}(\theta(\sigma_a + h(x^a))) = (\tilde{s} * m_-)|_{\sigma_a = h(x^a)},$$

where: $\theta(t) = 0$ for $t < 0$, $= t$ for $0 \leq t \leq 1$ and $= 1$ for $t > 1$; $h : X \to \mathbb{R}$ is smooth with $h(x) = -x^2$ for $x^2 \leq 1$, $= -1$ for $x^2 \geq 2$; and $\tilde{s}(\sigma) = \frac{1}{2} \max_{a \in L} \theta(\sigma_a)$. In particular, $S_0 - \frac{x^2}{2}$ is of the more general form

$$f(x) = u(\sigma)|_{\sigma_a = h(x^a)} \prod_{b \in L} h_b(x^b),$$

where $u : \mathbb{R}^{\#L} \to \mathbb{R}$, resp. $h_b : X \to \mathbb{R}$ are smooth bounded functions, whose derivatives are bounded, resp. compactly supported. Then $(x \cdot \nabla)f$ and $(c \cdot \nabla)f$, $(c \in X)$, are sums of functions of the same form. This proves (2.15). $\square$

3. Propagation estimates. We consider time dependent operators $A = A(t)$, $t \in [1, +\infty)$, on a Hilbert space $\mathcal{H}$ with $\sup_{t \geq 1} \|A(t)\| < +\infty$, which are weakly continuous in t. Let this be denoted by $A \in \mathcal{C}$. We say $P \in \mathcal{C}$ with $P(t) \geq 0$ satisfies a *propagation estimate* for a Hamiltonian $H = H^*$ if

$$(3.1) \qquad \int_1^\infty dt\, (\psi_t, P(t)\psi_t) \leq \text{const}\, \|\psi\|^2$$

for all $\psi \in \mathcal{H}$, where $\psi_t = e^{-itH}\psi$. It states that any Schrödinger trajectory ψ_t concentrates where $P(t)$ is small. A typical application of (3.1) is the comparison of two time evolutions:

LEMMA 3.1. *[14, 20] Let $R \in \mathcal{C}$ be (a sum of terms) of the form*

$$(3.2) \qquad R(t) = R_2(t)^* B(t) R_1(t)$$

in $\mathcal{C}$, where $R_i(t)^* R_i(t)$ satisfies (3.1) for H_i, $i = 1, 2$. Then the limit

$$\lim_{t \to +\infty} \int_1^t ds\, e^{isH_2} R(s) e^{-isH_1}$$

exists.

Proof. For any $t_1 \leq t_2$ we have

$$\left\| \int_{t_1}^{t_2} dt\, e^{itH_2} R(t) e^{-itH_1} \psi \right\| = \sup_{\|\varphi\|=1} \left| \int_{t_1}^{t_2} dt\, (\varphi_t, R(t)\psi_t) \right|,$$

where $\psi_t = e^{-itH_1}\psi$, $\varphi_t = e^{-itH_2}\varphi$. This is then estimated using

$$|(\varphi_t, R(t)\psi_t)| \leq \text{const}\, (\varphi_t, R_2(t)^* R_2(t)\varphi_t)^{1/2} (\psi_t, R_1(t)^* R_1(t)\psi_t)^{1/2},$$

$$(3.3) \qquad \left(\int_{t_1}^{t_2} dt\, |(\varphi_t, R(t)\psi_t)| \right)^2 \leq \text{const}\, \|\varphi\|^2 \cdot \int_{t_1}^{t_2} dt\, (\psi_t, R_1(t)^* R_1(t)\psi_t),$$

which is $o(1)$ as $t_1, t_2 \to +\infty$. $\square$

On the other hand, a typical way to derive (3.1) is by means of a *propagation observable* $\Phi \in \mathcal{C}$ with $\Phi(t) = \Phi(t)^*$. We assume the existence of the Heisenberg derivative $D\Phi \in \mathcal{C}$, i.e., of the time derivative of $\Phi(t)$ along Schrödinger trajectories:

$$\frac{d}{dt}(\psi_t, \Phi(t)\psi_t) = (\psi_t, D\Phi(t)\psi_t).$$

Formally, $D\Phi(t) = \mathrm{i}[H, \Phi(t)] + \partial\Phi/\partial t$.

LEMMA 3.2. *[15, 20] Let*

$$(3.4) \qquad\qquad D\Phi(t) \geq P(t) + R(t)$$

in $\mathcal{C}$, *where* $P(t) \geq 0$ *and the remainder* $R(t)$ *is of the form (3.2) with* $H_1 = H_2 = H$. *Then (3.1) holds true.*

Proof. From (3.3) we have $\int_1^\infty dt\,|(\psi_t, R(t)\psi_t)| \leq \mathrm{const}\,\|\psi\|^2$. Thus

$$(\psi_t, \Phi(t)\psi_t)\big|_1^T \geq \int_1^T dt\,(\psi_t, P(t)\psi_t) + \int_1^T dt\,(\psi_t, R(t)\psi_t),$$

where both the l.h.s. and the last term on the r.h.s. are bounded by $\mathrm{const}\,\|\psi\|^2$ uniformly in T. Since the first integrand is nonnegative, we conclude that $P(t)$ satisfies (3.1). $\square$

A first illustration of Lemma 3.2 is given by the following maximal velocity estimate.

THEOREM 3.3. *[21] Let* $\chi = \bar{\chi} \in C_0^\infty(\mathbb{R})$ *and* $\lambda > 0$ *large enough. Then*

$$(3.5) \qquad \int_1^\infty \frac{dt}{t}(\psi_t, \chi(H)F(\lambda \leq |x/t| \leq 2\lambda)\chi(H)\psi_t) \leq \mathrm{const}\,\|\psi\|^2$$

for all $\psi \in \mathcal{H}$, *where* $\psi_t = e^{-\mathrm{i}tH}\psi$ *and* $F(x \in A)$ *is the characteristic function of* $A \subset X$.

Proof. Let $h \in C_0^\infty(1/2, +\infty)$ with $h(r^2) = 1$ for $r \in [1, 2]$ and set $\tilde{h}(s) = \int_{-\infty}^s ds'\, h^2(s')$. We consider the propagation observable $\chi\Phi\chi$ where $\chi = \chi(H)$ and $\Phi = -\tilde{h}((x/\lambda t)^2)$. Then $D(\chi\Phi\chi) = \chi(D\Phi)\chi$ with

$$D\Phi = \frac{2}{t}\left(\frac{x}{\lambda t}\right)^2 h^2 - \frac{1}{\lambda t}\left(h^2\frac{x}{\lambda t}\cdot p + p\cdot\frac{x}{\lambda t}h^2\right),$$

where $h = h((x/\lambda t)^2)$. While the first term is $\geq t^{-1}h^2$, we have

$$\chi h^2\frac{x}{\lambda t}\cdot p\chi = \chi h\frac{x}{\lambda t}h\cdot p\chi = \chi h\tilde{\chi}\frac{x}{\lambda t}h\cdot p\chi + O(t^{-1}) = \chi h\tilde{\chi}p\cdot\frac{x}{\lambda t}h\chi + O(t^{-1}),$$

where we took $\tilde{\chi} \in C_0^\infty(\mathbb{R})$ with $\chi\tilde{\chi} = \chi$, set $\tilde{\chi} = \tilde{\chi}(H)$ and used $[h, \tilde{\chi}] = O(t^{-1})$, which is proved at the end of this section. This and $\tilde{\chi}p = O(1)$ show

$$\chi\left(h^2 \frac{x}{\lambda t} \cdot p + p \cdot \frac{x}{\lambda t}h^2\right)\chi \leq \text{const } \chi h^2\chi + O(t^{-1}),$$

and hence

$$\chi(D\Phi)\chi \geq \frac{1}{t}(1 - \text{const } \lambda^{-1})\chi h^2\chi + O(t^{-2}) \geq \frac{1}{2t}\chi h^2\chi + O(t^{-2})$$

provided $\lambda > 0$ is large enough. The claim now follows from (3.4). $\quad\square$

For the time being, let $S = S(x, t)$ be arbitrary. Motivated however by (1.7), we formally compute

$$(3.6) \quad DS = \tfrac{1}{2}(p \cdot \nabla S + \nabla S \cdot p) + \frac{\partial S}{\partial t},$$

$$D^2 S = p \cdot (\partial^2 S)p - \tfrac{1}{4}\Delta^2 S - \nabla S \cdot \nabla V + p \cdot \nabla\frac{\partial S}{\partial t} + \nabla\frac{\partial S}{\partial t} \cdot p + \frac{\partial^2 S}{\partial t^2}$$

$$(3.7') \qquad = (p - v) \cdot \partial^2 S(p - v)$$

$$(3.7'') \qquad -\tfrac{1}{4}\Delta^2 S + (p - v) \cdot (\mathcal{D}_v\nabla S) + (\mathcal{D}_v\nabla S) \cdot (p - v)$$

$$\qquad + \mathcal{D}_v^2 S - \nabla S \cdot (\nabla V + \mathcal{D}_v v),$$

where v is any vector field $v(x, t) \in X$ and $\mathcal{D}_v = v \cdot \nabla + \frac{\partial}{\partial t}$ is the corresponding material derivative, i.e., the time derivative along trajectories in X generated by v.

Let us now set

$$(3.8) \qquad S(x, t) = t^{-(1-2\delta)}S_0(x/t^\delta),$$

with S_0 given by (2.12). Then the term (3.7') is positive, because S_0 is convex. Since $\partial^2 S \approx \mathbf{1}_a/t$ on a-clustered configurations, that term may satisfy a propagation estimate if, on such configurations, v_a is asymptotic to p_a along Schrödinger trajectories. According to (1.6), suitable choices should be $v = \nabla S$ or, as we opt for,

$$v(x, t) = x/t.$$

In order to verify that (3.7'') is a remainder we shall use that the function $S(x, t)$ is a mild modification of $x^2/2t$, in the sense that

$$(3.9) \qquad \mathcal{D}^n\partial^\alpha(S - \tfrac{x^2}{2t}) = O(t^{-[n+1+(|\alpha|-2)\delta]})$$

uniformly in $x \in X$. Here $\mathcal{D} = \partial/\partial t$ or $\mathcal{D} = \mathcal{D}_v$, and ∂^α is any spatial derivative of order $|\alpha|$. This follows from $S - (x^2/2t) = t^{-(1-2\delta)}[S_0(y) - $

$y^2/2]_{y=x/t^\delta}$, from (2.15) and from the fact that functions of the form $h(x,t) = t^{-k}h_0(x/t^\delta)$ have $\partial^\alpha h = t^{-(k+|\alpha|\delta)}\partial^\alpha h_0|_{x/t^\delta}$ and

$$\frac{\partial h}{\partial t} = -t^{-(k+1)}[kh_0 + \delta(y \cdot \nabla)h_0]_{y=x/t^\delta}, \qquad \frac{x}{t} \cdot \nabla h = t^{-(k+1)}(y \cdot \nabla)h_0|_{y=x/t^\delta}.$$

Some of the operators in (3.6, 7) are actually unbounded. However, Lemma 3.2 will apply after multiplying them by some cutoffs. On their range, p and x/t will be bounded operators. Let $f \in C_0^\infty(X)$ be a fixed function with $0 \leq f \leq 1$, $f(x) = 0$ for $|x| \geq 2$, $f(x) = 1$ for $|x| \leq 1$ and $\mathrm{supp}\nabla f \subset \{x \in X \mid 1 < |x| < 2\}$.

THEOREM 3.2. *[10, 12, 21] Let* $\chi = \bar{\chi} \in C_0^\infty(\mathbb{R})$, $\lambda > 0$ *large enough and* $\mu^{-1} < \delta < 1$. *Then*

$$(3.10) \qquad \int_1^\infty dt\, (\psi_t, \chi f[(p-v) \cdot \partial^2 S(p-v)]f\chi\psi_t) \leq \mathrm{const}\, \|\psi\|^2$$

for all $\psi \in \mathcal{H}$, *where* $\psi_t = e^{-itH}\psi$, $\chi = \chi(H)$ *and* $f = f(x/\lambda t)$.

Proof. We consider the propagation observable $\chi f(DS)f\chi$, for which

$$(3.11) \quad D[\chi f(DS)f\chi] = \chi[f(D^2S)f + f(DS)(Df) + (Df)(DS)f]\chi,$$

where, as remarked, the contribution from (3.7′) is positive. Terms arising from

$$Df = \frac{1}{2\lambda t}(\nabla f \cdot (p - \tfrac{x}{t}) + (p - \tfrac{x}{t}) \cdot \nabla f)$$

are remainders of the form (3.2) thanks to (3.5):

$$(3.12) \qquad\qquad \chi f(DS)(Df)\chi = t^{-1}\tilde{\chi}gBg\tilde{\chi} + O(t^{-2}),$$

with $B(t) = O(1)$, where we took $\tilde{\chi} \in C_0^\infty(\mathbb{R})$, $g \in C_0^\infty(1 < |x| < 2)$ with $\tilde{\chi}\chi = \chi$, $g\nabla f = \nabla f$ and we set $\tilde{\chi} = \tilde{\chi}(H)$, $g = g(x/\lambda t)$. We thereby used (3.6), simple commutator estimates, as well as (3.9) in the form

$$(3.13) \quad \nabla S - \frac{x}{t} = O(t^{-(1-\delta)}), \qquad \frac{\partial S}{\partial t} + \frac{x^2}{2t^2} = O(t^{-2(1-\delta)}).$$

The contribution from (3.7″) to (3.11) is also a remainder: Functions of the form $h(x,t) = h_0(x/t)$ have $\mathcal{D}_{\sqsubseteq}\langle = 1$. In particular, $\mathcal{D}_v^2(x^2/2t) = (x^2/2t^2)\mathcal{D}_v^2 t = 0$ and $\mathcal{D}_v\nabla(x^2/2t) = \mathcal{D}_v v = 0$. From this and from (3.9) we obtain $\mathcal{D}_v^2 S = O(t^{-(3-2\delta)})$ and $\mathcal{D}_v\nabla S = O(t^{-(2-\delta)})$. Similarly, $\Delta^2(x^2/2t) = 0$ yields $\Delta^2 S = O(t^{-(1+2\delta)})$. In order to estimate $\nabla S \cdot \nabla V$, let $j_a = j_a(2x/t^\delta)$ and $j_a(x)$ given by (2.12). Then, by (2.13),

$$\nabla V = \sum_a j_a \nabla V^a + \nabla I - \sum_a (\nabla j_a)I_a,$$

where $I = \sum_a j_a I_a$ satisfies $I(p^2 + 1)^{-1} = O(t^{-\delta\mu})$ by (2.3, 1.4). Similarly, $\sum_a (\nabla j_a) I_a (p^2 + 1)^{-1} = O(t^{-\delta(\mu+1)})$, but more importantly $\nabla S \cdot j_a \nabla V^a = (\nabla S)_a \cdot j_a \nabla V^a = 0$ due to (2.14). Left to estimate is

$$\nabla S \cdot \nabla I = i((p \cdot \nabla S)I - I(\nabla S \cdot p)) - I\Delta S.$$

Using $(\nabla S \cdot p)f\chi = O(1)$ and $(p^2 + 1)f\chi = O(1)$ we find $\chi f(\nabla S \cdot \nabla V)f\chi = O(t^{-\delta\mu})$. The proof is completed by Lemma 3.2 because of $\min(3 - 2\delta, 2 - \delta, 1 + 2\delta, \delta\mu) > 1$. $\qquad\square$

Clearly, (3.10) also applies to H_a instead of H. As the above proof shows, the theorem also holds for $\mu \leq 1$, provided decay assumptions on ∇I_a are made. We conclude this section with some auxiliary estimates. The first of them has been used in the proof of Theorem 3.3:

LEMMA 3.3. *Let $\chi \in C_0^\infty(\mathbb{R})$ and $h \in C_0^\infty(X)$. Then*

$$(3.14) \qquad h\chi(H) - \chi(H)h = O(t^{-1}),$$

where $h = h(x/t)$. If in addition $\inf\{|z|_a \mid z \in \operatorname{supp} h\} > 0$, then

$$(3.15) \qquad (\chi(H) - \chi(H_a))h = O(t^{-1}).$$

Proof. We prove $h\chi(H) - \chi(H_a)h = O(t^{-1})$ under the additional hypothesis. It reduces to (3.14) for $a = \{0\}$ and together they yield (3.15). We estimate

$$h(H + i)^{-1} - (H_a + i)^{-1}h = (H_a + i)^{-1}(H_a h - hH)(H + i)^{-1} = O(t^{-1}),$$

by $(hH - H_a h)(H + i)^{-1} = (\tfrac{1}{2}[h, p^2] + hI_a)(H + i)^{-1} = O(t^{-1})$, due to (2.3, 1.4). For $\chi \in C_0^\infty(\mathbb{R})$ we set $\chi_1(x) = \chi(x)(x + i) \in C_0^\infty(\mathbb{R})$ and compute

$$(h\chi_1(H) - \chi_1(H_a)h)(H + i)^{-1}$$

$$= \int_{-\infty}^{\infty} dr\, \hat\chi_1(r) e^{-irH_a}(e^{irH_a} he^{-irH} - h)(H + i)^{-1}$$

$$= (-i) \int_{-\infty}^{\infty} dr\, \hat\chi_1(r) \int_0^r ds\, e^{i(s-r)H_a}(hH - H_a h)(H + i)^{-1} e^{-isH} = O(t^{-1}).$$

Then

$$h\chi(H) - \chi(H_a)h$$
$$= (h\chi_1(H) - \chi_1(H_a)h)(H + i)^{-1} + \chi_1(H_a)(h(H + i)^{-1} - (H_a + i)^{-1}h)$$
$$= O(t^{-1}).$$

$\qquad\square$

4. Deift-Simon wave operators. By (2.13) we have the decomposition

$$(4.1) \qquad DS = \sum_{a \in L} DS_a,$$

$$(4.2) \qquad S_a(x,t) = S(x,t) j_a(x/ut),$$

where $u > 0$ is arbitrary. Analogously to (3.9) we have

$$(4.3) \qquad \mathcal{D}^n \partial^\alpha (S_a - \tfrac{x^2}{2t} j_a(x/ut)) = O(t^{-[n+1+(|\alpha|-2)\delta]}).$$

This follows from (3.9), from $\mathcal{D}^n \partial^\alpha j_a(x/ut) = O(t^{-(n+|\alpha|)})$, and from the Leibniz rule.

THEOREM 4.1. *[24] Under the assumptions of Theorem 3.4, the limits*

$$(4.4) \qquad W_{a,\lambda} = \text{s}-\lim_{t \to +\infty} e^{itH_a} \chi(H_a) f(DS_a) f \chi(H) e^{-itH}$$

exist.

The proof will be based on (3.6, 7) with S replaced by S_a, but with $v = x/t$ as before. We indicate these equations by (3.6a, 3.7a) and begin with the following lemma:

LEMMA 4.1. *For any $a \in L$,*

$$(4.5) \qquad \pm \partial^2 S_a(x,t) \le \text{const} \, \partial^2 S(x,t)$$

on $|x/\lambda t| \le 2$, $ut^{1-\delta} \ge 2$.

Proof. From (3.8, 4.2) we get $t\partial^2 S = \partial^2 S_0(y)$ and

$$(4.6) \qquad \begin{aligned} t\partial^2 S_a &= j_a(z)\partial^2 S_0(y) + k^{-1}\nabla S_0(y) \otimes \nabla j_a(z) \\ &\quad + \nabla j_a(z) \otimes k^{-1}\nabla S_0(y) + k^{-2}S_0(y)\partial^2 j_a(z), \end{aligned}$$

where $k = ut^{1-\delta}$, $z = x/ut$, $y = x/t^\delta = kz$ and $\otimes$ is a dyadic product. Proving (4.5) thus amounts to show that the r.h.s. of (4.6) (and its negative) is bounded by $\text{const} \, \partial^2 S_0(y)$. It suffices to do this for $S_0(y)$ replaced by $S_0(\sigma; y)$, uniformly in $\sigma \in \Sigma$. The first term on the r.h.s. of (4.6) enjoys the bound. Due to (2.11, 14) the second equals

$$\sum_{b \in L} \chi_{\Gamma_b(\sigma)}(kz) z_b \otimes \nabla j_a(z) = \sum_{b \in L} \chi_{\Gamma_b(\sigma)}(kz) z_b \otimes \nabla_b j_a(z)$$

$$\le \text{const} \sum_{b \in L} \chi_{\Gamma_b(\sigma)}(kz) \mathbf{1}_b \le \text{const} \, \partial^2 S_0(\sigma; y)$$

for $|z| \le 2\lambda u$. The other two terms are dealt with similarly. $\square$

Proof of Theorem 4.1. Let $\chi = \chi(H)$, $\chi_a = \chi(H_a)$, $W(t) = f(DS_a)f$ and D be the Heisenberg derivative for H. Then

$$W_{a,\lambda} = e^{itH_a} \chi_a W(t) \chi e^{-itH} \Big|_{t=1} + \text{s}-\lim_{t \to +\infty} \int_1^t ds \, e^{isH_a} \chi_a (DW - iI_a W) \chi e^{-isH},$$

provided the limit exists. It will, if we show that $\chi_a I_a W \chi$ and

$$(4.7) \quad \chi_a D W \chi = \chi_a [f(D^2 S_a)f + f(DS_a)(Df) + (Df)(DS_a)f] \chi$$

are of the form (3.2). To verify this we need (3.6a, 4.3) and

$$\nabla(\tfrac{x^2}{2t} j_a) = \tfrac{x}{t} j_a(z) + \tfrac{x^2}{2t^2} u^{-1} \nabla j_a(z), \qquad \tfrac{\partial}{\partial t}(\tfrac{x^2}{2t} j_a) = -\tfrac{x^2}{2t^2}[j_a(z) + z \cdot \nabla j_a(z)],$$
(4.8)

where $z = x/ut$. From this and from (2.3) we obtain $(|x|_a+1)^{-\mu} f(DS_a)f\chi = O(t^{-\mu})$, hence $\chi_a I_a W \chi = O(t^{-\mu})$ by (1.4). Terms in (4.7) arising from Df are of the same form as the r.h.s. of (3.12), except for the left χ replaced by χ_a. To handle the contributions from (3.7''a), we note that $\mathcal{D}_v \nabla(\tfrac{x^2}{2t} j_a) = 0$ and $\mathcal{D}_v^2(\tfrac{x^2}{2t} j_a) = 0$. Then (4.3) yields $\mathcal{D}_v \nabla S_a = O(t^{-(2-\delta)})$, $\mathcal{D}_v^2 S_a = O(t^{-(3-2\delta)})$, as well as $f \Delta^2 S_a f = O(t^{-(1+2\delta)})$. Moreover,

$$\nabla S_a = [j_a(z)\nabla S + u^{-1}(S/t)\nabla j_a(z)]_{z=x/ut}$$

with j_a also satisfying (2.14), implies $\chi_a f(\nabla S_a \cdot \nabla V)f\chi = O(t^{-\delta\mu})$. Finally, we decompose the Hessian

$$\partial^2 S_a = (\partial^2 S_a)_+ - (\partial^2 S_a)_-, \qquad (\partial^2 S_a)_\pm \geq 0,$$

into its positive and negative parts. Then $(\partial^2 S_a)_\pm = (\partial^2 S_a)_\pm^{1/2}(\partial^2 S_a)_\pm^{1/2}$ are dominated by $\partial^2 S$ because of (4.5). This and (3.10) show that the contribution from (3.7'a) is of the form (3.2). $\qquad \square$

COROLLARY 4.3.

$$(4.9) \qquad i) \qquad\qquad \phi(H_a)W_{a,\lambda} = W_{a,\lambda}\phi(H)$$

for any bounded Borel function ϕ.

$$(4.10) \qquad ii) \quad W_{a,\lambda} = \text{s-}\lim_{t\to+\infty} e^{itH_a}\chi(H)f(DS_a)f\chi(H)e^{-itH}.$$

$$(4.11) \qquad iii) \qquad\qquad \|W_{\{0\},\lambda}\| \leq \frac{u^2}{2}\|\chi\|_\infty^2$$

for $u \leq \lambda$.

Proof. It suffices to prove (4.9) for $\phi(\omega) = e^{-is\omega}$. Then

$$e^{-isH_a}W_{a,\lambda}e^{isH} - W_{a,\lambda} = \text{s-}\lim_{t\to+\infty} e^{itH_a}\chi(H_a)[f(DS_a)f]_t^{t+s}\chi(H)e^{-itH} = 0,$$

since the expression is $O(t^{-1})$ in norm. Indeed,

$$\frac{\partial}{\partial t}f(DS_a)f = f\frac{\partial DS_a}{\partial t}f + f(DS_a)\frac{\partial f}{\partial t} + \frac{\partial f}{\partial t}(DS_a)f,$$

where $\partial f/\partial t = O(t^{-1})$ and $\chi_a f(\partial DS_a/\partial t)f\chi = O(t^{-1})$. The latter follows from (3.6a, 4.3, 8). Part (ii) follows from $DS_a = \nabla S_a \cdot p - (i/2)\Delta S_a + \partial S_a/\partial t$ and from (3.15). Since $z^2 \leq 1$ for $z \in \text{supp} j_{\{0\}}$, the factors $f = f(x/\lambda t)$ can be omitted in (4.4) when $a = \{0\}$. Hence

$$W_{\{0\},\lambda} = \text{s-}\lim_{t \to +\infty} \frac{1}{t} \int_1^t ds\, e^{isH} \chi(DS_{\{0\}}) \chi e^{-isH}$$

$$= \text{s-}\lim_{t \to +\infty} \frac{1}{t} [e^{isH} \chi S_{\{0\}} \chi e^{-isH}]_{s=1}^{s=t}.$$

Then (4.11) follows from $\overline{\lim}_{t \to +\infty} S_{\{0\}}(x,t)/t \leq u^2/2$ uniformly in x, where we used (4.3). □

5. The asymptotic observable W.

The limit [6, 24]

$$(5.1) \qquad W_\lambda = \text{s-}\lim_{t \to +\infty} e^{itH} \chi(H) f(DS) f\chi(H) e^{-itH}$$

$$= \text{s-}\lim_{t \to +\infty} e^{itH} \chi(H) f^2(DS) \chi(H) e^{-itH}$$

exists like the wave operators and (4.9) reads $[\phi(H), W_\lambda] = 0$. The second equality follows from $[DS, f] = O(t^{-1})$.

LEMMA 5.1. *The norm limit*

$$(5.2) \qquad\qquad W = \lim_{\lambda \to +\infty} W_\lambda$$

exists, and is given by

$$(5.3) \qquad (\varphi, W\psi) = \lim_{t \to +\infty} (\varphi, e^{itH} \chi(H)(DS)\chi(H) e^{-itH}\psi)$$

for $\varphi, \psi \in D(x^2)$. Moreover, $[\phi(H), W] = 0$ for any bounded Borel function ϕ.

Proof. We claim that

$$(5.4) \qquad\qquad \|x^2 \chi(H)\psi\| \leq \text{const} \|(x^2 + 1)\psi\|$$

for $\psi \in D(x^2)$ and given $\chi \in C_0^\infty(\mathbb{R})$. Since H and $p^2/2$ are relatively bounded w.r.t. one another, we have $e^{itH} p^2 e^{-itH} \leq \text{const}(p^2 + 1)$. From this and from

$$\frac{x^2}{2} e^{-itH} = e^{-itH} \frac{x^2}{2} + \int_0^t ds\, e^{-i(t-s)H} A e^{-isH},$$

$$x e^{-isH} = e^{-isH} x + \int_0^s du\, e^{-i(s-u)H} p e^{-iuH},$$

where $A = (p \cdot x + x \cdot p)/2 = p \cdot x + (i/2)\dim X$, we find

$$(5.5) \qquad \|\tfrac{x^2}{t^2}\psi_t\| + \|\tfrac{x}{t} \otimes p\psi_t\| \leq \text{const} \,\|(\tfrac{x^2}{t^2} + p^2 + 1)\psi\|$$

for $\psi \in D(p^2) \cap D(x^2)$, where $\psi_t = e^{-itH}\psi$. By following the pattern of the proof of Lemma 3.5 we obtain the boundedness of $[x^2, \chi(H)](x^2+1)^{-1}$, which proves (5.4). A further estimate we need is

$$(5.6) \quad \|(1-f)\tfrac{x}{t}\psi\| \leq \lambda^{-1}\|\tfrac{x^2}{t^2}\psi\|, \qquad \|(1-f)p\psi\| \leq \lambda^{-1}\|\tfrac{x}{t} \otimes p\psi\|,$$

where, as usual, $f = f(x/\lambda t)$. We then write $f^2 DS - DS = (f+1)(f-1)DS$ and $DS = \nabla S \cdot p - (i/2)\Delta S + \partial S/\partial t$, so that

$$(5.7) \quad \begin{aligned} &(\varphi, [f^2 DS - DS]\psi) \\ &\leq \text{const} \,(\|\tfrac{x}{t}\varphi\| + \|\varphi\|)(\|(1-f)\tfrac{x}{t}\psi\| + \|(1-f)p\psi\| + \|(1-f)\psi\|), \end{aligned}$$

for $t \geq 1$, where we used (3.13). All this yields

$$(5.8) \quad |(\varphi_t, \chi(H)[f^2 DS - DS]\chi(H)\psi_t)| \leq \text{const} \,\lambda^{-1}\|(\tfrac{x^2}{t^2}+1)\varphi\| \,\|(\tfrac{x^2}{t^2}+1)\psi\|$$

for $\varphi, \psi \in D(x^2)$, by using (5.4-7) in reverse order. As this bound tends to zero as $\lambda \to +\infty$ uniformly in $t \geq 1$, we conclude from (5.1) that for $\varphi, \psi \in D(x^2)$ the limit $\lim_{\lambda \to +\infty}(\varphi, W_\lambda \psi)$ exists, and is given by (5.3). Then (5.8) implies

$$|(\varphi, (W_\lambda - W)\psi)| \leq \text{const} \,\lambda^{-1}\|\varphi\| \,\|\psi\|,$$

proving norm convergence. $\qquad \square$

Let $S = \cup_{a \in L}\{\text{eigenvalues of } H^a\}$ be the set of eigenvalues and thresholds of H. According to the introduction, W is the asymptotic observable corresponding to dS/dt and should be positive at energies $E \notin S$. Let $E_\Omega(H)$ be the spectral projection of H on $\Omega \subset \mathbb{R}$.

LEMMA 5.2. *[24] Let* $\text{supp}\chi \subset \mathbb{R} \setminus S$. *Then* $W \geq \theta\chi(H)^2$ *for some* $\theta > 0$. *In particular, if* $\chi \upharpoonright \Omega = 1$, *then*

$$(5.9) \qquad\qquad W \geq \theta$$

as a map from $\text{Ran}\,E_\Omega(H)$ *to itself.*

Proof. We remark that the Mourre estimate [4, 11, 13] holds under the assumption (1.4). By compactness, it suffices to consider the case where it holds on $\text{supp }\chi$, i.e.,

$$\chi i[H, A]\chi \geq 2\theta\chi^2$$

for some $\theta > 0$, where $\chi = \chi(H)$. Using

$$e^{itH}\tfrac{x^2}{2}e^{-itH} = \tfrac{x^2}{2} + At + \int_0^t ds \int_0^s du \, e^{iuH}i[H, A]e^{-iuH}$$

on $D(p^2) \cap D(x^2)$, we find $(\psi_t, \chi \frac{x^2}{2} \chi \psi_t) \geq \theta t^2 (\psi, \chi^2 \psi) + O(t)$ for $\psi \in D(x^2)$ as $t \to +\infty$. From (5.3) we then get

$$
\begin{aligned}
(\psi, W\psi) &= \lim_{t \to +\infty} \frac{1}{t} \int_1^t ds \, (\psi_s, \chi(DS) \chi \psi_s) \\
&= \lim_{t \to +\infty} \frac{1}{t} (\psi_s, \chi S \chi \psi_s)|_{s=1}^{s=t} \\
&= \lim_{t \to +\infty} \frac{1}{t^2} (\psi_t, \chi \frac{x^2}{2} \chi \psi_t) \geq \theta(\psi, \chi^2 \psi),
\end{aligned}
$$

where we used (3.9).   □

Proof of Theorem 1.1. [5] The proof is by induction in $a \in L$ assuming (1.5) holds for H^a, $a > \{0\}$, w.r.t. the lattice $L^a = \{b \cap a^\perp \mid b \in L, \, b \geq a\}$. Equivalently, the induction assumption is

$$
(5.10) \qquad \left\| e^{-itH_a} \psi - \sum_{b \geq a} e^{-itH_b} (\mathbb{1} \otimes P^b) \psi^b \right\| \leq \varepsilon
$$

in the setting of (1.5). It is trivially satisfied for $a = X$.

Since (1.5) is obvious for $\psi = P^{\{0\}} \psi$, we may assume $P^{\{0\}} \psi = 0$, or even $\psi = E_\Omega(H)\psi$ for some compact $\Omega \subset \mathbb{R} \setminus S$. Indeed, the latter states are dense among the former, because S is closed and countable [4, 11, 13]. Take $\chi \in C_0^\infty(\mathbb{R})$ with $\chi \upharpoonright \Omega = 1$ and supp $\chi \subset \mathbb{R} \setminus S$. By (5.9), there is $\varphi = E_\Omega(H)\varphi$ with $\psi = W\varphi$. Given $\varepsilon > 0$, we pick u, $\lambda > 0$ small, resp. large enough so that

$$
(5.11) \qquad \|W_{\{0\}, \lambda} \varphi\| \leq \varepsilon, \qquad \|(W - W_\lambda)\varphi\| \leq \varepsilon
$$

by using (4.11, 5.2). Then, by (5.1, 4.1, 10),

$$
\begin{aligned}
e^{-itH} W_\lambda \varphi &= \chi(H) f(DS) f\chi(H) e^{-itH} \varphi + o(1) \\
&= \sum_a \chi(H) f(DS_a) f\chi(H) e^{-itH} \varphi + o(1) \\
&= \sum_a e^{-itH_a} W_{a, \lambda} \varphi + o(1)
\end{aligned}
$$

as $t \to +\infty$, and hence $\|e^{-itH} \psi - \sum_{a > \{0\}} e^{-itH_a} W_{a, \lambda} \varphi\| \leq 3\varepsilon$ for t large enough. Together with (5.10), this proves (1.5).   □

Acknowledgments. G.M.G. thanks W. Hunziker for very useful discussions. This work was done in part at the Institute for Mathematics and its Applications (IMA) of the University of Minnesota, where G.M.G. has profited from interesting discussions with P. Deift, J. Derezinski, I.M. Sigal, B. Simon and D. Yafaev, to all of whom he is indebted. His research at the IMA was supported in part with funds provided by the National Science Foundation. We thank M. Griesemer for pointing out a mistake in an earlier version of this work.

REFERENCES

[1] S. Agmon, *Lectures on exponential decay of solutions of second-order elliptic equations*, Mathematical notes, Vol. 29, Princeton University Press (1982).

[2] W.O. Amrein, A. Boutet de Monvel-Berthier, V. Georgescu, *Co-groups, commutator methods and spectral theory of N-body Hamiltonians*, Birkhäuser, (1996).

[3] A. Boutet de Monvel-Berthier, V. Georgescu, A. Soffer, N-body Hamiltonians with hard-core interactions. Rev. Math. Phys. 6, 515-596 (1994).

[4] H.L. Cycon, R.G. Froese, W. Kirsch, B. Simon, *Schrödinger operators*, Springer-Verlag (1987).

[5] P. Deift, B. Simon, A time-dependent approach to the completeness of multiparticle quantum systems. Commun. Pure Appl. Math. 30, 573-583 (1977).

[6] J. Derezinski, Algebraic approach to the N-body quantum long range scattering, Rev. Math. Phys. 3, 1-62 (1990).

[7] J. Derezinski, Asymptotic completeness for N-particle long-range quantum systems, Ann. Math. 138, 427-476 (1993).

[8] J. Derezinski, C. Gérard, *Scattering theory of classical and quantum N-particle systems, tracts and monographs in physics*, Springer-Verlag, to appear in 1997.

[9] V. Enss, "Geometric methods in spectral and scattering theory of Schrödinger operators", in *Rigorous atomic and molecular physics*, G. Velo, A.S. Wightman eds., Plenum (1981).

[10] V. Enss, "Introduction to asymptotic observables for multiparticle quantum scattering", in *Schrödinger operators, Aarhus 1985*, E. Balslev ed., Lecture Notes in Mathematics, Vol. 1218, Springer-Verlag (1986).

[11] R. Froese, I. Herbst, A new proof of the Mourre estimate. Duke Math. J. 49, 1075-1085 (1982).

[12] G.M. Graf, Asymptotic completeness for N-body short-range quantum systems: A new proof. Commun. Math. Phys. 132, 73-101 (1990).

[13] W. Hunziker, I.M. Sigal, "The general theory of N-body quantum systems", to appear. Preliminary version in *Mathematical quantum theory: II. Schrödinger operators*, J. Feldman et al. eds., AMS-publ. (1994).

[14] T. Kato, Wave operators and similarity for some non-self-adjoint operators. Math. Ann. 162, 258-279 (1966).

[15] T. Kato, Smooth operators and commutators. Stud. Math. 31, 535-546 (1968).

[16] E. Mourre, Absence of singular continuous spectrum for certain selfadjoint operators. Commun. Math. Phys. 78, 391-408 (1981).

[17] P. Perry, I.M. Sigal, B. Simon, Spectral analysis of N-body Schrödinger operators. Ann. Math. 114, 519-567 (1981).

[18] M. Reed, B. Simon: *Methods of modern mathematical physics, III. Scattering theory*, Academic Press, (1979).

[19] I.M. Sigal, A. Soffer, The N-particle scattering problem: asymptotic completeness for short-range systems. Ann. Math. 126, 35-108 (1987).

[20] I.M. Sigal, A. Soffer, Local decay and propagation estimates for time-dependent and time-independent Hamiltonians, Princeton University preprint (1988).

[21] I.M. Sigal, A. Soffer, Long-range many-body scattering. Asymptotic clustering for Coulomb-type potentials. Invent. Math. 99, 115-143 (1990).

[22] I.M. Sigal, A. Soffer, Asymptotic completeness for N-particle long-range scattering, J. AMS 7, 307-334 (1994).

[23] H. Tamura, Asymptotic completeness for N-body Schrödinger operators with short-range interactions, Commun. Part. Diff. Eq. 16, 1129-1154 (1991).

[24] D. Yafaev, Radiation condition and scattering theory for N-particle Hamiltonians. Commun. Math. Phys. 154, 523-554 (1993).

ON TRACE FORMULAS FOR SCHRÖDINGER-TYPE OPERATORS

F. GESZTESY* AND H. HOLDEN†

Abstract. We review a variety of recently obtained trace formulas for one- and multi-dimensional Schrödinger operators. Some of the results are extended to Sturm-Liouville and matrix-valued Schrödinger operators. Furthermore, we recall a set of trace formulas in one, two, and three dimensions related to point interactions as well as a uniqueness result for three-dimensional Schrödinger operators with spherically symmetric potentials.

Key words. Trace formulas, Sturm-Liouville and Schrödinger operators, Krein's spectral shift function.

1. Introduction. It is a well-established fact by now that trace formulas are of great importance in solving inverse spectral problems for Schrödinger operators. This is demonstrated in great detail in [7] in the context of short-range inverse scattering theory and in [9], [11], [26], [33], [38] in connections with the inverse periodic spectral problem. Historically these trace formulas originated in the works of Gelfand and Levitan [14] (see also [8], [12], [13]) for Schrödinger operators on a finite interval. Subsequent developments extended the range of validity of trace formulas in a variety of directions including algebro-geometric quasi-periodic finite gap potentials and certain classes of almost periodic potentials [6], [27], [30]–[32], [39]. Moreover, trace formulas proved to be a vital ingredient in descriptions of the isospectral manifold of quasi-periodic finite-gap potentials and some of their limiting cases as well as in the corresponding Cauchy problem for the Korteweg-de Vries equation. Due to the somewhat special nature of the potentials covered in the references cited thus far, it seemed natural to search for extensions of these trace formulas to a large class of potentials. This was the point of departure of our recent program which led to a trace formula for any continuous potential bound from below and subsequent generalizations to higher-order trace formulas in one dimension and certain multi-dimensional generalizations [15]–[19], [21]–[25], [37]. In the simplest case, the main new strategy is to compare the $L^2(\mathbb{R})$ Schrödinger operators $H = -\frac{d^2}{dx^2} + V$ and $H_y^D = -\frac{d^2}{dx^2} + V$, the corresponding operator with an additional Dirichlet boundary condition at the point $y \in \mathbb{R}$. The spectral characteristics of H and H_y^D, especially the Krein spectral shift function $\xi(\lambda, y)$ associated with the pair (H_y^D, H), then allows one to recover the potential $V(y)$.

---

* Department of Mathematics, University of Missouri, Columbia, MO 65211, USA
mathfg@mizzou1.missouri.edu

† Department of Mathematical Sciences, Norwegian Institute of Technology, University of Trondheim, N-7034 Trondheim, Norway
holden@imf.unit.no

In Section 2 we extend the results of [22] and [24] to Sturm-Liouville operators of the type $r^{-2}[-(p^2 f')' + qf]$ in $L^2(\mathbb{R}; r^2 dx)$ and consider general self-adjoint boundary conditions $\psi'(y) + \beta \psi(y) = 0$, $\beta \in \mathbb{R}$ in addition to the Dirichlet case $\beta = \infty$. Section 3 sketches an extension of the trace formula to matrix-valued Schrödinger operators in the Dirichlet case. Section 4 briefly reviews the multi-dimensional trace formulas in [25] and illustrates a possible abstract approach to some of these trace formulas in the special noninteracting case. In Section 5 we recall a different type of trace formula first derived in [17] in dimensions one, two, and three based on point interactions. Section 6 finally describes a recent uniqueness result for three-dimensional Schrödinger operators with spherically symmetric potentials originally proven in [19].

2. Trace formulas for Sturm-Liouville operators. Let $p, q, r \in C^\infty(\mathbb{R})$ be real-valued, $p, r > 0$ and q bounded from below.

We then define the self-adjoint Sturm-Liouville operator in $L^2(\mathbb{R}; r^2 dx)$ by

$$(2.1) \quad hf = \frac{1}{r^2}[-(p^2 f')' + qf],$$

$$f \in \mathcal{D}(h) = \{g \in L^2(\mathbb{R}; r^2 dx) | g, g' \in AC_{\text{loc}}(\mathbb{R}), hg \in L^2(\mathbb{R}; r^2 dx)\},$$

where $AC_{\text{loc}}(\Omega)$ denotes the set of locally absolutely continuous functions in $\Omega \subseteq \mathbb{R}$. In addition, we define the Dirichlet Sturm-Liouville operator

$$h_y^D f = \frac{1}{r^2}[-(p^2 f')' + qf],$$

$$(2.2) \quad f \in \mathcal{D}(h_y^D) = \{g \in L^2(\mathbb{R}, r^2 dx) | g, g' \in AC_{\text{loc}}(\mathbb{R}\backslash\{y\}),$$

$$\lim_{\epsilon \downarrow 0} g(y \pm \epsilon) = 0, \ h_y^D g \in L^2(\mathbb{R}; r^2 dx)\}.$$

In order to derive trace formulas we will compare the resolvents of h and h_y^D. Let $g(z, x, x')$ and $g_y^D(z, x, x')$ denote the Green's functions (i.e., the integral kernels of the resolvents) of h and h_y^D respectively,

$$(2.3) \quad g(z, x, x') = (h - z)^{-1}(x, x'), \quad g_y^D(z, x, x') = (h_y^D - z)^{-1}(x, x').$$

One verifies

$$(2.4) \quad g_y^D(z, x, x') = g(z, x, x') - \frac{g(z, x, y)g(z, y, x')}{g(z, y, y)},$$

and hence

$$(2.5) \quad \text{Tr}[(h_x^D - z)^{-1} - (h - z)^{-1}] = -\frac{d}{dz}\ln[g(z, x, x)].$$

To proceed further, we need a high-energy expansion, i.e., $z \to \infty$, of the diagonal Green's function $g(z, x, x)$. For that purpose we shall exploit the Liouville-Green transformation to find a Schrödinger operator

H which is unitarily equivalent to h and hence use known results for Schrödinger operators derived in [21], [22],[24].

Define the change of variable

$$(2.6) \qquad t = t(x) = \int_{x_0}^{x} dx' \, \frac{r(x')}{p(x')}$$

for an arbitrary but fixed point $x_0 \in \mathbb{R}$. Write

$$(2.7) \qquad P(t) = p(x(t)), \ Q(t) = q(x(t)), \ R(t) = r(x(t))$$

and introduce the unitary operator

$$(2.8) \quad U : L^2(\mathbb{R}; r^2 dx) \longrightarrow L^2(\mathbb{R}; dt)$$
$$(Uf)(t) = [P(t)R(t)]^{1/2} F(t), \ F(t) = f(x(t)), \ f \in L^2(\mathbb{R}; r^2 dx).$$

THEOREM 2.1. *([10], see also [20]) The operator* $H = U h U^{-1}$ *in* $L^2(\mathbb{R}; dt)$ *explicitly reads*

$$(2.9) \qquad Hf = -f'' + Vf,$$
$$f \in \mathcal{D}(H) = \{g \in L^2(\mathbb{R}; dt) \,|\, g, g' \in AC_{loc}(\mathbb{R}), Hg \in L^2(\mathbb{R}; dt)\},$$

where

$$V(t) = \frac{Q(t)}{R(t)^2} + \frac{1}{(R(t)P(t))^2} \left[\frac{1}{2}(R(t)P(t))(R(t)P(t))_{tt} - \frac{1}{4}((R(t)P(t))_t^2 \right]$$

$$(2.10) \qquad = \frac{q(x)}{r(x)} + \frac{p(x)}{2r(x)^3}(r(x)p(x))_{xx} + \frac{(r(x)p(x))_x}{2r(x)^2}\left(\frac{p(x)}{r(x)}\right)_x$$

$$- \frac{1}{4r(x)^4}(r(x)p(x))_x^2 := v(x), \quad x = x(t).$$

Furthermore,

$$(2.11) \qquad g(z, x, x') = \frac{G(z, t(x), t(x'))}{[r(x)p(x)r(x')p(x')]^{1/2}}, \ x, x' \in \mathbb{R}, \ z \in \mathbb{C},$$

where G is the Green's function of H. Moreover,

$$(2.12) \qquad H_u^D = U h_y^D U^{-1} = -\frac{d^2}{dt^2} + V$$

with V given by (2.10), is the Schrödinger operator with a Dirichlet boundary condition imposed at the point $u = \int_{x_0}^{y} dx[p(x)/r(x)]$. *Let G_u^D denote the Green's function of H_u^D. Then*

$$(2.13) \qquad g_y^D(z, x, x') = \frac{G_u^D(z, t(x), t(x'))}{[p(x)r(x)p(x')r(x')]^{1/2}}.$$

Hence we find, using known results for H [22], [24] that

$$(2.14) \qquad \text{Tr}[e^{-\tau h_x^D} - e^{-\tau h}] \underset{\tau \downarrow 0}{\sim} \sum_{\ell=0}^{\infty} s_\ell(x) \tau^\ell,$$

$$(2.15) \qquad \text{Tr}[(h_x^D - z)^{-1} - (h - z)^{-1}] \underset{\substack{|z| \to \infty \\ x \in \mathbb{C} \backslash C_\epsilon}}{\sim} \sum_{j=0}^{\infty} r_j(x) z^{-j-1},$$

where C_ϵ is a cone with apex at $E_0 := \inf\{\sigma(H)\}$ and opening angle $\epsilon > 0$. Recursion relations for s_ℓ and r_j are given by (cf. [22],[24])

$$(2.16) \qquad s_\ell(x) = (-1)^{\ell+1} \frac{r_\ell(x)}{\ell!}, \quad \ell \in \mathbb{N}_0,$$

$$r_0(x) = \frac{1}{2}, \ r_1(x) = \frac{1}{2} v(x),$$

$$(2.17) \qquad r_j(x) = j\gamma_j(x) - \sum_{\ell=1}^{j-1} \gamma_{j-\ell}(x) r_\ell(x), \ j = 2, 3, \ldots,$$

$$(2.18) \quad \gamma_0 = 1, \ \gamma_1 = \frac{1}{2} v,$$

$$\gamma_{j+1} = -\frac{1}{2} \sum_{\ell=1}^{j} \gamma_\ell \gamma_{j+1-\ell} + \frac{1}{2} \sum_{\ell=0}^{j} \left[v\gamma_\ell\gamma_{j-\ell} + \frac{1}{4}\gamma_{\ell,x}\gamma_{j-\ell,x} - \frac{1}{2}\gamma_{\ell,xx}\gamma_{j-\ell} \right],$$
$$j = 1, 2, \ldots.$$

Explicitly, one computes

$$(2.19) \qquad s_0 = -\frac{1}{2}, \ s_1(x) = \frac{1}{2} v(x), \quad \text{etc.}$$

The proof of (2.17) in [22] follows from the well-known differential equation for $\Gamma(z,t) = G(z,t,t)$, namely

$$(2.20) \qquad -2\Gamma_{tt}(z,t)\Gamma(z,t) + \Gamma_t(z,t)^2 + 4[V(t) - z]\Gamma(z,t)^2 = 1$$

and the asymptotic expansion

$$(2.21) \qquad \Gamma(z,t) \underset{\substack{|z| \to \infty \\ z \in \mathbb{C} \backslash C_\epsilon}}{\sim} \frac{i}{2} z^{-1/2} \sum_{j=0}^{\infty} \Gamma_j(t) z^{-j},$$

with $\Gamma_j(t)$ defined as in (2.18) but with $v(x)$ replaced by $V(t)$.

The next ingredient concerns the fact that $g(z, x, x)$ is a Herglotz function for all $x \in \mathbb{R}$, i.e., $g(\cdot, x, x)$: $\mathbb{C}_+ \to \mathbb{C}_+$ is analytic, $\mathbb{C}_+ = \{z \in \mathbb{C} \,|\, \text{Im}\, z > 0\}$. Hence g allows a representation [3]

$$(2.22) \quad g(z, x, x) = \exp\left\{ c(x) + \int_{\mathbb{R}} d\lambda \left[\frac{1}{\lambda - z} - \frac{\lambda}{1 + \lambda^2} \right] \xi(\lambda, x) \right\},$$

where $\xi(\lambda, x)$ is Krein's spectral shift function for the pair (h_x^D, h) [28], satisfying $0 \leq \xi(\lambda, x) \leq 1$, $\xi(\cdot, x) \in L_{loc}^1(\mathbb{R}; d\lambda)$, and $\int_{\mathbb{R}} d\lambda (1 + \lambda^2)^{-1} \xi(\lambda, x) < \infty$. Although it will not be subsequently used, for completeness we show how to obtain an expression for $c(x)$. Let $z = i$ in (2.22). By taking real parts of (2.22) one infers that

$$(2.23) \qquad c(x) = \operatorname{Re}\{\ln[g(i, x, x)]\}.$$

Fatou's lemma permits the explicit representation

$$(2.24) \qquad \xi(\lambda, x) = \frac{1}{\pi} \lim_{\epsilon \downarrow 0} \arg[g(\lambda + i\epsilon, x, x)] \text{ for a.e. } \lambda \in \mathbb{R}$$

and all $x \in \mathbb{R}$. We will normalize $\xi(\lambda, x)$ to be zero below the spectrum of h, i.e., $\xi(\lambda, x) = 0$ for $\lambda < E_0$. Using the spectral shift function, one can show that

$$(2.25) \qquad \operatorname{Tr}[F(h_x^D) - F(h)] = \int_{E_0}^{\infty} d\lambda F'(\lambda) \xi(\lambda, x)$$

whenever $F \in C^2(\mathbb{R})$, $(1 + \lambda^2) F^{(j)} \in L^2((0, \infty))$, $j = 1, 2$ and $F(\lambda) = (\lambda - z)^{-1}$, $z \in \mathbb{C} \setminus [E_0, \infty)$.

In particular,

$$(2.26) \qquad \operatorname{Tr}[e^{-\tau h_x^D} - e^{-\tau h}] = -\tau \int_{E_0}^{\infty} d\lambda e^{-\tau\lambda} \xi(\lambda, x), \quad \tau > 0,$$

$$\operatorname{Tr}[(h_x^D - z)^{-1} - (h - z)^{-1}] = -\int_{E_0}^{\infty} d\lambda \frac{\xi(\lambda, x)}{(\lambda - z)^2}, \quad z \in \mathbb{C} \setminus \{\sigma(h_x^D) \cup \sigma(h)\}.$$

(2.27)

Combining (2.14) and (2.26) we obtain the general trace formula for Sturm-Liouville operators

$$(2.28) \qquad 2s_1(x) = v(x) = E_0 + \lim_{\tau \downarrow 0} \int_{E_0}^{\infty} d\lambda e^{-\tau\lambda} [1 - 2\xi(\lambda, x)].$$

The Abelian regularization cannot be removed in general, see [18].

Higher-order trace formulas are given in the next theorem.

THEOREM 2.2. *One infers*

$$s_0(x) = -\tfrac{1}{2},$$
$$s_\ell(x) = \frac{(-1)^{\ell-1}}{\ell!} \left\{ \frac{E_0^\ell}{2} + \ell \lim_{t \downarrow 0} \int_{E_0}^{\infty} d\lambda e^{-t\lambda} \lambda^{\ell-1} \left[\tfrac{1}{2} - \xi(\lambda, x) \right] \right\}, \quad \ell \in \mathbb{N}.$$

(2.29)

From the high-energy behavior of the Green's function we find that

$$(2.30) \qquad p(x) r(x) = i\{ \lim_{z \downarrow -\infty} [\sqrt{z} g(z, x, x)] \}^{-1}.$$

In contrast to the Schrödinger case, the spectral shift function $\xi(\lambda, x)$ does not contain all the information necessary to construct both p and q in the Sturm-Liouville case, given the weight r. From (2.11) and (2.24) we see that in fact the spectral shift functions Ξ and ξ of H and h respectively, are identical in the sense that $\xi(\lambda, x) = \Xi(\lambda, t(x))$. For a given V we may construct $\Xi(\lambda, t)$ associated with (H_t^D, H). By choosing any positive $p \in C^\infty(\mathbb{R})$ we may define the Sturm-Liouville operator h using (2.10) (or (2.11) for the Green's function). By construction, the pair (h_x^D, h) will have $\xi(\lambda, x)$ as the corresponding spectral shift function.

The behavior of $\xi(\lambda, x)$ is particularly simple in spectral gaps of h. Since p, q, and r are real-valued, $g(\lambda+i0, x, x)$ is real-valued for $\lambda \in \mathbb{R} \backslash \sigma(h)$. More precisely, suppose $(\lambda_1, \lambda_2) \subset \mathbb{R} \backslash \sigma(h)$ and assume that $\mu(x) \in (\lambda_1, \lambda_2)$ is an eigenvalue of h_x^D. Then one has

$$(2.31) \qquad \xi(\lambda, x) = \begin{cases} 0, & \lambda_1 < \lambda < \mu(x) \\ 1, & \mu(x) < \lambda < \lambda_2. \end{cases}$$

Next, assume that p, q, and r are periodic, i.e.,

$$(2.32) \quad p(x + a) = p(x), \ q(x + a) = q(x), \ r(x + a) = r(x), \quad x \in \mathbb{R}$$

for some $a > 0$. Then Floquet theory implies that

$$(2.33) \quad \sigma(h) = \bigcup_{n=1}^\infty [E_{2(n-1)}, E_{2n-1}], \quad E_0 < E_1 \leq E_2 < E_3 \leq \cdots$$

and

$$\sigma(h_x^D) = \sigma(h) \cup \{\mu_n(x)\}_{n \in \mathbb{N}}, \quad E_{2n-1} \leq \mu_n(x) \leq E_{2n}, \ x \in \mathbb{R}, \ n \in \mathbb{N}.$$
(2.34)

In the periodic case $g(\lambda + i0, x, x)$ is purely imaginary on the spectrum, and hence

$$(2.35) \qquad \xi(\lambda, x) = \begin{cases} 0, & \lambda < E_0, \ \mu_n(x) < \lambda < E_{2n}, \ n \in \mathbb{N} \\ 1, & E_{2n-1} < \lambda < \mu_n(x), \ n \in \mathbb{N} \\ \frac{1}{2}, & E_{2(n-1)} < \lambda < E_{2n-1}, \ n \in \mathbb{N} \end{cases}$$

Combining (2.29) and (2.35) we obtain the following result.

THEOREM 2.3. Let $p, q, r \in C^\infty(\mathbb{R})$, $p, r > 0$ be periodic, $p(x + a) = p(x)$, $q(x + a) = q(x)$, $r(x + a) = r(x)$ for some $a > 0$. Then

$$2(-1)^{\ell+1} \ell! s_\ell(x) = E_0^\ell + \sum_{n=1}^\infty [E_{2n-1}^\ell + E_{2n}^\ell - 2\mu_n(x)^\ell], \ \ell \in \mathbb{N}, \ x \in \mathbb{R}.$$

(2.36)

In particular,

$$(2.37) \qquad 2s_1(x) = v(x) = E_0 + \sum_{n=1}^\infty [E_{2n-1} + E_{2n} - 2\mu_n(x)].$$

Finally, we turn to the case where the Dirichlet boundary condition is replaced by a family of (Robin-type) self-adjoint boundary conditions. Define

$$(2.38) \qquad h_{\beta,y} f = \frac{1}{r^2}[-(p^2 f')' + qf],$$

$$f \in \mathcal{D}(h_{\beta,y}) = \{g \in L^2(\mathbb{R};r^2 dx) \,|\, g, g' \in AC([y,\pm R]), \quad R > 0,$$
$$\lim_{\epsilon \downarrow 0}[g'(y \pm \epsilon) + \beta g(y \pm \epsilon)] = 0, \ h_{\beta,y} g \in L^2(\mathbb{R};r^2 dx)\}.$$

($\beta = 0$ corresponds to a Neumann boundary condition at y.)

$h_{\beta,y}$ is unitarily equivalent (using the operator U in (2.8)) to the Schrödinger operator

$$(2.39) \qquad H_{\nu(\beta,u),u} = -\frac{d^2}{dt^2} + V,$$

$$\mathcal{D}(H_{\nu(\beta,u),u}) = \{g \in L^2(\mathbb{R};dt) \,|\, g, g' \in AC([u,\pm R]), \quad R > 0,$$
$$\lim_{\epsilon \downarrow 0}[g'(u \pm \epsilon) + \nu(\beta,u)g(u \pm \epsilon)] = 0, \ H_{\nu(\beta,u),u} g \in L^2(\mathbb{R};dt)\},$$

where V is given by (2.10), the boundary condition is located at

$$(2.40) \qquad u(y) = \int_{x_0}^{y} dx \frac{r(x)}{p(x)},$$

and $\nu(\beta,u)$ depends on u as well as on β, viz.,

$$(2.41) \quad \nu = \nu(\beta,u) = \left[\frac{p}{r}\beta - \frac{(pr)_x}{2r^2}\right]\Big|_{x=y} = \left[\frac{P}{R}\beta - \frac{(PR)_t}{2PR}\right]\Big|_{t=u}.$$

The Green's function of $h_{\beta,y}$ is given by

$$g_{\beta,y}(z,x,x') = (h_{\beta,y} - z)^{-1}(x,x')$$
$$(2.42) \qquad = g(z,x,x') - \frac{(\beta + \partial_2)g(z,x,y)(\beta + \partial_1)g(z,y,x')}{(\beta + \partial_1)(\beta + \partial_2)g(z,y,y)},$$

where we abbreviate

$$\partial_1 g(z,y,x') = \partial_x g(z,x,x')|_{x=y}, \ \partial_2 g(z,x,y) = \partial_{x'} g(z,x,x')|_{x'=y}, \text{ etc.}$$
(2.43)
In this case $-(\beta + \partial_1)(\beta + \partial_2)g(z,y,y)$ is a Herglotz function such that $\text{Im}[(\beta + \partial_1)(\beta + \partial_2)g(\lambda + i0, y, y)] < 0$ for $-\lambda > 0$ large enough. Krein's spectral shift function for the pair $(h_{\beta,x}, h)$ then reads

$$(2.44) \ \xi_\beta(\lambda, x) = \frac{1}{\pi}\lim_{\epsilon \downarrow 0}\{\arg[(\beta + \partial_1)(\beta + \partial_2)g(\lambda + i\epsilon, x, x)]\} - 1,$$

$$\beta \in \mathbb{R}, \ x \in \mathbb{R}, \ \lambda \in \mathbb{R},$$

and it satisfies

(2.45) $\xi_\beta(\lambda, x) = 0$ for $\lambda < \zeta_{\beta,0}(x) := \inf\{\sigma(h_{\beta,x})\},$

(2.46) $\operatorname{Tr}[F(h_{\beta,x}) - F(h)] = \int_{\zeta_{\beta,0}(x)}^{\infty} d\lambda F'(\lambda)\xi_\beta(\lambda, x)$

for functions F as in (2.25). In particular, we find

(2.47) $\operatorname{Tr}[e^{-\tau h_{\beta,x}} - e^{-\tau h}] \underset{\tau\downarrow 0}{\sim} \sum_{\ell=0}^{\infty} s_{\beta,\ell}(x)\tau^\ell,$

where

(2.48) $s_{\beta,\ell}(x) = (-1)^{\ell+1}\dfrac{r_{\beta,\ell}(x)}{\ell!}, \quad \ell \in \mathbb{N}_0,$

with (cf. [22],[24]),

(2.49)

$$r_{\beta,0}(x) = -\frac{1}{2}, \; r_{\beta,1}(x) = \nu(\beta, u(x))^2 - \frac{1}{2}v(x),$$

$$r_{\beta,j}(x) = j\gamma_{\beta,j-1}(x) - \sum_{\ell=1}^{j-1}\gamma_{\beta,j-\ell-1}(x)r_{\beta,\ell}(x), \; j = 2, 3, \ldots,$$

$$\gamma_{\beta,-1} = 1, \; \gamma_{\beta,0} = \nu^2 - \frac{1}{2}v, \; \gamma_{\beta,1} = \frac{1}{2}\nu^2 v + \frac{1}{2}\nu v_x - \frac{1}{8}v^2 + \frac{1}{8}v_{xx},$$

$$\gamma_{\beta,2} = -\frac{1}{16}v^3 + \frac{3}{8}\nu^2 v^2 + \frac{3}{16}v_x(4\nu v + v_x) + \frac{1}{8}v_{xx}(v - \nu^2)$$

$$-\frac{1}{8}\nu v_{xxx} - \frac{1}{64}v_{xxxx},$$

$$\gamma_{\beta,j+1} = \frac{1}{8}\sum_{\ell=1}^{j}[2(v - \nu^2)\gamma_{\beta,\ell-1}\gamma_{\beta,j-\ell,xx} - (v - \nu^2)\gamma_{\beta,\ell-1,x}\gamma_{\beta,j-\ell,x}$$

$$-4\gamma_{\beta,\ell}\gamma_{\beta,j-\ell+1} - 4v(v - \nu^2)\gamma_{\beta,\ell-1}\gamma_{\beta,j-\ell} - 2v_x\gamma_{\beta,\ell-1}\gamma_{\beta,j-\ell,x}$$

$$+\gamma_{\beta,\ell-1}\gamma_{\beta,j-\ell}] + \frac{1}{8}\sum_{\ell=0}^{j}[\gamma_{\beta,\ell,x}\gamma_{\beta,j-\ell,x} - 2\gamma_{\beta,\ell}\gamma_{\beta,j-\ell,xx}$$

(2.50) $-4(\nu^2 - 2v)\gamma_{\beta,\ell}\gamma_{\beta,j-\ell}], \qquad j = 2, 3, \ldots.$

Explicitly, one computes

(2.51) $s_{\beta,0}(x) = \dfrac{1}{2}, \; s_{\beta,1}(x) = \nu(\beta, u(x))^2 - \dfrac{1}{2}v(x),$ etc.

The proof of (2.49) in [22] is based on the differential equation for $\Gamma_\nu(z,t) = (\nu + \partial_1)(\nu + \partial_2)G(z,t,t)$, namely

(2.52) $2[V(t) - \nu^2 - z]\Gamma_{\nu,tt}(z,t)\Gamma_\nu(z,t) - [V(t) - \nu^2 - z]\Gamma_{\nu,t}(z,t)^2$

$$-2V_t(t)\Gamma_{\nu,t}(z,t)\Gamma_\nu(z,t) - 4\{[V(t) - z][V(t) - \nu^2 - z] - \nu V_t(t)\}\Gamma_\nu(z,t)^2$$
$$= -[V(t) - \nu^2 - z]^3$$

and the asymptotic expansion

(2.53)
$$\Gamma_\nu(z,t) \underset{\substack{|z| \to \infty \\ z \in \mathbb{C} \backslash C_\epsilon}}{\sim} \frac{i}{2} z^{-1/2} \sum_{j=-1}^{\infty} \Gamma_{\nu,j}(t) z^{-j},$$

with $\Gamma_{\nu,j}(t)$ defined as in (2.50) with β replaced by ν and $v(x)$ by $V(t)$.
    The analog of Theorem 2.2 now reads

(2.54)
$$s_{\beta,\ell}(x) = \frac{(-1)^\ell}{\ell!} \left\{ \frac{\zeta_{\beta,0}(x)^\ell}{2} + \ell \lim_{\tau \downarrow 0} \int_{\zeta_{\beta,0}(x)}^{\infty} d\lambda e^{-\tau\lambda} \lambda^{\ell-1} \left[-\frac{1}{2} + \xi_\beta(\lambda, x) \right] \right\},$$
$$\ell \in \mathbb{N},$$

and, in particular,

(2.55)
$$s_{\beta,1}(x) = \nu(\beta, u(x))^2 - \frac{1}{2} v(x)$$
$$= -\frac{1}{2} \zeta_{\beta,0}(x) - \lim_{\tau \downarrow 0} \int_{\zeta_{\beta,0}(x)}^{\infty} d\lambda e^{-\tau\lambda} \left[-\frac{1}{2} + \xi_\beta(\lambda, x) \right].$$

Our last example in this section will be the periodic case, assuming (2.32) to hold.
    In this case

$$\sigma(h_{\beta,x}) = \sigma(h) \cup \{\zeta_{\beta,n}(x)\}_{n \in \mathbb{N}_0},$$

$$\zeta_{\beta,0}(x) \leq E_0, \ E_{2n-1} \leq \zeta_{\beta,n}(x) \leq E_{2n}, \ x \in \mathbb{R}, \ n \in \mathbb{N},$$
(2.56)
with $\sigma(h)$ given as in (2.33). The spectral shift function now reads

(2.57) $\xi_\beta(\lambda, x) = \begin{cases} 0, & \lambda < \zeta_{\beta,0}(x), \ E_{2n-1} < \lambda < \zeta_{\beta,n}(x), \ n \in \mathbb{N} \\ -1, & \zeta_{\beta,n}(x) < \lambda < E_{2n}, \ n \in \mathbb{N}_0 \\ -\frac{1}{2}, & E_{2(n-1)} < \lambda < E_{2n-1}, \ n \in \mathbb{N} \end{cases}$

and the trace formula (2.55) in the periodic case now equals

$$2(-1)^\ell \ell! s_{\beta,\ell}(x) = 2\zeta_{\beta,0}(x)^\ell - E_0^\ell + \sum_{n=1}^{\infty} [2\zeta_{\beta,n}(x)^\ell - E_{2n-1}^\ell - E_{2n}^\ell],$$
(2.58)
$$\ell \in \mathbb{N}, \ x \in \mathbb{R}.$$

In the case $\ell = 1$ we find

$$-2s_{\beta,1}(x) = v(x) = \frac{q(x)}{r(x)} + \frac{p(x)}{2r(x)^3}(r(x)p(x))_{xx} + \frac{(r(x)p(x))_x}{2r(x)^2} \left(\frac{p(x)}{r(x)} \right)_x,$$

$$(2.59) \qquad -\frac{(r(x)p(x))_x^2}{4r(x)^4} = 2\left(\frac{p(x)}{r(x)}\beta - \frac{(p(x)r(x))_x^2}{2r(x)^2}\right)^2$$

$$+2\zeta_{\beta,0}(x) - E_0 + \sum_{n=1}^{\infty}[2\zeta_{\beta,n}(x) - E_{2n-1} - E_{2n}].$$

Subtracting this equation from (2.37) yields

$$-\left(\frac{p(x)}{r(x)}\beta - \frac{(p(x)r(x))_x^2}{2r(x)^2}\right)^2$$

$$(2.60) \qquad = E_0 - \zeta_{\beta,0}(x) + \sum_{n=1}^{\infty}[E_{2n-1} + E_{2n} - \mu_n(x) - \zeta_{\beta,n}(x)].$$

3. Matrix-valued Schrödinger operators. In this section we extend the trace formula (2.28) to self-adjoint matrix-valued Schrödinger operators. General background on matrix-valued differential expressions can be found, e.g., in [1], [40]. Unlike all other sections in this contribution, the material below is in a preliminary stage with more details appearing elsewhere.

Let H in $L^2(\mathbb{R})^m \cong L^2(\mathbb{R}) \otimes \mathbb{C}^m$ be a self-adjoint operator defined by

$$(3.1) \qquad Hf = -I_m f'' + Qf,$$
$$f \in \mathcal{D}(H) = \{g \in L^2(\mathbb{R})^m \,|\, g_j, g_j' \in AC_{loc}(\mathbb{R}), 1 \le j \le m;$$
$$Hg \in L^2(\mathbb{R})^m\},$$

where $f = (f_1, \ldots, f_m)^T$, I_m denotes the identity in $\mathbb{C}^m$, and $Q = (Q_{j,k})_{1 \le j,k \le m}$ denotes a self-adjoint matrix satisfying

$$(3.2) \qquad Q_{j,k} \in C(\mathbb{R}) \text{ bounded from below, } 1 \le j, k \le m.$$

Closely associated with the equation

$$(3.3) \qquad Hf = zf$$

is the first-order $2m \times 2m$ system

$$(3.4) \qquad L(z)(f, f')^T = 0,$$

where $(f, f')^T = (f_1, \ldots, f_m, f_1', \ldots, f_m')^T$ and

$$(3.5) \qquad L(z) = I_{2m}\frac{d}{dx} - A(z), \quad A(z) = \begin{pmatrix} 0 & I_m \\ Q-z & 0 \end{pmatrix},$$

with I_{2m} the identity in $\mathbb{C}^{2m}$. If $\Psi(z,x)$ denotes a fundamental matrix for $L(z)$, that is,

$$(3.6) \qquad L(z)\Psi(z) = 0,$$

or equivalently,

(3.7) $$\Psi'(z, x) = A(z, x)\Psi(z, x),$$

then $\tilde{\Psi}(z, x)$ defined by

(3.8) $$\tilde{\Psi}(z, x) = \Psi(z, x)^{-1}$$

satisfies the adjoint system

(3.9) $$\tilde{\Psi}'(z, x) = -\tilde{\Psi}(z, x)A(z, x).$$

Moreover, the fundamental matrices $\Psi(z, x)$ and $\tilde{\Psi}(z, x)$ are of the form

$$\Psi(z, x) = \begin{pmatrix} \psi_1(z, x) & \psi_2(z, x) \\ \psi_1'(z, x) & \psi_2'(z, x) \end{pmatrix}, \quad \tilde{\Psi}(z, x) = \begin{pmatrix} \tilde{\psi}_2'(z, x) & -\tilde{\psi}_2(z, x) \\ -\tilde{\psi}_1'(z, x) & \tilde{\psi}_1(z, x) \end{pmatrix},$$
(3.10)

and one verifies that

(3.11)
$$-\psi_j''(z, x) + Q(x)\psi_j(z, x) = z\psi_j(z, x),$$
$$-\tilde{\psi}_j''(z, x) + \tilde{\psi}_j(z, x)Q(x) = z\tilde{\psi}_j(z, x), \quad j = 1, 2.$$

In particular, assuming $\psi_j(z)$, $\tilde{\psi}_j(z)$ to be unique solutions of (3.11) (up to right resp. left multiplication of matrices constant with respect to x) satisfying

(3.12)
$$\psi_{\frac{1}{2}}(z, \cdot): \quad = \quad \psi_\pm(z, \cdot) \in L^2([R, \pm\infty))^{m \times m},$$
$$\tilde{\psi}_{\frac{1}{2}}(z, \cdot): \quad = \quad \tilde{\psi}_\pm(z, \cdot) \in L^2([R, \pm\infty))^{m \times m},$$
$$R \in \mathbb{R}, z \in \mathbb{C} \setminus \sigma(H),$$

the Green's matrix $G(z, x, x')$ of H becomes

(3.13) $$G(z, x, x') = \begin{cases} \psi_+(z, x)\tilde{\psi}_-(z, x'), & x \geq x' \\ \psi_-(z, x)\tilde{\psi}_+(z, x'), & x \leq x' \end{cases}$$

and hence the resolvent of H is given by

(3.14) $$((H - z)^{-1}f)(x) = \int_{\mathbb{R}} dx' \, G(z, x, x')f(x'), \, f \in L^2(\mathbb{R})^m, \, z \in \mathbb{C} \setminus \sigma(H).$$

Since

(3.15) $$- \psi_j''(\bar{z}, x)^* + \psi_j(\bar{z}, x)^*Q(x) = z\psi_j(\bar{z}, x)^*, \, j = 1, 2,$$

$\tilde{\psi}_j(z, x)$ are of the type

(3.16) $$\tilde{\psi}_j(z, x) = A_{j,1}(z)\psi_1(\bar{z}, x)^* + B_{j,2}(z)\psi_2(\bar{z}, x)^*, \, j = 1, 2$$

for matrices $A_{j,k}(z)$, $B_{j,k}(z)$, $1 \le j, k \le 2$ in $\mathbb{C}^m$ constant with respect to x. Introducing the "Wronskian" $W(\phi, \psi)(x)$ of $m \times m$ matrices ϕ and ψ by

$$(3.17) \qquad W(\phi, \psi)(x) = \phi(x)\psi'(x) - \phi'(x)\psi(x),$$

one verifies that

$$(3.18) \qquad \frac{d}{dx} W(\phi(\bar{z})^*, \psi(z))(x) = 0$$

for solutions $\psi(z, x)$ and $\phi(\bar{z}, x)^*$ of

$$(3.19) \qquad -\psi''(z, x) + [Q(x) - z]\psi(z, x) = 0, \quad -\phi''(\bar{z}, x)^* + \phi(\bar{z}, x)^*[Q(x) - z] = 0.$$

Relations (3.8), (3.12), (3.15), and (3.16) then yield

$$(3.20) \qquad \tilde{\psi}_\pm(z, x) = \pm W(\psi_\pm(\bar{z})^*, \psi_\mp(z))^{-1} \psi_\pm(\bar{z}, x)^*$$

and hence

$$(3.21) \qquad \begin{aligned} G(z, x, x) &= -\psi_+(z, x) W(\psi_-(\bar{z})^*, \psi_+(z))^{-1} \psi_-(\bar{z}, x)^* \\ &= \psi_-(z, x) W(\psi_+(\bar{z})^*, \psi_-(z))^{-1} \psi_+(\bar{z}, x)^*. \end{aligned}$$

The corresponding matrix-valued Dirichlet Schrödinger operator H_y^D in $L^2(\mathbb{R})^m$ then reads

$$(3.22) \qquad \begin{aligned} H_y^D f &= -I_m f'' + Qf, \\ f \in \mathcal{D}(H_y^D) &= \{g \in L^2(\mathbb{R})^m \,|\, g_j \in AC_{\text{loc}}(\mathbb{R}), g_j' \in AC_{\text{loc}}(\mathbb{R} \setminus \{y\}), \\ &\quad \lim_{\epsilon \downarrow 0} g_j(y \pm \epsilon) = 0, H_y^D g \in L^2(\mathbb{R})^m \} \end{aligned}$$

and its Green's matrix $G_y^D(z, x, x')$, the analog of (2.4), is given by

$$(3.23) \quad G_y^D(z, x, x') = G(z, x, x') - G(z, x, y)G(z, y, y)^{-1}G(z, y, x').$$

The analog of (2.5) then becomes

$$(3.24)$$

$$\begin{aligned} \text{Tr}[(H_x^D - z)^{-1} - (H - z)^{-1}] &= -\text{Tr}[G(z, \cdot, x)G(z, x, x)^{-1}G(z, x, \cdot)] \\ &= -\text{Tr}[G(z, x, x)^{-1}G(z, x, \cdot)G(z, \cdot, x)] \\ &= -\text{Tr}_{\mathbb{C}^m}\{G(z, x, x)^{-1}[\frac{d}{dz}G(z, x, x)]\} \\ &= -\frac{d}{dz}\text{Tr}_{\mathbb{C}^m}\{\ln[G(z, x, x)]\} = -\frac{d}{dz}\ln\{\det_{\mathbb{C}^m}[G(z, x, x)]\}, \end{aligned}$$

where we used cyclicity of the trace,

$$(H - z)^{-2}(x, x')_{j,k} = \frac{d}{dz} G(z, x, x')_{j,k}$$

(3.25)
$$= \sum_{\ell=1}^{m} \int_{\mathbb{R}} dx'' \, G(z, x, x'')_{j,\ell} G(z, x'', x')_{\ell,k} ,$$

and $\mathrm{Tr}_{\mathbb{C}^m}[\ln(M)] = \ln[\det_{\mathbb{C}^m}(M)]$ for matrices M in $\mathbb{C}^m$. Moreover, $\mathrm{Tr}(\cdot)$ and $\mathrm{Tr}_{\mathbb{C}^m}(\cdot)$ in (3.24) denote the trace in $L^2(\mathbb{R})^m$ and $\mathbb{C}^m$, respectively.

Introducing the matrix-valued Green's kernel diagonal with respect to x (cf. (3.21))

(3.26)
$$\Gamma(z, x) = G(z, x, x),$$

the matrix analog of (2.20) reads

$$-\Gamma(z, x)\Gamma_{xx}(z, x) - \Gamma_{xx}(z, x)\Gamma(z, x) + \Gamma_x(z, x)^2 + \Gamma(z, x)^2 Q(x)$$
(3.27)
$$+ Q(x)\Gamma(z, x)^2 + 2\Gamma(z, x)Q(x)\Gamma(z, x) - 4z\Gamma(z, x)^2 = I_m$$

and considerations along the lines of (2.20), (2.21) then yield

(3.28)
$$\Gamma(z, x) \underset{\substack{|z| \to \infty \\ z \in \mathbb{C} \backslash C_\epsilon}}{\sim} \frac{i}{2} z^{-1/2} \sum_{j=0}^{\infty} \Gamma_j(x) z^{-j},$$

with

(3.29)
$$\Gamma_0(x) = I_m, \quad \Gamma_1(x) = \frac{1}{2} Q(x), \quad \text{etc.}$$

Similarly,

(3.30)
$$-\frac{d}{dz} \ln[G(z, x, x)] \underset{\substack{|z| \to \infty \\ z \in \mathbb{C} \backslash C_\epsilon}}{\sim} \sum_{j=0}^{\infty} R_j(x) z^{-j-1},$$

where

(3.31)
$$R_0(x) = \frac{1}{2} I_m, \quad R_1(x) = \frac{1}{2} Q(x), \quad \text{etc.}$$

Next, define for all $x \in \mathbb{R}$ the analog of (2.24) by

$$\Xi(\lambda, x) = \frac{1}{\pi} \lim_{\epsilon \downarrow 0} \mathrm{Im}\{\ln[G(\lambda + i\epsilon, x, x)]\} \text{ for a.e. } \lambda \in \mathbb{R},$$
(3.32) $\Xi(\lambda, x) = 0, \lambda < E_0 := \inf\{\sigma(H)\},$

where $\mathrm{Im}(M)$, $\mathrm{Re}(M)$, in obvious notation, abbreviate

(3.33)
$$\mathrm{Im}(M) = \frac{1}{2i}(M - M^*), \quad \mathrm{Re}(M) = \frac{1}{2}(M + M^*)$$

for matrices M in $\mathbb{C}^m$. It follows from the results in [5] that

(3.34) $$0 \leq \Xi(\lambda, x) \leq I_m \quad \text{for a.e. } \lambda \in \mathbb{R}.$$

In the following denote by $C_{R,\epsilon}$ the counter-clockwise oriented contour

$$C_{R,\epsilon} = \{z = E_0 + \epsilon e^{i\phi} \mid \frac{3\pi}{2} \geq \phi \geq \frac{\pi}{2}\} \cup \{z = E_0 + \lambda + i\epsilon \mid 0 \leq \lambda \leq R\}$$
$$\cup \ \{z = E_0 + Re^{i\phi} \mid \arctan(\epsilon/R) \leq \phi \leq 2\pi - \arctan(\epsilon/R)\}$$
(3.35) $$\cup \ \{z = E_0 + \lambda - i\epsilon \mid 0 \leq \lambda \leq R\}, \ R > \epsilon > 0.$$

Applying the residue theorem, taking into account that $G(z, x, x)$, $x \in \mathbb{R}$, is analytic in $z \in \mathbb{C} \setminus \sigma(H)$ and $\det[G(z, x, x)] \neq 0$ for $z \in \mathbb{C} \setminus \sigma(H)$ (cf. (3.23)), then yields

(3.36)

$$\{\ln[G(z, x, x)]\}_{j,k} = \frac{1}{2\pi i} \oint_{C_{R,\epsilon}} dz' \frac{\{\ln[G(z', x, x)]\}_{j,k}}{z' - z}$$

$$= \frac{1}{2\pi i} \oint_{C_{R,\epsilon}} dz' \{\ln[G(z', x, x)]\}_{j,k} \frac{z'}{1 + z'^2}$$

$$+ \frac{1}{2\pi i} \oint_{C_{R,\epsilon}} dz' \{\ln[G(z', x, x)]\}_{j,k} \left[\frac{1}{z' - z} - \frac{z'}{1 + z'^2} \right]$$

$$= \text{Re}\{\ln[G(i, x, x)]\}_{j,k}$$

$$+ \frac{1}{\pi} \int_{E_0}^{R} d\lambda \text{Im}\{\ln[G(\lambda + i0, x, x)]\}_{j,k} \left[\frac{1}{\lambda - z} - \frac{\lambda}{1 + \lambda^2} \right] + o(\epsilon) + o(R^{-1})$$

$$\xrightarrow[R\to\infty, \epsilon\downarrow 0]{} \text{Re}\{\ln[G(i, x, x)]\}_{j,k}$$

$$+ \int_{E_0}^{\infty} d\lambda \, \Xi(\lambda, x)_{j,k} \left[\frac{1}{\lambda - z} - \frac{\lambda}{1 + \lambda^2} \right], \quad 1 \leq j, k \leq m.$$

Thus

(3.37) $$\frac{d}{dz} \ln[G(z, x, x)] = \int_{E_0}^{\infty} d\lambda \, \Xi(\lambda, x)(\lambda - z)^{-2},$$

and the matrix analog of (2.28) then reads

(3.38) $$Q(x) = E_0 I_m + \lim_{z \to i\infty} \int_{E_0}^{\infty} d\lambda \, z^2 (\lambda - z)^{-2} [I_m - 2\Xi(\lambda, x)],$$

where we used a resolvent instead of a heat kernel regularization.

Defining

(3.39) $$\xi(\lambda, x) = \text{Tr}_{\mathbb{C}^m}[\Xi(\lambda, x)],$$

one infers from (3.24) that

$$(3.40) \quad \text{Tr}[(H_x^D - z)^{-1} - (H - z)^{-1}] = -\int_{E_0}^{\infty} d\lambda \xi(\lambda, x)(\lambda - z)^{-2}$$

and that

$$\xi(\lambda, x) = \frac{1}{\pi} \lim_{\epsilon \downarrow 0} \arg\{\det_{\mathbb{C}^m}[G(\lambda + i\epsilon, x, x)]\},$$

$$(3.41) \qquad\qquad 0 \le \xi(\lambda, x) \le m \quad \text{for a.e. } \lambda \in \mathbb{R}.$$

Further details and applications of this formalism to inverse spectral problems will appear elsewhere.

4. Multi-dimensional trace formulas. First, reporting on recent work in [25], we attempt to extend the leading behavior in (2.14),

$$(4.1) \qquad 2\text{Tr}[e^{-\tau H} - e^{-\tau H_x^D}] = 1 - \tau V(x) + o(\tau) \quad \text{as } \tau \downarrow 0$$

to arbitrary space dimensions $\nu \in \mathbb{N}$. The key to such an extension is an appropriate combination of Dirichlet and Neumann boundary conditions on various hyperplanes through the point $x \in \mathbb{R}^\nu$ taking into account that (4.1) is equivalent to

$$(4.2) \qquad \text{Tr}[e^{-\tau H_x^N} - e^{-\tau H_x^D}] = 1 - \tau V(x) + o(\tau) \quad \text{as } \tau \downarrow 0,$$

where $H_x^N = H_x^0$ denotes the operator (2.39) with a Neumann boundary condition at $x \in \mathbb{R}$. We start by introducing proper notations. In the following let V be a real-valued continuous function on $\mathbb{R}^\nu$ bounded from below and define the self-adjoint operator

$$(4.3) \qquad\qquad\qquad H = -\Delta \dot{+} V$$

as a form sum in $L^2(\mathbb{R}^\nu)$. Next, let $A \subseteq \{1, ..., \nu\}$ and denote by $|A|$ the number of elements of A. Moreover, let $B_\alpha^{(x)}, \alpha \subseteq \{1, ..., \nu\}$ be the 2^ν blocks obtained by removing the hyperplanes $\mathcal{P}_j^{(x)} = \{y \in \mathbb{R}^\nu \mid y_j = x_j\}$ from $\mathbb{R}^\nu$, that is, $B_\alpha^{(x)} = \{y \in \mathbb{R}^\nu \mid y_\ell > x_\ell \text{ if } \ell \in \alpha, y_\ell < x_\ell \text{ if } \ell \notin \alpha\}$ and denote by $\mathcal{P}_\nu$ the power set of $\{1, ..., \nu\}$. The operator $H_{A;x}$ is then defined to be $-\Delta + V$ on $\bigoplus_{\alpha \in \mathcal{P}_\nu} L^2(B_\alpha^{(x)})$ with Dirichlet boundary conditions on $\{P_j^{(x)}\}_{j \in A}$ and Neumann boundary conditions on $\{P_j^{(x)}\}_{j \notin A}$.

THEOREM 4.1. *[25] Define* $C_\tau = \sum_{A \in \mathcal{B}_\nu} (-1)^{|A|} e^{-\tau H_{A;0}}, \tau > 0$. *Then the integral kernel of* C_τ *is given by*

$$(4.4) \quad C_\tau(x, x') = \begin{cases} 2^\nu e^{-\tau H}(x, -x'), & x, x' \text{ in the same orthant} \\ 0, & \text{otherwise.} \end{cases}$$

Moreover, $C_\tau, \tau > 0$ is a trace class operator in $L^2(\mathbb{R}^\nu)$ and

$$(4.5) \qquad \cdot \ Tr(C_\tau) = 2^\nu \int_{\mathbb{R}^\nu} d^\nu x \, e^{-tH}(x, -x), \quad \tau > 0.$$

The proof of (4.4) in [25] is based on the method of images while the trace class property of C_τ and (4.5) follow from the direct sum decomposition of C_τ in $\bigoplus_{\alpha \in \mathcal{P}_\nu} L^2(B_\alpha^{(x)})$.

Applying a Feynman-Kac-type analysis then yields the following ν-dimensional generalization of (4.2).

THEOREM 4.2. *[25]*

$$(4.6) \qquad Tr\Big(\sum_{A \in \mathcal{P}_\nu} (-1)^{|A|} e^{-\tau H_{A;x}} \Big) = 1 - \tau V(x) + o(\tau) \text{ as } \tau \downarrow 0.$$

While Theorem 4.2 represents a multidimensional trace formula for Schrödinger operators associated with unbounded regions in $\mathbb{R}^\nu$, one can also prove new trace formulas for Schrödinger operators defined in boxes. One obtains, e.g.,

THEOREM 4.3. *[25] Let V be continuous on $[0,1]^\nu$. For $A \subseteq \{1, ..., \nu\}$, let H_A be $-\Delta + V$ on $L^2([0,1]^\nu)$ with Dirichlet boundary conditions on the hyperplanes with $x_j = 0$ or 1 and $j \in A$ and Neumann boundary conditions on the hyperplanes with $x_j = 0$ or 1 and $j \notin A$. Let $\langle V \rangle$ be the average of V at the 2^ν corners of $[0,1]^\nu$. Then*

$$(4.7) \qquad \sum_{A \in \mathcal{P}_\nu} (-1)^{|A|} Tr(e^{-\tau H_A}) = 1 - \tau \langle V \rangle + o(\tau) \text{ as } \tau \downarrow 0.$$

This result holds also for rectangular boxes $\times_{j=1}^\nu [a_j, b_j]$ but the rectangular symmetry is crucial in the proof of [25]. Similarly, one can prove

THEOREM 4.4. *[25] Let V be continuous on $[0,1]^\nu$. For $A \subseteq \{1, ..., \nu\}$ let $\tilde{H}_A$ be $-\Delta + V$ on $L^2([0,1]^\nu)$ with Dirichlet boundary conditions on the hyperplanes with $x_j = 0$ for $j \in A$ and Neumann boundary conditions on the hyperplanes with $x_j = 0$ for $j \notin A$ or $x_k = 1$ for all $k \in \{1, ..., \nu\}$. Then*

$$(4.8) \qquad \sum_{A \in \mathcal{P}_\nu} (-1)^{|A|} Tr(e^{-\tau \tilde{H}_A}) = 2^{-\nu}[1 - \tau V(0) + o(\tau)] \text{ as } \tau \downarrow 0.$$

Finally, we mention an Abelianized version of a trace formula that Lax [29] derived formally in two dimensions.

THEOREM 4.5. *[25] Let V be a continuous periodic function on $\mathbb{R}^2$ with $V(x_1 + n_1, x_2 + n_2) = V(x_1, x_2)$ for all $(x_1, x_2, n_1, n_2) \in \mathbb{R}^2 \times \mathbb{Z}^2$. Let $H_P, H_A, H_{AP}, H_{PA}, H_N$, and H_D be the operators $-\Delta + V$ on $L^2([0,1]^2)$ with periodic, antiperiodic, AP, PA, Neumann, and Dirichlet boundary*

conditions respectively, where AP (resp. PA) means antiperiodic in the
x_1 *(resp.* x_2*) direction and periodic in the* x_2 *(resp.* x_1*) direction. Then*

$$\text{Tr}[e^{-\tau H_P} + e^{-\tau H_A} + e^{-\tau H_{AP}} + e^{-\tau H_{PA}} - 2e^{-\tau H_N} - 2e^{-\tau H_D}]$$

$$(4.9) \quad = -1 + \tau V(0) + o(\tau) \text{ as } \tau \downarrow 0.$$

For a different kind of two-dimensional trace formula for $V(x)$ comparing the heat kernels for $H = -\Delta + V$ and $H_0 = -\Delta$ with Dirichlet boundary conditions on a rectangular box, see [34]. Trace formulas for heat kernels of multi-dimensional Schrödinger operators in the short-range case have also recently been derived in [4].

Finally, we illustrate a possible new abstract approach to the trace formulas (4.7) based on certain commutation (supersymmetric) techniques in the noninteracting case where $V(x) = 0$, $x \in \mathbb{R}^\nu$. We need a bit of notation. Let $\mathcal{H}$ be a (complex separable) Hilbert space, F a closed densely defined linear operator in $\mathcal{H}$ and define the self-adjoint operators

$$(4.10) \qquad\qquad H_1 = F^* F , \ H_2 = F F^*$$

in $\mathcal{H}$ and

$$(4.11) \qquad\qquad Q = \begin{pmatrix} 0 & F^* \\ F & 0 \end{pmatrix}, \ P = \begin{pmatrix} 1 & 0 \\ 0 & -1 \end{pmatrix}$$

in $\mathcal{H} \oplus \mathcal{H}$. Moreover, we denote by tr(.) the trace in $\mathcal{H}$, by Tr(.) the trace in $\mathcal{H} \oplus \mathcal{H}$, and by $\mathcal{B}(\mathcal{H})$ (resp. $\mathcal{B}_1(\mathcal{H})$) the set of bounded (resp. trace class) operators on $\mathcal{H}$.

LEMMA 4.6. *One infers that*

$$(4.12) \qquad\qquad QP + PQ = 0,$$

$$(4.13) \qquad\qquad Q^2 = \begin{pmatrix} H_1 & 0 \\ 0 & H_2 \end{pmatrix},$$

$$(4.14) \qquad F e^{-tH_1} \supseteq e^{-tH_2} F, \ F^* e^{-tH_2} \supseteq e^{-tH_1} F^*.$$

Proof. While (4.12) and (4.13) are obvious, (4.14) follows from

$$Q e^{-tQ^2} \supseteq e^{-tQ^2} Q.$$

LEMMA 4.7. *Assume* $B \in \mathcal{B}(\mathcal{H} \oplus \mathcal{H})$ *is bounded and commutes with* Q, *i.e.,* $QB \supseteq BQ$. *Suppose* e^{-tQ^2}, $Q^2 e^{-tQ^2} \in \mathcal{B}_1(\mathcal{H} \oplus \mathcal{H})$, $t > 0$. *Then*

$$(4.15) \qquad\qquad \frac{d}{dt} \text{Tr}[P e^{-tQ^2} B] = 0.$$

Proof.

$$\frac{d}{dt}\text{Tr}[Pe^{-tQ^2}B] = -\text{Tr}[PQ^2e^{-tQ^2}B]$$

(4.16) $$= \text{Tr}[PQe^{-tQ^2}QB] = \ldots = \text{Tr}[PQ^2e^{-tQ^2}B]$$

using commutativity of Q and B and anticommutativity of Q and P in (4.12) and cyclicity of the trace. The fact that Q is unbounded is offset by the trace class hypotheses in Lemma 4.7. In fact, rewriting

$$-\text{Tr}[PQ^2e^{-tQ^2}B] = -\text{Tr}[PQ(1+|Q|)^{-1}Q(1+|Q|)e^{-tQ^2}B]$$

enables one to prove 4.16 in a trivial manner by reshuffling $Q(1+|Q|)^{-1} \in \mathcal{B}(\mathcal{H} \oplus \mathcal{H})$ as opposed to Q in (4.16).

Next we introduce the closed densely defined linear operators F_n, $1 \leq n \leq \nu$ in $\mathcal{H}$ and define $H_{1,n} = F_n^*F_n$, $H_{2,n} = F_nF_n^*$, $1 \leq n \leq \nu$ as in (4.10). Moreover, assume

$$e^{-tH_{j,n}}, \qquad H_{j,n}e^{-tH_{j,n}} \in \mathcal{B}_1(\mathcal{H}), \qquad 1 \leq n \leq \nu$$

and

$$[F_m, F_n] \subseteq 0, \qquad [F_m, F_n^*] \subseteq 0, \qquad m \neq n$$

implying

$$[H_{j,m}, H_{\ell,n}] \subseteq 0, \qquad j, \ell = 1, 2, \qquad m \neq n.$$

We also denote

(4.17) $$Q_n = \begin{pmatrix} 0 & F_n^* \\ F_n & 0 \end{pmatrix}, \qquad 1 \leq n \leq \nu$$

in $\mathcal{H} \oplus \mathcal{H}$ as in (4.11) and define for any $A \in \mathcal{P}_\nu$ (the power set of $\{1, 2, \ldots, \nu\}$) the self-adjoint operator

(4.18) $$H_A^0 = \sum_{n \in A} H_{1,n} + \sum_{n \notin A} H_{2,n}.$$

Then an abstract version of (4.7) in the noninteracting case reads as follows.

THEOREM 4.8.

(4.19) $$\sum_{A \in \mathcal{P}_\nu} (-1)^{|A|} tr(e^{-tH_A^0}) = \sum_{A \in \mathcal{P}_\nu} (-1)^{|A|} \dim Ran[P_{H_A^0}(\{0\})],$$

where $P_{H_A^0}(\Omega)$, $\Omega \subseteq \mathbb{R}$, denote the spectral projections of H_A^0.

Proof. One computes

$$\frac{d}{dt} \sum_{A \in \mathcal{P}_\nu} (-1)^{|A|} \text{tr}(e^{-tH_A^0}) = -\sum_{A \in \mathcal{P}_\nu} (-1)^{|A|} \text{tr}(H_A^0 e^{-tH_A^0})$$

$$(4.20) \qquad\qquad = -\sum_{n=1}^{\nu} \text{Tr}[PQ_n^2 e^{-tQ_n^2} B_{\nu,n}] = 0$$

by (4.15), where

$$B_{1,1} = \begin{pmatrix} 1 & 0 \\ 0 & 1 \end{pmatrix},$$

$$B_{\nu,n} = \begin{pmatrix} b_{\nu,n} & 0 \\ 0 & b_{\nu,n} \end{pmatrix}, \quad b_{\nu,n} = -\prod_{\substack{m=1 \\ m \neq n}}^{\nu} (e^{-tH_{2,m}} - e^{-tH_{1,m}}), \quad \nu \geq 2$$

(4.21)

are bounded and commute with Q_n. Thus the left-hand side in (4.19) is independent of t and taking $t \uparrow \infty$ then determines the right-hand side of (4.19).

Identifying $A_n = 1 \otimes \ldots \otimes 1 \otimes \frac{\partial}{\partial x_n}\big|_D \otimes 1 \otimes \ldots \otimes 1$ in $L^2([0,1])^\nu)$ with

$$(4.22) \qquad\qquad \frac{\partial'}{\partial x_n}\Big|_D = \overline{\frac{d}{dx}\Big|_{C_0^\infty((0,1))}}, \quad 1 \leq n \leq \nu$$

in $L^2([0,1])$ then yields (4.7) in the case $V(x) = 0$ since only the zero-energy eigenvalue of the Neumann operator $H_\emptyset^0$ contributes on the right-hand side of (4.19). More generally, if A_n has the tensor product structure

$$A_n = 1 \otimes \ldots \otimes 1 \otimes a_n \otimes 1 \otimes \ldots \otimes 1$$

in $\mathcal{H} = \mathcal{H}_1 \otimes \ldots \otimes \mathcal{H}_\nu$, then clearly $[A_m, A_n] \subseteq 0$ and one evaluates

$$(4.23) \qquad \sum_{A \in \mathcal{P}_\nu} (-1)^{|A|} \text{tr}(e^{-tH_A^0}) = \prod_{n=1}^{\nu} \text{tr}(e^{-ta_n a_n^*} - e^{-ta_n^* a_n}).$$

In the special case (4.22), where $a_n = \frac{\partial}{\partial x_n}\big|_D$, one confirms that

$$\text{tr}\left(e^{-ta_n a_n^*} - e^{-ta_n^* a_n}\right) = 1, \quad 1 \leq n \leq \nu.$$

5. Trace formulas and point interactions in dimensions one, two, and three.

In this section we describe a different kind of multi-dimensional trace formula based on point interactions [2] and hence rank-one perturbations of resolvents first derived in [17] in a slightly different form. Since point interactions (also called contact interactions or δ-interactions) are limited to $\nu = 1, 2, 3$ space dimensions, so will be our approach below.

Assuming V to be real-valued, continuous and bounded from below on $\mathbb{R}^\nu, \nu = 1, 2, 3$, we introduce $H = -\Delta \dotplus V$ as in (4.3). The resolvent of the self-adjoint Hamiltonian $H_{\alpha,x}$, modeling H plus a point interaction centered at $x \in \mathbb{R}^\nu$ (whose strength is parameterized in terms of $\alpha \in \mathbb{R}$), is defined as follows (see, e.g., [2], [42])

$$(5.1) \quad (H_{\alpha,x} - z)^{-1} = (H - z)^{-1} + D_{\alpha,x}(z)^{-1}(\overline{G(z, x, .)}, .)G(z, ., x),$$
$$z \in \mathbb{C}\backslash\{\sigma(H_{\alpha,x}) \cup \sigma(H)\},$$

where

$$(5.2) \quad D_{\alpha,x}(z) = \begin{cases} -\alpha^{-1} - \Gamma_\nu(z, x), & \nu = 1, \ \alpha \in \mathbb{R} \cup \{\infty\}, \alpha \neq 0 \\ \alpha - \Gamma_\nu(z, x), & \nu = 2, 3, \ \alpha \in \mathbb{R}, \end{cases}$$

$$\Gamma_1(z, x) = G(z, x, x), \quad \Gamma_2(z, x) = \lim_{|\epsilon|\downarrow 0}[G(z, x, x + \epsilon) - (2\pi)^{-1}\ln(|\epsilon|)],$$

$$(5.3) \quad \Gamma_3(z, x) = \lim_{|\epsilon|\downarrow 0}[G(z, x, x + \epsilon) - (4\pi |\epsilon|)^{-1}],$$

and $G(z, x, x')$ denotes the Green's function of H. In analogy to (2.5) one then computes

$$(5.4) \quad \text{Tr}[(H_{\alpha,x} - z)^{-1} - (H - z)^{-1}] = -\frac{d}{dz}\ln[D_{\alpha,x}(z)].$$

Krein's spectral shift function for the pair $(H_{\alpha,x}, H)$ is then introduced via

$$(5.5) \quad \text{Tr}[(H_{\alpha,x} - z)^{-1} - (H - z)^{-1}] = -\int_{E_{\alpha,x,0}}^{\infty} d\lambda \frac{\xi_{\alpha,x}(\lambda)}{(\lambda - z)^2},$$

with $E_{\alpha,x,0} = \inf\{\sigma(H_{\alpha,x}) \cup \sigma(H)\}$ and the normalization

$$(5.6) \quad \xi_{\alpha,x}(\lambda) = 0, \quad \lambda < E_{\alpha,x,0}.$$

$\xi_{\alpha,x}(\lambda)$ is related to $D_{\alpha,x}(z)$ as $\xi(\lambda, x)$ is to $g(z, x, x)$ in 2.24. The high-energy expansion (see, e.g., [35], [41])

$$\lim_{|\epsilon|\downarrow 0}[G(z, x, x + \epsilon) - G^{(0)}(z, x, x + \epsilon)]$$

$$(5.7) \qquad = -V(x) \begin{cases} 1/4(-z)^{3/2} + o(z^{-3/2}), & \nu = 1 \\ -1/4\pi z + o(z^{-1}), & \nu = 2 \\ 1/8\pi(-z)^{1/2} + o(z^{-1/2}), & \nu = 3 \end{cases}$$

then yields

$$D_{\alpha,x}(z) = \begin{cases} -\alpha^{-1} - \frac{i}{2}z^{-1/2} - \frac{i}{4}V(x)z^{-3/2} + o(z^{-3/2}), & \nu = 1 \\ (2\pi)^{-1}\ln((-iz)^{1/2}) + \tilde{\alpha} - (4\pi)^{-1}V(x)z^{-1} + o(z^{-1}), & \nu = 2 \\ -i(4\pi)^{-1}z^{1/2} + \alpha + i(8\pi)^{-1}V(x)z^{-1/2} + o(z^{-1/2}), & \nu = 3 \end{cases}$$

(5.8)

where

$$\tilde{\alpha} = \alpha + (2\pi)^{-1}\gamma - (2\pi)^{-1}\ln(2)$$

with $\gamma = .5772\ldots$ being Euler's constant. A combination of (5.4), (5.5), and (5.8) then implies the following trace formula.

THEOREM 5.1. *[17]*

$\nu = 1$:

$$V(x) = \begin{cases} \lim_{z\downarrow-\infty}\left\{-z - 2\int_{\inf[\sigma(H)]}^{\infty} d\lambda z^2(\lambda - z)^{-2}\xi_{\infty,x}(\lambda)\right\}, & \alpha = \infty, \\[2mm] \frac{1}{6}\alpha^2 + \lim_{z\downarrow-\infty}\left\{-\frac{2}{3}z + \frac{i}{3}\alpha z^{1/2}\right. \\[2mm] \qquad \left. + \frac{8i}{3}\alpha^{-1}z^{5/2}\int_{E_{\alpha,x,0}}^{\infty} d\lambda(\lambda - z)^{-2}\xi_{\alpha,x}(\lambda)\right\}, & \alpha \in \mathbb{R}\backslash\{0\} \end{cases}$$

(5.9)

$\nu = 2$:

$$\begin{aligned} V(x) &= \lim_{z\downarrow-\infty}\left\{-z + 4\pi[(2\pi)^{-1}\ln(-iz^{1/2}) + \tilde{\alpha}]\right. \\[2mm] &\qquad \left. \times \int_{E_{\alpha,x,0}}^{\infty} d\lambda z^2(\lambda - z)^{-2}\xi_{\alpha,x}(\lambda)\right\}. \end{aligned}$$

(5.10)

$\nu = 3$:

$$\begin{aligned} V(x) &= 16\pi^2\alpha^2 + \lim_{z\downarrow-\infty}\left\{\quad - z + 4\pi i\alpha z^{1/2}\right. \\[2mm] &\qquad \left. + 2\int_{E_{\alpha,x,0}}^{\infty} d\lambda z^2(\lambda - z)^{-2}\xi_{\alpha,x}(\lambda)\right\}. \end{aligned}$$

(5.11)

Using the systematic high-energy expansion of $\lim_{|\epsilon|\downarrow 0}[G(z,x,x+\epsilon) - G^{(0)}(z,x,x+\epsilon)]$ in terms of (multi-dimensional) KdV invariants (see, e.g., [35], [41]) one can extend Theorem 5.1 to higher-order trace relations in analogy to 2.29 and 2.55.

In the special case where $V^{(0)} \equiv 0$, one obtains explicitly,

$$(5.12) \qquad D_\alpha^{(0)}(z) = \begin{cases} -\alpha^{-1} - (-4z)^{-1/2}, & \nu = 1 \\[1mm] \tilde{\alpha} + (2\pi)^{-1}\ln((-z)^{1/2}), & \nu = 2 \\[1mm] \tilde{\alpha} + (4\pi)^{-1}(-z)^{1/2}, & \nu = 3, \end{cases}$$

and

$$(5.13) \qquad \text{Tr}[(H_{\alpha,x}^{(0)} - z)^{-1} - (H^{(0)} - z)^{-1}] = -\int_{E_{\alpha,0}^{(0)}}^{\infty} d\lambda \frac{\xi_\alpha^{(0)}(\lambda)}{(\lambda - z)^2}.$$

Here, for $\nu = 1$,

$$\xi_\alpha^{(0)}(\lambda) = \begin{cases} 0, & \lambda < -\alpha^2/4 \\ -1, & -\alpha^2/4 < \lambda < 0 \\ a_\alpha(\lambda), & \lambda > 0 \end{cases} \begin{cases} 0, & \lambda < 0 \\ \frac{1}{2}, & \lambda > 0 \end{cases} \begin{cases} 0, & \lambda < 0 \\ 1 + a_\alpha(\lambda), & \lambda > 0 \end{cases}$$

$$\alpha < 0 \qquad\qquad\qquad \alpha = 0 \qquad\qquad\quad \alpha \in (0, \infty]$$

(5.14)

writing $a_\alpha(\lambda) = -\pi^{-1}\arctan(|\alpha|/2\lambda^{1/2})$, and, for $\nu = 2$,

$$\xi_\alpha^{(0)}(\lambda) = \begin{cases} 0, & \lambda < -e^{-4\pi\tilde\alpha} \\ -1, & -e^{-4\pi\tilde\alpha} < \lambda \le 0 \\ -\pi^{-1}\arctan[\pi/(4\pi\tilde\alpha + \ln(\lambda))] - 1, & 0 \le \lambda \le e^{-4\pi\tilde\alpha} \\ -\pi^{-1}\arctan[\pi/(4\pi\tilde\alpha + \ln(\lambda))], & \lambda \ge e^{-4\pi\tilde\alpha} \end{cases}$$

(5.15)

and, finally, for $\nu = 3$,

$$\xi_\alpha^{(0)}(\lambda) = \begin{cases} 0, & \lambda < -(4\pi\alpha)^2 \\ -1, & -(4\pi\alpha)^2 < \lambda < 0 \\ A_\alpha(\lambda), & \lambda > 0 \end{cases} \begin{cases} 0, & \lambda < 0 \\ -\frac{1}{2}, & \lambda > 0 \end{cases} \begin{cases} 0, & \lambda < 0 \\ A_\alpha(\lambda), & \lambda > 0 \end{cases}$$
$$\qquad\qquad \alpha < 0 \qquad\qquad\qquad \alpha = 0 \qquad\qquad \alpha > 0$$

(5.16)

writing $A_\alpha(\lambda) = -\pi^{-1}\arctan(\lambda^{1/2}/4\pi|\alpha|)$, and

(5.17) $\quad E_{\alpha,0}^{(0)} = \begin{cases} -\alpha^2/4, & \alpha < 0 \\ 0, & \alpha \in [0,\infty] \end{cases} \begin{cases} -e^{-4\pi\tilde\alpha} \end{cases} \begin{cases} -(4\pi\alpha)^2, & \alpha < 0 \\ 0, & \alpha \ge 0 \end{cases}.$
$$\qquad\qquad\qquad \nu = 1 \qquad\qquad\quad \nu = 2 \qquad\qquad \nu = 3$$

6. A uniqueness result for three-dimensional Schrödinger Operators. Finally, we briefly sketch a uniqueness result in the context of three-dimensional Schrödinger operators with spherically symmetric potentials originally derived in [19]. Consider the potential $V : \mathbb{R}^3 \to \mathbb{R}$,

(6.1) $\qquad V(x) = v(|x|), \qquad v \in L^1([0, R)])$ for all $R > 0$

and define the self-adjoint Schrödinger operator H in $L^2(\mathbb{R}^3)$ associated with the differential expression $-\Delta + v(|x|)$ by decomposition with respect to angular momenta. This represents H as an infinite direct sum of half-line operators in $L^2((0,\infty); r^2 dr)$ associated with differential expressions of the type

(6.2) $\quad \hat\tau_\ell = -\dfrac{d^2}{dr^2} - \dfrac{2}{r}\dfrac{d}{dr} + \dfrac{\ell(\ell+1)}{r^2} + v(r), \qquad r = |x| > 0, \qquad \ell \in \mathbb{N}_0.$

A simple unitary transformation (see, e.g., [36], Appendix to Sect. X.1) reduces (6.2) to

(6.3) $$\tau_\ell = -\dfrac{d^2}{dr^2} + \dfrac{\ell(\ell+1)}{r^2} + v(r)$$

and associated Hilbert space $L^2((0,\infty); dr)$. Next, let $G(z, x, x')$, $x \ne x'$ denote the Green's function of H and define $H_{\alpha,0}$ in $L^2(\mathbb{R}^3)$, $\alpha \in \mathbb{R}$ as in (5.1) (with $x = 0$) and the corresponding Krein spectral shift function $\xi_{\alpha,0}(\lambda)$ as in (5.5), i.e.,

(6.4) $\qquad \xi_{\alpha,0}(\lambda) = \lim_{\epsilon\downarrow 0} \pi^{-1}\mathrm{Im}\{\ln[D_{\alpha,0}(\lambda + i\epsilon)]\} \qquad$ a.e.

Then the following uniqueness result holds.

THEOREM 6.1. *[19] Define* $H_j, H_{j,\alpha_j,0}$, $\alpha_j \in \mathbb{R}$ *associated with* $-\Delta +$ $v_j(|x|)$, $x \in \mathbb{R}^3$, $j = 1, 2$ *as above and introduce Krein's spectral shift function* $\xi_{j,\alpha_j,0}(\lambda)$ *for the pair* $(H_{j,\alpha_j,0}, H_j)$, $j = 1, 2$. *Then the following are equivalent:*

(i) $\xi_{1,\alpha_1,0}(\lambda) = \xi_{2,\alpha_2,0}(\lambda)$ *for a.e.* $\lambda \in \mathbb{R}$.

(ii) $\alpha_1 = \alpha_2$ *and* $V_1(x) = V_2(x)$ *for a.e.* $x \in \mathbb{R}^3$.

The proof of this result in [19] is based on detailed Weyl-m-function investigations associated with the angular momentum channel $\ell = 0$.

Acknowledgments. We are indebted to B. Simon and Z. Zhao for joint collaborations which led to most of the results presented in this contribution and to B. M. Levitan for discussions which inspired Section 3. We are also grateful to G. Stolz for discussions on matrix-valued differential expressions. F.G. would like to thank A. Friedman and the Institute for Mathematics and its Applications, IMA, University of Minnesota, USA and the Department of Mathematical Sciences, NTH, University of Trondheim, Norway for the great hospitality extended to him during a month long stay at each institution in the summer of 1995. F.G. also thanks B. Simon and D. Truhlar for their kind invitation to the IMA workshop "Multiparticle Quantum Scattering with Applications to Nuclear, Atomic and Molecular Physics". Moreover, we are both indebted to T. Hoffmann-Ostenhof for the kind invitation to the Erwin Schrödinger International Institute for Mathematical Physics, ESI, Vienna, Austria for a period in June and July of 1995. Financial support by the IMA, the Norwegian Research Council, and the ESI is gratefully acknowledged.

REFERENCES

[1] Z. S. Agranovich and V. A. Marchenko, *The Inverse Problem of Scattering Theory*, Gordon and Breach, New York, 1963.

[2] S. Albeverio, F. Gesztesy, R. Høegh-Krohn, and H. Holden, *Solvable Models in Quantum Mechanics*, Springer, New York, 1988.

[3] N. Aronszajn and W. F. Donoghue, On exponential representations of analytic functions in the upper half-plane with positive imaginary part, *J. Anal. Math.*, 5: 321–388 (1957).

[4] R. Banuelos and A. Sa Barreto, On the heat trace of Schrödinger operators, preprint 1995.

[5] R. W. Carey, A unitary invariant for pairs of self-adjoint operators, *J. reine angew. Math.*, *283*: 294–312 (1976).

[6] W. Craig, The trace formula for Schrödinger operators on the line, *Commun. Math. Phys.*, *126*: 379–407 (1989).

[7] P. Deift and E. Trubowitz, Inverse scattering on the line, *Commun. Pure Appl. Math.*, *32*: 121–251 (1979).

[8] L. A. Dikii, Trace formulas for Sturm-Liouville differential operators, *Amer. Math. Soc. Transl. Ser. (2), 18:* 81–115 (1961).

[9] B. A. Dubrovin, Periodic problems for the Korteweg-de Vries equation in the class of finite band potentials, *Funct. Anal. Appl., 9:* 215–223 (1975).

[10] N. Dunford and J. T. Schwartz, *Linear Operators II: Spectral Theory, Self-Adjoint Operators in Hilbert Space,* Interscience, New York, 1988.

[11] H. Flaschka, On the inverse problem for Hill's operator, *Arch. Rat. Mech. Anal., 59:* 293–309 (1975).

[12] I. M. Gelfand, On identities for the eigenvalues of a second-order differential operator, *Uspehi Mat. Nauk, 11:1:* 191–198 (1956) (Russian). English translation in *Izrail M. Gelfand, Collected Papers Vol. I* (S. G. Gindikin, V. W. Guillemin, A. A. Kirillov, B. Kostant, S. Sternberg, eds.), Springer, Berlin, 1987, pp. 510–517.

[13] I. M. Gelfand and L. A. Dikii, Asymptotic behaviour of the resolvent of Sturm-Liouville equations and the algebra of the Korteweg-de Vries equations, *Russ. Math. Surv., 30:5:* 77–113 (1975).

[14] I. M. Gelfand and B. M. Levitan, On a simple identity for the eigenvalues of a second-order differential operator, *Dokl. Akad. Nauk SSSR, 88:* 593–596 (1953) (Russian). English translation in *Izrail M. Gelfand, Collected Papers Vol. I* (S. G. Gindikin, V. W. Guillemin, A. A. Kirillov, B. Kostant, S. Sternberg, eds.), Springer, Berlin, 1987, pp. 457–461.

[15] F. Gesztesy, New trace formulas for Schrödinger operators, *Evolution Equations,* G. Ferreyra, G. Ruiz Goldstein, F. Neubrander (eds.), Marcel Dekker, New York, 1995, pp. 201–221.

[16] F. Gesztesy and H. Holden, Trace formulas and conservation laws for nonlinear evolution equations, *Rev. Math. Phys., 6:* 51–95 (1994).

[17] F. Gesztesy and H. Holden, On new trace formulae for Schrödinger operators, *Acta Appl. Math., 39:* 315–333 (1995).

[18] F. Gesztesy and B. Simon, The ξi function, *Acta Math.,* to appear.

[19] F. Gesztesy and B. Simon, Uniqueness theorems in inverse spectral theory for one-dimensional Schrödinger operators, *Trans. Amer. Math. Soc., 348:* 349–373 (1996).

[20] F. Gesztesy and K. Unterkofler, Isospectral deformations for Sturm-Liouville and Dirac-type operators and associated nonlinear evolution equations, *Rep. Math. Phys., 31:* 113–137 (1992).

[21] F. Gesztesy, H. Holden, and B. Simon, Absolute summability of the trace relation for certain Schrödinger operators, *Commun. Math. Phys., 168:* 137–161 (1995).

[22] F. Gesztesy, R. Ratnaseelan, and G. Teschl, The KdV hierarchy and associated trace formulas, to appear in the proceedings of the International Conference on Applications of Operator Theory (I. Gohberg, P. Lancaster, P. N. Shivakumar, eds.), *Operator Theory: Advances and Applications,* Birkhäuser.

[23] F. Gesztesy, H. Holden, B. Simon, and Z. Zhao, Trace formulae and inverse spectral theory for Schrödinger operators, *Bull. Amer. Math. Soc., 29:* 250–255 (1993).

[24] F. Gesztesy, H. Holden, B. Simon, and Z. Zhao, Higher order trace relations for Schrödinger operators, *Rev. Math. Phys., 7:* 893–922 (1995).

[25] F. Gesztesy, H. Holden, B. Simon, and Z. Zhao, A trace formula for multidimensional Schrödinger operators, *J. Funct. Anal.,* to appear.

[26] H. Hochstadt, On the determination of a Hill's equation from its spectrum, *Arch. Rat. Mech. Anal., 19:* 353–362 (1965).

[27] S. Kotani and M. Krishna, Almost periodicity of some random potentials, *J. Funct. Anal., 78:* 390–405 (1988).

[28] M. G. Krein, Perturbation determinants and a formula for the traces of unitary and self-adjoint operators, *Sov. Math. Dokl., 3:* 707–710 (1962).

[29] P. D. Lax, Trace formulas for the Schroedinger operator, *Commun. Pure Appl.*

Math., 47: 503–512 (1994).

[30] B. M. Levitan, On the closure of the set of finite-zone potentials, Math. USSR Sbornik, 51: 67–89 (1985).

[31] B. M. Levitan, Inverse Sturm-Liouville Problems, VNU Science Press, Utrecht, 1987.

[32] V. A. Marchenko, Sturm-Liouville Operators and Applications, Birkhäuser, Basel, 1986.

[33] H. P. McKean and P. van Moerbeke, The spectrum of Hill's equation, Invent. Math., 30: 217–274 (1975).

[34] V. Papanicolaou, Trace formulas and the behavior of large eigenvalues., SIAM J. Math. Anal., 26: 218–237 (1995).

[35] A. M. Perelomov, Schrödinger equation spectrum and Korteweg–de Vries type invariants, Ann. Inst. H. Poincaré, Sect. A, 24: 161–164 (1976).

[36] M. Reed and B. Simon, Methods of Modern Mathematical Physics, II. Fourier Analysis, Self-Adjointness, Academic Press, New York, 1975.

[37] B. Simon, Spectral analysis of rank one perturbations and applications, Proc. Mathematical Quantum Theory II: Schrödinger Operators, J. Feldman, R. Froese, L. M. Rosen (eds.), CRM Proceedings and Lecture Notes Vol. 8, Amer. Math. Soc., Providence, RI, 1995, pp. 109–149.

[38] E. Trubowitz, The inverse problem for periodic potentials, Commun. Pure Appl. Math., 30: 321–337 (1977).

[39] S. Venakides, The infinite period limit of the inverse formalism for periodic potentials, Commun. Pure Appl. Math., 41: 3–17 (1988).

[40] J. Weidmann, Spectral Theory of Ordinary Differential Operators, Lecture Notes in Mathematics, Vol. 1258, Springer, Berlin, 1987.

[41] S. F. J. Wilk, Y. Fujiwara, and T. A. Osborn, N-body Green's function and their semiclassical expansion, Phys. Rev., 24A: 2187–2202 (1981).

[42] J. Zorbas, Perturbation of self-adjoint operators by Dirac distributions, J. Math. Phys., 21: 840–847 (1980).

MULTIPARTICLE QUANTUM SYSTEMS IN CONSTANT MAGNETIC FIELDS*

I. ŁABA†

Abstract. This article is an introduction to the spectral and scattering theory of Hamiltonians of quantum N-particle systems in a constant magnetic field, developed recently by C. Gérard and the author. For such systems, both the effects due to the interactions between particles and those caused by the magnetic field play an important role, and it is the interplay between them that accounts for the richness and complexity of the theory.

We begin with a detailed discussion of the basic properties of the magnetic Hamiltonians, including an analogue of the center of mass separation and the characteristics of bound states, and then proceed to present, in the simplest possible settings, the main ideas in the proof of asymptotic completeness for wide classes of systems.

1. Introduction. This article is an introduction to the spectral and scattering theory of Hamiltonians of N-particle systems in a constant magnetic field, developed in [GL1–3]. It is meant to be accessible to scattering theorists who have not thought about magnetic fields, as well as to those mathematicians who wish to learn more about multiparticle magnetic Hamiltonians, but have not worked on the problem of asymptotic completeness, do not necessarily intend to do so, and are not familiar with the relevant notation, "standard arguments", and folklore.

The subject of our interest is a Hamiltonian of the form:

$$H = \frac{1}{2}(D - A)^2 + V.$$

where $D = -i\nabla$, A is a vector potential corresponding to a uniform magnetic field $\vec{B}$, and V is a scalar (electric) potential. For multiparticle systems, $V = \sum V_{ij}$ is the sum of the pair potentials.

Neither an N-particle potential V nor a constant magnetic field can be treated as small perturbation of the free Hamiltonian $H_0 = \frac{1}{2}D^2$ — introducing either one of them changes the structure of the essential spectrum in a fundamental way, and the behaviour of the physical system in question will be qualitatively different. Thus, for systems of N interacting particles one has to deal with the so-called *multichannel scattering*, in which the different scattering channels correspond to subsets of the system forming bound states. Although partial results were available earlier, it was only in the last two decades that the multiparticle scattering theory underwent a period of rapid development, which culminated in the proof of asymptotic completeness for N-body systems.

The spectral properties of the Hamiltonian $\frac{1}{2}(D - A)^2$ of a single charged particle in a constant magnetic field were analyzed in the 1920's by

* Supported by NSF under Grant DMS-9501033.
† Department of Mathematics, UCLA, Los Angeles, CA 90095 U.S.A.

L. Landau. They are rather different from those of the free Hamiltonian, and reflect the well known from physics fact that the dynamics of such a particle is a superposition of free propagation along the field and stable bounded motion in the directions transversal to it.

In [GL1-3], we considered the case of N-particle systems in 3 dimensions in a constant magnetic field $\vec{B}$. As it often happens, the whole turns out to be more than merely the sum of its ingredients. Compared to either one of the cases $\vec{B} \equiv 0$ or $V \equiv 0$, the multiparticle magnetic Schrödinger operators display a richer and more complex structure, due to the interplay between the geometry of the potential and of the magnetic field.

This paper begins with a detailed discussion of the basic properties of magnetic Hamiltonians analyzed in [GL1], such as the reduction procedure (an analogue of the center of mass separation for standard N-body systems), and the characteristics of bound states. Some of the explanations (especially those in Sections 2 and (3) would be unnecessary or even inappropriate in a research article; however, I did not think that they should be omitted in a paper addressed to a wider audience.

The main new ideas of [GL2] and [GL3] are described in Sections 6, 9, and 10. In an attempt to keep the forest from being obscured by the trees, I present them in the simplest possible settings. For instance, the assumption (V) is rather restrictive – in [GL1-3] we prove our results for a much larger class of potentials. The center of orbit observable is introduced and used for the first time for a 1-particle Hamiltonian (Section 6), the geometry of charged systems is explained in the 2-body case (Section 10), and the proof of the Mourre estimate in the dispersive case is given under the assumption that the spectrum of the reduced Hamiltonians has a particularly simple structure (Section 9). Moreover, a number of technical arguments has been omitted.

The notation in this paper is usually as consistent with that of [GL1-3] as possible, with the obvious exception of places where those articles are not consistent with one another in that respect. However, there is one conscious departure from a convention used both in [GL1-3] and generally in the literature – I replaced the subscripts and superscripts "a_{max}" and "a_{min}" by "$\Diamond$" and "$\sharp$". I found the latter esthetically more appealing, and I hope that the reader will agree on that.

We found several errors in [GL1] after its publication; in this presentation they are pointed out and corrected.

In our work we adopt the time-dependent approach to proving asymptotic completeness, which goes back to the work of Enss [E1], [E2], and has been particularly successful in the N-body scattering. Together with the method of positive commutators (discussed in Appendix A), it led to the proof of asymptotic completeness for N-body systems ([SS1], [Gr], [De1], [SS5]; see [DG] for a comprehensive review of the relevant literature).

Prior to our work ([GL1-3]), scattering theory for multiparticle systems in a constant magnetic field was almost nonexistent. Asymptotic completeness for a single particle in a constant magnetic field and an external short-range or Coulomb potential was shown in [AHS1] ; a different proof was given in [S]. The separation of the center of mass in the presence of a magnetic field was considered in [AHS2] by Avron, Herbst and Simon, who also worked out explicitly the spectral properties of the reduced Hamiltonians in the 2-body case. The general long-range one-body problem and long-range two-particle systems with total charge zero were studied in [L1], [L2], and [I1]. In [I2], Iwashita obtained the Mourre estimate for reduced 3−particle Hamiltonians in certain special cases. Localization of the essential spectrum for N-particle Hamiltonians with a constant magnetic field in the "charged" case was discussed in the recent article [VZ]. There are also several papers on one-particle scattering in a magnetic field vanishing at infinity ([BP], [C], [E3], [LT1,2], [Ni], [NR]). Asymptotic completeness for N-body Stark Hamiltonians was proved recently in [HMS1], [HMS2], and E. Skibsted [Sk] considered subsequently the case of 3-particle systems in constant electric and magnetic fields.

2. Notation. The Hamiltonian of a system of N interacting particles of masses m_i and electric charges q_i in a constant magnetic field $\vec{B} = (0, 0, 2b)$, $b > 0$, in $\mathbb{R}^3$ is given by:

$$(2.1) \qquad H = \sum_1^N \frac{1}{2m_i}(D_i - q_i J x_i)^2 + \sum_{i<j} V_{ij}(x_i - x_j),$$

where J is the vector potential associated with the field. We will assume that $b > 0$ and use the transversal gauge in which J is the skew-symmetric matrix

$$(2.2) \qquad J = \begin{pmatrix} 0 & -b & 0 \\ b & 0 & 0 \\ 0 & 0 & 0 \end{pmatrix}.$$

We consider systems such that all of the particles are charged, *i.e.*, $q_i \neq 0$ for $1 \leq i \leq N$.

The coordinates in the configuration space $X = \mathbb{R}^{3N}$ will be denoted by

$$x = (x_1, \ldots, x_N), \quad x_i \in \mathbb{R}^3.$$

The magnetic field $\vec{B}$ "does not act" in the direction parallel to it: the Lorenz force $\vec{F} = q\vec{v} \times \vec{B}$ is always perpendicular to the field. Correspondingly, the dynamics of the systems in the plane transversal to the field will be entirely different from that along the third axis. It is therefore natural to introduce notation which emphasises that distinction. We will write:

$$x_i = (y_i, z_i),$$

where $(y_i, z_i) = (y_i^{(1)}, y_i^{(2)}, z_i) \in \mathbb{R}^2 \times \mathbb{R}$ are the coordinates of the i-th particle in the plane transversal to $\vec{B}$ and along the direction of $\vec{B}$ respectively.

Let:

$$Y = \{x \in X \mid z_i = 0, \ 1 \le i \le N\},$$

$$Z = \{x \in X \mid y_i = 0, \ 1 \le i \le N\} = \{x \in X \mid Ax = 0\}.$$

Then $X = Y \oplus^\perp Z$, and the projections of a vector $x \in X$ onto these subspaces will be denoted by $y \in Y$ and $z \in Z$ respectively. We equip X with the scalar product

$$\langle x, \bar{x} \rangle = \sum_1^N m_i x_i \bar{x}_i.$$

The dual spaces to X, Y, Z will be denoted by X', Y', and Z'. The duality is given by

$$\langle \xi, x \rangle = \sum_1^N \xi_i x_i,$$

and the scalar product on X' is

$$\langle \xi, \bar{\xi} \rangle = \sum_1^N m_i^{-1} \xi_i \bar{\xi}_i,$$

The N-particle vector potential A is defined by:

$$Ax = (q_1 J x_1, \ldots, q_N J x_N).$$

A is a antisymmetric mapping $A : X \to X'$, which we will also consider as an antisymmetric bilinear form on $X \times X$.

We can now rewrite H as

$$(2.3) \quad H = \frac{1}{2}(D - Ax)^2 + V(x) = \frac{1}{2}D_z^2 + \frac{1}{2}(D_y - Ay)^2 + V(x),$$

where

$$V(x) = \sum_{i<j} V_{ij}(x_i - x_j).$$

For simplicity, we will assume that the pair potentials satisfy

$$|D^\alpha V_{ij}(x_i - x_j)| \le C_\alpha (1 + |x_i - x_j|)^{-|\alpha|-\mu} \text{ for all } \alpha, \ |\alpha| \ge 0, \quad (V)$$

for some $\mu > 0$. This hypothesis is much more restrictive in terms of regularity than those used in [GL1–3], where, in particular, we allow Coulomb singularities. Kato's inequality (see *e.g.*, [CFKS, Section 1.3]) implies that a multiplicative operator which is Δ-bounded with the relative bound a is also $\frac{1}{2}(D - Ax)^2$-bounded with a relative bound of at most a, and the arguments used to replace boundedness by relative boundedness are essentially the same as in the case $\vec{B} = 0$.

We put

$$V_{ij,L} = 0 \text{ if } \mu > 1, \ V_{ij,L} = V_{ij} \text{ if } 0 < \mu \le 1,$$

$$V_L = \sum_{i,j} V_{ij,L}.$$

The cases $\mu > 1$ and $0 < \mu \le 1$ will be referred to as short-range and long-range respectively. Selfadjointness of H follows immediately from that of $(D - Ax)^2$.

Let us introduce the center of mass coordinates. We can write

$$X = X_\diamond \oplus^\perp X^\diamond,$$

where

$$X_\diamond = \{x \in X \mid x_i = x_j \text{ for all } i, j\},$$

and

$$X^\diamond = \left\{ x \in X \mid \sum_{i=1}^N m_i x_i = 0 \right\},$$

are the configuration spaces of the center of mass and of the relative motion of the particles respectively. It is easy to check that $X_\diamond$ and $X^\diamond$ are orthogonal in the above scalar product. We will write:

$$Y_\diamond = X_\diamond \cap Y, \ Z_\diamond = X_\diamond \cap Z,$$

$$Y^\diamond = X^\diamond \cap Y, \ Z^\diamond = X^\diamond \cap Z.$$

Note that the projections onto $X_\diamond$, $X^\diamond$, and onto Y, Z commute, so that:

$$X_\diamond = Y_\diamond \oplus Z_\diamond, \ X^\diamond = Y^\diamond \oplus Z^\diamond.$$

The symbols

$$x_\diamond, \ y_\diamond, \ z_\diamond,$$

$$x^\diamond, \ y^\diamond, \ z^\diamond,$$

will stand for the orthogonal projections of $x \in X$ on the above spaces. The spaces dual to $X_\diamond$ and $X^\diamond$ are:

$$X'_\diamond = \left\{ \xi \in X' \mid \frac{\xi_i}{m_i} = \frac{\xi_j}{m_j} \text{ for all } i, j \right\},$$

$$X^{\Diamond\prime} = \left\{ x \in X \mid \sum_{i=1}^{N} \xi_i = 0 \right\}.$$

We have:

(2.4) $$|x|^2 = |x_\Diamond|^2 + |x^\Diamond|^2, |\xi|^2 = |\xi_\Diamond|^2 + |\xi^\Diamond|^2,$$

and

$$\langle \xi, x \rangle = \langle \xi_\Diamond, x_\Diamond \rangle + \langle \xi^\Diamond, x^\Diamond \rangle.$$

Spaces dual to $Y_\Diamond$, $Z_\Diamond$, ..., will be denoted by $Y'_\Diamond$, $Z'_\Diamond$, etc.

The symbols $A^\Diamond{}_\Diamond$, $A_\Diamond{}^\Diamond$, $A_{\Diamond\Diamond}$, $A^{\Diamond\Diamond}$ will stand for the restrictions of the bilinear form A to $X^\Diamond \times X_\Diamond$, $X_\Diamond \times X^\Diamond$, $X_\Diamond \times X_\Diamond$, $X^\Diamond \times X^\Diamond$ respectively ($e.g.$, $A^\Diamond{}_\Diamond x_\Diamond = (Ax_\Diamond)^\Diamond$, where the first projection is taken in X and the second one in X').

In order to carry out the calculations in the next section, we introduce the following notation. We will write

$$x = \begin{pmatrix} y^{(1)} \\ y^{(2)} \\ z \end{pmatrix}, \ y^{(i)} = \begin{pmatrix} y_1^{(i)} \\ y_2^{(i)} \\ \vdots \\ y_N^{(i)} \end{pmatrix}, \ z = \begin{pmatrix} z_1 \\ z_2 \\ \vdots \\ z_N \end{pmatrix}$$

(note the order of coordinates). In this basis, we have:

(2.5) $$x_\Diamond = \begin{pmatrix} T & 0 & 0 \\ 0 & T & 0 \\ 0 & 0 & T \end{pmatrix} x, \ \xi_\Diamond = \begin{pmatrix} S & 0 & 0 \\ 0 & S & 0 \\ 0 & 0 & S \end{pmatrix} \xi,$$

(2.6) $$x^\Diamond = x - x_\Diamond, \ \xi^\Diamond = \xi - \xi_\Diamond,$$

where

$$T = \frac{1}{M} \begin{pmatrix} m_1 & m_2 & \cdots & m_N \\ m_1 & m_2 & \cdots & m_N \\ \vdots & \vdots & \ddots & \vdots \\ m_1 & m_2 & \cdots & m_N \end{pmatrix} = \Omega\mathcal{M},$$

$$S = T^T = \frac{1}{M} \begin{pmatrix} m_1 & m_1 & \cdots & m_1 \\ m_2 & m_2 & \cdots & m_2 \\ \vdots & \vdots & \ddots & \vdots \\ m_N & m_N & \cdots & m_N \end{pmatrix} = \mathcal{M}\Omega,$$

$$M = \frac{1}{M} \begin{pmatrix} m_1 & 0 & \cdots & 0 \\ 0 & m_2 & \cdots & 0 \\ \vdots & \vdots & \ddots & \vdots \\ 0 & 0 & \cdots & m_N \end{pmatrix}, \quad \Omega = \begin{pmatrix} 1 & 1 & \cdots & 1 \\ 1 & 1 & \cdots & 1 \\ \vdots & \vdots & \ddots & \vdots \\ 1 & 1 & \cdots & 1 \end{pmatrix},$$

and $M = \sum_{i=1}^{N} m_i$. Indeed, it is easy to check that $x_\diamond$, $x^\diamond$, $\xi_\diamond$, and $\xi^\diamond$, given by (2.5) and (2.6), are in $X_\diamond$, $X^\diamond$, $X'_\diamond$, and $X^{\diamond'}$ respectively.

The matrix of the mapping $A : X \to X'$ in the above bases is

$$(2.7) \quad A = \begin{pmatrix} 0 & -Q & 0 \\ Q & 0 & 0 \\ 0 & 0 & 0 \end{pmatrix}, \quad \text{where } Q = b \begin{pmatrix} q_1 & 0 & \cdots & 0 \\ 0 & q_2 & \cdots & 0 \\ \vdots & \vdots & \ddots & \vdots \\ 0 & 0 & \cdots & q_N \end{pmatrix}.$$

Finally, $F(\cdot \in \Omega)$ will stand for a smoothed out characteristic function of Ω, equal to 0 outside Ω and to 1 in a slightly smaller set.

3. Center of mass separation. Let

$$K = D + Ax$$

be the generator of the *magnetic translations*:

$$e^{i(x',K)} u(x) = e^{i(x',Ax)} u(x + x'), \quad \forall x' \in X.$$

It is easy to check that

$$(3.1) \qquad \left[K, \frac{1}{2}(D - Ax)^2 \right] = 0.$$

One deduces from (3.1) and from the translational invariance of the potential:

$$V(x + x') = V(x), \quad x' \in X_\diamond,$$

that the total pseudomomentum

$$K_\diamond := D_{x_\diamond} + (Ax)_\diamond$$

satisfies $[H, iK_\diamond] = 0$.

Due to the invariance of H under magnetic translations, the spectrum of H is infinitely degenerate. In the next two subsections we describe a procedure, analogous to the separation of the center of mass motion for standard N-body Hamiltonians (without a magnetic field), which leads to the removal of that degeneracy. We will use that the pseudomomentum $K_\diamond$ is a constant of motion, and the details of the construction will depend on the properties of $K_\diamond$.

Let $Q = \sum q_i$ be the total electric charge of the system. We have

$$\left[\sum_{i=1}^{N}(D_{y_i^{(1)}} - q_i by_i^{(2)}), \sum_{i=1}^{N}(D_{y_i^{(2)}} + q_i by_i^{(1)})\right] = 2Qb.$$

Hence the two orthogonal to the field components of $K_\diamond$ commute if $Q = 0$, and have (up to a multiplicative factor) the Heisenberg commutation relations if $Q \neq 0$. In [GL1], we construct a unitary mapping $U_\diamond$ such that

$$U_\diamond K_\diamond U_\diamond^* = D_{x_\diamond}, \text{ if } Q = 0,$$

and

$$U_\diamond K_\diamond U_\diamond^* = (\kappa^{-1}D_{y_\diamond^{(1)}}, \kappa^{-1}y_\diamond^{(1)}, D_{z_\diamond}), \text{ if } Q > 0,$$

where the factor $\kappa > 0$ is given explicitly in Lemma 3.2 (if $Q < 0$, we need to interchange the order of $D_{y_\diamond^{(1)}}$ and $y_\diamond^{(1)}$). $U_\diamond$ will be the unitary operator *implementing* the symplectic transformation constructed in Subsection 3.1.

In the neutral case ($Q = 0$), the transformed Hamiltonian $\tilde{H}_\diamond = U_\diamond H U_\diamond$ will commute with $D_{x_\diamond}$ (it will not depend explicitly on $x_\diamond$). We will write it as a direct integral

$$\tilde{H}_\diamond = \int^\oplus \tilde{H}(\xi_\diamond)d\xi_\diamond.$$

The reduced Hamiltonian will be $\tilde{H}(\xi_\diamond)$ with the value of $\xi_\diamond$ fixed, acting on $L^2(X^\diamond)$. Note that the spectrum of it will depend on $\xi_\diamond$.

If the system is charged, $\tilde{H}_\diamond$ will not depend explicitly either on $y_\diamond^{(1)}$ or on $D_{y_\diamond^{(1)}}$, so that we can consider it as an operator on $Y_\diamond^{(2)} \times Z_\diamond \times X^\diamond$. This, combined with the separation of the center of mass along the z axis (which in both cases is identical to that for standard, *i.e.*, without a magnetic field, N-body systems), removes the degeneracy of the spectrum of H. In the charged case the spectrum of the reduced Hamiltonian will not depend on the transversal to the field components of the pseudomomentum.

3.1. Symplectic transformations.
LEMMA 3.1. *If $Q = 0$, the following linear transformation is symplectic:*

(3.2)
$$\chi_\diamond : T^*X \to T^*X$$
$$(x, \xi) \mapsto (\tilde{x}, \tilde{\xi}),$$

where

$$
(3.3) \quad
\begin{cases}
\tilde{\eta}_\diamond = \eta_\diamond + (Ax)_\diamond, \\[4pt]
\tilde{\zeta}_\diamond = \zeta_\diamond, \\[4pt]
\tilde{y}_\diamond = y_\diamond, \\[4pt]
\tilde{z}_\diamond = z_\diamond, \\[4pt]
\tilde{\xi}^\diamond = \xi^\diamond - A^\diamond{}_\diamond x_\diamond, \\[4pt]
\tilde{x}^\diamond = x^\diamond.
\end{cases}
$$

LEMMA 3.2. *Assume that the system is charged, and suppose that* $Q > 0$. *Let* $\nu = \sum_{i=1}^{N} m_i^2 / M^2$ *and* $\kappa = (2\nu b |Q|)^{-1/2}$. *Then the following transformation is symplectic:*

$$
(3.4) \quad
\begin{aligned}
\chi_\diamond : T^*X &\to T^*X \\
(x, \xi) &\mapsto (\tilde{x}, \tilde{\xi}),
\end{aligned}
$$

where

$$
(3.5) \quad
\begin{cases}
(\tilde{\eta}_\diamond^{(1)}, \tilde{y}_\diamond^{(1)}) = \kappa(\eta_\diamond + (Ax)_\diamond), \\[4pt]
(\tilde{\eta}_\diamond^{(2)}, \tilde{y}_\diamond^{(2)}) = \kappa(\eta_\diamond - A_\diamond{}_\diamond x_\diamond + A_\diamond{}^\diamond x^\diamond), \\[4pt]
\tilde{z}_\diamond = z_\diamond, \\[4pt]
\tilde{\zeta}_\diamond = \zeta_\diamond, \\[4pt]
\tilde{\xi}^\diamond = \xi^\diamond - A^\diamond{}_\diamond x_\diamond, \\[4pt]
\tilde{x}^\diamond = x^\diamond.
\end{cases}
$$

If $Q < 0$, *the above transformation with* $\tilde{y}_\diamond$ *and* $\tilde{\eta}_\diamond$ *interchanged is symplectic.*

Proof of Lemmas 3.1 and 3.2. Let us rewrite the above transformations in coordinates. If $Q = 0$ (the case considered in Lemma 3.1), we have

$$\begin{cases} S\tilde{\eta}^{(1)} = S\eta^{(1)} - SQy^{(2)}, \\ S\tilde{\eta}^{(2)} = S\eta^{(2)} + SQy^{(1)}, \\ T\tilde{y}^{(i)} = Ty^{(i)}, \\ (1 - S)\tilde{\eta}^{(1)} = (1 - S)\eta^{(1)} + (1 - S)QTy^{(2)}, \\ (1 - S)\tilde{\eta}^{(2)} = (1 - S)\eta^{(2)} - (1 - S)QTy^{(1)}, \\ (1 - T)\tilde{x} = (1 - T)x, \\ \tilde{\zeta} = \zeta, \tilde{z} = z. \end{cases}$$

If, as in Lemma 3.2, $Q > 0$ (the case $Q < 0$ is similar), we have

$$\begin{cases} S\tilde{\eta}^{(1)} = \kappa(S\eta^{(1)} - SQy^{(2)}), \\ T\tilde{y}^{(1)} = \kappa(S\eta^{(2)} + SQy^{(1)}), \\ S\tilde{\eta}^{(2)} = \kappa(S\eta^{(2)} - SQTy^{(1)} + SQ(1 - T)y^{(1)}), \\ T\tilde{y}^{(2)} = \kappa(S\eta^{(1)} + SQTy^{(2)} - SQ(1 - T)y^{(2)}), \\ (1 - S)\tilde{\eta}^{(1)} = (1 - S)\eta^{(1)} + (1 - S)QTy^{(2)}, \\ (1 - S)\tilde{\eta}^{(2)} = (1 - S)\eta^{(2)} - (1 - S)QTy^{(1)}, \\ (1 - T)\tilde{x} = (1 - T)x, \\ \tilde{\zeta} = \zeta, \tilde{z} = z. \end{cases}$$

To prove that the above transformations are symplectic, it suffices to show that they preserve the Poisson brackets involving $y^{(i)}$ and $\eta^{(i)}$. We first note that if $A = (a_{ij})$ and $B = (b_{ij})$ are two $N \times N$ matrices,

$$\{A\eta^{(i)}, By^{(i)}\} = \sum_{i,j} a_{ij}b_{ij}, \quad \{A\eta^{(i)}, By^{(j)}\} = 0 \text{ if } i \neq j.$$

Using this formula, we verify the identities:

$$\{S\eta^{(i)}, Ty^{(i)}\} = 1,$$

$$\{S\eta^{(i)}, (1-T)y^{(i)}\} = \{(1-S)\eta^{(i)}, Ty^{(i)}\} = 0,$$

$$\{S\eta^{(i)}, Q(1-T)y^{(i)}\} = \{(1-S)\eta^{(i)}, QTy^{(i)}\} = 0,$$

$$\{S\eta^{(i)}, SQy^{(i)}\} = \{S\eta^{(i)}, SQTy^{(i)}\} = \{\eta^{(i)}, SQTy^{(i)}\} = \nu b Q,$$

$$\{S\eta^{(i)}, QTy^{(i)}\} = \{\eta^{(i)}, SQy^{(i)}\} = \tfrac{b}{M}\sum m_i q_i.$$

From this, the invariance of the Poisson bracket under the above transformations follows by direct computation. $\square$

Let us now check how the symbol of the Hamiltonian is transformed by the above mappings.

Case $Q = 0$. If $Q = 0$, we have

(3.6) $$(Ax_\diamond)_\diamond = 0.$$

Indeed, this follows from (2.7) and the identity $SQT = 0$ if $Q = 0$, which can be verified by direct computation. (3.6) implies that $(Ax)_\diamond = (Ax^\diamond)_\diamond$. Using this and (2.4), we obtain that

$$|\xi - Ax|^2 = |\zeta|^2 + |\eta_\diamond - (Ax)_\diamond|^2 + |\eta^\diamond - (Ax)^\diamond|^2$$

$$= |\zeta|^2 + |\eta_\diamond + (Ax)_\diamond - 2(Ax^\diamond)_\diamond|^2 + |\eta^\diamond - (Ax_\diamond)^\diamond - (Ax^\diamond)^\diamond|^2$$

$$= |\tilde{\zeta}|^2 + |\tilde{\eta}_\diamond - 2(A\tilde{x}^\diamond)_\diamond|^2 + |\tilde{\eta}^\diamond - (A\tilde{x}^\diamond)^\diamond|^2.$$

Hence the symbol of H in the new coordinates is

(3.7) $$h(\tilde{\xi}, \tilde{x}) = \frac{1}{2}|\tilde{\zeta}|^2 + \frac{1}{2}|\tilde{\eta}_\diamond - 2(A\tilde{x}^\diamond)_\diamond|^2 + \frac{1}{2}|\tilde{\eta}^\diamond - (A\tilde{x}^\diamond)^\diamond|^2 + V(\tilde{x}^\diamond).$$

Case $Q > 0$. We compute:

$$|\xi - Ax|^2 = |\zeta|^2 + |\eta_\diamond - (Ax)_\diamond|^2 + |\eta^\diamond - (Ax)^\diamond|^2$$

$$= |\zeta|^2 + |\eta_\diamond - (Ax_\diamond)_\diamond + (Ax^\diamond)_\diamond - 2(Ax^\diamond)_\diamond|^2$$

$$+|\eta^\diamond - (Ax_\diamond)^\diamond - (Ax^\diamond)^\diamond|^2$$

$$= |\tilde{\zeta}|^2 + |\kappa^{-1}(\tilde{\eta}_\diamond^{(2)}, \tilde{y}_\diamond^{(2)}) - 2(A\tilde{x}^\diamond)_\diamond|^2 + |\tilde{\eta}^\diamond - (A\tilde{x}^\diamond)^\diamond|^2.$$

In this case, the symbol of H becomes

(3.8) $$h(\tilde{\xi}, \tilde{x}) = \frac{1}{2}|\tilde{\zeta}|^2 + R(\tilde{\eta}_\diamond^{(2)}, \tilde{y}_\diamond^{(2)}, \tilde{x}^\diamond) + \frac{1}{2}|\tilde{\eta}^\diamond - (A\tilde{x}^\diamond)^\diamond|^2,$$

where

(3.9) $$R(\tilde{\eta}_\diamond^{(2)}, \tilde{y}_\diamond^{(2)}, \tilde{x}^\diamond) = \frac{1}{2}|\kappa^{-1}(\tilde{\eta}_\diamond^{(2)}, \tilde{y}_\diamond^{(2)}) - 2(A\tilde{x}^\diamond)_\diamond|^2.$$

3.2. Reduced Hamiltonians. We will say that a unitary operator

$$U : L^2(X) \to L^2(X)$$

implements a linear symplectic transformation $\chi : T^*X \to T^*X$, if for any linear form $a(x, \xi)$ on T^*X one has

(3.10) $U^* a(x, D) U = (a \circ \chi)(x, D).$

It is well known that if (3.10) holds for linear a, the same is true for a much larger class of symbols if one uses the Weyl quantization, *e.g.*, we have

(3.11) $U^* a^w(x, D) U = (a \circ \chi)^w(x, D), \ \forall a \in S(T^*X),$

if

$$a^w(x, D) u(x) = (2\pi)^{-n} \int \int e^{i(x-y,\xi)} a\left(\frac{x+y}{2}, \xi\right) u(y) dy d\xi.$$

Any linear symplectic transformation χ can be implemented by a unitary map U which is unique up to a phase factor (see *e.g.*, [Hö, Thm 18.5.9]). Here we will only use the Weyl quantization and denote $a^w(x, D)$ simply by $a(x, D)$.

Case $Q = 0$. Let $U_\diamond$ be a unitary transformation implementing $\chi_\diamond$. Explicitly:

$$U_\diamond u(x) = e^{i(x_\diamond, Ax)} u(x).$$

One has

(3.12) $U_\diamond H U_\diamond^* = \tilde{H},$

where

(3.13)
$$\tilde{H} = \frac{1}{2} D_{z_\diamond}^2 + \frac{1}{2}(D_{y_\diamond} - 2A_\diamond{}^\diamond x^\diamond)^2 +$$
$$\frac{1}{2}(D^\diamond - A^\diamond{}^\diamond x^\diamond)^2 + V(x^\diamond) \qquad = \frac{1}{2} D_{z_\diamond}^2 + \tilde{H}^\diamond.$$

The operator $\tilde{H}^\diamond$ acting on $L^2(Y_\diamond \times X^\diamond)$ does not depend explicitly on $y_\diamond$, hence it commutes with $D_{y_\diamond}$. It can therefore be written as

$$\tilde{H}^\diamond = \int_{Y'_\diamond}^{\oplus} \tilde{H}^\diamond(\eta_\diamond) d\eta_\diamond,$$

where

$$\tilde{H}^\diamond(\eta_\diamond) = \frac{1}{2}(\eta_\diamond - 2A_\diamond{}^\diamond x^\diamond)^2 + \frac{1}{2}(D^\diamond - A^\diamond{}^\diamond x^\diamond)^2 + V(x^\diamond)$$

acts on $L^2(X^\diamond)$. Thus in the neutral case reduction of the Hamiltonian means fixing a value of $\eta_\diamond$ and studying $\tilde{H}^\diamond(\eta_\diamond)$ as an operator on $L^2(X^\diamond)$.

The following proposition is easy to prove.

PROPOSITION 3.1. *Assume that hypothesis (V1) holds. Then the mapping*

$$Y'_\diamond \ni \eta_\diamond \mapsto \tilde{H}^\diamond(\eta_\diamond)$$

is continuous in the norm resolvent sense.

Case $Q > 0$. As in the neutral case, we can implement the linear symplectic transformation $\chi_\diamond$ by a unitary map $U_\diamond$. We have

(3.14) $U_\diamond H U_\diamond^\ast = \tilde{H},$

where

(3.15)
$$\tilde{H} = \frac{1}{2}D_{z_\diamond}^2 + \tilde{H}^\diamond,$$
$$\tilde{H}^\diamond = R_\diamond(D_{y^{(2)}_\diamond}, y^{(2)}_\diamond, x^\diamond) + \frac{1}{2}(D^\diamond - A^{\diamond\diamond}x^\diamond)^2 + V(x^\diamond),$$

and $R_\diamond$ is defined in (3.9).

Note that $\tilde{H}^\diamond$ does not act on the $y^{(1)}_\diamond$ and $z_\diamond$ variables. We can therefore consider it as an operator acting only on $L^2(Y^{(2)}_\diamond \times X^\diamond)$.

Example 1: $N = 1$. Let us first consider the one-particle magnetic Hamiltonian

$$H = \frac{1}{2}(D - Ax)^2 \text{ on } L^2(\mathbb{R}^3)$$

(we omit the subscript 1 and assume that $q > 0$). In this case $U_\diamond$ is the well known unitary transformation mapping a 2-dimensional magnetic Hamiltonian onto a 1-dimensional harmonic oscillator. To see that, we note that for $N = 1$ we have $X_\diamond = X$ and $X^\diamond = \{0\}$, so that the transformation $\chi_\diamond$ from Lemma 3.2 has the form

$$\begin{cases} (\tilde{\eta}^{(1)}, \tilde{y}^{(1)}) = \kappa(\eta + Ax), \\ (\tilde{\eta}^{(2)}, \tilde{y}^{(2)}) = \kappa(\eta - Ax), \\ \tilde{z} = z, \ \tilde{\zeta} = \zeta, \end{cases}$$

where $\kappa = (2bq)^{-1/2}$. Under this transformation, the symbol h of H becomes

$$h = \frac{1}{2}|\zeta|^2 + \frac{1}{2}(\eta - Ax)^2 = \frac{1}{2}|\tilde{\zeta}|^2 + \frac{1}{2}\kappa^{-2}(|\tilde{\eta}^{(2)}|^2 + |\tilde{y}^{(2)}|^2)$$
$$= \frac{1}{2}|\tilde{\zeta}|^2 + bq(|\tilde{\eta}^{(2)}|^2 + |\tilde{y}^{(2)}|^2).$$

The corresponding quantum Hamiltonian is

(3.16) $$\tilde{H} = \frac{1}{2}|D_z|^2 + bq(|D_{y^{(2)}}|^2 + |y^{(2)}|^2).$$

The spectrum of the reduced Hamiltonian, *i.e.*, of $\tilde{H}^{\diamond} = \tilde{H} - \frac{1}{2}D_z^2$ acting on $L^2(Y^{(2)})$, consists of the *Landau levels*

$$\Lambda_n = \frac{bq}{m}(2n+1), \ n = 0, 1, 2, \ldots.$$

The infinite degeneracy of Λ_n as eigenvalues of $H^{\diamond} = H - \frac{1}{2}D_z^2$ is due to the fact that $\tilde{H}^{\diamond}$ does not act on the $y^{(1)}$ variable. To remove that degeneracy, we consider the Hamiltonian $\tilde{H}^{\diamond}$ acting only on $Y^{(2)}$.

Example 2: a neutral pair ($N = 2$, $q_1 + q_2 = 0$). In this case the reduction procedure was worked out first by Avron, Herbst and Simon [AHS2]. Recall that the reduced Hamiltonians $\tilde{H}^{\diamond}(\eta_{\diamond})$, acting on $L^2(X^{\diamond})$, are defined by:

$$\tilde{H}^{\diamond}(\eta_{\diamond}) = \frac{1}{2}(\eta_{\diamond} - 2A_{\diamond}{}^{\diamond}x^{\diamond})^2 + \frac{1}{2}(D^{\diamond} - A^{\diamond}{}^{\diamond}x^{\diamond})^2 + V^{\diamond}(x^{\diamond}).$$

The spectral properties of $\tilde{H}(\eta_{\diamond})$ are analyzed (under various assumptions on the potential) in [AHS2], [L2], [GL3]. Here we state, without proof, Theorem 6.1 of [GL3]. Let

(3.17) $$\tau = \left\{ \sum_1^2 \frac{|q_i|}{m_i}(2n_i + 1)b, n_i \in \mathbb{N} \right\},$$

be the set of Landau levels of the pair of particles (see Example 1 above), and let

$$\nu_0 = \inf \tau = \sum_1^2 \frac{|q_i|}{m_i} b$$

be the lowest Landau level. To simplify the notation, we will denote the parameter $\eta_{\diamond} \in Y'_{\diamond}$ by k.

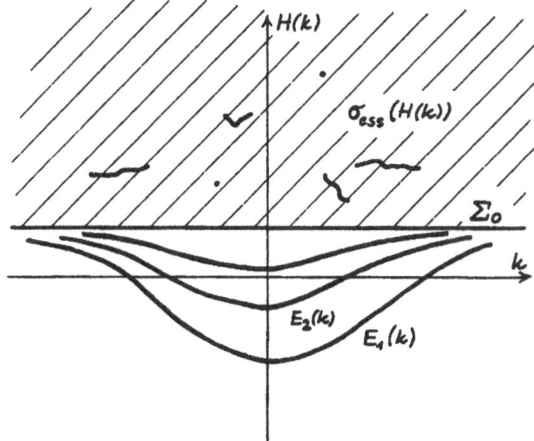

FIG. 3.1. *Modified spheres in 2 dimensions.*

THEOREM 3.3. *Assume that (V) holds. Then:*
i) $\tilde{H}^{\diamond}(k)$ *is selfadjoint with domain* $D = \{u \in L^2(X^{\diamond})|D^2_{x_{\diamond}}u, (y^{\diamond})^2u \in L^2(X^{\diamond})\}$;
ii) the essential spectrum $\sigma_{\text{ess}}(\tilde{H}^{\diamond}(k))$ *is equal to* $[\nu_0, +\infty)$;
iii) the eigenvalues of $\tilde{H}^{\diamond}(k)$ *can accumulate only at* τ;
iv) if $E(k)$ *is a discrete eigenvalue of* $\tilde{H}^{\diamond}(k)$, *then* $\lim_{k\to\infty} E(k) = (\nu_0)$;
v) the map

$$Y'_{\diamond} \ni k \mapsto \tilde{H}^{\diamond}(k)$$

is analytic in norm resolvent sense.

The spectrum of $\tilde{H}^{\diamond}(k)$ is pictured in Figure 3.1. Note that, apart from the discrete eigenvalues E_i below the continuous spectrum, $\tilde{H}^{\diamond}(k)$ may have embedded eigenvalues, whose behaviour can be rather irregular and difficult to control.

4. Bound and scattering states. Recall that for standard N-particle Hamiltonians

$$H = \sum_{i=1}^{N} \frac{1}{2m_i}D_i^2 + \sum_{i<j} V_{ij}(x_i - x_j),$$

bound and scattering states are defined as eigenfunctions and states from the continuous spectral subspace respectively of the Hamiltonian with the center of mass motion removed. More precisely, one writes

$$H = \frac{1}{2}D^2_{x_{\diamond}} + H^{\diamond},$$

where $H^\diamond$ acts only on the relative coordinates $x^\diamond$ of the particles. The subspaces of bound and scattering states are then defined as

$$\mathcal{H}_{\text{bound}} = \mathcal{H}_{\text{pp}}(H^\diamond) \otimes L^2(X_\diamond),$$

$$\mathcal{H}_{\text{scatt}} = \mathcal{H}_{\text{cont}}(H^\diamond) \otimes L^2(X_\diamond).$$

In fact, one often considers only $H^\diamond$ instead of H, so that $\mathcal{H}_{\text{bound}}$ and $\mathcal{H}_{\text{scatt}}$ are subspaces of $L^2(X^\diamond)$. The motion of the center of mass of the system, for both bound and scattering states, is described by the free Hamiltonian $\frac{1}{2} D^2_{x_\diamond}$.

For magnetic N-particle Hamiltonians the center of mass motion cannot be simply separated in a way similar to that described above; yet, common sense dictates that there should be a physically meaningful notion of bound and scattering states. We will define them using the reduction procedure from the previous section. The cases of neutral and charged systems will be considered separately. The behaviour of bound states will depend on the total charge of the system. We will show that, while bound states of charged systems perform bounded motion in directions transversal to the field (as single charged particles do), a neutral system in a bound state can travel to infinity across the field.

We emphasize here that the motion of the center of mass of bound states of neutral systems in directions perpendicular to the field has nothing

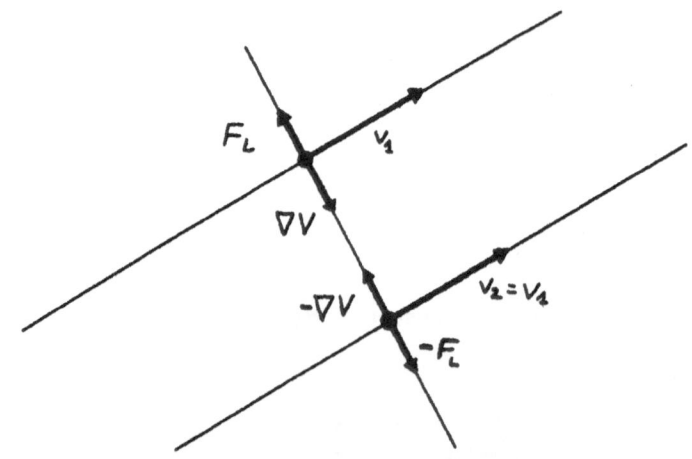

FIG. 4.1. *Two classical particles.*

to do with the free motion occurring in the absence of external forces (as in the standard case discussed above). Rather, it is due to the cancellations between the forces exerted on the particles by the magnetic field and by one another. The following simple example from classical mechanics illustrates well the mechanism at work.

Example: Two classical particles. Consider two classical particles whose electric charges are $q_1 = e$ and $q_2 = -e$ (e.g., a proton and an electron), interacting via a Coulomb potential $V = -e^2/|x_1 - x_2|$, moving in a constant magnetic field. The equations of motion have a family of particularly simple solutions, corresponding to the two particles moving across the field along parallel lines with the same velocity (Figure 4.1; the magnetic field is perpendicular to the plane in which the particles are moving).

The trajectories are given by:

$$y_1(t) = \vec{v}t + \vec{b}_0, \ z_1 = 0,$$

$$y_2(t) = \vec{v}t + \vec{b}_0 + \vec{b}, \ z_2 = 0,$$

where the vector $\vec{b}$ is perpendicular to the velocity $\vec{v}$. (For simplicity, we consider the solutions for which there is no motion along the field). Indeed, the Lorenz forces acting on the two particles are

$$\vec{F}_{L,1} = \vec{F}_L = e\vec{v} \times \vec{B}, \ \vec{F}_{L,2} = -\vec{F}_L = -e\vec{v} \times \vec{B},$$

and the Coulomb forces are

$$\vec{F}_{C,1} = \vec{F}_C = \nabla V(\vec{b}), \ \vec{F}_{C,2} = -\vec{F}_C = -\nabla V(\vec{b}).$$

The forces $\vec{F}_L$ and $\vec{F}_C$ act in the same direction, and, given a vector $\vec{v}$, one can find $\vec{b}$ orthogonal to it such that $\vec{F}_L = -\vec{F}_C = -\nabla V(\vec{b})$, so that the Coulomb and Lorentz forces cancel out.

4.1. Charged systems. If the system is charged, we have

(4.1)
$$U_\Diamond H U_\Diamond^\star = \tilde{H},$$

where

(4.2)
$$\tilde{H} = \frac{1}{2}D_{z_\Diamond}^2 + R_\Diamond(D_{y_\Diamond^{(2)}}, y_\Diamond^{(2)}, x^\Diamond) + \frac{1}{2}(D^\Diamond - A^{\Diamond\Diamond}x^\Diamond)^2 + V(x^\Diamond)$$
$$= \frac{1}{2}D_{z_\Diamond}^2 + \tilde{H}^\Diamond.$$

We will consider $\tilde{H}^\Diamond$ as an operator acting only on $L^2(Y_\Diamond^{(2)} \times X^\Diamond)$.

DEFINITION 4.1. *Assume that $Q \neq 0$. Let*

$$\mathcal{H}_{\text{scatt}} = U_\Diamond^\star(\mathcal{H}_{\text{cont}}(\tilde{H}^\Diamond) \otimes L^2(Y_\Diamond^{(1)} \times Z_\Diamond)),$$

$$\mathcal{H}_{\text{bound}} = U_{\diamond}^{*}\big(\mathcal{H}_{\text{pp}}(\tilde{H}^{\diamond}) \otimes L^{2}(Y_{\diamond}^{(1)} \times Z_{\diamond})\big).$$

We have:

$$L^{2}(X) = \mathcal{H}_{\text{scatt}} \oplus \mathcal{H}_{\text{bound}}.$$

Note that the factor $L^{2}(Y_{\diamond}^{(1)})$ plays no significant role in the above definition and can be omitted if $\tilde{H}^{\diamond}$ is considered as an operator on $Y_{\diamond}$. (The factor $L^{2}(Z_{\diamond})$ was left out in [GL1] and [GL2] by mistake, both in the charged and the neutral case.)

We will show that a state in $\mathcal{H}_{\text{bound}}$ describes a bounded cluster of particles in the sense that all relative coordinates $x^{\diamond}$ of the particles are bounded uniformly in time. Moreover, similar bounds hold for all coordinates y transversal to the field, *i.e.*, the motion of the system across the field is essentially restricted to a bounded area. The center of mass of the system moves freely in the z direction.

PROPOSITION 4.1. *Let $Q \neq 0$ and $u \in \mathcal{H}_{\text{bound}}$. We have:*

$$(i) \quad \lim_{R \to \infty} \sup_{t \in \mathbb{R}} \Big\| F\Big(\frac{|x^{\diamond}|}{R} \geq 1\Big) e^{-itH} u \Big\| = 0,$$

$$(ii) \quad \lim_{R \to \infty} \sup_{t \in \mathbb{R}} \Big\| F\Big(\frac{|y_{\diamond}|}{R} \geq 1\Big) e^{-itH} u \Big\| = 0.$$

Proof. It suffices to prove the proposition for $u \in \mathcal{H}_{\text{bound}}$ such that

$$U_{\diamond} u = v = v_{1} \cdot v_{2},$$

where $v_{1} \in L^{2}(X^{\diamond} \times Y_{\diamond}^{(2)})$ is an eigenfunction of $\tilde{H}^{\diamond}$, corresponding to an eigenvalue E, and $v_{2} \in L^{2}(Y_{\diamond}^{(1)} \times Z_{\diamond})$. For such u, we have:

$$e^{-itH} u = U_{\diamond}^{*} e^{-it\tilde{H}} v$$

$$= U_{\diamond}^{*}(e^{-itE} v_{1} \cdot e^{-itD_{z_{\diamond}}^{2}/2} v_{2}).$$

Hence, if Φ is any one of the observables

$$x^{\diamond}, y_{\diamond}^{(1)}, D_{y_{\diamond}^{(1)}}, y_{\diamond}^{(2)}, D_{y_{\diamond}^{(2)}},$$

we have

$$\lim_{R \to \infty} \sup_{t \in \mathbb{R}} \Big\| F\Big(\frac{|\Phi|}{R} \geq 1\Big) e^{-it\tilde{H}} v \Big\| = 0.$$

Inverting the transformation $U_{\diamond}$ and using the invariance of the Weyl calculus under linear symplectic transformations, we obtain the proposition. $\square$

4.2. Neutral systems. Recall that if the system is neutral, we have:

$$(4.3) \qquad\qquad U_\diamond H U_\diamond^* = \tilde{H},$$

where

$$\tilde{H} = \frac{1}{2}D_{z_\diamond}^2 + \frac{1}{2}(D_{y_\diamond} - 2A_\diamond{}^\diamond x^\diamond)^2 + \frac{1}{2}(D^\diamond - A^{\diamond\diamond} x^\diamond)^2 + V(x^\diamond)$$

$$= \frac{1}{2}D_{z_\diamond}^2 + \tilde{H}^\diamond.$$

(4.4)

The transformed Hamiltonian $\tilde{H}^\diamond$ can be written as a direct integral:

$$\tilde{H}^\diamond = \int_{Y_\diamond'}^{\oplus} \tilde{H}^\diamond(\eta_\diamond)\, d\eta_\diamond,$$

where

$$\tilde{H}^\diamond(\eta_\diamond) = \frac{1}{2}(\eta_\diamond - 2A_\diamond{}^\diamond x^\diamond)^2 + \frac{1}{2}(D^\diamond - A^{\diamond\diamond} x^\diamond)^2 + V(x^\diamond)$$

acts on $L^2(X^\diamond)$. The bound and scattering states will be defined as direct integrals of eigenfunctions and states from the continuous spectral subspaces respectively of the reduced Hamiltonians $\tilde{H}^\diamond(\eta_\diamond)$.

DEFINITION 4.2. *Assume that $Q = 0$. Let*

$$\mathcal{H}_{\text{scatt}} = U_\diamond^* \left(L^2(Z_\diamond) \otimes \int_{Y_\diamond'}^{\oplus} \mathcal{H}_{\text{cont}}(\tilde{H}^\diamond(\eta_\diamond))\, d\eta_\diamond \right),$$

$$\mathcal{H}_{\text{bound}} = U_\diamond^* \left(L^2(Z_\diamond) \otimes \int_{Y_\diamond'}^{\oplus} \mathcal{H}_{\text{pp}}(\tilde{H}^\diamond(\eta_\diamond))\, d\eta_\diamond \right).$$

We have:

$$L^2(X) = \mathcal{H}_{\text{scatt}} \oplus \mathcal{H}_{\text{bound}}.$$

As in the charged case, a state in $\mathcal{H}_{\text{bound}}$ describes a bounded cluster of particles, *i.e.*, one has uniform in time bounds on all relative coordinates $x^\diamond$ of the particles. The motion of the center of mass in the z direction is described by the free Hamiltonian $\frac{1}{2}D_{z_\diamond}^2$. The dynamics in the plane transversal to the field is more complicated. The effective kinetic energy of the motion of a bound state $u = U_\diamond^* v$, where $v = Pv$,

$$P = 1_{L^2(Z_\diamond)} \otimes \int^{\oplus} P(\eta_\diamond)d\eta_\diamond,$$

and $P(\eta_\diamond)$ is an eigenprojection of $\tilde{H}^\diamond(\eta_\diamond)$ corresponding to the eigenvalue $E(\eta_\diamond)$, is:

$$\frac{1}{2}D_{z_\diamond}^2 + E(\eta_\diamond),$$

i.e., one has

$$e^{-itH}u = U_\diamond^* \int^{\oplus} e^{-it(\frac{1}{2}\zeta_\diamond^2 + E(\eta_\diamond))}v(\eta_\diamond, \zeta_\diamond)d\eta_\diamond d\zeta_\diamond.$$

Proposition 4.2*(ii)* below states that for such states, the average velocity of the center of mass motion across the field is $\nabla E(\eta_\diamond)$.

PROPOSITION 4.2. *Assume that $Q = 0$. Then the following holds.*
i) for any $u \in \mathcal{H}_{\text{bound}}$ we have

$$\lim_{R \to \infty} \sup_{t \in \mathbb{R}} \left\| F\left(\frac{|x^\diamond|}{R} \geq 1\right) e^{-itH}u \right\| = 0.$$

ii) let $u \in \mathcal{H}_{\text{bound}}$ satisfy

$$U_\diamond u = \left(1_{L^2(Z_\diamond)} \otimes P(D_{y_\diamond})\right) U_\diamond u,$$

and assume that

$$Y_\diamond' \ni \eta_\diamond \mapsto (P(\eta_\diamond), \nabla E(\eta_\diamond)) \in B(L^2(X^\diamond)) \times \mathbb{R}$$

is a Lipschitz function. Then one has

$$\lim_{t \to +\infty} \left(g\left(\frac{y_\diamond}{t}\right) - g(\nabla E(D_{y_\diamond} + A_\diamond \diamond x^\diamond))\right) e^{-itH}u = 0 \text{ for } g \in C_0^\infty(Y_\diamond).$$

Proof. i). Follows directly from the definition of $\mathcal{H}_{\text{bound}}$ and from the fact that $U_\diamond$ conserves $x^\diamond$.

Let us prove *ii).* By density, we need only consider functions of the form $u = U_\diamond^*(v_1 \cdot v_2)$, where $v_1 \in L^2(Z_\diamond)$ and $v_2 = P(D_{y_\diamond})v_2 \in L^2(X^\diamond \times Y_\diamond)$. We have:

$$e^{-itH}u = U_\diamond^*\left(e^{-it\frac{1}{2}D_{z_\diamond}^2}v_1 \cdot e^{-itE(D_{y_\diamond})}v_2\right).$$

Let $g \in C_0^\infty(Y_\diamond)$. We will show that for any $\phi \in L^2(X^\diamond \times Y_\diamond)$,

$$(4.5) \quad \lim_{t \to \infty} \|e^{itE(D_{\eta\diamond})}g\left(\frac{y_\diamond}{t}\right)e^{-itE(D_{\eta\diamond})}\phi - g(\nabla E(D_{y_\diamond}))\phi\| = 0.$$

This implies $ii)$ by the invariance of the Weyl calculus under linear symplectic transformations.

It suffices to show (4.5) for $\phi \in C_0^\infty(X^\diamond \times Y_\diamond)$. In the proof below the variables $x^\diamond$ will be only parameters, so that we will omit the subscript $\diamond$ and write D, y instead of $D_{y_\diamond}$ and $y_\diamond$. $\mathcal{F}$ and $\check{}$ will denote the Fourier transform only in the $y_\diamond$ variables. We have:

$$\mathcal{F}\left(e^{itE(D)}g\left(\frac{y}{t}\right)e^{-itE(D)}\phi\right)$$

$$= e^{itE(\eta)}\mathcal{F}\left(g\left(\frac{y}{t}\right)e^{-itE(D)}\phi\right)$$

$$= e^{itE(\eta)}\left(\mathcal{F}\left(g\left(\frac{y}{t}\right)\right) * e^{-itE(\eta)}\hat{\phi}\right)$$

$$= e^{itE(\eta)}\int t^2\hat{g}((\eta - \lambda)t)e^{-itE(\lambda)}\hat{\phi}(\lambda)d\lambda$$

$$= \int e^{it(E(\eta)-E(\lambda))}t^2\hat{g}((\eta - \lambda)t)\hat{\phi}(\lambda)d\lambda,$$

where we used that $\mathcal{F}(g(y/t)) = t^2\hat{g}(\eta t)$. Substituting $(\eta - \lambda)t = \theta$, we see that the last integral is equal to:

$$(4.6) \quad \int \exp\left(i\theta\frac{E(\eta) - E(\eta - \theta/t)}{\theta/t}\right)\hat{g}(\theta)\hat{\phi}(\eta - \theta/t)d\theta.$$

As $t \to \infty$, the integrand in (4.6) converges to

$$e^{i\theta\nabla E(\eta)}\hat{g}(\theta)\hat{\phi}(\eta).$$

Using the rapid decay of $\hat{g}$ and the dominated convergence theorem, we see that (4.6) converges in L^2 as $t \to \infty$ to

$$\int e^{i\theta\nabla E(\eta)}\hat{g}(\theta)\hat{\phi}(\eta)d\theta = g(\nabla E(\eta))\hat{\phi}(\eta) = \mathcal{F}(g(\nabla E(D))\phi),$$

which proves (4.5). $\square$

5. Two-particle systems. In this section we will analyze the scattering states of a 2-particle Hamiltonian

$$(5.1) \quad H = \sum_{i=1}^{2}\frac{1}{2m_i}(D_i - q_iJx_i)^2 + V(x_1 - x_2),$$

with the potential $V = V_{12}$ satisfying the hypothesis (V). We will be interested in the large time asymptotics of the solutions $e^{-itH}u$ of the time-dependent Schrödinger equation for $u \in \mathcal{H}_{\text{scatt}}$. The main result presented

in this section is the asymptotic completeness of the modified wave operators for Hamiltonians of the form (5.1) (Theorem 5.1). Roughly speaking,
it states that for scattering states of H, the large time behaviour of the
system is similar to that of noninteracting particles.

Let us first briefly discuss the spectral properties of the "free" magnetic
Hamiltonian:

$$H_{\parallel} = \sum_{i=1}^{2} \frac{1}{2m_i}(D_i - q_i J x_i)^2 = \frac{1}{2}D_z^2 + H^{\sharp},$$

where

$$H^{\sharp} = \sum_{i=1}^{2} \frac{1}{2m_i}(D_{y_i} - q_i J y_i)^2.$$

The spectrum of $H^{\sharp}$ is the set τ of the Landau levels of the two particles
(see Subsection 3.2, Example 1):

$$\tau = \left\{ \sum_{1}^{2} \frac{|q_i|}{m_i}(2n_i + 1)b, n_i = 0, 1, 2, \ldots \right\}.$$

Moreover, $H^{\sharp}$ and D_z commute. Thus the dynamics of two noninteracting
particles is a superposition of bounded motion $e^{-itH^{\sharp}}$ in the directions
perpendicular to the magnetic field and free propagation $e^{-itD_z^2}$ along the
field.

Recall also that if no magnetic field is present ($\vec{B} = 0$), the asymptotic
evolution of the system is given by

$$e^{-iS(t,D)},$$

where $S(t, D)$ is essentially given by the corresponding equations of classical
mechanics. In the short-range case (V satisfies the Hypothesis (V) with
$\mu > 1$) $S(t, D) = \frac{1}{2}D^2 t$ is the usual free dynamics, and a modified free
evolution (an approximate solution of the Hamilton-Jacobi equation) is
used for long-range interactions, such as Coulomb potentials.

Since the magnetic field does not act in the z direction, we expect that
the propagation along it will not differ significantly from the case $\vec{B} = 0$.
This part of the problem can be treated using the standard techniques of
scattering theory.

Entirely different will be the dynamics of the system in the directions
transversal to the magnetic field and the methods that we will employ to
analyze it. We will prove that, for scattering states of H, the field binds the
charged particles in these directions, as it would if no interaction between
them was present.

To state our result, we first have to define the modified dynamics mentioned above. It is well known that a long-range interaction between particles escaping to infinity with nonzero relative velocities cannot be entirely

neglected as $t \to \infty$, so that one has to modify the free evolution appearing in the definition of the wave operators. For simplicity, we have chosen the Dollard modifier [Do], in which an approximate solution of the Hamilton-Jacobi equation is used. Since, as we will prove, the particles can separate only along the field, it suffices to modify the dynamics in that direction. Let

$$(5.2) \qquad I(t, x) = F\left(\frac{|z^{\diamond}|\log t}{t} \geq 1\right) V(x).$$

Then $I(t, x)$ satisfies

$$(5.3) \qquad |\partial_x^\alpha I(t, x)| \leq C_{\alpha, \epsilon}\langle t \rangle^{-|\alpha|-\mu+\epsilon}, \quad |\alpha| \geq 0.$$

for all $\epsilon > 0$. Define

$$S(t, \zeta) = \int\limits_0^t \frac{1}{2}|\zeta|^2 + I(s, s\zeta)ds.$$

Note that $S(t, D_z)$ and $H^\sharp$ commute and $S(t, \zeta) = \frac{1}{2}t|\zeta|^2$ if the potential V is short-range. Then we have the following theorem:

THEOREM 5.1. *Let H be given by (5.1), where V satisfies the hypothesis (V) with $\mu > 1/2$. Let P_{scatt} denote the orthogonal projection onto $\mathcal{H}_{\text{scatt}}$. Then:*

i) the modified wave operators

$$\Omega^\pm = s - \lim_{t \to \pm\infty} e^{itH} e^{-iS(t, D_z) - itH^\sharp} P_{\text{scatt}}$$

exist;

ii) the system is asymptotically complete:

$$\text{Ran}\Omega^\pm = \mathcal{H}_{\text{scatt}}.$$

The above theorem is a special case ($N = 2$) of the general results of [GL1] and [GL2] for short-range and long-range interactions respectively ([GL1, Theorem 6.8], [GL2, Theorem 4.5]; see also [L2] for a proof in the case of 2-particle neutral systems). It is not our goal to present here a complete proof of it – especially since some of the arguments are identical to those used in the well understood 2-body (or 1-body, after the center of mass motion has been separated) problem with $H = -\Delta + V(x)$. Instead, we concentrate on the new features of the problem with a magnetic field and the methods developed in solving it. Several technical arguments are omitted; we refer the reader to [GL1], [GL2] for proofs.

Let us first sketch the general outline of the proof of Theorem 5.1. In this and the next section we will discuss in more detail two parts of it:

the Mourre estimate, and the analysis of the dynamics of the particles in transversal directions using the *center of orbit* observable.

In Subsection 5.1 we prove the Mourre estimate for the reduced Hamiltonians with $A = \frac{1}{2}(zD_z + D_z z)$ as the conjugate operator. A point that we want to emphasize is that if the system is neutral, the dependence of the spectrum of the reduced Hamiltonian $\tilde{H}^\diamond(\eta_\diamond)$ on the parameter $\eta_\diamond$ has to be taken into account, so that our estimates will be localized not only in energy, but also in pseudomomentum.

Once we have this, the propagation along the z axix can be treated by standard arguments borrowed from the 2-body scattering theory. In Subsection 5.2 we show that for states in $\mathcal{H}_{scatt}$ the two particles separate in the direction of the field with nonzero relative velocity, *i.e.*, that for u in a dense subset of $\mathcal{H}_{scatt}$, there is an $\epsilon > 0$ such that for any function $f \in C_0^\infty(-\epsilon, \epsilon)$ we have

$$\lim_{t \to \infty} \left\| f\left(\frac{z}{t}\right) e^{-itH} u \right\| = 0.$$

This readily yields asymptotic completeness in the short-range case.

For long-range interactions, this allows us to consider, instead of a 2-particle system, two systems, each consisting of one particle in a magnetic field with an additional external time-dependent force (the interaction with the other particle) vanishing as $t^{-\mu}$ as $t \to \infty$. More precisely, we obtain the existence and completeness of the wave operators intertwining $e^{-itH} P_{scatt}$ and the evolution generated by the time-dependent Hamiltonian

$$H_\parallel + I(x, t),$$

where $I(x, t)$ is given by 5.2. Writing $H_\parallel$ as a sum of two 1-particle Hamiltonians, we see that the evolution of the i-th particle ($i = 1, 2$) is governed by a time-dependent Hamiltonian of the form:

$$H_i(t) = \frac{1}{2}(D_i - Ax_i)^2 + I(x, t)$$

($I(x, t)$ depends on the coordinates of the other particle).

The analysis of the dynamics generated by such Hamiltonians consists of two steps. We first prove that the transversal to the field coordinates y of the particles grow no faster than t^α for any $\alpha > 1/2$. In Section 6 we will do this for more general 1-particle time-dependent Hamiltonians using the *center of orbit* observable. This allows us to replace H by $H^\parallel + H_z$, where $H_z = \frac{1}{2}|D_z|^2 + I(0, z)$. Since $H^\parallel$ commutes with H_z, it suffices to apply well known results on one-particle scattering (see *e.g.*, [Do], [E1], [Sig]) to the Hamiltonian H_z to obtain asymptotic completeness.

Note that if the system is neutral, one can bypass the study of time-dependent Hamiltonians and use instead a simpler argument involving the conservation of energy and pseudomomentum (see [L2]). However, it is the argument presented here that generalizes to N-particle systems.

5.1. The Mourre estimate. Given two selfadjoint operators B, H on a Hilbert space $\mathcal{H}$ and $c \in \mathbb{R}$, we will say that $B > c$ at $H = E$ if there is a $\delta > 0$ such that for any function $f \in C_0^\infty((E - \delta, E + \delta))$ we have

$$(5.4) \qquad f(H)Bf(H) > cf^2(H).$$

We will say that H satisfies the *Mourre estimate* with the conjugate operator A (which is required to be selfadjoint) at $H = E$ if there is a constant $c > 0$ and a compact operator K such that

$$[H, iA] > c + K \text{ at } H = E.$$

The importance of the Mourre estimate in spectral and scattering theory for multiparticle systems is well known (see e.g., [Mo], [Pe], [PSS], [SS1–5], [CFKS], [DG], [HuSi]). One can use it to obtain such results as the absence of singular continuous spectrum, the limiting absorption principle, local decay and minimal velocity estimates. It also plays a key role in the existing proofs of asymptotic completeness for N-body systems (see *e.g.*, [SS1], [Gr], [De1]).

Let

$$A = \frac{1}{2}(zD_z + D_z z).$$

In this subsection, we will show that the reduced Hamiltonians satisfy the Mourre estimate with A as the conjugate operator at all nonthreshold energy levels (*i.e.*, at all $E \notin \tau$). This and Lemma A.1 will imply that a strict Mourre estimate (with $K = 0$) holds away from the point spectrum of the reduced Hamiltonians. In the neutral case, the latter will depend on the value of the pseudomomentum, and our estimates will have to be localized accordingly.

To obtain the compactness of the error K, we will need the following lemma, which is an easy consequence of Kato's inequality (see *e.g.*, [AHS1], [CFKS]). A proof is given in [AHS2]; note, however, that those authors use a somewhat different notion of reduced Hamiltonians. We will denote by Q the total charge $Q = q_1 + q_2$ of the pair.

LEMMA 5.2. *Assume that the potential V satisfies the condition (V), and let $V_\alpha = (|x| + 1)^{-|\alpha|} D^\alpha V$. Then, for $0 \leq |\alpha| \leq 2$, V_α is relatively compact with respect to the reduced Hamiltonians, i.e., $V_\alpha(\tilde{H}^\Diamond(\eta_\Diamond) + i)^{-1}$ is compact on $L^2(X^\Diamond)$ if $Q = 0$, and $V_\alpha(\tilde{H}^\Diamond + i)^{-1}$ is compact on $L^2(X^\Diamond \times Y_\Diamond^{(2)})$ if $Q \neq 0$.*

Remark. To have the relative compactness of the potential, we need to use the reduced Hamiltonians. $V(x_1 - x_2)$ does not vanish at infinity as $x_1, x_2 \to \infty$ if $x_1 - x_2$ remains bounded, and therefore is not relatively compact with respect to H. (Neither is it $-\Delta$-compact.)

Case. $Q = 0$. Recall that the transformed Hamiltonian $\tilde{H}^\diamond$ can be written as a direct integral

$$\tilde{H}^\diamond = \int_{Y'_\diamond}^{\oplus} \tilde{H}(\eta_\diamond) d\eta_\diamond,$$

where

$$\tilde{H}^\diamond(\eta_\diamond) = \frac{1}{2}(\eta_\diamond - 2A_\diamond{}^\diamond x^\diamond)^2 + \frac{1}{2}(D^\diamond - A^{\diamond\diamond}x^\diamond)^2 + V^\diamond(x^\diamond)$$

acts on $L^2(X^\diamond)$ for each $\eta_\diamond \in Y'_\diamond$. To simplify the notation, we will denote the parameter $\eta_\diamond$ by k.

LEMMA 5.3. *Let $f_\delta \in C_0^\infty(\mathbb{R})$ be a cutoff function supported in $[E - \delta, E + \delta]$. Then:*

$$f_\delta(\tilde{H}^\diamond(k))[\tilde{H}^\diamond(k), iA]f_\delta(\tilde{H}^\diamond(k))$$

$$= f_\delta(\tilde{H}^\diamond(k))D_{z^\diamond}^2 f_\delta(\tilde{H}^\diamond(k)) + K_1(k),$$

where $K_1(k)$ is compact on $L^2(X^\diamond)$ and the mapping $k \mapsto K_1(k)$ is norm continuous. In particular, $f_\delta(\tilde{H}^\diamond(k))[\tilde{H}^\diamond(k), iA]f_\delta(\tilde{H}^\diamond(k))$ is a norm continuous function of k.

Proof. Since

$$[\tilde{H}^\diamond(k), iA] = D_{z^\diamond}^2 - \langle z^\diamond, \nabla_{z^\diamond} V \rangle,$$

we have:

$$K_1(k) = -f_\delta(\tilde{H}^\diamond(k))\langle z^\diamond, \nabla_{z^\diamond} V \rangle f_\delta(\tilde{H}^\diamond(k)).$$

This operator is compact on $L^2(X)$ by Lemma 5.2. Continuity of $K_1(k)$ follows from the norm resolvent continuity of $\tilde{H}^\diamond(k)$ (Theorem 3.3(v)). $\square$

We will denote by

$$d(E) = \text{dist}(E, \tau),$$

the distance from E to the threshold set τ. Note that we will not follow the usual convention, according to which $d(E)$ would stand for the distance only to the thresholds to the left of E. Later, this will allow us to avoid certain technical complications due to the lack of continuity of the latter function.

LEMMA 5.4. *Let $E \in \mathbb{R}$. For any $\epsilon > 0$ there is a $\delta > 0$ such that for any function $f_\delta \in C_0^\infty((E - \delta, E + \delta))$ we have:*

$$(5.5) \quad f_\delta(\tilde{H}^\diamond(k))D_{z^\diamond}^2 f_\delta(\tilde{H}^\diamond(k)) \geq (2d(E) - \epsilon)f_\delta^2(\tilde{H}^\diamond(k)) + K_2(k),$$

where $K_2(k)$ is compact on $L^2(X^\diamond)$ and the mapping $k \mapsto K_2(k)$ is norm continuous.

Proof. Fix $\epsilon > 0$. Let $H_\sharp^\diamond = H^\diamond - V(x^\diamond)$, $\tilde{H}_\sharp^\diamond = U_\diamond^* H_\sharp^\diamond U_\diamond = \tilde{H}^\diamond - V(x^\diamond)$, and $\tilde{H}_\sharp^\diamond(k) = \tilde{H}^\diamond(k) - V(x^\diamond)$.

Since $D_{z\diamond}^2 = 2(H_\sharp^\diamond - H^\sharp)$, D_z commutes with $H^\sharp$, and $\sigma(H^\sharp) = \tau$, for sufficiently small $\delta > 0$ we have

$$f_\delta(H_\sharp^\diamond) D_{z\diamond}^2 f_\delta(H_\sharp^\diamond) \geq (2d(E) - \epsilon) f_\delta(H_\sharp^\diamond)^2.$$

Since D_z commutes with $U_\diamond$, this implies that:

$$f_\delta(\tilde{H}_\sharp^\diamond) D_{z\diamond}^2 f_\delta(\tilde{H}_\sharp^\diamond) \geq (2d(E) - \epsilon) f_\delta(\tilde{H}_\sharp^\diamond)^2.$$

Hence, for the same δ and all k:

$$(5.6) \qquad f_\delta(\tilde{H}_\sharp^\diamond(k)) D_{z\diamond}^2 f_\delta(\tilde{H}_\sharp^\diamond(k)) \geq (2d(E) - \epsilon) f_\delta(\tilde{H}_\sharp^\diamond(k))^2.$$

Using that $f(\tilde{H}^\diamond(k)) - f(\tilde{H}_\sharp^\diamond(k))$ is compact on $L^2(X)$ for any function $f \in C_0^\infty(\mathbb{R})$ and for any k (which follows from Lemma 5.2), we obtain the lemma. $\square$

The following Mourre estimate for $\tilde{H}^\diamond(k)$ is an immediate consequence of Lemmas 5.3 and 5.4.

LEMMA 5.5. *For any $k \in Y_\diamond'$, $E \in \mathbb{R} \setminus \tau$, and $\epsilon > 0$, there is an operator $K_\epsilon(k)$ compact on $L^2(X^\diamond)$ such that*

$$[\tilde{H}^\diamond(k), iA^\diamond] > 2d(E) - \epsilon + K_\epsilon(k) \text{ at } \tilde{H}^\diamond(k) = E.$$

Case $Q \neq 0$. In this case the following lemma holds:

LEMMA 5.6. *For any $E \in \mathbb{R} \setminus \tau$ and $\epsilon > 0$ there is an operator K_ϵ compact on $L^2(X^\diamond \times Y_\diamond^{(2)})$ such that*

$$[\tilde{H}^\diamond, iA] > 2d(E) - \epsilon + K_\epsilon \text{ at } \tilde{H}^\diamond = E.$$

The proof of Lemma 5.6 is exactly the same as that of Lemma 5.5, except that there is no dependence on k to be taken into account.

5.2. Minimal velocity estimate. We will now use the Mourre estimate proved in the previous subsection to show that for the scattering states of H the two particles separate with nonzero relative velocity in the z direction, i.e., $|z_1 - z_2|$ grows linearly in t. More precisely, we want to show that for each $u \in \mathcal{H}_0$, where $\mathcal{H}_0$ is a dense subset of $\mathcal{H}_{\text{scatt}}$, there is an $\epsilon > 0$ such that for any function $f \in C_0^\infty(-\epsilon, \epsilon)$ we have

$$(5.7) \qquad \lim_{t \to \infty} \left\| f\left(\frac{z}{t}\right) e^{-itH} u \right\| = 0.$$

Note that in [GL1,2] we have to use the *asymptotic velocity* observable to obtain an analogous result for N-particle systems. For $N = 2$, (5.7) is easier to prove and sufficient for our purposes.

Case $Q = 0$. We will first define the set $\mathcal{H}_0$. The following lemma is proved in [GL1] (Proposition 6.3):

LEMMA 5.7. *The set $\mathcal{H}_0$ of vectors $u \in L^2(X)$ such that for some $\epsilon = \epsilon(u) > 0$ we have:*

$$U_\diamond u = \sum_{i=1}^{k} v_i,$$
$$v_i = \chi_i(D_{y_\diamond}) f_i(\tilde{H}^\diamond(k)) v_i,$$

for some $\chi_i \in C_0^\infty(Y_\diamond')$ and $f_i \in C_0^\infty(\mathbb{R})$ satisfying

$$(5.8) \quad \text{dist} \left(\text{supp} f_i, \sigma_{\text{pp}}(\tilde{H}^\diamond(k)) \cup \tau \right) \geq \epsilon > 0 \text{ for all } k \in \text{supp} \chi_i$$

are dense in $\mathcal{H}_{\text{scatt}}$.

PROPOSITION 5.1. *If $u \in \mathcal{H}_0$, where $\mathcal{H}_0$ is defined in Lemma 5.7, there is an $\epsilon > 0$ such that (5.7) holds for any $f \in C_0^\infty((-\epsilon, \epsilon))$.*

Proposition 5.1 will follow by well known arguments (see *e.g.*, [Pe], [SS2], [Sig], [DG]) from the following lemma:

LEMMA 5.8. *For any $\chi \in C_0^\infty(Y_\diamond')$ and $f \in C_0^\infty(\mathbb{R})$ satisfying (5.8),*

$$(5.9) \qquad f(H)\chi(k)i[H^\diamond, A^\diamond]f(H)\chi(k) > \epsilon f^2(H)\chi^2(k).$$

Proof. We write the left hand side of (5.9) as

$$U_\diamond \left(\int^\oplus f(\tilde{H}^\diamond(k))\chi(k)i[\tilde{H}^\diamond(k), A^\diamond]f(\tilde{H}^\diamond(k))\chi(k)dk \right) U_\diamond^*.$$

Lemmas 5.5 and A.1 imply that if dist $\left(E, \sigma_{\text{pp}}(\tilde{H}^\diamond(k)) \cup \tau \right) \geq \epsilon > 0$ for all $k \in \text{supp}\chi$, then for each k

$$f_{E,\delta}(\tilde{H}^\diamond(k))\chi(k)i[\tilde{H}^\diamond(k), A^\diamond]f_{E,\delta}(\tilde{H}^\diamond(k))\chi(k)$$
$$(5.10) \qquad\qquad\qquad \geq \frac{\epsilon}{2} f_{E,\delta}^2(\tilde{H}^\diamond(k))\chi^2(k),$$

for all $f_{E,\delta} \in C_0^\infty((E - \delta, E + \delta))$, provided that $\delta > 0$ is small enough. Moreover, since the left hand side of (5.10) is a norm-continuous function of k (see Lemma 5.3), we can use Lemma A.2 to conclude that δ can be chosen the same for all $k \in \text{supp}\chi$.

We cover supp$f$ with intervals $(E - \delta, E + \delta)$ chosen as above, pick a finite subcovering of it and a partition of unity associated with that subcovering. This will allow us to write

$$f = \sum_{i=1}^{N_0} f_{E_i, \delta_i},$$

where (5.10) is satisfied for each f_{E_i, δ_i}. Integrating (5.10) over suppχ for all i and adding up the obtained inequalities, we obtain (5.9). $\square$

Case $Q \neq 0$. We first deduce from Lemmas 5.6 and A.1, and the general Mourre theory (see Appendix A), that $\sigma_{pp}(\tilde{H}^\Diamond)$ can accumulate only at thresholds τ and that

$$[\tilde{H}^\Diamond, iA] > d(E) \text{ at } \tilde{H}^\Diamond = E,$$

if E is not an eigenvalue of $\tilde{H}^\Diamond$. Hence the set $\tau \cup \sigma_{pp}(\tilde{H}^\Diamond)$ is closed and discrete, and for any compact interval Δ supported away from it we have:

$$E_\Delta(\tilde{H}^\Diamond)[\tilde{H}^\Diamond, iA]E_\Delta(\tilde{H}^\Diamond) > cE_\Delta^2(\tilde{H}^\Diamond)$$

for some $c > 0$ (this follows from Lemma A.2). By the same standard arguments as in the neutral case, this implies (5.7) for u such that $E_\Delta(\tilde{H}^\Diamond)u = u$ for an interval Δ as above.

6. Center of orbit – 1-particle Hamiltonians. Consider a single particle in a constant magnetic field and an external time-dependent potential, described by the Hamiltonian

$$(6.1) \qquad H(t) = H + W(t, x), \quad H = \frac{1}{2m}(D - Ax)^2 + V(x),$$

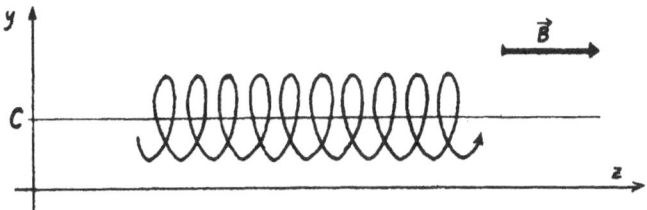

FIG. 6.1. *The classical observable* $c = \frac{1}{2}A^{-1}(\eta + Ay)$.

where $A = qJ$, J is given by (2.2), $V(x)$ and $W(t, x)$ satisfy

$$|D^\alpha V(x)| \leq \text{const}(1 + |x|)^{-|\alpha|-\mu}, \quad |\alpha| \leq 2, \quad \mu > 1/2,$$

$$(6.2)$$

$$|D^\alpha W(t, x)| \leq \text{const}(1 + |t|)^{-|\alpha|-\mu}, \quad |\alpha| \leq 2, \quad \mu > 1/2.$$

In particular, this includes the one-particle time-dependent Hamiltonians mentioned in Section 5 (in this case $V = 0$). We consider the more general case $V \neq 0$ to present an easier version of the argument which was used in [GL2] to prove asymptotic completeness for N-particle systems. The main ideas behind that generalization will be discussed in Section 10.

We denote by $U(t)$ the evolution generated by $H(t)$:

$$i\frac{d}{dt}U(t)u = H(t)U(t)u, \ U(0) = I.$$

Let also $u_t = U(t)u$. We will show that the transversal coordinates y of the particle can grow no faster than t^α, $\alpha > 1/2$, along the evolution $U(t)$. To do this, we need the *center of orbit* observable.

Note that the mapping A restricted to Y is invertible. We will denote by $A^{-1} : Y' \to Y$ the inverse of $A : Y \to Y'$.

DEFINITION 6.1. *The center of orbit observable is defined by:*

$$C = \frac{1}{2}A^{-1}(D_y + Ay).$$

A trajectory of a classical charged particle in a constant magnetic field, if no other forces are present, is a superposition of uniform motion along the field and circular motion in the transversal plane. C is the center of that circle. (For a quantum particle, the two components of C do not commute.)

LEMMA 6.1. *(i)* $C - y$ *is* H-*bounded;*
(ii) C *commutes with* $\frac{1}{2}(D - Ax)^2$.

These properties describe the fact that a constant magnetic field binds charged particles in the two directions orthogonal to it. The part *(i)* implies that once we fix the energy interval, estimates on C and on y are equivalent. On the other hand, because of *(ii)*, bounds on C are much easier to prove.

Proof. To show *(i)*, we compute:

$$C - y = \frac{1}{2}A^{-1}(D_y + Ay) - y = \frac{1}{2}A^{-1}(D_y + Ay - 2Ay)$$

$$= \frac{1}{2}A^{-1}(D_y - Ay),$$

which is $(D - Ay)^2$-bounded.

(ii) follows from the fact that $C = \frac{1}{2}A^{-1}K$ and $K = D + Ay$ commutes with $(D - Ay)^2$. $\square$

Due to Lemma 6.1 *(i)*, to show that $\||y|u_t\| = O(t^\alpha)$, it suffices to prove a similar bound on $\||C|u_t\|$. To obtain it, we will use Lemma 6.1 *(ii)* and the decay of V at infinity. Suppose that the propagation takes place in $|y| > t^\alpha$. Then the Heisenberg derivative of C can be estimated by

(6.3)
$$[H(t), C] = i[V + W, C] = A^{-1}\nabla(V + W)$$

$$= O((1 + |x|)^{-\mu-1}) = O((1 + |t|)^{-\alpha(1+\mu)}).$$

Integrating this from 0 to t we obtain that $\|Cu_t\| = O(t^{1-\alpha(1+\mu)})$ as $t \to \infty$. If $\alpha > 1/2$, we have

$$1 - \alpha(1 + \mu) < 1 - \alpha < \alpha,$$

so that $\|Cu_t\| = O(t^\alpha)$ and, consequently, $\|yu_t\| = O(t^\alpha)$.

This reasoning can be made rigorous, and the method works not only for one-particle Hamiltonians given by (6.1), but also for Hamiltonians of the form (2.1) for arbitrary N, under more general assumptions on potentials, provided that the system has no neutral subsystems [GL2]. However, the above argument is rather sloppy the way it is stated. In particular, it has one obvious logical flaw: we had no right to assume at the beginning that propagation takes place *only* in $|y| > t^\alpha$, and not both in and out of that region. Thus, a key point is that, instead of C, one considers a time-dependent observable $\mathcal{F}_t(C)$ such that, as $t \to \infty$,

$$(6.4) \qquad i[H(t), \mathcal{F}_t(C)] = i[V + W, \mathcal{F}_t(C)] = O(t^{-(\mu+1)\alpha}).$$

For 1-particle Hamiltonians, one can take $\mathcal{F}_t(C) = \mathcal{F}(\frac{C}{t^\alpha})$, where $\mathcal{F} \in C_0^\infty(Y)$ satisfies $\mathcal{F} \equiv 1$ in a neighbourhood of 0. Then on $\operatorname{supp}\nabla\mathcal{F}(\frac{C}{t^\alpha})$ one has $|C| > \operatorname{const} \cdot t^\alpha$. Once we restrict our attention to a bounded energy interval, this is essentially equivalent to $|y| > \operatorname{const} \cdot t^\alpha$. In that region, we have:

$$i[H(t), C] = A^{-1}\nabla(V + W) = O(t^{-\alpha(1+\mu)}),$$

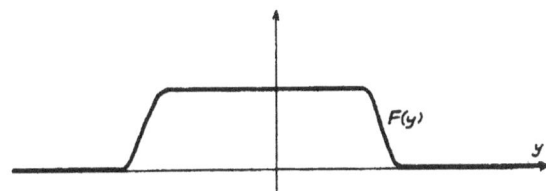

FIG. 6.2. *The function $\mathcal{F}(y)$.*

so that, using the commutator expansion from Lemma 6.2, we see that the Heisenberg derivative of $\mathcal{F}_t(C)$ satisfies:

$$
\begin{aligned}
(6.5) \quad D_t\mathcal{F}\left(\frac{C}{t^\alpha}\right) &= \frac{\partial}{\partial t}\mathcal{F}\left(\frac{C}{t^\alpha}\right) + i\left[H(t), \mathcal{F}\left(\frac{C}{t^\alpha}\right)\right] \\
&= -\alpha\frac{C}{t^{\alpha+1}}\nabla\mathcal{F}\left(\frac{C}{t^\alpha}\right) + i\left[H(t), \frac{C}{t^\alpha}\right]\nabla\mathcal{F}\left(\frac{C}{t^\alpha}\right) + O(t^{-2\alpha}) \\
&= -\alpha\frac{C}{t^{\alpha+1}}\nabla\mathcal{F}\left(\frac{C}{t^\alpha}\right) + O(t^{-(\mu+2)\alpha}) + O(t^{-2\alpha}).
\end{aligned}
$$

If $\alpha > 1/2$, both remainder terms in (6.5) are integrable in t. As we will see, it turns out that to handle the first term on the right hand side of

(6.5), it suffices to pick $\mathcal{F}$ such that $y \cdot \nabla \mathcal{F}(y) \leq 0$. We will show that for such functions

$$\text{s-}\lim_{t\to\infty} U(t)^{\star} \mathcal{F}\left(\frac{C}{t^{\alpha}}\right) U(t) = 1$$

for all $\alpha > \frac{1}{2}$.

For N-particle Hamiltonians, the construction of an appropriate observable $\mathcal{F}_t$ is much more difficult; the main ideas behind it are described in Section 10.

We will now present a 1-particle version of Theorem 7.6 of [GL2]. To simplify further the exposition, we will consider only one component of C at a time, so that certain technical arguments concerning the functional calculus for the noncommuting observable C will not be necessary. Henceforth, C will stand for $C^{(i)}$, with $i = 1$ or 2 fixed throughout the rest of this section, and y will denote $y^{(i)}$ with the same i. We omit the superscript (i) to keep the notation as simple as possible.

We will need several technical results, which are proved in [GL2, Section 7]. (In fact, in [GL2] they are shown for functions of the noncommuting observable C; if we follow the above convention and consider functions of only one component of C, the proofs become easier.)

The first lemma deals with the properties of C. Part (i) is a consequence of Lemma 6.1. The commutator expansion in (ii) for functions satisfying (6.2) follows from the composition theorem for pseudodifferential operators; for less regular potentials suitable functional calculus would have to be used (as in [GL2]). (iii) follows immediately from Lemma 7.4(ii) and the functional calculus of [HeSj] (see the formula (7.4) of [GL2]).

LEMMA 6.2. (i) For any $\chi \in C_0^{\infty}(\mathbb{R})$ and $F \in C_0^{\infty}(Y)$, $F_1 \in C^{\infty}(Y) \cap L^{\infty}(Y)$, such that $F_1 \equiv 1$ on $\operatorname{supp} F$, we have

$$\chi(H)\left(F\left(\frac{C}{t^{\alpha}}\right) - F\left(\frac{C}{t^{\alpha}}\right) F_1\left(\frac{y}{t^{\alpha}}\right)\right) = O(t^{-2\alpha}).$$

(ii) Assume that (6.2) holds. Then for $F \in C_0^{\infty}(Y)$, we have:

$$\left[V, iF\left(\frac{C}{t^{\alpha}}\right)\right] = \frac{1}{t^{\alpha}}[V, iC]\nabla_c F\left(\frac{C}{t^{\alpha}}\right) + O(t^{-2\alpha}).$$

(iii) Let $F \in C_0^{\infty}(Y)$ and $\chi \in C_0^{\infty}(\mathbb{R})$. Then:

$$\left[\chi(H), F\left(\frac{C}{t^{\alpha}}\right)\right] = O(t^{-\alpha}).$$

The next lemma is proved in [GL2] (Lemma 7.5; the intertwining property (6.7), which was not demonstrated there, follows directly from (6.6)).

LEMMA 6.3. Let $\chi \in C_0^{\infty}(\mathbb{R})$. Then:

(i) $D_t \chi(H) = O(t^{-1-\mu})$;
(ii) the norm limit

$$(6.6) \qquad \gamma(\chi) = \lim_{t \to +\infty} U(t)^* \chi(H) U(t)$$

exists. Moreover, there is an unbounded self-adjoint operator H^+ with a dense domain such that the limit in (6.6) equals $\chi(H^+)$, and one has

$$(6.7) \qquad U(t) \chi(H^+) = \chi(H) U(t).$$

The following theorem states that no propagation takes place in $|C| > t^\alpha$ for $\alpha > 1/2$.

THEOREM 6.4. *Let $\mathcal{F} \in C_0^\infty(Y)$ satisfy $y\mathcal{F}'(y) \le 0$ and $\mathcal{F} \equiv 1$ in a neighbourhood of 0, and let $\alpha > \frac{1}{2}$. Then:*

$$\text{s-} \lim_{t \to \infty} U^*(t) \mathcal{F}\left(\frac{C}{t^\alpha}\right) U(t) = 1.$$

Proof. Throughout the proof, we will assume that $t > 1$. We will show that for $1 > \alpha > \frac{1}{2}, R > 1$, the limit

$$(6.8) \qquad P_{\alpha,R} = \text{s-} \lim_{t \to \infty} U^*(t) \mathcal{F}\left(\frac{C}{Rt^\alpha}\right) U(t)$$

exists. Moreover, $[H^+, P_{\alpha,R}] = 0$, and

$$(6.9) \qquad \text{s-} \lim_{R \to +\infty} P_{\alpha,R} = 1.$$

This implies the theorem in the following way. Let $\alpha > \frac{1}{2}$, and let us pick β such that $\alpha > \beta > \frac{1}{2}$ and $\beta < 1$. Then for any $R > 0$ we have

$$\mathcal{F}\left(\frac{C}{Rt^\beta}\right) \le \mathcal{F}\left(\frac{C}{t^\alpha}\right),$$

when t is sufficiently large, so that

$$P_{\beta,R} \le P_{\alpha,1}.$$

Hence $P_{\alpha,1} \ge \text{s-}\lim_{R \to \infty} P_{\beta,R} = 1$. Since, by definition, $0 \le P_{\alpha,1} \le 1$, we obtain that $P_{\alpha,1} = 1$, which is what we had to show.

Let us now prove (6.8) and (6.9). Consider the propagation observable:

$$M_R = \chi(H) \mathcal{F}\left(\frac{C}{Rt^\alpha}\right) \chi(H).$$

We have

$$D_t M_R = \chi(H) D_t \mathcal{F}\left(\frac{C}{Rt^\alpha}\right)\chi(H) + D_t\chi(H)\mathcal{F}\left(\frac{C}{Rt^\alpha}\right)\chi(H) + \text{hc.}$$

To estimate the last two terms, we note that by Lemma 6.3(i),

$$(6.10) \qquad D_t\chi(H) = O(R^0 t^{-1-\mu}).$$

Next, we have:

$$(6.11) \qquad \partial_t \mathcal{F}\left(\frac{C}{Rt^\alpha}\right) = -\alpha\frac{C}{Rt^{\alpha+1}}\mathcal{F}'\left(\frac{C}{Rt^\alpha}\right) \geq 0.$$

Let us now estimate the term $[H(t), i\mathcal{F}(\frac{C}{Rt^\alpha})]$. By Lemma 6.1(ii),

$$\left[H, \mathcal{F}\left(\frac{C}{Rt^\alpha}\right)\right] = \left[V, \mathcal{F}\left(\frac{C}{Rt^\alpha}\right)\right].$$

Let $F \in C_0^\infty(Y)$ be a function such that $F \equiv 0$ in a neighbourhood of 0 and $F \equiv 1$ on $\text{supp}\,\mathcal{F}'$. Using Lemma 6.2(i) and (ii), we see that

$$\left\|\chi(H)\left[V, \mathcal{F}\left(\frac{C}{Rt^\alpha}\right)\right]\chi(H)\right\|$$

$$(6.12) \qquad \begin{aligned} &\leq R^{-1}t^{-\alpha}\|\chi(H)|\nabla V|\mathcal{F}'\left(\frac{C}{Rt^\alpha}\right)\chi(H)\| + O(R^{-2}t^{-2\alpha}) \\ &= R^{-1}t^{-\alpha}\|\chi(H)|\nabla V|F\left(\frac{y}{Rt^\alpha}\right)\mathcal{F}'\left(\frac{C}{Rt^\alpha}\right)\chi(H)\| + O(R^{-2}t^{-2\alpha}) \\ &= O(R^{-2-\mu}t^{-(2+\mu)\alpha}) + O(R^{-2}t^{-2\alpha}) = O(R^{-2}t^{-2\alpha}). \end{aligned}$$

Moreover, by our assumption on the decay of $W(t,x)$ we have

$$(6.13) \qquad \left[W(t,x), i\mathcal{F}\left(\frac{C}{Rt^\alpha}\right)\right] = O(R^{-1}t^{-1-\mu-\alpha}) + O(R^{-2}t^{-2\alpha}).$$

Combining (6.12) and (6.13), we obtain that:

$$(6.14) \qquad \chi(H)\left[H(t), \mathcal{F}\left(\frac{C}{Rt^\alpha}\right)\right]\chi(H) = O(R^{-1}t^{-2\alpha}).$$

This, together with (6.10), implies that:

$$D_t M_R = \Phi(t) + r(t),$$

where

$$(6.15) \qquad \begin{aligned} &\Phi(t) = \chi(H)\left(\partial_t\mathcal{F}\left(\frac{C}{Rt^\alpha}\right)\right)\chi(H) \geq 0, \\ &\|r(t)\| = O(R^0 t^{-2\alpha}) \in L^1([1,\infty)). \end{aligned}$$

We first note that (6.15) implies that

$$D_t M_R = |D_t M_R| + O(R^0 t^{-\alpha}).$$

By definition, we have $0 \leq M_R(t) \leq 1$ for all R and t. Hence

$$\int_1^T |D_t M_R| \leq \int_1^T D_t M_R + \int_1^T O(R^0 t^{-2\alpha})$$
$$= M_R(T) - M_R(1) + O(1) = O(1),$$

uniformly in R and T. It follows that

$$\int_1^\infty D_t M_R < \infty,$$

so that the limit

(6.16) $$\lim_{t \to \infty} \langle U^\star(t) M_R(t) U(t) u, u \rangle$$

exists for $u \in L^2(X)$.

We will now prove (6.8). Let $u \in L^2(X)$. By Lemma 6.3, it suffices to consider u of the form

$$u = \chi^2(H^+) u$$

for a cutoff function $\chi \in C_0^\infty(\mathbb{R})$. By Lemma 6.2(iii) and (6.7), we have

$$U^\star(t) \mathcal{F}\left(\frac{C}{Rt^\alpha}\right) U(t) u$$
$$= U^\star(t) \chi(H) \mathcal{F}\left(\frac{C}{Rt^\alpha}\right) \chi(H) U(t) u + o(1).$$

The existence of the limit (6.8) follows then from the existence of (6.16). The fact that $P_{\alpha,R}$ commutes with H^+ follows directly from (6.6) and (6.8). To prove (6.9), we first note that since $0 \leq \mathcal{F} \leq 1$, we have:

(6.17) $$0 \leq P_{\alpha,R} \leq 1.$$

Let $\chi \in C_0^\infty(\mathbb{R})$. Using (6.7), we write:

$$\chi(H^+) P_{\alpha,R} \chi(H^+) u$$

(6.18) $$= U^\star(t_0) \chi(H) \mathcal{F}(\frac{C}{Rt_0^\alpha}) \chi(H) U(t_0) u$$
$$+ \int_{t_0}^{+\infty} U^\star(t) D_t(M_R) U(t) u \, dt.$$

We deduce from (6.15) that

$$D_t M_R \geq -c_0 R^0 t^{-2\alpha}.$$

hence, given an $\epsilon > 0$, we can choose t_0 such that the integral in (6.18) is greater than $-\epsilon$ for all $R > 1$. Since $\mathcal{F}(y) \equiv 1$ near 0,

$$s-\lim_{R\to+\infty} \mathcal{F}\left(\frac{C}{Rt_0^\alpha}\right) = 1.$$

Hence for $u \in L^2(X)$ and any $\epsilon > 0$ we have:

$$\|\chi(H^+)u\|^2 \geq (\chi(H^+)P_{\alpha,R}\chi(H^+)u, u) \geq \|\chi(H^+)u\|^2 - 2\epsilon\|u\|^2,$$

provided that R is sufficiently large, which implies that

$$s-\lim_{R\to+\infty} \chi(H^+)P_{\alpha,R}\chi(H^+) = \chi^2(H^+).$$

By density, this yields (6.9). $\square$

The above theorem yields the estimate which was needed in the proof of asymptotic completeness in [GL2]. However, using methods similar to those described above (and Lemma 9.4*(ii)* of [GL3]), one can deduce from it the following corollary.

COROLLARY 6.5. *Let $\mathcal{F}$ and α be as in Theorem 6.4. Then:*

$$s-\lim_{t\to\infty} U^*(t)\mathcal{F}\left(\frac{y}{t^\alpha}\right) U(t) = 1.$$

7. Channel Hamiltonians.

We now return to the study of N-particle Hamiltonians of the form (2.1) for general N. Bound states of such Hamiltonians and their properties were described in Section 4; the remainder of this paper will be devoted to the analysis of the scattering states.

We first introduce some of the standard notation used in the N-body theory. $\mathcal{A}$ will stand for the set of all cluster decompositions, *i.e.*, partitions $a = (C_1, \ldots, C_k)$ of $\{1, \ldots, N\}$ into disjoint non-empty sets C_i called clusters. The number of clusters of a will be denoted by $\sharp a$. We will say that a is a refinement of b and write $a \leq b$ if all clusters of a are subsets of clusters of b. The relation $\leq$ defines a natural lattice structure on $\mathcal{A}$ with the maximal and minimal elements

$$a_{\max} = (\{1, \ldots, N\}), \quad a_{\min} = (\{1\}, \ldots, \{N\}).$$

We will write $a < b$ if $a \leq b$ and $a \neq b$.

Given a cluster decomposition a, X can be written as

$$X = X_a \oplus^\perp X^a,$$

where

$$X_a = \{x \in X \mid x_i = x_j \text{ for all } i, j \text{ such that } (ij) \le a\},$$

and

$$X^a = \left\{ x \in X \mid \sum_{i \in C_k} m_i x_i = 0 \text{ for all } C_k \in a \right\}$$

are the configuration spaces of the centers of mass of the clusters in a, and of the relative motion within clusters. In subscripts and superscripts, we will replace $a_{\max}$ and $a_{\min}$ by $\Diamond$ and $\sharp$ respectively. It is easy to check that

$$a \le b \text{ iff } X_b \subset X_a,$$

$$X_\Diamond = \{x \in X \mid x_i = x_j \text{ for all } i, j\}, \ X_\sharp = X.$$

We denote $Y_a = Y \cap X_a$, $Y^a = Y \cap X^a$, etc. The symbols x_a, x^a, y_a, etc., will stand for the orthogonal projections of $x \in X$ on the corresponding spaces. $A^a{}_a, A_a{}^a, A_a{}_a, A^a{}^a$ are the restrictions of the bilinear form A to $X^a \times X_a, X_a \times X^a, X_a \times X_a, X^a \times X^a$ respectively.

We will write $(ij) \le a$ if i, j are in the same cluster of a. Then we have:

$$H = H_a + I_a,$$

where

$$H_a = \frac{1}{2}(D - Ax)^2 + V^a(x^a), \ V^a(x^a) = \sum_{(ij) \le a} V_{ij}(x_i - x_j),$$

is the Hamiltonian of noninteracting clusters, and

$$I_a = V(x) - V^a(x^a) = \sum_{(ij) \nleq a} V_{ij}(x_i - x_j)$$

is the intercluster interaction.

Let us consider a cluster decomposition a equal to:

$$a = (C_1, \cdots C_{\sharp a}).$$

We denote by

$$Q_{C_l} = \sum_{i \in C_l} q_i$$

the total charge of the cluster C_l, we call the cluster C_l *neutral* if $Q_{C_l} = 0$ and *charged* if $Q_{C_l} \neq 0$. The numbers of charged and neutral clusters of a will be denoted by $\#_c a$ and $\#_n a$ respectively.

Let

$$X[C_l] = \{x \in X \mid x_k = 0 \text{ if } k \notin C_l\}$$

be the configuration space of the particles in C_l (isomorphic to $\mathbb{R}^{3n_l}$, where n_l is the number of particles in the cluster C_l). The Hamiltonian H_a can be written as

$$(7.1) \qquad\qquad H_a = \sum_{l=1}^{\#a} H[C_l],$$

where $H[C_l]$ is the Hamiltonian of the cluster C_l:

$$H[C_l] = \sum_{I \in C_l} \frac{1}{2m_i} (D_i - q_i J x_i)^2 + \sum_{i,j \in C_l} V_{ij}(x_i - x_j).$$

We can treat $H[C_l]$ as a n_l-particle Hamiltonian acting on $L^2(\mathbb{R}^{3n_l})$, and repeat the procedure of center of mass separation described in Section 3 with $H[C_l]$ instead of H, *i.e.*, apply to $H[C_l]$ the unitary transformation implementing the symplectic map described in Lemma 3.1 or Lemma 3.2, depending on whether the cluster C_l is neutral or charged. Combining such transformations for all clusters C_l in a, we will obtain the reduced channel Hamiltonian $\tilde{H}_a = \sum \tilde{H}[C_l]$.

If we wanted to write out explicitly the symplectic mappings in question for all clusters C_l, the notation would be unmanageable. Instead, we note that all of those mappings for neutral clusters are of the same form, so that we can combine them into one. We do the same for all charged clusters. Thus, we write

$$H_a = H_a^n + H_a^c,$$

where

$$H_a^n = \sum_{C_l \text{ neutral}} H[C_l] \quad \text{and} \quad H_a^c = \sum_{C_l \text{ charged}} H[C_l].$$

The spaces X_a and X^a can be decomposed accordingly:

$$X_a^n = \bigoplus_{\text{neutral } C_l} X_{C_l}, \quad X_a^c = \bigoplus_{\text{charged } C_l} X_{C_l},$$

$$X^{a,n} = \bigoplus_{\text{neutral } C_l} X^{C_l}, \quad X^{a,c} = \bigoplus_{\text{charged } C_l} X^{C_l},$$

where

$$X_{C_l} = \{x \in X \mid x_i = x_j \text{ if } (i,j) \in C_l, \ x_k = 0 \text{ if } k \notin C_l\}$$

and

$$X^{C_l} = \left\{ x \in| \sum_{i \in C_l} m_i x_i = 0, \; x_k = 0 \text{ if } k \notin C_l \right\}$$

are the configuration spaces of the center of mass of the cluster C_l and of the relative motion of the particles in C_l respectively. We have:

$$X_a = X_a^n \oplus X_a^c,$$

$$X^a = X^{a,n} \oplus X^{a,c}.$$

We denote by x_a^n, $x^{a,n}$, etc., the orthogonal projections on the above spaces. In particular, note that $y_\Diamond = y_\Diamond^n$ or $y_\Diamond = y_\Diamond^c$, depending on whether the system is neutral or charged. Y_a^n, Y_a^c, etc., are defined as the intersections of the corresponding subspaces of X with Y. Furthermore, to avoid triple superscripts (as in $y^{(1)a,c}$), we will write $y^{(1)} = \bar{y}$, $y^{(2)} = \hat{y}$. Similarly, let $\eta = (\bar{\eta}, \hat{\eta})$.

There exist bijective linear mappings

$$\bar{M}_a^c : Y_a^{c\prime} \to T^\star \mathbb{R}^{\natural c a}$$

$$\hat{M}_a^c : Y_a^{c\prime} \to T^\star \mathbb{R}^{\natural c a},$$

such that the following linear transformation is symplectic:

$$\chi_a : (x, \xi) \mapsto (x', \xi'),$$

where

$$\begin{cases} x' = (x_a', x^{a\prime}), \; x_a' = (\bar{y}_a^{c\prime}, \hat{y}_a^{c\prime}, y_a^{n\prime}, z_a'), \\ \xi' = (\xi_a', \xi^{a\prime}), \; \xi_a' = (\bar{\eta}_a^{c\prime}, \hat{\eta}_a^{c\prime}, \eta_a^{n\prime}, \zeta_a'), \end{cases}$$

and

$$\begin{cases} (\bar{\eta}_a^{c\prime}, \bar{y}_a^{c\prime}) = \bar{M}_a^c(\eta_a^c + (Ax)_a^c), \\ (\hat{\eta}_a^{c\prime}, \hat{y}_a^{c\prime}) = \hat{M}_a^c(\eta_a^c + A_a{}^{ac}x^a - A_{a\,a}^c x_a), \\ \eta_a^{n\prime} = (\eta_a^n + (Ax)_a^n), \\ y_a^{n\prime} = y_a^n, \\ z_a' = z_a, \\ \zeta_a' = \zeta_a, \\ \xi^{a\prime} = \xi^a - A^a{}_a x_a, \\ x^{a\prime} = x^a. \end{cases}$$

Let U_a be the unitary transformation implementing χ_a. Then

$$U_a H_a U_a^\star = \tilde{H}_a = \frac{1}{2} D_{z_a}^2 + \tilde{H}^a,$$

where

(7.2)
$$\begin{aligned}\tilde{H}^a = R_a^{\mathrm{w}}(D_{\hat{y}_a^c}, \hat{y}_a^c, x^a) + \frac{1}{2}(D_{y_a^{\mathrm{n}}} - 2A_{a,\mathrm{n}}{}^a x^a)^2 \\ + \frac{1}{2}(D_{x^a} - A^a{}^a x^a)^2 + V^a(x^a),\end{aligned}$$

and $R_a(\hat{\eta}_a^c, \hat{y}_a^c, x^a)$ is a second order polynomial equal to:

$$R_a(\hat{\eta}_a^c, \hat{y}_a^c, x^a) = \frac{1}{2}\left((\hat{M}_a^c)^{-1}(\hat{\eta}_a^c, \hat{y}_a^c) - 2A_{a,c}{}^a x^a\right)^2.$$

We also define:

$$H^a = H_a - \frac{1}{2}D_{z_a}^2 = U_a^\star \tilde{H}^a U_a.$$

The transformed Hamiltonian $\tilde{H}^a$ can be written as a direct integral:

$$\tilde{H}^a = \int_{Y_a^{\mathrm{n}\prime}}^{\oplus} \tilde{H}^a(\eta_a^{\mathrm{n}}) d\eta_a^{\mathrm{n}},$$

where

$$\tilde{H}^a(\eta_a^{\mathrm{n}}) = R_a^{\mathrm{w}}(D_{\hat{y}_a^c}, \hat{y}_a^c, x^a) + \frac{1}{2}(\eta_a^{\mathrm{n}} - 2A_{a,\mathrm{n}}{}^a x^a)^2 + \frac{1}{2}(D_{x^a} - A^a{}^a x^a)^2 + V^a(x^a)$$

acts on $L^2(\hat{Y}_a^c \times X^a)$.

8. Asymptotic completeness – N particles. We will now state the results of [GL1–3] on the problem of asymptotic completeness for N-particle systems in constant magnetic fields. Our goal is to describe the large time asymptotic behaviour of the solutions of the Schrödinger equation

$$i\frac{\partial u}{\partial t} = Hu,$$

where H is a Hamiltonian of the form (2.1). We saw in Section 4 that the Hilbert space $\mathcal{H} = L^2(X)$ can be decomposed into two subspaces:

$$\mathcal{H} = \mathcal{H}_{\mathrm{bound}} \oplus \mathcal{H}_{\mathrm{scatt}},$$

where $\mathcal{H}_{\mathrm{bound}}$ is the subspace of bound states for which all relative coordinates of the particles are bounded uniformly in time, i.e.,

$$\lim_{R \to \infty} \sup_{t \in \mathbb{R}} \|F(|x_i - x_j| > R)e^{-itH}u\| = 0$$

for all i, j, and $\mathcal{H}_{\mathrm{scatt}}$ is the subspace of scattering states.

Our task is to analyze the large time behaviour of $u_t = e^{-itH}u$ for $u \in \mathcal{H}_{\text{scatt}}$. Physical considerations suggest that as $t \to \infty$, the system breaks up into two or more stable and asymptotically independent subsystems, so that:

$$\|u_t - \sum_a u_{a,t}\| \to 0 \text{ as } t \to \infty,$$

where $u_{a,t}$ is the asymptotic evolution of noninteracting clusters in a decomposition $a = \{C_1, \ldots, C_n\}$.

Let us first discuss the case when the potentials V_{ij} are short-range (the modification needed for long-range interactions is described below). Under this assumption, we expect that

$$u_{a,t} = e^{-itH_a}u_a,$$

where H_a is the Hamiltonian of noninteracting clusters, and u_a is its bound state. Using the decomposition (7.1), we can write

$$u_{a,t} = e^{-itH_a}u_a = \bigotimes_l e^{-itH[C_l]}u_{C_l},$$

where $H[C_l]$ is the Hamiltonian of the cluster C_l (a precise definition is given in the previous section), and u_{C_l} is a bound state of the cluster.

The behaviour of $e^{-itH[C_l]}u_{C_l}$ depends on the electric charge Q_{C_l} of the cluster. Consider first the case when the system has no neutral subsystems, so that all possible clusters are charged. Then the asymptotic evolution of a scattering state of the system is a superposition of free (classical) motion of the centers of mass of subsystems in the direction of the field and bounded motion within subsystems, including the motion of their centers of mass across the field. The external and internal dynamics are decoupled, $i.e.$, the pseudodifferential operators describing them commute. This is illustrated in Figure 8.1: the system splits into asymptotically noninteracting subsystems, which travel with constant velocities along the field and perform bounded motion in other directions. If neutral clusters are present, the behaviour of the system changes qualitatively since such clusters can, and usually will, travel with nonzero average velocity across the field. Such motion is described in more detail in Section 4.

Asymptotic completeness for systems such that all proper subsystems are charged, $i.e.$,

$$\sum_{k=1}^{n} q_{i_k} \neq 0 \text{ for any } \{i_1, \ldots, i_n\} \subset \{1, \ldots, N\}, \ n < N, \qquad (Q\spadesuit)$$

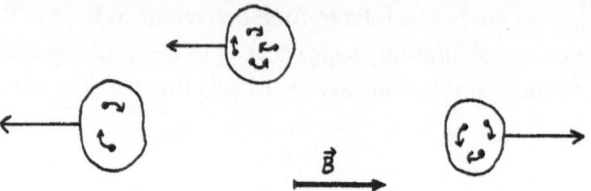

FIG. 8.1. *Charged clusters travelling along the field.*

(the total charge $Q = \sum_{i=1}^{N} q_i$ of the system may be arbitrary) was proved in [GL1] and [GL2] for short-range and long-range interactions respectively. In [GL3], we considered the case when neutral subsystems are allowed, *i.e.*, the only assumption on the electric charges of the particles was that all of them be nonzero:

$$q_i \neq 0, \ i = 1, \ldots, N. \tag{$Q\clubsuit$}$$

However, we have been able to handle only 3-particle systems, and we needed the additional assumptions described later in this section. Since the motion of neutral clusters is governed by an effective dispersive Hamiltonian, we call this case *dispersive*.

To state precisely the results of [GL1–3], we need to introduce some notation. Recall that for a cluster decomposition a, H_a^c and H_a^n denote the parts of $H_a = \sum H[C_l]$ corresponding to charged and neutral clusters respectively (see Section 7 for a definition), and let

$$\tilde{H}_a^c = U_a H_a^c U_a^*, \ \tilde{H}_a^n = U_a H_a^n U_a^*.$$

We can write the reduced Hamiltonian $\tilde{H}^a(\eta_a^n)$ as a sum:

$$\tilde{H}^a(\eta_a^n) = \tilde{H}^{a,n}(\eta_a^n) + \tilde{H}^{a,c},$$

where

$$(8.1) \quad \tilde{H}^{a,n}(\eta_a^n) = \frac{1}{2}(\eta_a^n - 2A_{a,n}{}^a x^a)^2 + \frac{1}{2}(D_{x^{a,n}} - A^{a\,a} x^{a,n})^2 + V^{a,n}(x^{a,n})$$

acts on $L^2(X^{a,n})$, and

$$(8.2) \quad \tilde{H}^{a,c} = R_a^w(D_{\hat{y}_a^c}, \hat{y}_a^c, x^a) + \frac{1}{2}(D_{x^{a,c}} - A^{a\,a} x^{a,c})^2 + V^{a,c}(x^{a,c})$$

acts on $L^2(\hat{Y}_a^c \times X^{a,c})$. Here we denote by

$$V^{a,c} = \sum_{(ij) \in \text{ charged } C_l} V_{ij},$$

$$V^{a,n} = \sum_{(ij) \in \text{ neutral } C_l} V_{ij},$$

the internal potentials of charged and neutral clusters respectively.

The channel identification operators can now be defined as follows:

$$(8.3) \qquad \Pi_a := U_a^* \left[E_{\mathrm{pp}}(\tilde{H}^{a,c}) \otimes \int_{Y_a^{n\prime}}^{\oplus} E_{\mathrm{pp}}(\tilde{H}^{a,n}(\eta_a^n))d\eta_a^n \right] U_a.$$

Note that

$$E_{\mathrm{pp}}(\tilde{H}^{a,c}) \otimes E_{\mathrm{pp}}(\tilde{H}^{a,n}(\eta_a^n)) = E_{\mathrm{pp}}(\tilde{H}^a(\eta_a^n)).$$

The states from $\mathrm{Ran}\Pi_a$ are those for which each cluster in a is in a bound state. In particular, we have:

$$\mathcal{H}_{\mathrm{bound}} = \mathrm{Ran}\Pi_{\diamond}.$$

Using the decomposition (7.1), $i.e.$, writing $H_a = \sum H[C_l]$, and applying to each $H[C_l]$ Proposition 4.2 or 4.4, depending on whether the cluster C_l is charged or neutral, we obtain the following straightforward generalization of [GL1, Proposition 6.6]:

PROPOSITION 8.1. $Let\ u \in \mathrm{Ran}\Pi_a$. $Then$:

$$\lim_{R\to\infty} \sup_{t\in\mathbb{E}} \left\| F\left(\frac{|x^a|}{R} \geq 1 \right) e^{-itH_a} u \right\| = 0,$$

$$\lim_{R\to\infty} \sup_{t\in\mathbb{E}} \left\| F\left(\frac{|y_a^c|}{R} \geq 1 \right) e^{-itH_a} u \right\| = 0.$$

As we mentioned above, the hypothesis (V) on the potentials does not suffice for our proof of asymptotic completeness for 3-particle systems containing neutral subsystems. We now state the additional assumptions that we need to handle this case. The first one is that the interactions are of Coulomb type.

DEFINITION 8.2. $The\ interactions\ V_{ij}\ are\ said\ to\ be\ of$ Coulomb type if:

$$V_{ij,L}(x_i - x_j) = q_i q_j V^l(x_i - x_j),$$

$where\ V^l\ is\ the\ same\ for\ all\ i, j$.

The key property of the Coulomb interactions which we needed in [GL3] is that if C_l is a neutral cluster, then $\sum_{i\in C_l} V ik, L(x_i - x)$ vanishes if $x_i = x_j$ for $(ij) \in C_l$. Note that Coulomb potentials $V_{ij} = -q_i q_j |x_i - x_j|^{-1}$ satisfy this condition, but not the assumption (V) because of the singularity at the origin. They are, however, allowed in [GL3].

Next, we will need certain implicit assumptions on the regularity of the eigenvalues of the reduced Hamiltonians of the neutral subsystems. It follows from Theorem 3.3(iii) and (v) that these conditions are satisfied for all simple discrete eigenvalues of $\tilde{H}^{a,n}(\eta_a^n)$. They fail to hold if the eigenvalues of $\tilde{H}^{a,n}(\eta_a^n)$ cross, and they seem very difficult to check for the possible eigenvalues of $\tilde{H}^{a,n}(\eta_a^n)$ embedded in the essential spectrum. Nonetheless, we believe that they allow us to understand the behaviour of a generic physical 3–body system with neutral clusters (see Section 11, Problem 1). Similar conditions were used by Dereziński in his study of dispersive Hamiltonians [De2].

We recall that $C^{n,1}(X)$ denotes the space of functions f in $C^n(X)$ such that $\partial_x^\alpha f$ is Lipschitz for $|\alpha| = n$. For a cluster decomposition a containing a neutral pair (i_1, i_2), we will denote by

$$\tau_a^n = \left\{ \sum_{j=i_1, i_2} \frac{|q_j|}{m_j}(2n_j + 1)b, \; n_j \in \mathbb{N} \right\}$$

the set of the Landau levels of the neutral pair, and by Γ_a — the set

$$\Gamma_a = \{(\lambda, \eta_a^n) \mid \lambda \in \sigma_{\mathrm{pp}}(\tilde{H}^{a,n}(\eta_a^n)), \lambda \notin \tau_a^n\} \subset \mathbb{R} \times Y_a^{n'}.$$

Hypothesis (H). There is a locally finite covering of Γ_a by open sets $\mathcal{V}_j = I_j \times \Omega_j$, $I_j \subset \mathbb{R}$, $\Omega_j \subset Y_a^{n'}$, $C^{1,1}$ projections $P_j : \Omega_j \to B(\mathcal{H})$, and $C^{2,1}$ functions $\omega_j : \Omega_j \to \mathbb{R}$, such that:

i) $\Gamma_a \cap \mathcal{V}_j = \{(\lambda, \eta_a^n) \mid \eta_a^n \in \Omega_j, \; \lambda = \omega_j(\eta_a^n)\}$;

ii) $E_{\{\omega_j(\eta_a^n)\}}(\tilde{H}^{a,n}(\eta_a^n)) = P_j(\eta_a^n)$ for $\eta_a^n \in \Omega_j$;

iii) The set τ_{a,Ω_j} of the critical values of ω_j on Ω_j is finite.

Let us now introduce the modified free dynamics that we will use to construct the wave operators in the long-range case. As in Section 5, we will use the Dollard modifiers. Since in the 3-particle dispersive case we assume that the potentials are of Coulomb type, the effective intercluster potential between a neutral pair and a third particle will be of short-range (see the remark after Definition 8.1). This is the reason why we only need to modify the dynamics in the direction of the magnetic field. Let us denote

$$|z|_a = \min_{(ij) \leq a} |z_i - z_j|,$$

and

(8.4) $$I_a(t, x) = F\left(\frac{|z|_a \log t}{t} \geq 1\right) Ia, L(x).$$

Then $I_a(t, x)$ satisfies

(8.5) $$|\partial_x^\alpha I_a(t, x)| \leq C_{\alpha, \epsilon} \langle t \rangle^{-|\alpha| - \mu + \epsilon}, \; |\alpha| \leq 1.$$

for all $\epsilon > 0$. Define

$$S_a(t, \zeta_a) = \int\limits_0^t \frac{1}{2}\zeta_a^2 + I_a(s, 0, s\zeta_a)ds.$$

Note that $S_a(t, D_{z_a})$ and H^a commute and:

$$S_a(t, \zeta_a) = \frac{1}{2}t\zeta_a^2, \text{ if } a \in \mathcal{A}^n.$$

The results of [GL1–3] on the problem of asymptotic completeness (proved there under weaker assumptions on the potentials) can be summarized in the following theorem.

THEOREM 8.1. *Let H be given by (2.1), where V satisfies (V) with $\mu > \sqrt{3} - 1$. Moreover, assume that one of the following sets of hypotheses holds: either*

(a) $(Q\spadesuit)$,

or

(b) $N = 3$, $(Q\clubsuit)$, the potentials V_{ij} are of Coulomb type (see the definition above), and the hypothesis (H) is satisfied for all cluster decompositions a containing neutral pairs.

Then:

(i) the modified wave operators

$$\Omega_a^+ := s - \lim_{t \to \infty} e^{itH} e^{-iS_a(t, D_{z_a}) - itH^a} \Pi_a$$

exist and their ranges are mutually orthogonal.

(ii) the system is asymptotically complete:

$$\bigoplus_{a \neq a_{\max}} \mathrm{Ran}\Omega_a^+ = \mathcal{H}_{\mathrm{scatt}}.$$

In the proof of the above theorem, we often rely on the methods of the standard N-particle scattering theory; this particularly applies to the analysis of the dynamics of the system in the direction of the magnetic field, where we frequently either simply invoke the known results or repeat, with only minor modifications, the arguments used in their proofs. Thus, the Mourre theory ([Mo], [PSS], [FH], [De2]), the construction of asymptotic velocity due to Dereziński [De3] (and based on the geometrical ideas of Graf [Gr]), and a minimal velocity estimate due to Sigal and Soffer [SS3], all play an important role in our work. At the last stage of the proof of asymptotic completeness in the long-range N-body charged case, one of the crucial steps is to apply Dereziński's arguments [De1] to the propagation along the field. (Hence the assumption on the rate of decay of the pair potentials.) We use a variety of tools and techniques developed for standard N-body systems, including those of [SS1–5], [G], and many other authors.

Graf's idea of constructing observables which depend only on some of the variables in certain regions of the configuration space inspired our geometrical analysis of the dynamics of the system in the long-range case, carried out in [GL2] and explained here in Section 10.

Here we will concentrate on the new problems encountered in the presence of a magnetic field and the methods developed in [GL1-3] in order to deal with them. A detailed and comprehensive discussion of the subject is well beyond the scope of this paper. Rather, we will present a few intermediate results, in whose proofs the main ideas of [GL1-3] find an application.

Section 9 is devoted to the proof of the Mourre estimate. The "charged" N-body case is considered in Subsection 9.1, where we present the results of Section 5 of [GL1]. The point that we want to make is that the center of mass separation is convenient, but not indispensable, in scattering theory: instead, we use the unitary transformations constructed in Sections 3 and 7. In Subsection 9.2, we demonstrate one of the main new ideas of [GL3] by proving the Mourre estimate in the 3-particle case for a scattering channel corresponding to unbounded motion of a bound state of a neutral pair across the field. Finally, geometrical arguments used in [GL2] to prove asymptotic completeness in the long-range case are discussed in Section 10.

9. The Mourre estimate. Throughout this section, we will use freely the notation and results of Appendix A.

9.1. The charged case. We will first assume that $(Q\spadesuit)$ holds, *i.e.*, all clusters consisting of less than N particles are charged. Our main result in this subsection is the Mourre estimate for $\tilde{H}^\diamond$ with the conjugate operator

$$A = \frac{1}{2}(\langle z, D_z \rangle + \langle D_z, z \rangle).$$

For a cluster decomposition a, we define its *threshold set*:

$$\tau_a = \bigcup_{b < a} \sigma_{pp}(\tilde{H}^b),$$

and put $d_a(\lambda) = \text{dist}(\lambda, \tau_a)$. We will also need the following notion of a−compact operators (see [PSS]).

DEFINITION 9.1. *An operator K^a acting on $L^2(\bar{Y}_a^c \times \hat{Y}_a^c \times Y_a^n \times X^a)$ is called $a-$ compact for $a \in \mathcal{A}$ if*

$$K^a = 1_{\bar{Y}_a^c} \otimes \int_{Y_a^{n\prime}}^{\oplus} K^a(\eta_a^n) d\eta_a^n,$$

where the operator $K^a(\eta_a^n) \in \mathcal{B}(L^2(\hat{Y}_a^c \times X^a))$ *satisfies*

$$K^a(\eta_a^n) \text{ is compact,}$$

$$\eta_a^n \to K^a(\eta_a^n) \text{ is norm continuous,}$$

$$\lim_{\eta_a^n \to \infty} K^a(\eta_a^n) = 0.$$

If a contains no neutral clusters, which we assume to be true for all $a \neq a_{\max}$, we have simply:

$$K^a = 1_{\hat{Y}_a^c} \otimes \tilde{K}^a,$$

where $\tilde{K}^a$ is compact on $L^2(\hat{Y}_a^c \times X^a)$.

THEOREM 9.1. *Assume that hypotheses (V) and (Q♠) hold. Then:*

(i) for each $\lambda \in \mathbb{R}$ *and* $\epsilon > 0$ *there is an* $a_{\max}$*-compact operator* $K^\diamond = K^\diamond(\epsilon)$ *such that:*

(9.1) $$[\tilde{H}^\diamond, iA] \geq 2d_\diamond(\lambda) - \epsilon - K^\diamond \text{ at } \tilde{H}^\diamond = \lambda;$$

(ii) the set $\tau_\diamond$ *is closed and countable.*

Remark. In [GL1] we show that for any $\epsilon > 0$ there is a $a_{\max}$-compact operator $K^\diamond = K^\diamond(\epsilon)$ such that:

$$[\tilde{H}^\diamond, iA] \geq 2\tilde{d}(\lambda) - \epsilon - K^\diamond \text{ at } \tilde{H}^\diamond = \lambda,$$

where

$$\tilde{d}(\lambda) = \text{dist}(\lambda, \tau_\diamond \cap (-\infty, \lambda])$$

is the distance from λ to the highest threshold to the left of it. This constant is optimal. Here, in order to simplify the proof, we do not pay attention to the value of the constant. For the intended application in the proof of asymptotic completeness, it suffices that it is positive if $\lambda \notin \tau_\diamond$.

Proof of Theorem 9.1. The proof below is a somewhat rearranged version of the one given by Froese-Herbst (see [FH], [CFKS, Sect 4.5]) for standard N-body Hamiltonians. We will use the lattice structure on the set $\mathcal{A}$ of all cluster decompositions to prove by induction in $a \in \mathcal{A}$ that:

(i') for any $\lambda \in \mathbb{R}$ *and* $\epsilon > 0$ *there is an* a*-compact operator* $K^a = K^a(\epsilon)$ *such that:*

$$[\tilde{H}^a, iA] \geq 2d_a(\lambda) - \epsilon - K^a \text{ at } \tilde{H}^a = \lambda;$$

(ii') the set τ_a *is closed and countable.*

This will imply the theorem, which states that the above conditions hold for $a = a_{\max}$.

To start the induction, we note that

$$\tilde{H}_{\parallel}^{\diamond} = \frac{1}{2}D_{z\diamond}^2 + \tilde{H}^{\parallel}$$

is unitarily equivalent to

$$H_{\parallel}^{\diamond} = \frac{1}{2}D_{z\diamond}^2 + \sum_1^N \frac{1}{2m_i}(D_{y_i} - q_i J y_i)^2.$$

This shows that the spectrum of $\tilde{H}^{\parallel}$ is purely discrete and consists of the Landau levels of noninteracting particles:

$$\sigma(\tilde{H}^{\parallel}) = \left\{ \sum_1^M \frac{|q_i|}{m_i}(2n_i + 1)b, \; n_i \in \mathbb{N} \right\}.$$

Since $[\tilde{H}_{\parallel}^{\diamond}, iA] = D_{z\diamond}^2$, the conditions (i')–(ii') are trivially satisfied for $a = a_{\min}$.

Assume now that (i')–(ii') hold with a replaced by b for all $b < a$. We will show that they also hold for a. We first note that (i')–(ii') for all $b < a$ implies (ii') for a, since the threshold set τ_a of $\tilde{H}^a$ consists of eigenvalues of $\tilde{H}^b$ for $b < a$, which, by (i') for b and a result of Mourre [Mo], can accumulate only at τ_b. Part (i') is an immediate consequence of Lemmas 9.2 and 9.3. □

Let $b < a$. Since $X^b \subset X^a$, we can write:

$$X^a =: X^b \oplus X_b^a.$$

We denote by x_b^a the orthogonal projection of $x^a \in X^a$ on X_b^a. Let also:

$$H_b^a = H_b - \frac{1}{2}D_{z_a}^2.$$

LEMMA 9.2. *Suppose that* $[\tilde{H}_b^a, iA] \geq c$ *at* $\tilde{H}_b^a = \lambda$ *for all* $b < a$, *where*

$$\tilde{H}_b^a = U_a H_b^a U_a^\star = U_a H_b U_a^\star - \frac{1}{2}D_{z_a}^2.$$

Then for any $\epsilon > 0$ *there is an* a-*compact operator* $K^a = K^a(\epsilon)$ *such that*

$$[\tilde{H}^a, iA] \geq c - \epsilon + K^a(\epsilon) \; at \; \tilde{H}^a = \lambda.$$

Proof. The proof is based on the following channel expansion: there is a partition of unity:

(9.2) $$1 = \sum_{b \leq a}(q_b^a(x^a))^2,$$

such that, if we denote by $q^a_{b,R}$ the operator of multiplication by $q^a_b(\frac{x^a}{R})$, we have:

$$f(\tilde{H}^a)[\tilde{H}^a, iA^a]f(\tilde{H}^a)$$

(9.3)
$$= \sum_{b \leq a} q^a_{b,R} f(\tilde{H}^a_b)[\tilde{H}^a_b, iA^a]f(\tilde{H}^a_b)q^a_{b,R} + o(R^0),$$
$$f^2(\tilde{H}^a) = \sum_{b \leq a} q^a_{b,R} f^2(\tilde{H}^a_b)q^a_{b,R} + o(R^0),$$

for any $f \in C^\infty_0(\mathbb{R})$. The partition q^a_b is a standard $N-$body partition of unity on X^a (an explicit construction is given e.g., in [Gr]), such that

$$\operatorname{supp} q^a_b \subset \{x \in X^a \mid |x^b| \leq \epsilon_0, |x^c| \geq \epsilon_1, \forall c \leq a, c \not\leq b\}$$

for some constants ϵ_0, ϵ_1. Then (9.3) follows easily from the hypothesis (V).

Fix $f \in C^\infty_0(\mathbb{R})$ such that $\operatorname{supp} f \subset [\lambda - 2\delta, \lambda + 2\delta]$ and $f \equiv 1$ on $[\lambda - \delta, \lambda + \delta]$. Given $\epsilon > 0$, we can find sufficiently large R such that the terms $o(R^0)$ is less that $\epsilon/2$. Hence, to prove the lemma, it suffices to show that $q^a_{a,R}f(\tilde{H}^a)$ is $a-$compact. Indeed, it follows from Kato's inequality that

$$F\left(\frac{|x^a|}{R} \leq 1\right)\left(R_a(D_{y^c_a}, y^c_a, x^a) + \frac{1}{2}(D^a - A^{a\,a}x^a)^2\right)^{-1}$$

is compact on $L^2(\hat{Y}^c_a \times X^a)$. If $a_{\max}$ is a charged cluster, the same argument shows that

$$q^\diamond_{\diamond,R}f(\tilde{H}^\diamond)$$

is $a_{\max}-$compact. If $a_{\max}$ is neutral, it is also straightforward to verify that

$$K(\eta_\diamond) = F\left(\frac{|x^\diamond|}{R} \leq 1\right)\left(\frac{1}{2}(\eta_\diamond - 2(Ax)_\diamond)^2 + \frac{1}{2}(D^\diamond - A^{\diamond\,\diamond}x^\diamond)^2\right)^{-1}$$

is compact on $L^2(X^\diamond)$, and that

$$\lim_{\eta_\diamond \to \infty} \|K(\eta_\diamond)\| = 0.$$

This completes the proof of the lemma. $\square$

LEMMA 9.3. *Assume that (i')–(ii') hold for a cluster decomposition* $b < a$. *Then for all* $\lambda \in \mathbb{R}$

$$[\tilde{H}^a_b, iA] \; a\text{-}\geq 2d_a(\lambda) \; at \; \tilde{H}^a_b = \lambda.$$

Proof. First, we claim that $\tilde{H}_b^a$ is unitarily equivalent to $\frac{1}{2}D_{z_b^a}^2 + \tilde{H}^b$. Indeed, this follows from the fact that:

$$U_a H_b U_a^\star = \frac{1}{2}D_{z_a}^2 + \tilde{H}_b^a,$$

and

$$U_b H_b U_b^\star = \frac{1}{2}D_{z_b}^2 + \tilde{H}^b = \frac{1}{2}D_{z_a}^2 + \frac{1}{2}D_{z_b^a}^2 + \tilde{H}^b.$$

Fix $\lambda \in \mathbb{R}$, and let $f_\delta \in C_0^\infty((\lambda - \delta, \lambda + \delta))$ for $0 < \delta < 1$. We have

$$
\begin{aligned}
(9.4) \quad & f_\delta\left(\frac{1}{2}D_{z_b^a}^2 + \tilde{H}^b\right)\left(\left[\frac{1}{2}D_{z_b^a}^2 + \tilde{H}^b, iA\right]\right) f_\delta\left(\frac{1}{2}D_{z_b^a}^2 + \tilde{H}^b\right) \\
&= f_\delta\left(\frac{1}{2}D_{z_b^a}^2 + \tilde{H}^b\right)(D_{z_b^a}^2 + [\tilde{H}^b, iA]) f_\delta\left(\frac{1}{2}D_{z_b^a}^2 + \tilde{H}^b\right) \\
&= \int_{z_b^{a'}}^{\oplus} f_\delta\left(\frac{1}{2}|\zeta_b^a|^2 + \tilde{H}^b\right)(|\zeta_b^a|^2 + [\tilde{H}^b, iA]) f_\delta\left(\frac{1}{2}|\zeta_b^a|^2 + \tilde{H}^b\right) d\zeta_b^a.
\end{aligned}
$$

Since $\tilde{H}^b$ is bounded from below, the integrand in (9.4) can be nonzero only for ζ_b^a in a compact set $\Omega = \{|\zeta_b^a| \leq C\}$, the same for all $\delta < 1$.

By *(i')–(ii')* for $\tilde{H}^b$, Lemma A.1, and the virial theorem, we have at $\tilde{H}^b = \nu$:

$$[\tilde{H}^b, iA] \ a \geq 2d_b(\nu) \geq 2d_a(\nu) \text{ if } \nu \notin \sigma_{pp}(\tilde{H}^b),$$

$$[\tilde{H}^b, iA] \ a \geq 0 = 2d_a(\nu) \text{ if } \nu \in \sigma_{pp}(\tilde{H}^b).$$

Hence $[\tilde{H}^b, iA] \ a \geq 2d_a(\lambda - \frac{1}{2}|\zeta_b^a|^2)$ at $\tilde{H}^b = \lambda - \frac{1}{2}|\zeta_b^a|^2$. By Lemma A.2*(ii)*, this holds uniformly for $\zeta_b^a \in \Omega$. Therefore, given $\epsilon > 0$, for $\delta = \delta(\epsilon)$ sufficiently small the right hand side of (9.4) is greater or equal than:

$$
\int_\Omega^{\oplus} f_\delta\left(\frac{1}{2}|\zeta_b^a|^2 + \tilde{H}^b\right)\left(|\zeta_b^a|^2 + 2d_a\left(\lambda - \frac{1}{2}|\zeta_b^a|^2\right)\right) f_\delta\left(\frac{1}{2}|\zeta_b^a|^2 + \tilde{H}^b\right) d\zeta_b^a
$$

$$
\geq 2d_a(\lambda) \int_\Omega^{\oplus} f_\delta^2\left(\frac{1}{2}|\zeta_b^a|^2 + \tilde{H}^b\right) d\zeta_b^a
$$

$$
\geq 2d_a(\lambda) f_\delta^2\left(\frac{1}{2}D_{z_b^a}^2 + \tilde{H}^b\right),
$$

where we used the triangle inequality:

$$d_a(\lambda - s) + s \geq d_a(\lambda) \text{ for } s \geq 0.$$

This proves the lemma. $\square$

9.2. The dispersive case. Let us now consider a 3-particle system for which the assumptions (V), $(Q\clubsuit)$, and (H) for all cluster decompositions containing a neutral pair are satisfied. As in the previous subsection, the Mourre estimate for $\tilde{H}^{\diamond}$ will be obtained from estimates for $\tilde{H}_a^{\diamond}$ for all cluster decompositions $a < a_{max}$. However, this time the conjugate operator will be obtained by gluing together operators $A + B_a$, where B_a depends both on a and the energy level λ.

Consider a cluster decomposition $a = \{(1,2),(3)\}$ in which the pair $(1,2)$ is neutral: $q_1 + q_2 = 0$. The transformed Hamiltonian $\tilde{H}_a^{\diamond}$ has the form

$$\tilde{H}_a^{\diamond} = \frac{1}{2}D_{z_a^{\diamond}} + \tilde{H}^{a,c} + \tilde{H}^{a,n},$$

where $\tilde{H}^{a,c}$ is a 1-particle reduced Hamiltonian, and $\tilde{H}^{a,n}$ is the transformed Hamiltonian of a neutral pair of particles with the center of mass motion along the z axis removed (see Subsection 3.2, Examples 1 and 2 respectively). The spectrum of $\tilde{H}^{a,c}$ consists of the Landau levels of the third particle:

$$\sigma_{pp}(\tilde{H}^{a,c}) = \tau_a^c = \left\{ \frac{|q_3|}{m_3}(2j+1)b \mid j = 0,1,\ldots \right\}.$$

We will denote $\Lambda_j = \frac{|q_3|}{m_3}(2j+1)b$. The Hamiltonian $\tilde{H}^{a,n}$ can be written as a direct integral

$$\tilde{H}^{a,n} = \int_{\eta_a^n \in Y_a^{n'}}^{\oplus} \tilde{H}^{a,n}(\eta_a^n).$$

The spectrum of $\tilde{H}^{a,n}(\eta_a^n)$ is described in Theorem 3.3. For simplicity, we will denote the parameter η_a^n by k.

We will first define the threshold set for H_a. Let

$$\tau_a = \left\{ \sum_1^3 \frac{|q_i|}{m_i}(2n_i+1)b, n_i \in \mathbb{N} \right\}$$

be the set of Landau levels of all particles. The Mourre estimate will fail to hold at these energy levels. We will also have to exclude other energy levels, which are defined below.

DEFINITION 9.2. *The secondary threshold set $\tilde{\tau}_a$ is the set*

$$\tilde{\tau}_a = \{\Lambda + \kappa \mid \Lambda \in \tau_a^c, \ \kappa \in \tilde{\tau}_a^n\},$$

where $\tau_a^c = \{\Lambda_j \mid j \in \mathbb{N}\}$, $\Lambda_j = \frac{|q_3|}{m_3}(2j+1)b$, is the set of the Landau levels of the third particle, and

$$\tilde{\tau}_a^n = \bigcup_j \tau_{a,\Omega_j},$$

where τ_{a,Ω_j} are defined in hypothesis (H).

The set $\tilde{\tau}_a^n$ consists of critical values of the effective kinetic energy of the motion of the bound states of the neutral pair in a. The following lemma is proved in [GL3].

LEMMA 9.4. *The set $\tau_a \cup \tilde{\tau}_a$ is discrete and closed.*

We can now state and prove the main result of this section — the Mourre estimate for $\tilde{H}_a^\diamond$.

THEOREM 9.5. *Assume that (V), $(Q\clubsuit)$, and (H) hold. Let $\lambda \in \mathbb{R} \backslash (\tau_a \cup \tilde{\tau}_a)$. Then there is an operator*

$$\tilde{B}_a(\lambda) = \frac{1}{2}\left(\langle m(\lambda; D_{z_a^\diamond}, D_{y_a^n}), y_a^n \rangle + \langle y_a^n, m(\lambda; D_{z_a^\diamond}, D_{y_a^n})\rangle\right)$$

where $m(\zeta_a^\diamond, \eta_a^n)$ is a $C^{1,1}$ function, and $\alpha_0(\lambda) > 0$, such that

$$(9.5) \qquad [\tilde{H}_a^\diamond, iA^\diamond + c\tilde{B}_a] \geq \min(1, c)\alpha_0(\lambda) \text{ at } \tilde{H}_a^\diamond = \lambda.$$

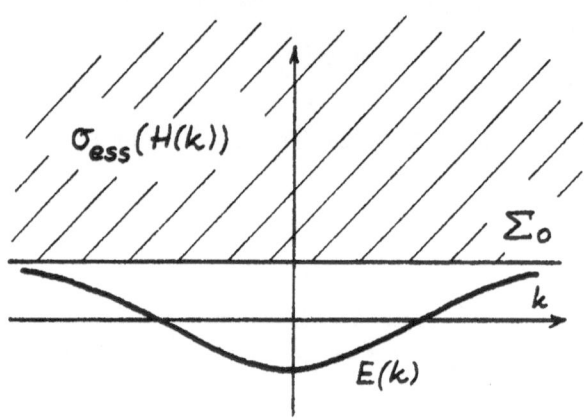

FIG. 9.1. *The spectrum of $H(k)$, simplified.*

The proof of Theorem 9.5 is given in [GL3]. Here we prove it under the following simplifying assumption.

Assumption (E). For each k, $\tilde{H}^{a,n}(k)$ has only one discrete and simple eigenvalue $E(k)$.

The eigenprojections corresponding to $E(k)$ will be denoted by $P(k)$. It follows from Theorem 3.3(iv), (v), that $E(k)$ is analytic in k and that $\lim_{|k|\to\infty} E(k) = \nu_0 = \inf \sigma_{ess} \tilde{H}^{a,n}(k)$.

Let us first explain the main idea of the proof. We write $\tilde{H}^{a,n}$ as a direct integral

$$\tilde{H}^{a,n} = \int_{Y_\diamond'}^{\oplus} H(k)\,dk.$$

If, as before, we let $A = \langle z, D_z \rangle$, we may have

$$\inf_{\tilde{H}_a^\diamond = \lambda} [\tilde{H}_a^\diamond, iA] = \inf_{\tilde{H}_a^\diamond = \lambda} D_{z_a}^2{}_\diamond = 0,$$

whenever $\lambda - \Lambda_j = E(k)$ for some $j \in \mathbb{N}$ and $k \in \mathbb{R}$. The set of such values of λ is not discrete – in fact, it is the interval $[\min E(k), \nu_0)$. Let

$$P = \int^\oplus P(k) dk.$$

Since A commutes with k, we have by the virial theorem:

$$P[\tilde{H}^{a,\mathrm{n}}, iA]P = \int^\oplus P(k)[H(k), A]P(k) dk = 0.$$

The same argument shows that one cannot expect a Mourre estimate with a conjugate operator commuting with k to hold for $\tilde{H}_a^\diamond$. (Recall that to prove a strict Mourre estimate, without a compact error, for a neutral pair of particles in Section 5, we had to localize it in the pseudomomentum k. We cannot do that here, since k is not a constant of motion for the three-particle Hamiltonian H.)

We will therefore have to add to A a term of the form

$$\tilde{B}_a = \frac{1}{2} P(\langle \nabla_k E(k), D_k \rangle + \langle D_k, \nabla_k E(k) \rangle) P.$$

We have:

$$P[\tilde{H}_a^\diamond, i(A + \tilde{B}_a)]P = \int^\oplus (|\zeta_a^\diamond|^2 + |\nabla E(k)|^2) P(k) dk,$$

and its infimum at $\tilde{H}_a^\diamond = \lambda$, $\lambda < \nu_0$, is strictly positive unless $\lambda - \Lambda_j$ is a critical value of $E(k)$ for some $j \in \mathbb{N}$. The set of such values λ is excluded as the secondary threshold set $\tilde{\tau}_a$.

We will denote by

$$\tau_a^\mathrm{n} = \left\{ \sum_1^2 \frac{|q_i|}{m_i} (2n_i + 1) b, n_i \in \mathbb{N} \right\},$$

the set of Landau levels for the pair of particles $(1, 2)$, and by

$$d(E) = \mathrm{dist}(E, \tau_a),$$

$$d^\mathrm{n}(E) = \mathrm{dist}(E, \tau_a^\mathrm{n}),$$

the distances from E to the threshold sets τ_a and τ_a^n. Note that $d(E)$ and $d^n(E)$ are continuous functions of E. Moreover,

$$(9.6) \qquad d(E + s) \leq d(E) + s \text{ for all } E \in \mathbb{R},\ s > 0,$$

and

$$(9.7) \qquad d^n(E - \Lambda_j) \geq d(E).$$

We will need the following lemma, which is proved in [GL3]:

LEMMA 9.6. *i) Let $E \in \mathbb{R}$ and $\epsilon > 0$. Then there is a constant $C > 0$ such that*

$$(9.8) \qquad [\tilde{H}^{a,n}(\eta_a^n), iA^a] > 2d^n(E) - \epsilon \text{ at } \tilde{H}^{a,n}(\eta_a^n) = E$$

uniformly for $|\eta_a^n| > C$.

ii) If $I \subset \mathbb{R}$ is a compact interval, for any $\epsilon > 0$ there is a $C > 0$ such that (9.8) holds uniformly for $|\eta_a^n| > C$ and $E \in I$.

Proof of Theorem 9.5. We will fix the value of λ as in the statement of the theorem. To simplify the notation, we will write $H(k)$ and ζ instead of $\tilde{H}^{a,n}(k)$ and $\zeta_a^\diamond$, and, except when it is necessary, we will not display the dependence of constants, operators, etc., on λ.

Let f_δ denote a function in $C_0^\infty([\lambda - \delta, \lambda + \delta])$ such that $f_\delta \equiv 1$ on $[\lambda - \delta/2, \lambda + \delta/2]$. We have

$$f_\delta(\tilde{H}_a^\diamond)[\tilde{H}_a^\diamond, iA]f_\delta(\tilde{H}_a^\diamond)$$

$$= \sum_{j=0}^\infty \int^\oplus f_\delta\left(\frac{1}{2}\zeta^2 + \Lambda_j + H(k)\right)(\zeta^2 + [H(k), iA])$$

$$f_\delta\left(\frac{1}{2}\zeta^2 + \Lambda_j + H(k)\right)d\zeta dk,$$

(recall that $\Lambda_j = \frac{|q_3|}{m_3}(2j + 1)b$, $j \in \mathbb{N}$, are the Landau levels of the third particle). Since $H(k)$ are bounded from below uniformly in k, $f_\delta(\frac{1}{2}\zeta^2 + \Lambda_j + H(k))$ is nonzero only for finitely many values of j ($j = 1, 2, \ldots, N_0$), and the sum on the right-hand side is in fact finite. For $1 \leq j \leq N_0$, let $\lambda_j = \lambda - \Lambda_j$ and $f_j(s) = f_\delta(s + \Lambda_j)$ (so that f_j is supported in $(\lambda_j - \delta, \lambda_j + \delta)$). If δ is small enough, the supports of f_j's are disjoint for different values of j. Note that if $\lambda \notin \tau_a$, $\lambda_j \notin \tau_a^n$ for all j.

For any function $f \in C_0^\infty([\lambda_j - 1, \lambda_j + 1])$ we have $f(H(k) + \frac{1}{2}\zeta^2) = 0$ if $|\zeta|$ is large enough ($|\zeta| \geq C_1$). By Lemma 9.6ii), for any $\epsilon > 0$ there is a constant $C_2 = C_2(\epsilon)$ such that for $1 \leq j \leq N_0$,

$$[H(k), iA] \geq 2d^n\left(\lambda_j - \frac{1}{2}\zeta^2\right) - \epsilon \geq 2d\left(\lambda - \frac{1}{2}\zeta^2\right) - \epsilon \text{ at } H(k) = \lambda_j - \frac{1}{2}\zeta^2$$

uniformly for $|k| > C_2(\epsilon)$ and $|\zeta| < C_1$ (the second inequality follows from (9.7)). Let $C(\epsilon) = \max(C_1, C_2(\epsilon))$. Then, if δ is sufficiently small, we have

$$
\int\limits_{|(\zeta,k)| \geq C(\epsilon)}^{\oplus} f_j\left(\frac{1}{2}\zeta^2 + H(k)\right)\left(\zeta^2 + [H(k), iA]\right) f_j\left(\frac{1}{2}\zeta^2 + H(k)\right) d\zeta dk
$$

$$
\geq \int\limits_{|(\zeta,k)| \geq C(\epsilon)}^{\oplus} f_j\left(\frac{1}{2}\zeta^2 + H(k)\right)\left(\zeta^2 + 2d\left(\lambda - \frac{1}{2}\zeta^2\right) - \epsilon\right) f_j\left(\frac{1}{2}\zeta^2 + H(k)\right) d\zeta dk
$$

$$
\geq \int\limits_{|(\zeta,k)| \geq C_0}^{\oplus} (2d(\lambda) - \epsilon) f_j^2\left(\frac{1}{2}\zeta^2 + H(k)\right) d\zeta dk,
$$

where we have used that by (9.6),

$$
(9.9) \qquad d\left(\lambda - \frac{1}{2}\zeta^2\right) + \frac{1}{2}\zeta^2 \geq d(\lambda).
$$

Let $\epsilon = d(\lambda) > 0$, then for $C_0 = C(d(\lambda))$ we have:

$$
\int\limits_{|(\zeta,k)| \geq C_0}^{\oplus} f_j\left(\frac{1}{2}\zeta^2 + H(k)\right)\left(\zeta^2 + [H(k), iA]\right) f_j\left(\frac{1}{2}\zeta^2 + H(k)\right) d\zeta dk
$$

$$
\geq \int\limits_{|(\zeta,k)| \geq C_0}^{\oplus} d(\lambda) f_j^2\left(\frac{1}{2}\zeta^2 + H(k)\right) d\zeta dk,
$$

(9.10)

Note that $\lambda_j - \frac{1}{2}|\zeta|^2 \geq \nu_0 = \inf \sigma_{pp}(H(k))$ if and only if $|\zeta|^2 \leq \nu_j = 2(\lambda_j - \nu_0)$. By Lemmas 5.5 and A.1, and by (9.7), for $|\zeta|^2 \leq \nu_j$ and all k we have

$$
(9.11) \qquad
\begin{aligned}
[H(k), iA] \; a\text{-}&\geq 2d^n\left(\lambda_j - \frac{1}{2}\zeta^2\right) \\
&\geq 2d\left(\lambda - \frac{1}{2}\zeta^2\right) \text{ at } H(k) = \lambda_j - \frac{1}{2}\zeta^2.
\end{aligned}
$$

By Lemma A.2, for any $0 < \epsilon < d(\lambda)$

$$
[H(k), iA] \geq 2d\left(\lambda - \frac{1}{2}\zeta^2\right) - \epsilon
$$

at $H(k) = \lambda_j - \frac{1}{2}\zeta^2$ uniformly on the compact set $M_j = \{|\zeta|^2 \leq \nu_j, |k| \leq C_0\}$. (The continuity assumptions of Lemma A.2 are satisfied by Theorem

3.3(v), Lemma 5.3, and Hypothesis (H).) Hence for sufficiently small δ,

$$\int\limits_{M_j}^{\oplus} f_j \left(\frac{1}{2}\zeta^2 + H(k)\right) (\zeta^2 + [H(k), iA]) f_j \left(\frac{1}{2}\zeta^2 + H(k)\right) d\zeta dk$$

$$(9.12) \geq \int\limits_{M_j}^{\oplus} \left(\zeta^2 + 2d\left(\lambda - \frac{1}{2}\zeta^2\right) - \epsilon\right) f_j^2 \left(\frac{1}{2}\zeta^2 + H(k)\right) d\zeta dk$$

$$\geq d(\lambda) \int\limits_{M_j}^{\oplus} f_j^2 \left(\frac{1}{2}\zeta^2 + H(k)\right) d\zeta dk.$$

Let

$$\Sigma_j = \left\{(\zeta, k) | \lambda_j = \frac{1}{2}\zeta^2 + E(k)\right\}.$$

Note that Σ_j is compact, $\Sigma_j \cap \Sigma_l = \emptyset$ for $j \neq l$, and $\Sigma_j \subset \{|\zeta|^2 \geq \nu_0\}$.

Let $(\zeta_0, k_0) \in \Sigma_j$. Since $\lambda_j \notin \tilde{\tau}_a^n$, we can choose a neighbourhood $\mathcal{V}$ of (ζ_0, k_0) such that on $\mathcal{V} \cap \{\zeta = 0\}$ we have:

$$(9.13) \qquad\qquad |\nabla E(k)| > 0.$$

In particular, (9.13) implies that $\zeta^2 + |\nabla E(k)|^2 > 0$ for all $(\zeta, k) \in \Sigma_j$. Since Σ_j is compact for each j, there are open neighbourhoods $\mathcal{V}_j$ of Σ_j and a constant $\theta_0 > 0$ such that $\mathcal{V}_j \cap \mathcal{V}_l = \emptyset$ if $j \neq l$, $\mathcal{V}_j \subset \{|\zeta|^2 < \nu_j\}$, and for $1 \leq j \leq N_0$:

$$(9.14) \qquad \inf\{\zeta^2 + |\nabla E(k)|^2 \mid (\zeta, k) \in \mathcal{U}_j\} > \theta_0.$$

By the virial theorem, we have:

$$P(k)[H(k), iA]P(k) = 0.$$

Hence, if $\delta > 0$ is sufficiently small,

$$\int\limits_{\mathcal{V}_j}^{\oplus} f_j \left(\frac{1}{2}\zeta^2 + H(k)\right) (\zeta^2 + [H(k), iA]) f_j \left(\frac{1}{2}\zeta^2 + H(k)\right) d\zeta dk$$

$$(9.15) \qquad \geq \int\limits_{\mathcal{V}_j}^{\oplus} \zeta^2 f_j^2 \left(\frac{1}{2}\zeta^2 + H(k)\right) P(k) d\zeta dk.$$

We define

$$\tilde{B}_a = \frac{1}{2} P(\langle \nabla E(k), D_k \rangle + \langle D_k, \nabla E(k) \rangle) P,$$

where $P = \int^{\oplus} P(k)dk$. Using the identity

(9.16)
$$[H(k), iP(k)D_kP(k)]$$
$$= [E(k), iP(k)D_kP(k)] = P(k)\nabla E(k)P(k),$$

we obtain:

(9.17)
$$f(\tilde{H}_a^{\diamond})[\tilde{H}_a^{\diamond}, ic\tilde{B}_a]f(\tilde{H}_a^{\diamond})$$
$$\geq \sum_{j=1}^{N_0} \int_{\mathcal{V}_j}^{\oplus} f_j^2 \left(\frac{1}{2}\zeta^2 + H(k)\right) P(k)c|\nabla E(k)|^2 d\zeta dk.$$

By (9.14),

(9.18)
$$f_j^2 \left(\frac{1}{2}\zeta^2 + H(k)\right)(\zeta^2 + c|\nabla E(k)|^2)P(k)$$
$$\geq \min(1, c)\alpha_0 f_j^2 \left(\frac{1}{2}\zeta^2 + H(k)\right) P(k),$$

where $\alpha_0 = \min(\theta_0/2, d(\lambda))$. Collecting (9.10), (9.12), (9.15), (9.17), and using (9.18), we obtain:

$$f_\delta(\tilde{H}_a^{\diamond})[\tilde{H}_a^{\diamond}, i(A + c\tilde{B}_a)]f_\delta(\tilde{H}_a^{\diamond}) \geq \min(1, c)\alpha_0 f_\delta^2(\tilde{H}_a^{\diamond}),$$

which completes the proof of the theorem. □

One point that requires a clarification is why we need the constant c in Theorem 9.5. The Mourre estimate for $H^{\diamond}$ is obtained in [GL3] by gluing together the estimates with conjugate operators as in Theorem 9.5 for the cluster decompositions containing a neutral pair, and with A as a conjugate operator for cluster decompositions in which all clusters are charged. Since different conjugate operators are used for different cluster decompositions, this is not merely a standard procedure involving a partition of unity and localization in almost commuting variables.

Our construction is based on a channel expansion ([GL3, Proposition 7.7]), which states that the error terms due to the auxiliary conjugate operators $\tilde{B}_a$ are essentially localized in the free region, where all particles are separated. (It is crucial for its proof that we deal with a 3-body system, so that the regions of the configuration space corresponding to different 2-cluster decompositions are disjoint.) We then choose the constant c sufficiently small, so that in the free region the commutator is dominated by the positive term $[H_{\parallel}^{\diamond}, iA] = D_{z\diamond}^2$, and use Theorem 9.5 to show that the terms corresponding to $a \in \mathcal{A}$, $\sharp a = 2$, are positive.

10. N-body long-range case: geometrical methods. In this section we present a geometrical analysis of the dynamics of the system in the transversal directions, which will allow us to generalize the argument from Section 6 showing that there is no propagation in the region

$$|y| > t^\alpha, \ \alpha > \frac{1}{2},$$

under the assumption that the system has no neutral subsystems, i.e.,

$$\sum_{k=1}^{n} q_{i_k} \neq 0 \text{ for all } \{i_1, \ldots, i_n\} \subset \{1, \ldots, N\}, \ n \leq N. \qquad (A\spadesuit)$$

In particular, we assume that the total charge Q of the system is nonzero. In this section, we restrict our attention to the dynamics in the transversal directions.

As in the 1-particle case, we define the *center of orbit* observable by:

$$C = \frac{1}{2}A^{-1}(D + Ay).$$

Lemma 6.1 holds in the N-particle case, and the proof is identical to that in Section 6. Hence to estimate the growth of y by t^α it suffices to prove similar bounds on C.

We want to use the same method as in Section 6, *i.e.*, find a cutoff function $\mathcal{F}$ for which (6.5) and (6.4) are satisfied, so that we could prove an analogue of Theorem 6.4. This time, however, we cannot simply take any $\mathcal{F} \in C_0^\infty(Y)$ equal to 1 in a neighbourhood of 0. In Section 6 we used that for a potential V satisfying (6.2), if $y \geq t^\alpha$, $\nabla V(y) = O(t^{-1-\mu})$. This is not true for an N-particle potential $V = \sum V_{ij}$ which does not vanish at infinity in all directions.

The starting point is, therefore, the observation that for N-body systems there is a system of subspaces of the configuration space along which the total potential does not vanish: the hyperplanes $\{y_i = y_j\}$ and all their intersections, which are denoted here by Y_a (see Section 7). We expect unbounded motion of the system across the field only along these subspaces. The heuristic argument here is that, if the particles move apart from one another, they become asymptotically independent and should behave the way noninteracting particles do: remain in a bounded region in the directions transversal to the field. This idea can be translated into a rigorous mathematical proof: in [GL3, Section 9], we use the center of orbit observable to show that there is no propagation in the "free region" (away from X_a for all $a \neq a_{\min}$) for 3-body systems satisfying the weaker assumption $(Q\clubsuit)$ on the charges. An extension to the N-body case is straightforward.

Let us first sketch a 2-particle version of the argument used in [GL2].

Example: 2-particle systems. Consider the case $N = 2$. The potential $V(x_1 - x_2)$ does not vanish at infinity only if $y_1 - y_2$ is bounded (i.e., along the plane $Y_\diamond = \{y_1 = y_2\}$, so that we expect propagation only in that region.

On the other hand, we have the following bound on the *center of charge* of the system:

$$\left\| \sum q_i y_i |u_t\| \right\| = O(1) \text{ as } t \to \infty.$$

To see this, we note that the total pseudomomentum $\sum (D_i + q_i J y_i)$ is a constant of motion, and, by the assumption (V) on the potentials, $\sum (D_i - q_i J y_i)$ is H-bounded. This implies uniform bounds on $\sum q_i J y_i$, and, since $J : Y_i \to Y_i'$ is invertible, on $\sum q_i y_i$.

Thus, we have reasonable grounds to believe that $y_1 - y_2$ and $q_1 y_1 + q_2 y_2$ are bounded along the evolution. The condition that the system is charged $(q_1 \neq -q_2)$ means that $q_1 y_1 + q_2 y_2$ is not a multiple of $y_1 - y_2$, hence if both of them are bounded, y_1 and y_2 must be bounded as well.

We now want to generalize this argument to an arbitrary number of particles.

Center of charge coordinates. We introduce an inner product on Y with coefficients given by the electric charges of the particles:

$$q(y, \bar{y}) = \sum q_i y_i \bar{y}_i.$$

Note that, since we allow both positively and negatively charged particles, q need not be positive definite.

For each of the subspaces Y_a we define the subspace $Y^{a,q}$ orthogonal to it in the metric q:

$$Y^{a,q} = \{y \in Y \mid q(y, \bar{y}) = 0 \text{ for all } \bar{y} \in Y_a\}$$

The condition that there are no neutral subsystems means that the metric q restricted to each of the subspaces Y_a is non-degenerate: $Y_a \cap Y^{a,q} = \{0\}$ for all Y_a. (This is not true if neutral subsystems are present; the nonempty intersections correspond to the possible propagation of these subsystems to infinity).

In the case considered here, we can write Y as:

$$Y = Y_a \oplus_q Y^{a,q},$$

and for any $y \in Y$ we have a unique decomposition $y = y_a^q + y^{a,q}$.

Example ($N = 2$) continued. There is only one subspace $Y_\diamond = \{y_1 = y_2\}$ along which V does not vanish. Its orthogonal complement $Y^{\diamond,q}$ is given by

$$Y^{\diamond,q} = \{y \mid q_1 y_1 + q_2 y_2 = 0\}.$$

If $Q = q_1 + q_2 \neq 0$, these subspaces intersect only at 0. The projection $y_\diamond^q$ is the center of charge of the system:

$$y_\diamond^q = (Q^{-1}(q_1 y_1 + q_2 y_2), Q^{-1}(q_1 y_1 + q_2 y_2)).$$

A geometric construction. This is the most difficult technically part of the proof. We construct a function $\mathcal{F} \in C_0^\infty(Y)$ with the following properties:

   (i) $\mathcal{F} \equiv 1$ in a neighbourhood of 0;
   (ii) near each of the subspaces Y_a $\mathcal{F}(y)$ depends only on the projection y_a^q of y on that subspace;
   (iii) $y \cdot \nabla \mathcal{F}(y) \leq 0$.

A similar geometrical analysis in the usual (no magnetic field) N-body scattering, with center of mass instead of center of charge coordinates, is due to Graf [Gr]. The difference between his construction and ours comes from the fact while that the scalar product $\langle y, \bar{y} \rangle = \sum m_i y_i \bar{y}_i$, defining the center of mass coordinates, is positive definite, q is not. We can explain this by drawing pictures of the supports of the functions in question. In each case, their boundaries are obtained from a sphere $S = \{\langle x, x \rangle = 1\}$ by modifying it in neighbourhoods of the intersections with Y_a's, so that the obtained surface is in these neighbourhoods orthogonal to Y_a in the given inner product. However, since Y^a is parallel to the hyperplane tangent to S at the points where Y_a intersects it, Graf's construction could be accomplished by flattening S around these intersections. On the other hand, the subspace $Y^{a,q}$, to which our surface is to be parallel, can be almost perpendicular to the tangent to S at the intersections with Y_a, and a similar procedure of flattening S, or variations thereof, will not work. Instead, we attach to S long and thin extensions, which can then be sliced off in the desired direction. The 1-dimensional spheres (circles) modified as in Graf's construction and in ours are shown in Figure 10.1(a) and (b) respectively.

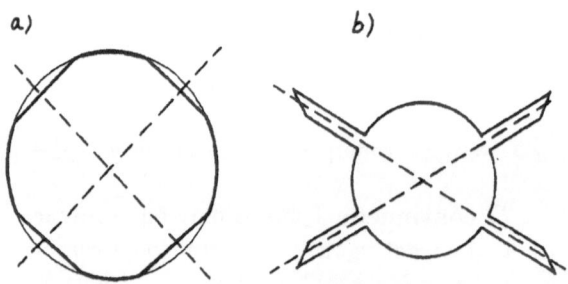

FIG. 10.1. *Modified spheres in 2 dimensions.*

The main difficulty for 4 or more particles is to make the deformations along different Y_a's compatible at their intersections (see Figure 10.2). A picture of our set for 4 particles is given in Figure 10.3. For $N > 4$ we just proceed by induction.

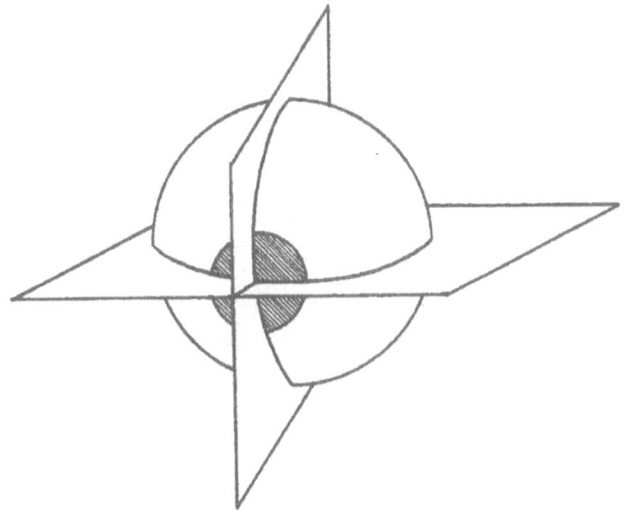

FIG. 10.2. *Planes intersecting a sphere in 3 dimensions.*

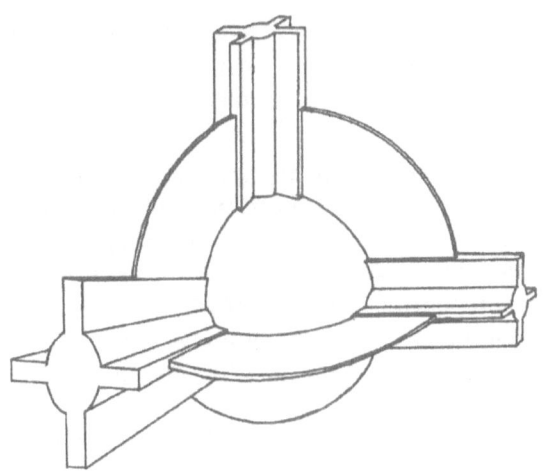

FIG. 10.3. *The modified sphere in 3 dimensions.*

Proof of the estimate on y. Our main observable will be $\mathcal{F}(\frac{C}{t^\alpha})$, where $\mathcal{F}$ is the function we have just constructed, and C is, as before, the center of orbit. Consider the Heisenberg derivative of $\mathcal{F}(\frac{C}{t^\alpha})$. Using the properties *(i)-(ii)* of $\mathcal{F}$ and the commutator expansion of Lemma 6.2*(ii)*, we obtain that:

$$DF\left(\frac{C}{t^\alpha}\right) = \frac{\partial}{\partial t}F\left(\frac{C}{t^\alpha}\right) + i\left[H, F\left(\frac{C}{t^\alpha}\right)\right]$$

$$= -\alpha\frac{C}{t^{\alpha+1}}\nabla F\left(\frac{C}{t^\alpha}\right) + A^{-1}t^{-\alpha}\nabla V \cdot \nabla F\left(\frac{C}{t^\alpha}\right) + O(t^{-2\alpha}).$$

Now comes the crucial step: the projections C_a^q have been constructed in such a way that

$$[V_{ij}, C_a^q] = 0$$

for all pair potentials V_{ij} which do not vanish at infinity along Y_a.

Example $(N = 2)$ **continued.** $Y_\diamond = \{y_1 = y_2\}$, and $C_\diamond^q$ is the center of charge of the system, with the coordinates of the particles replaced by the centers of their orbits. When we compute it, we see that the electric charges cancel out:

$$(C_\diamond^q)_i = Q^{-1}(q_1 C_1 + q_2 C_2) = Q^{-1}\sum_{i=1,2} q_i \cdot \frac{1}{2}q_i^{-1}J^{-1}(D_i + q_i J y_i)$$

$$= \frac{1}{2}Q^{-1}J^{-1}\sum_{i=1,2}(D_i + q_i J y_i).$$

This is a multiple of the pseudomomentum and, of course, commutes with $V(x_1 - x_2)$.

Once we have this, we can apply essentially the same one-body argument as in Section 6 and use the decay of V *in the directions orthogonal to Y_a's* to estimate the Heisenberg derivative of $F(\frac{C}{t^\alpha})$. This will imply the needed estimate on y.

11. Open problems. While the problem of asymptotic completeness in the "charged" case (for systems containing no neutral proper subsystems, *i.e.*, satisfying $(Q\spadesuit)$) was solved in [GL1] and [GL2], we are only beginning to understand the general case. The problems listed below are just a few of the many remaining open questions. Although some are stated as "prove that ...", it might be more appropriate to ask under what conditions the given conclusions hold, since we expect that additional assumptions may be needed.

Here I will list only questions most directly related to our work on N-particle scattering. It should also be possible to carry out a more detailed analysis of the properties and behaviour of bound states of specific physical systems; however, I do not feel competent to discuss that subject in detail.

1. Check for which 3-particle systems Hypothesis (H) holds. As explained in Section 8, to prove the Mourre estimate and asymptotic completeness for 3-particle systems we need certain implicit assumptions on the regularity of the eigenvalues of the reduced Hamiltonians $H(k)$ of the 2-particle neutral subsystems. It is not difficult to show that these assumptions are satisfied for the discrete simple eigenvalues of $H(k)$. However, we do not know how to control the behaviour of the possible eigenvalues of $H(k)$ embedded in the continuous spectrum to which one cannot apply the usual methods of perturbation theory. Neither can we show that the discrete eigenvalues of $H(k)$ do not cross, in which case the Hypothesis (H) would also fail.

2. Prove asymptotic completeness for 3-particle systems with general long-range potentials. In [GL3], to prove asymptotic completeness for 3-particle systems with neutral subsystems, we had to assume that the potentials were of Coulomb type. We do not know whether a similar result would hold without this assumption.

The final step in the proof of asymptotic completeness is the analysis of the dynamics of clusters moving apart with nonzero relative velocities. Consider thus a decomposition consisting of a neutral pair $(1, 2)$ and a charged particle (3). The evolution of the neutral pair is governed by an effective time-dependent Hamiltonian

$$H(t) = H[(1,2)] + I(t,x),$$

where $H[(1,2)] = \sum_1^2 \frac{1}{2m_i}(D_i - q_i J x_i)^2 + V_{12}(x_1 - x_2)$, and $I(t,x)$ is the interaction of the pair with the third particle. It is here that the assumption that the potentials are of Coulomb type is needed: it implies that if the pair is in a bound state and the distance between the two clusters grows linearly in time, $I(t,x)$ is short-range, i.e., $I(t,x) = O(t^{-1-\epsilon})$. Hence, as $t \to \infty$, $I(t,x)$ can be neglected and the asymptotic evolution of the pair is given simply by $e^{-itH[(1,2)]}$.

For standard (without a magnetic field) multiparticle Schrödinger operators, in the long-range case the asymptotic dynamics of the centers of mass of clusters is not free, but has to be modified. More precisely, one shows that for large time the evolution of the system is approximated by

$$e^{iS_a(D_a,t)}e^{-itH^a},$$

where $S_a(\xi_a, t)$ is an approximate solution of the classical Hamilton-Jacobi equation, and e^{-itH^a} is the internal dynamics of the clusters. Since S_a and H^a commute, the internal and external dynamics are independent.

The motion of the center of mass of a neutral cluster across a constant magnetic field is coupled to the internal structure of the cluster (see Section 4). If an additional long-range time-dependent force (the intercluster interaction) is present, we expect that, as for standard N-particle systems, the center of mass dynamics has to be modified. However, it seems impossible to do this while leaving the internal structure of the cluster unchanged.

3. Include an external electric potential. Many physical systems, including atoms and molecules with infinitely heavy nuclei, can be described by a Hamiltonian $H_e = H + V_e$, where H has the same form as in [GL1–3], and $V_e = \sum_i V_e(x_i)$ is an additional electric potential vanishing at infinity.

The operator $H_e = H + V_e$ may have discrete spectrum, with eigenfunctions corresponding to all particles remaining in a bounded region of the configuration space. All other states could be labelled as scattering states; this includes those for which all particles of finite mass travel to infinity, while remaining close together.

4. Investigate the case when the magnetic field is constant only at infinity. More precisely, consider a magnetic field $\vec{B} = \vec{B}_0 + \vec{B}_1$, where B_0 is constant and B_1 vanishes at infinity. There are several results ([C], [LT1], [LT2], [E3], [BP], [Ni], [NR]) on asymptotic completeness and related questions for a single particle in a magnetic field vanishing as $x \to \infty$. It may be possible to combine the methods used in these articles with ours to handle multiparticle systems in magnetic fields $\vec{B}$ of the above form.

The results of [LT1], [LT2], [E3] suggest that at least the case of a short-range additional field $\vec{B}_1$ (such that the vector potential $A_1 = \frac{1}{2}(\vec{B}_1 \times x)$ vanishes as $|x|^{-1/2-\epsilon}$ at infinity) and systems containing no proper neutral subsystems (as in [GL1], [GL2]) should be similar to that of a constant field. If A_1 vanishes more slowly, the asymptotic evolution may have to be modified. Moreover, one will certainly have to assume something about the regularity of $\vec{B}_1$.

5. Include the case when neutral particles are present. In [GL1] we claimed that neutral particles were allowed in the short-range charged case. That, however, was a mistake. We did not take it into account that the projections on X_a and on Z do not commute if Z is defined as in that paper, and, in fact, our results hold only if all particles are charged.

It should be easier to analyze a system containing neutral particles than one in which neutral clusters consisting of charged particles are present – a neutral particle is a neutral cluster but has no internal structure. Thus, for example, it should not be very difficult to handle the 2-body case with electric charges $q_1 \neq 0$, $q_2 = 0$. However, for $N \geq 3$ substantial new problems arise.

6. Prove the Mourre estimate for N-particle systems containing neutral clusters. To do this, one would probably need more detailed information about the bound states of the neutral subsystems.

7. Prove asymptotic completeness for general N-body systems in constant magnetic fields. It certainly would be reasonable to assume that the potentials are of Coulomb type (see Problem 2 above). In any case, I do not have much to say about the problem in this generality at this time.

A. Appendix: The Mourre theory. Here we explain the notation and abstract results used in the main text in the proof of the Mourre estimate. We will follow Section 5 of [GL3]. The arguments involved are due to other authors, most notably Mourre [Mo], Perry, Sigal and Simon [PSS], and Froese and Herbst [FH]; ours is only this particular way of stating them.

DEFINITION A.1. Let A, B and H be self-adjoint operators on a Hilbert space $\mathcal{H}$, and let $c \in \mathbb{R}$. We will say that:

(i) $B \geq c$ at $H = E_0$ if there is a $\delta > 0$ such that for any function $f \in C_0^\infty(\mathbb{R})$ supported in $(E_0 - \delta, E_0 + \delta)$ we have

$$\text{(A.1)} \qquad\qquad f(H)Bf(H) \geq cf^2(H);$$

(ii) B is almost positive $(B \; a\text{-}\geq 0)$ at $H = E_0$ if $B \geq -\epsilon$ at $H = E_0$ for any $\epsilon > 0$;

(iii) $B \geq A$ (resp., $B \; a\text{-}\geq A$) at $H = E_0$ if $B - A \geq 0$ (resp., $B - A \; a\text{-}\geq 0$) at $H = E_0$.

We will write

$$\inf_{H=E} B = \sup\{c \in \mathbb{R} \mid B \geq c \text{ at } H = E\}.$$

The following lemma is an abstract version of an argument which was used in [Mo] (part (i)) and [FH] (part (ii)). A proof is given in [GL3, Section 5].

LEMMA A.1. Let B and H be self-adjoint operators on $\mathcal{H}$ such that B is H-bounded, let c be a positive constant, and fix $\lambda \in \mathbb{R}$. Suppose that for any $\epsilon > 0$ there is a compact operator K_ϵ such that

$$\text{(A.2)} \qquad\qquad B \geq c - \epsilon + K_\epsilon \text{ at } H = \lambda.$$

(i) If λ is not an eigenvalue of H, we have $B \; a\text{-}\geq c$ at $H = \lambda$.

(ii) Suppose that λ is an eigenvalue of H. Let $P = E_{\{\lambda\}}(H)$ be the corresponding spectral projection, and let $c_0 = \|PBP\|$. Then

$$B \; a\text{-}\geq c(1 - P) - 2c_0 P \text{ at } H = \lambda.$$

Equation (A.2) with $B = i[H, A]$ for a selfadjoint operator A (called a conjugate operator) and $c > 0$ is known as the *Mourre estimate*. Among Mourre's results [Mo] are the following:

(i) if $[H, iA] \geq c + K$ at $H = \lambda$, where $c > 0$ and K is compact, then $\sigma_{pp}(H)$ cannot accumulate at λ;

(ii) assume that $ad_A^\alpha(H+i)^{-1}$ is bounded for $\alpha = 1, 2$. If $[H, iA] \geq +K$ at $H = \lambda$ for all $\lambda \in \Delta$, with $c > 0$ and K compact, then H has no singular continuous spectrum in Δ.

The Mourre estimate has many other applications in spectral and scattering theory; see *e.g.*, [CFKS], [Pe], [HuSi] for reviews. Although the origins of the method of positive commutators and its applications to N-particle scattering can be traced back to earlier work of Kato, Putnam, and Lavine, prior to Mourre's work it could be used only for repulsive potentials. For N-body systems, the Mourre estimate was first proved in [PSS]; later, a simpler proof was given in [FH]. The Mourre theory was further developed in numerous papers, including [SS1-3] and [BG].

Lemma A.1 implies that if H satisfies the Mourre estimate at $H = \lambda$ with the conjugate operator A, and λ is not an eigenvalue of H, then

$$\inf_{H=\lambda} [H, iA] > 0.$$

It is this inequality that plays a key role in scattering theory. We refer to it as the *strict Mourre estimate*.

In multiparticle scattering theory we often have to prove local inequalities for operators which can be written as direct integrals. Definition A.2 and Lemma A.2 will greatly facilitate handling such cases. Similar arguments were used in [PSS] and [FH].

DEFINITION A.2. *Let $B(k, E)$ and $H(k)$ be measurable families of self-adjoint operators on $\mathcal{H}$ for $k \in \Omega_1 \subset \mathbb{R}^m$, $E \in I \subset \mathbb{R}$. Let*

$$H = \int_{\Omega_1}^{\oplus} H(k)dk, \ B(E) = \int_{\Omega_1}^{\oplus} B(k, E)dk.$$

We will say that:

(i) $B \geq c$ locally in H uniformly for $(k, E) \in \Omega \subset \Omega_1 \times I$, if there is a $\delta > 0$ such that for any $f \in C_0^\infty(\mathbb{R})$ supported in $(-\delta, \delta)$ we have

$$f(H(k) - E)B(k, E)f(H(k) - E) \geq cf^2(H(k) - E) \text{ for all } (k, E) \in \Omega;$$

(ii) B a-≥ 0 locally in H uniformly for $(k, E) \in \Omega \subset \Omega_1 \times I$, if for any $\epsilon > 0$ $B \geq -\epsilon$ locally in H uniformly for $(k, E) \in \Omega$.

LEMMA A.2. *Let $B(k, E)$ and $H(k)$ be as in Definition A.2. Assume that the mapping $\Omega_1 \ni k \to H(k)$ is continuous in the norm resolvent sense, and that the mapping $\Omega_1 \times I \ni (k, E) \to f(H(k))B(k, E)f(H(k))$ is norm continuous for any $f \in C_0^\infty(\mathbb{R})$. Let Ω be an open subset of $\Omega_1 \times I$.*

(i) If $B(k_0, E_0)$ a-≥ 0 at $H(k_0) = E_0$ for some $(k_0, E_0) \in \Omega$, then for any $\epsilon > 0$ there is a neighbourhood $\mathcal{U}$ of (k_0, E_0) such that $B \geq -\epsilon$ locally in H uniformly for $(k, E) \in \mathcal{U}$.

(ii) If B a-≥ 0 locally in H pointwise on Ω (i.e., $B(k, E)$ a-≥ 0 at $H(k) = E$ for any $(k, E) \in \Omega$), then B a-≥ 0 locally in H uniformly on any compact subset of Ω.

Proof. Since (ii) follows from (i) by a standard covering argument, we will only prove the latter.

Let $\epsilon > 0$. Then there is a $\delta_0 > 0$ such that for any C_0^∞ function f_0 supported in $(E_0 - \delta_0, E_0 + \delta_0)$ we have

$$f_0(H(k_0))B(k_0, E_0)f_0(H(k_0)) \geq -\frac{\epsilon}{2}f_0^2(H(k_0)).$$

Moreover, we can choose f_0 so that $f_0 = 1$ on $(E_0 - \delta_0/2, E_0 + \delta_0/2)$. By our assumptions on continuity, there is an open neighbourhood $\mathcal{U}_1$ of k_0 and $\delta_1 > 0$ such that for all $k \in \mathcal{U}_1$, $|E - E_0| < \delta_1$,

$$(A.3) \qquad f_0(H(k))B(k, E)f_0(H(k)) \geq -\frac{\epsilon}{2}f_0^2(H(k)) - \frac{\epsilon}{2}.$$

Let $\delta_2 = \min(\delta_0, \delta_1)$, $\Delta = (E_0 - \delta_2/10, E_0 + \delta_2/10)$, and $\delta = \delta_2/10$. Then for any $E \in \Delta$ and for any function f supported in $(E - \delta, E + \delta)$ we have $f f_0 = f$. Multiplying (A.3) from both sides by f, we obtain that for all $k \in \mathcal{U}_1$,

$$f(H(k))B(k, E)f(H(k)) \geq -\epsilon f^2(H(k)).$$

Hence we can take $\mathcal{U} = \mathcal{U}_1 \times \Delta$. $\square$

Acknowledgement. I wish to take this opportunity to thank C. Gérard for a fruitful and most enjoyable collaboration, in which the results presented here were obtained and my understanding of scattering theory took its present shape.

REFERENCES

[Ag] S. Agmon: *Lectures on exponential decay of solutions of second order elliptic equations*, Princeton University Press, Princeton 1982.

[AHS1] J. Avron, I. Herbst, B. Simon: Schrödinger operators with magnetic fields I: General interactions, Duke Math. J. 45 (1978), 847–884.

[AHS2] J. Avron, I. Herbst, B. Simon: Schrödinger operators with magnetic fields II: Separation of the center of mass in homogeneous magnetic fields, Ann. Phys. 114 (1978), 431–451.

[BG] A. Boutet de Monvel, I.V. Georgescu: Spectral and scattering theory by the conjugate operator method, Algebra and Analysis Vol 4 (1992) 73–116.

[BP] A. Boutet de Monvel, R. Purice: Limiting absorption principle for Schrödinger Hamiltonians with magnetic fields. Comm. in P.D.E. 19 (1994), 89–117.

[C] M. Combescure: Long range scattering for the two-body Schrödinger equation with "Hörmander-like" potentials. Comm. P. D. E. 11 (1986), 123–165.

[CFKS] H.L. Cycon, R. Froese, W. Kirsch, B. Simon: Schrödinger operators with applications to Quantum Mechanics and Global Geometry, Texts and Monographs in Physics, Springer-Verlag 1987.

[Do] J. Dollard: Asymptotic convergence and Coulomb interaction, J. Math. Phys. Vol 5, (1964), p 729–738.

[De1] J. Dereziński: Asymptotic Completeness for N-particle Long-range Quantum Systems, Ann. Math. 138 (1993), 427–476.

[De2] J. Dereziński: The Mourre estimate for dispersive N-body Schrödinger operators, Trans. of AMS 317, (1990) p. 773–798.

[De3] J. Dereziński: Algebraic approach to the N-body long range scattering. Rev.
       Math. Phys. 3 (1991), 1–62.

[DG]   J. Dereziński, C. Gérard: Asymptotic completeness of N−particle systems,
       chapter 3, preprint, Erwin Schrödinger Institute, 1993; chapter 5, preprint,
       Ecole Polytechnique, 1994.

[E1]   V. Enss: Asymptotic completeness for quantum-mechanical potential scattering,
       I. Short-range potentials, Comm. Math. Phys. 61 (1978), 285–291

[E2]   V. Enss: Quantum scattering theory for two- and three-body systems with
       potentials of short and long range, in: Schrödinger Operators, S.Graffi ed.,
       Springer LN Math. 1218, Berlin, 1986.

[E3]   V. Enss: Quantum scattering with long-range magnetic fields. Operator Theory:
       Adv. and Appl. 57 (1993), 61–70.

[FH]   R. Froese, I.Herbst: A new proof of the Mourre estimate, Duke Math. J. 49
       (1982), 1075–1085.

[G]   C. Gérard: Sharp propagation estimates for N-particle systems. Duke Math. J.
       67 (1992), 483–515.

[GL1] C. Gérard, I. Łaba: Scattering theory for N−particle systems in constant magnetic fields, Duke Math. J. 76 (1994), 433–465.

[GL2] C. Gérard, I. Łaba: Scattering theory for N−particle systems in constant magnetic fields, II. Long-range interactions, Comm. in P.D.E. 20 (1995), 1791–
       1830.

[GL3] C. Gérard, I. Łaba: Scattering theory for 3−particle systems in constant magnetic fields: dispersive case, Annales de l'Institut Fourier (Grenoble), 46
       (1996), 801–876.

[Gr]   G.M. Graf: Asymptotic completeness for N−body short range quantum systems : A new proof, Comm. in Math. Phys. 132 (1990), 73–101.

[HeSj] B. Helffer, J. Sjöstrand: Equation de Schrödinger avec champ magnétique et
       équation de Harper, Springer-Verlag Lectures Notes in Physics n° 345 (1989)
       118–197.

[HMS1] I. Herbst, J. S. Møller, E. Skibsted: Spectral analysis of N-body Stark Hamiltonians, preprint, Aarhus Universitet, 1994.

[HMS2] I. Herbst, J.S. Møller, E. Skibsted: Asymptotic completeness for N-body Stark
       Hamiltonians, preprint, Aarhus Universitet, 1994.

[Hö]   L. Hörmander: *The analysis of linear partial differential operators*, volume III
       Springer-Verlag (1985).

[HuSi] W. Hunziker, I.M. Sigal: The General Theory of N-Body Quantum Systems,
       Centre de Recherches Mathématiques Proceedings and Lecture Notes, vd.
       8, American Mathematical Society (1995).

[I1]   H. Iwashita: On the long-range scattering for one- and two-particle Schrödinger
       operators with constant magnetic fields, Preprint, Nagoya Institute of Technology, 1993.

[I2]   H. Iwashita: Spectral theory for 3−particle quantum systems with constant
       magnetic fields, Preprint, Nagoya Institute of Technology, 1993.

[L1]   I. Łaba: Long-range one-particle scattering in a homogeneous magnetic field,
       Duke Math. J. 70 (1993), 283–303.   .

[L2]   I. Łaba: Scattering for hydrogen-like systems in a constant magnetic field,
       Comm. in P.D.E. 20 (1995), 741–762.

[LT1] M. Loss, B. Thaller: Scattering of particles by long-range magnetic fields. Ann.
       Phys. 176 (1987), 159–180

[LT2] M. Loss, B. Thaller: Short-range scattering in long-range magnetic fields: the
       relativistic case. J. Diff. Equ. 73(1988), 225–236

[Mo]   E. Mourre: Absence of singular continuous spectrum for certain selfadjoint operators, Comm. in Math. Phys. 78 (1981), 519–567.

[Ni]   F. Nicoleau: Théorie de la diffusion pour l'opérateur de Schrödinger en présence d'un champ magnétique, Thèse de Doctorat de l'Université de Nantes, 1991.

[NR] F. Nicoleau, D. Robert: Théorie de la diffusion quantique pour des perturbations à longue et courte portée du champ magnétique. Ann. Fac. Sci. de Toulouse 12 (1991), 185–194.

[Pe] P. Perry: Scattering Theory by the Enss Method. Harwood Academic Press, London, 1983.

[PSS] P. Perry, I.M. Sigal, B. Simon: Spectral analysis of N–body Schrödinger operators, Ann. Math. 114 (1981), 519–567.

[RS] M. Reed, B. Simon: Methods of Modern Mathematical Physics, Academic Press.

[Sig] I.M. Sigal: On the long-range scattering. Duke Math. J. 60(2) (1990), 473–496.

[SS1] I.M. Sigal, A. Soffer: The N–particle scattering problem: asymptotic completeness for short-range quantum systems, Ann. Math. 125 (1987), 35–108.

[SS2] I.M. Sigal, A. Soffer: Long-range many-body scattering. Asymptotic clustering for Coulomb-type potentials, Invent. Math. 99 (1990), 115–143.

[SS3] I.M. Sigal, A. Soffer: Local decay and velocity bounds, preprint, Princeton University, 1988.

[SS4] I.M. Sigal, A. Soffer: Asymptotic completeness for four-body Coulomb systems, Duke Math. J. 71 (1993), 243–298.

[SS5] I.M. Sigal, A. Soffer: Asymptotic completeness of N particle long range scattering, J.AMS. 7 (1994), 307–334.

[S] B. Simon: Phase space analysis of simple scattering systems: extensions of some work of Enss, Duke Math. J. 46 (1979), 119–168.

[Sk] E. Skibsted: On the asymptotic completeness for particles in constant electromagnetic fields. preprint, Aarhus Universitet, 1995.

[VZ] S.A. Vugalter, G.M. Zhislin: Localization of the essential spectrum of the energy operators of n-particle quantum systems in a magnetic field. Theor. and Math. Phys. 97(1) (1993), 1171–1185.

NEW CHANNELS OF SCATTERING FOR TWO-AND THREE-BODY QUANTUM SYSTEMS WITH LONG-RANGE POTENTIALS

D. YAFAEV*

Abstract. We consider a system of three one-dimensional particles with one of pair potentials $V^\alpha(x^\alpha)$ decaying at infinity as $|x^\alpha|^{-\rho}$, $0 < \rho < 1/2$. It is shown that such a system can possess channels of scattering not included in the usual list of channels called the asymptotic completeness.

A similar result holds for the two-particle Schrödinger operator with a long-range potential if the usual condition on its derivatives is relaxed.

1. An aim of the scattering theory is to find the asymptotics as $t \to \infty$ of the solution $u(t) = \exp(-iHt)f$ of the time-dependent Schrödinger equation with a Hamiltonian

$$(1) \qquad H = -2^{-1}\Delta + V(x), \ V(x) = \overline{V(x)},$$

in the space $\mathcal{H} = L_2(\mathbb{R}^d)$. If f is an eigenvector of H, i.e. $Hf = \lambda f$, then obviously $u(t) = \exp(-i\lambda t)f$. Suppose now that f is orthogonal to the subspace $\mathcal{H}^{(p)}$ spanned by all eigenvectors. In the two-body short-range case when $V(x) = O(|x|^{-\rho})$, $\rho > 1$, the asymptotics of $u(t)$ is the same as that for the free system, that is

$$(2) \qquad \exp(-iHt)f = \exp(-iH_0 t)f_0 + o(1)$$

for some $f_0 \in \mathcal{H}$ and $H_0 = -2^{-1}\Delta$. The symbol $o(1)$ means a function such that its norm in the space $\mathcal{H}$ tends to zero as $t \to \infty$. One can rewrite (2) in an equivalent way as

$$(3) \qquad \exp(-iHt)f)(x) = (U_0(t)g)(x) + o(1),$$

where

$$(U_0(t)g)(x) = \exp(i\Phi_0(x,t))t^{-d/2}g(x/t),$$

$\Phi_0(x,t) = x^2(2t)^{-1}$, $g = \exp(i\pi d/4)\hat{f}_0$ and $\hat{f}_0 = Ff_0$ is the Fourier transform of f_0.

The relation (3) holds true (see e.g.[1]) also for long-range potentials satisfying the condition

$$(4) \qquad |D^\kappa V(x)| \le C(1 + |x|)^{-\rho - |\kappa|}, \quad \rho > 0,$$

for $|\kappa| = 0, 1, 2$. In this case the phase function $\Phi_0(x,t)$ is a (perhaps, approximate) solution of the eikonal equation

$$(5) \qquad \partial\Phi_0/\partial t + 2^{-1}|\nabla\Phi_0|^2 + V = 0.$$

---

* Département de Mathématiques, Université de Rennes-1, Campus Beaulieu, 35042 Rennes, France.

The asymptotics (3) shows that, if f belongs to the absolutely continuous subspace $\mathcal{H}^{(ac)} = \mathcal{H} \ominus \mathcal{H}^{(p)}$ of the operator H, then the solution $(\exp(-iHt)f)(x)$ "lives" in the region where $|x| \sim t$. The elements f and g in (3) are connected by the equality $f = W_0 g$, where

$$W_0 = s - \lim_{t \to \infty} \exp(iHt)U_0(t)$$

is the wave operator. This limit exists, is isometric and its range coincides with the subspace $\mathcal{H}^{(ac)}$. Clearly, $W_0 F$ is the usual (modified) wave operator relating H_0 and H.

The situation is more complicated in the many-body case when

$$H = -2^{-1}\Delta + \sum_\alpha V^\alpha(x^\alpha), \quad 1 \le \alpha \le \alpha_0 < \infty,$$

and x^α are orthogonal projections of $x \in \mathbb{R}^d$ on some given subspaces $X^\alpha \subset \mathbb{R}^d$. Set $X_\alpha = \mathbb{R}^d \ominus X^\alpha$. The three-body case is distinguished by the assumption $X_\alpha \cap X_\beta = \{0\}$ for $\alpha \ne \beta$. Suppose first that all pair potentials V^α are short-range, i.e. that they satisfy the condition $V^\alpha(x^\alpha) = O(|x^\alpha|^{-\rho})$ for some $\rho > 1$. The wave operator W_0 (for $\Phi_0(x,t) = x^2(2t)^{-1}$) again exists and the subspace $\mathcal{H}_0 := R(W_0) \subset \mathcal{H}^{(ac)}$. The evolution (2) (or (3)) for $f \in \mathcal{H}_0$ corresponds to the process where all particles are asymptotically free.

In general, $\mathcal{H}_0 \ne \mathcal{H}^{(ac)}$ and the orthogonal complement $\mathcal{H}^{(ac)} \ominus \mathcal{H}_0$ is determined by the point spectra of pair Hamiltonians $H^\alpha = -2^{-1}\Delta_{x^\alpha} + V^\alpha$ acting in the spaces $L_2(X^\alpha)$. Let us introduce their eigenvalues $\lambda^{\alpha,k}$ and eigenvectors $\psi^{\alpha,k}$. For every couple $\{\alpha, k\}$ there exists the wave operator

$$(6) \qquad W_{\alpha,k} = s - \lim_{t \to \infty} \exp(iHt)U_{\alpha,k}(t),$$

where

$$(U_{\alpha,k}(t)g)(x) = \psi^{\alpha,k}(x^\alpha)\exp(i\Phi_{\alpha,k}(x_\alpha,t))t^{-d_\alpha/2}g(x_\alpha/t), \quad d_\alpha = \dim X_\alpha,$$

and $\Phi_{\alpha,k}(x_\alpha,t) = x_\alpha^2(2t)^{-1} - \lambda^{\alpha,k}t$. The range $\mathcal{H}_{\alpha,k} := R(W_{\alpha,k}) \subset \mathcal{H}^{(ac)}$ of $W_{\alpha,k}$ is orthogonal to $\mathcal{H}_0$. The asymptotics

$$(7) \quad (\exp(-iHt)f)(x) = (U_{\alpha,k}(t)g)(x) + o(1), \quad f = W_{\alpha,k}g \in \mathcal{H}_{\alpha,k},$$

corresponds to the channel of scattering where the pair α of particles forms a bound state described by the wave function $\psi^{\alpha,k}$ with the energy $\lambda^{\alpha,k}$ and the third particle is asymptotically free.

The basic result of the three-particle scattering theory, called the asymptotic completeness, is that the sum of all subspaces $\mathcal{H}_0$ and $\mathcal{H}_{\alpha,k}$ exhausts the absolutely continuous subspace $\mathcal{H}^{(ac)}$ of the operator H. Actually, $\mathcal{H}^{(ac)}$ can be decomposed into the orthogonal sum

$$(8) \qquad \mathcal{H}^{(ac)} = \mathcal{H}_0 \oplus \left(\bigoplus_{\alpha,k} \mathcal{H}_{\alpha,k} \right).$$

This result was first obtained by L. D. Faddeev [2] under some additional assumptions. The optimal formulation is due to V. Enss [3].

For three-body systems with long-range pair potentials V^α the wave operators W_0 and $W_{\alpha,k}$ again exist (for suitable functions $\Phi_0(x,t)$ and $\Phi_{\alpha,k}(x_\alpha, t)$) if $V^\alpha(x^\alpha)$ satisfy the condition (4) as functions of x^α. For $f \in \mathcal{H}_0$ the asymptotics (3) and for $f \in \mathcal{H}_{\alpha,k}$ the asymptotics (7) are fulfilled. We emphasize that for $f \in \mathcal{H}_{\alpha,k}$ the solution $\exp(-iHt)f$ is localized in the variable x^α and, on the contrary, $|x_\alpha| \sim t$. Thus such solutions play an intermediary role between those for $f \in \mathcal{H}^{(p)}$ and $f \in \mathcal{H}_0$. As shown by V. Enss [4,5], the asymptotic completeness (8) holds if $\rho > \sqrt{3} - 1$ in the condition (4) on functions $V^\alpha(x^\alpha)$. Note also that in the case $\rho > 1/2$ the relation (8) was established in the papers [6,7] under some additional assumptions.

We do not discuss here results on N body systems for $N > 3$. Let us mention only that the condition $\rho > \sqrt{3} - 1$ reappeared in the paper by J. Dereziński [8], where asymptotic completeness was considered for arbitrary N by a method different from [4,5].

2. Our goal is to construct new channels of scattering for the Schrödinger operator (1) which arise due to eigenvalues generated by a potential $V(x)$ on "cross-sections" of the problem. We suppose that

$$\mathbb{R}^d = X_1 \oplus X^1, \quad \dim X_1 = d_1, \ \dim X^1 = d^1, \ d_1 + d^1 = d,$$

but we do not make any special assumptions about a potential $V(x) = V(x_1, x^1)$. Let us introduce the operator

$$(9) \qquad H^1(x_1) = -2^{-1}\Delta_{x^1} + V(x_1, x^1)$$

acting in the space $L_2(X^1)$. Suppose that the operator $H^1(x_1)$ has an eigenvalue $\lambda(x_1)$ and denote by $\psi(x_1)$ the corresponding normalized eigenfunction. In the particular case when $V(x)$ does not depend on x_1 the operator H describes a three-body system with only one non-trivial pair interaction and H^1 is the Hamiltonian of this pair. In this case the wave operator (6) exists (for $\lambda^1 = \lambda$, $\psi^1 = \psi$ and $\Phi_1(x_1, t) = x_1^2(2t)^{-1} - \lambda t$).

We are looking for a generalization of (6) *in the case when* $\lambda(x_1) \to 0$ *as* $|x_1| \to \infty$. In interesting situations the function $\lambda(x_1)$ decreases slower than $|x_1|^{-1}$. Let us consider it as an "effective" potential energy and associate to the long-range potential $\lambda(x_1)$ the phase function $\Phi(x_1, t)$. This means that Φ is (an approximate) solution of the eikonal equation (5) with λ in place of V. We prove under some assumptions the existence of the wave operator

$$(10) \qquad \mathcal{W} = s - \lim_{t \to \infty} \exp(iHt)\mathcal{U}(t),$$

where

$$(\mathcal{U}(t)g)(x) = \psi(x_1, x^1) \exp(i\Phi(x_1, t)) t^{-d_1/2} g(x_1/t).$$

The range $\mathfrak{H} := R(W) \subset \mathcal{H}^{(ac)}$ of W is orthogonal to the subspace $\mathcal{H}_0$ whenever the wave operator W_0 exists. The restriction of the operator H on the subspace $\mathfrak{H}$ has the absolutely continuous spectrum which coincides with the positive half-axis.

Our construction (its details can be found in [9]) of the wave operator W is similar but more straightforward than that in [10]. The basic difference is that now we make calculations in the coordinate representation whereas they were done in the momentum representation in [10].

The existence of the wave operator (10) requires rather special assumptions which are naturally formulated in terms of eigenfunctions $\psi(x_1, x^1)$ of the operator $H^1(x_1)$. It turns out that typically the asymptotics of $\psi(x_1, x^1)$ as $\lambda(x_1) \to 0$ has a certain self-similarity:

$$(11) \qquad \psi(x_1, x^1) \sim |x_1|^{-\sigma d^1/2} \Psi(|x_1|^{-\sigma} x^1)$$

for some $\Psi \in L_2(X^1)$ and $\sigma > 0$. We prove the existence of W if the asymptotics (11) is fulfilled for $\sigma < 1/2$. In this case the solution

$$(\exp(-iHt)f)(x) = (\mathcal{U}(t)g)(x) + o(1), \quad f = Wg \in \mathfrak{H},$$

of the time-dependent Schrödinger equation "lives" in a parabolic region where $|x_1| \sim t$, $|x^1| \sim t^\sigma$. This implies orthogonality of the subspaces $\mathfrak{H}$ and $\mathcal{H}_0$.

The study of the asymptotics of eigenfunctions $\psi(x_1)$ as the corresponding eigenvalues $\lambda(x_1)$ tend to zero could be a cumbersome problem. Therefore it is important that $\psi(x_1, x^1)$ can be chosen as an *approximate* solution of the equation $H^1(x_1)\psi(x_1) - \lambda(x_1)\psi(x_1) = 0$. In our applications this allows us to construct $\psi(x_1)$ in terms of eigenfunctions of some auxiliary operator with the *discrete spectrum*. Actually, it is not required that V have a negative part. A function $\psi(x_1)$ can be generated by a *positive* minimum (in the variable x^1) of the potential $V(x_1, x^1)$. In this case $\psi(x_1)$ corresponds to a resonant state of the operator $H^1(x_1)$.

3. In applications to two-body systems (this is discussed more thoroughly in [11]) our construction of new channels works if the condition (4) on derivatives of $V(x)$ is relaxed. A typical example of such a potential is given by the equality

$$(12) \quad V(x_1, x^1) = -v\left(<x_1>^q + <x^1>^q\right)^{-p/q}, \quad p \in (0, 1), \quad q \in (0, 2),$$

$<x> = (1 + x^2)^{1/2}$, $v > 0$. For potentials (12) the bound (4) is fulfilled for arbitrary κ outside of any conical neighbourhood of the planes X_1 and X^1. This suffices for the existence of the wave operator W_0. If $1 \leq q < 2$, then the bound (4) is satisfied (uniformly in directions of x) for $|\kappa| = 1$ but is violated for $|\kappa| = 2$. If $0 < q < 1$, then (4) is violated already for $|\kappa| = 1$. Let $\Psi(x^1)$ be any eigenfunction of the auxiliary operator

$K = -2^{-1}\Delta_{x^1} + v\rho q^{-1}|x^1|^q$ with the discrete spectrum. We prove that there exists an approximate eigenfunction of the operator (9) satisfying (11) for

$$\sigma = (\rho + q)(2 + q)^{-1}.$$

Thus in the case (12), the wave operator (10) exists if $q < 2(1 - \rho)$.

We emphasize that the wave operator $W = W_n$ can be constructed in terms of every eigenfunction, $\Psi = \Psi_n$, of the operator K. Furthermore, one can interchange the roles of variables x_1 and x^1. This gives us a new set of wave operators W^m. The ranges of all wave operators W_n, W^m and W_0 are orthogonal one to another.

Notice that the potential (12) is radial if $q = 2$. In this case $\sigma = (2 + \rho)/4 > 1/2$ and the wave operator (10) does not exist. This, of course, should have been expected, since the wave operator W_0 is complete now. This example shows that the relation (11) for $\sigma > 1/2$ does not ensure the existence of the wave operator W.

4. For some three-body systems (see [9], for more details) with sufficiently slowly decreasing pair potentials the "additional" wave operator W may exist even if pair potentials satisfy the condition (4) for all κ. Actually, we consider one-dimensional particles with pair potentials satisfying (4) for $\rho \in (0, 1/2)$. We pick up some α, say $\alpha = 1$, and prove the existence of the wave operator (10). For initial data $f \in \mathfrak{H}$ the solution $u(t) = \exp(-iHt)f$ of the Schrödinger equation "lives" in the region where $|x_1| \sim t$ and $|x^1| \sim t^\sigma$ for $\sigma = (\rho + 1)/3 \in (1/3, 1/2)$. Thus solutions $u(t)$ for $f \in \mathfrak{H}$ are intermediary between those for $f \in \mathcal{H}_0$ and $f \in \mathcal{H}_{1,k}$. The subspace $\mathfrak{H}$ is orthogonal to $\mathcal{H}_0$ and $\mathcal{H}_{\alpha,k}$ for all α and k. The existence of the wave operator W contradicts, of course, the asymptotic completeness (8).

We require that one of pair potentials

(13) $$V^\alpha(x^\alpha) = v|x^\alpha|^{-\rho}, \quad \alpha \neq 1, \; 0 < \rho < 1/2,$$

for $\pm v > 0$ and large positive values of $\mp x^\alpha$. Other pair potentials can be short-range. Thus we consider two cases "+" and "−" which differ by the sign of the long-range interaction. In both cases "±" the function Ψ intervening in the asymptotics (11) is an eigenfunction of the auxiliary operator $K = -2^{-1}(d/dx^1)^2 + v\rho x^1$ in the space $L_2(\mathbb{R}_+)$ with the Dirichlet boundary condition at $x^1 = 0$. Thus Ψ is the Airy function. Note that the operator (10) can be constructed in terms of every eigenfunction $\Psi = \Psi_n$ of the operator K. We emphasize that in the case "+" the operator (9) does not have eigenvalues and the functions $\Psi_n(x_1)$ correspond to its resonant states.

From the point of view of quantum mechanical interpretation we construct channels of scattering where particles of some chosen pair are relatively close to one another and the third particle is far away. This pair is bound by a potential depending on the position of the third particle but its bound state is evanescent as $t \to \infty$. The third particle moves freely in some effective long-range potential created by the pair of the first two particles. This means, in particular, that its distance from the center of mass of this pair grows as t.

One can imagine that, for example, the mass of the first particle is infinite and it is fixed at the origin. Positions of the second and third particles are given by coordinates x^1 and x_1, respectively. The first and second particles interact by a potential $V^1(x^1)$ which is positive and decreases not too rapidly as $x^1 \to -\infty$:

$$V^1(x^1) \geq c|x^1|^{-r}, \quad c > 0, \quad r \in (0, 2).$$

This potential barrier retains the second particle on the positive half-axis. We may suppose that the third particle does not interact with the first one. On the contrary, a potential $V^2(x^2) = V^2(x^1 - x_1)$ of interaction between the second and third particles is long-range and satisfies (13). In the case "+" (the case "−"), the third particle moves to plus (minus) infinity and $|x_1|$ grows as t for $t \to \infty$. In both cases the second particle is jammed to the first one either by a repulsion from the third particle (in the case "+") or by an attraction to it (in the case "−"). In the long run the second particle escapes to plus infinity but x^1 behaves as t^σ, $\sigma \in (1/3, 1/2)$, for large t.

Note that the existence of new channels for three-particle systems automatically implies the same phenomenon for systems of more than three particles. It suffices to take a system where all particles but three are free and the system of these three distinguished particles possesses a described channel.

REFERENCES

[1] L. Hörmander, *The analysis of linear partial differential operators* IV, Springer-Verlag, 1985.

[2] L.D. Faddeev, *Mathematical Aspects of the Three Body Problem in Quantum Scattering Theory*, Trudy MIAN, **69**, 1963. (Russian)

[3] V. Enss, Completeness of three-body quantum scattering, in: *Dynamics and processes*, P. Blanchard and L. Streit, eds., Springer Lecture Notes in Math., **1031** (1983), 62–88.

[4] V. Enss, Quantum scattering theory of two-body and three-body systems with potentials of short and long range, in: *Schrödinger operators*, S. Graffi, ed., Springer Lecture Notes in Math., **1159** (1985), 39–176.

[5] V. Enss, Long-range scattering of two- and three-body quantum systems, in: *Journées EDP*, Saint Jean de Monts (1989), 1–31.

[6] X.P. Wang, On the three body long range scattering problems, Reports Math. Phys., **25** (1992), 267–276.

[7] C. Gérard, Asymptotic completeness of 3-particle long-range systems, Invent. Math., **114** (1993), 333–397.

[8] J. Dereziński, Asymptotic completeness of long-range quantum systems, Ann. Math., **138** (1993), 427–473.

[9] D.R. Yafaev, New channels of scattering for in three-body quantum systems with long-range potentials, Duke Math. J., **82**, no. 3 (1996), 553–584.

[10] D.R. Yafaev, On the break-down of completeness of wave operators in potential scattering, Comm. Math. Phys., **65** (1979), 167–179.

[11] D.R. Yafaev, New channels in two-body long-range scattering, Algebra and Analysis, **8**, no. 1 (1996), 211–233. (This journal is published in Russian. Its regular English translation is called "Saint-Petersburg Math. J.").

STATE-OF-STATE TRANSITION PROBABILITIES AND CONTROL OF LASER-INDUCED DYNAMICAL PROCESSES BY THE (T, T') METHOD

NIMROD MOISEYEV*

Abstract. The time evolution operator for general time-dependent (not necessarily time-periodic) Hamiltonians is given by the (t, t') method as, $\hat{u}(x, t', t) = \exp\{-\frac{i}{\hbar}(\hat{p}_{t'} + H(x, t')(t - t_0))\}$, where $\hat{p}_{t'} \equiv (\hbar/i)\partial/\partial t'$ and t' serves as an additional coordinate. Therefore the inner product is defined in the generalized Hilbert space of x and t'. The physical solution is given by $\psi(x, t', t)|_{t'=t}$ where $\psi(x, t', t) = \hat{u}(x, t', t)\Psi_0$.

A brief review of the (t, t') method is given emphasizing two advantages of the method: (I) it enables the extension of time-independent scattering theory to time-dependent Hamiltonian; (II) it enables the derivation of time-independent-like closed-form simple expression of the time evolution operator which can be described by global type very high order polynomial expansions.

It is shown how one can use the (t, t') method to control the state-to-state transition probabilities in laser-induced processes.

1. Introduction. For sufficiently high intensity fields, the interaction between an atomic or molecular system, $\hat{H}_M$, and an electromagnetic field can be well described semiclassically by a time-dependent formalism [1]. That is,

$$(1.1) \qquad \hat{H}(t) = \hat{H}_M + \hat{\mu}f(t)$$

where $\hat{\mu}$ is the dipole moment operator. For a monochromatic laser, $f(t) = g(t)\cos wt$, where w is the frequency of the laser and $g(t)$ provides the laser pulse shape. Recent theoretical and experimental studies have shown that the dynamics can be controlled by carefully designing the shape of the laser pulse or by using multicolor lasers where the relative phases are carefully matched [2].

In order to predict the state-to-state transition probabilities when the Hamiltonian is time-dependent, the time-dependent Schrödinger equation should be solved

$$(1.2) \qquad \hat{H}(t)\psi(t) = \hbar i \frac{\partial\psi}{\partial t}$$

for the initial condition

$$(1.3) \qquad \psi(t = 0) = \phi_0 \ .$$

Since the Hamiltonian is time-dependent the propagator, $\hat{u}(t \leftarrow 0)$ that propagates from time 0 to time t does not have a closed-formed expression

---

* Department of Chemistry and Minerva Center for Nonlinear Physics, Technion – Israel Institute of Technology, Haifa 32 000, Israel.

and is given by the following Magnus expansion,

$$\hat{u}(t \leftarrow 0) = \exp\left\{-i/\hbar \int_0^t \hat{H}(t')dt' - 1/2\hbar \int_0^t dt' \int_0^{t'} dt''[\hat{H}(t'), \hat{H}(t'')] + \ldots\right\}$$
(1.4)

where

$$(1.5) \qquad\qquad \psi(t) = \hat{u}(t \leftarrow 0)\phi_0 \ .$$

Due to the complexity of problems in which the Hamiltonian is time-dependent, $\psi(t)$ cannot be propagated analytically to $t = \infty$, and there is no available time-independent scattering theory which provides closed-form expressions for the state-to-state transition probabilities. As shown by Peskin and Moiseyev [3], however, an important result of the (t, t') method is the extension of time-independent scattering theory to time-dependent Hamiltonians.

2. Brief review of the (t, t') method. Let us first define a Floquet-like operator $\hat{\mathcal{H}}$:

$$(2.1) \qquad\qquad \hat{\mathcal{H}}(x, t') = \hat{p}_{t'} + \hat{H}(x, t') \ .$$

$H(x, t')$ is the physical time-dependent Hamiltonian where t is replaced by

$$(2.2) \qquad\qquad 0 \leq t' \leq T \ .$$

T is the period of the field when the Hamiltonian is time-periodic, or for the most general case it is any time that is much larger than the duration of the laser pulse. The momentum-like operator $\hat{p}_{t'}$ is given by

$$(2.3) \qquad\qquad \hat{p}_{t'} = \frac{\hbar}{i}\frac{\partial}{\partial t'} \ .$$

As first discussed by Howland in 1974 the inner product in the generalized Hilbert space is defined as [4]:

$$(2.4) \qquad << f|g >> = \frac{1}{T}\int_0^T dt' \int_{-\infty}^{\infty} dx \int f^*(x, t')g(x, t')$$

where f and g are two functions in the generalized Hilbert space. Here t' serves as an additional coordinate and therefore the dimensionality of the problem is increased by one. The time-dependent Schrödinger equation with time-dependent Hamiltonian is replaced by another differential equation

$$(2.5) \qquad\qquad \hat{\mathcal{H}}(x, t')\Phi(x, t', t) = i\hbar\frac{\partial \Phi(x, t', t)}{\partial t}$$

Since $\hat{\mathcal{H}}$ is a time-independent operator (here we refer to t as time and not to t'), then

$$(2.6) \qquad \Phi(x, t', t) = e^{-i\hat{\mathcal{H}}(x, t')(t-t_0)/\hbar} \, \Phi(x, t', t_0)$$

where t_0 is the initial time and $\Phi(x, t', t_0)$ is t'-independent and is equal to the initial wavepacket $\psi(x, t_0) = \phi_0$. (See Eq. (1.3)).

The physical solution of the time-dependent Schrödinger equation is a cut in $\{t, t'\}$ plane:

$$(2.7) \qquad \Psi(x, t) = \Phi(x, t', t) \Big|_{t'=t} .$$

For time-periodic Hamiltonian

$$(2.8) \qquad t' = t(\text{mod} \quad T)$$

and therefore the number of grid points along the t'-axis which are required to get converged results when Eq. (2.5) is numerically solved is reduced.

Simple proof of Eq. (2.7).

The simple proof of Peskin and Moiseyev [3] is based on the chain-rule:

$$(2.9) \qquad \frac{d\psi(u, v)}{du} = \frac{\partial \psi}{\partial u} + \frac{\partial v}{\partial u} \cdot \frac{\partial \psi}{\partial v}$$

It is clear that

$$(2.10) \qquad \frac{d\psi(u, v = u)}{du} = \frac{d\psi(u, v)}{du}\Big|_{v=u} = \frac{\partial \psi}{\partial u}\Big|_{v=u} + \frac{\partial \psi}{\partial v}\Big|_{v=u}$$

Although it may look embarrassing, for the sake of clarity let us look at simple examples: When

$$\psi = u^3$$

then $\partial \psi/\partial u = 3u^2$. Now let us define ψ as

$$\psi = u^2 v$$

such that $\psi(u, v = u) = u^3$. It is obvious that $\partial \psi/\partial u = 2uv$ and $\partial \psi/\partial v = u^2$. Consequently,

$$\frac{\partial \psi}{\partial u}\Big|_{v=u} + \frac{\partial \psi}{\partial v}\Big|_{v=u} = 2uu + u^2 = 3u^2$$

We can construct $\psi(u, v = u) = u^3$ in a different way:

$$\psi = \frac{u^4}{v}$$

Therefore,

$$\frac{\partial \psi}{\partial u} = \frac{4u^3}{v} \quad , \qquad \frac{\partial \psi}{\partial v} = -\frac{u^4}{v^2}$$

and as before

$$\frac{\partial \psi}{\partial u}\bigg|_{v=u} + \frac{\partial \psi}{\partial v}\bigg|_{v=u} = \frac{4u^3}{u} - \frac{u^4}{u^2} = 3u^2$$

Let us substitute in Eq. (2.10)

$$\begin{aligned} u &= t \\ v &= t' \ . \end{aligned}$$

Then we get that

$$(2.11) \qquad \frac{d\psi(t'=t,\,t)}{dt} = \frac{\partial \psi(t,\,t')}{\partial t}\bigg|_{t'=t} + \frac{\partial \psi(t,\,t')}{\partial t'}\bigg|_{t'=t} \ .$$

Since (here we "ignore" the fact that ψ is a function of x as well)

$$\hat{H}(t)\psi(t) = i\hbar\frac{d\psi(t)}{dt}$$

it is obvious that

$$(2.12) \qquad \hat{H}(t')\psi(t',\,t)\bigg|_{t'=t} = i\hbar\frac{d}{dt}\psi(t'=t,\,t)$$

By substituting Eq. (2.11) into Eq. (2.12) one immediately gets that

$$\left(-i\hbar\frac{\partial}{\partial t'} + H(t')\right)\psi(t',\,t)\bigg|_{t'=t} = i\hbar\frac{\partial}{\partial t}\psi(t',\,t)\bigg|_{t'=t}$$

Therefore the solution of Eq. (2.6) at $t' = t$, i.e. Eq. (2.7), provides the physical desired solution.

3. Powerful numerical algorithm for propagation. Let the time evolution operator for general time-dependent Hamiltonian be given on a grid of x and t':

$$(3.1) \qquad \psi(x,t) = e^{-i\hat{\mathcal{H}}(x,\,t')(t-t_0)/\hbar}\phi_0(x)\bigg|_{t'=t}$$

where $\exp(-i\hat{\mathcal{H}}t/\hbar)$ can be expanded in Chebyschev, Newton, Faber or any other global polynomial expansion. To operate with $\hat{\mathcal{H}}$ on a given function the FFT method can be applied both with x and t'. This enabled Peskin, Kosloff and Moiseyev [5] to get numerical results of a given accuracy with a computational effort reduced by 6 orders of magnitude (10^6) by using the (t, t') method to construct the propagator.

When basis set methods were used to construct the Floquet like matrix $\mathcal{H}(x, t')$ it was found out that about three Fourier basis functions, $\{\exp(inwt'); n = -1, 0, +1\}$, are needed to get the same quasi-energies as obtained by direct diagonalization of $\hat{\mathcal{H}}$ when hundreds or even thousands of Fourier basis functions are used (i.e. the Floquet matrix is extremely large).

In order to understand this remarkable efficiency of the (t, t') method, let us write Eq. (2.12) within the framework of the finite matrix representation approach:

$$\vec{\psi}(t) = u(t \leftarrow t_0)\vec{\phi}_0$$

where

(3.2)
$$\vec{\psi}(t) = \sum_{n=-\infty}^{+\infty} e^{iwnt} \left[e^{-i\mathcal{H}(t-t_0)/\hbar} \right]_{n,0} \vec{\phi}_0$$

and $\mathcal{H}$ is a Hamiltonian matrix

(3.3)
$$(\mathcal{H})_{(i, n'), (j, n)} = \langle\langle i, n' | \hat{\mathcal{H}} | j, n \rangle\rangle$$

constructed of

(3.4)
$$|i\rangle = \varphi_i(x) \quad ; \quad i = 1, 2, 3, \ldots$$

Which are the basis functions for the x-coordinate and of

(3.5)
$$|n\rangle = e^{iwnt'} \quad .$$

Which are the basis functions for the t'-coordinate where w is taken as

(3.6)
$$w = \frac{2\pi}{T}$$

T is the duration of the laser pulse or the periodicity of the Hamiltonian when the Hamiltonian is time-periodic. The initial state vector is a projection of the initial state $\Psi(x, 0)$ on the $|i>$ basis functions,

(3.7)
$$\left(\vec{\phi}_0\right)_i = \langle i | \Psi(0) \rangle$$

The time-dependent solution is given by

(3.8)
$$\Psi(x, t) = \sum_i \left[\vec{\psi}(t)\right]_i \varphi_i(x)$$

Let us carry out the propagation up to $t = m\tau$ when τ is a small time step. In such a case the time evolution matrix, u (given in Eq. (3.2)), can be written as:

(3.9) $u(m\tau \leftarrow 0) = u(m\tau \leftarrow (m-1)\tau)\ldots u(2\tau \leftarrow \tau)u(\tau \leftarrow 0)$

where

(3.10) $u(\tau \leftarrow 0) = \sum\limits_{n=-\infty}^{\infty} e^{iwn\tau}\left[e^{-i\mathcal{H}\tau/\hbar}\right]_{n,0}$

(3.11) $u(2\tau \leftarrow \tau) = \sum\limits_{n=-\infty}^{\infty} e^{iw2n\tau}\left[e^{-i\mathcal{H}\tau/\hbar}\right]_{n,0}$

and for the m time step:

(3.12) $u(m \leftarrow (m-1)\tau) = \sum\limits_{n=-\infty}^{\infty} e^{iwnm\tau}\left[e^{-i\mathcal{H}\tau/\hbar}\right]_{n,0}$.

The conclusion is clear – the exponential term

(3.13) $\exp\left\{-i\mathcal{H}\tau/\hbar\right\}_{n,0}$

should be calculated only once! If τ is small enough, then

(3.14) $\exp\left\{-i\mathcal{H}\tau/\hbar\right\}_{n,0} \cong 1 - i\tau[\mathcal{H}]_{n,0}$

when $(\mathcal{H})_{n,0}$ is a matrix which is composed of the elements

(3.15) $\langle\!\langle i, n|\hat{\mathcal{H}}|j, 0\rangle\!\rangle$, $i, j = 1, 2, \ldots N$.

The key point in this analysis is the realization that $\mathcal{H}$ is a band matrix. For cw (continuous wave) lasers where the molecular/field interaction is given by $\hat{\mu}\epsilon_0 \cos wt$, the Floquet matrix $\mathcal{H}$ is a three-block-diagonal matrix. In such a case the matrix elements $(\mathcal{H})_{n,0}$ are non-zero for only three values of n, namely $n = -1, 0, +1$. They are identically equal to zero if $n \neq 0, \pm 1$. From time-periodic Hamiltonians (cw lasers) it is clear that for a sufficiently small time step, τ, (see Eq. 3.14)

(3.16) $\exp\left\{-i\mathcal{H}\tau/\hbar\right\}_{n,0} \simeq 0$

if $n \neq 0, \pm 1$. Therefore the $-\infty$ to $+\infty$ summations in Eqs. (3.10-3.12) (where time-evolution matrices are constructed) can be replaced by summations from -1 to $+1$! Here we show that the exact leading term is obtained if $\mathcal{H}$ is a 3×3 block matrix rather than a $\infty \times \infty$ 1! For non-periodic Hamiltonians the Floquet-like matrix $\mathcal{H}$ is not a 3-block-diagonal matrix, but it is still a band matrix. In their molecular experiment realization of the kicked chaotic rotor, Blumel, Fishman and Smilansky [6] found out that a δ-like function is well described by 10-15 Fourier functions. This implies that for very short living laser pulses the Floquet-like matrix $\mathcal{H}$ should be constructed of, at the most, 15×15 "Floquet-channels" rather than the 3 used for time-periodic Hamiltonians.

4. Quasi-energies for non-periodic Hamiltonians. For time-periodic Hamiltonians the quasi stationary solutions which are known as Floquet states are given by

$$(4.1) \qquad \Psi(x,t) = e^{-iEt/\hbar}\phi(x,t)$$

where $\phi(x,t)$ is a time-periodic function,

$$(4.2) \qquad \phi(x,t) = \phi(x,t+T)$$

E is the quasi-energy which is defined modulo $\hbar w$ when $w = 2\pi/T$. E and $\phi(x,t)$ are respectively the eigenvalue and eigenfunction of the Floquet operator $\mathcal{H}$.

$$(4.3) \qquad \mathcal{H}\phi(x,t) = E\phi(x,t) \ .$$

The (t,t') approach extends this concept of quasi-energy state to a general not necessarily time-periodic Hamiltonians. Even when $\mathcal{H}(x,t') \neq \mathcal{H}(x,t'+T)$ but when for $t' \geq T$, $\mathcal{H}$ is a time-independent operator (this is the situation when the laser is turned off) any solution of the time-dependent Schrödinger equation can be expanded in the basis of the eigenfunctions of $\mathcal{H}$. That is

$$(4.4) \qquad \Psi(x,t) = \psi(x,t',t)\Big|_{t'=t}$$

when

$$(4.5) \qquad \psi(x,t',t) = \sum_{\alpha} C_\alpha e^{-iE_\alpha t/\hbar}\phi_\alpha(x,t')$$

and

$$(4.6) \qquad \mathcal{H}(x,t')\phi_\alpha(x,t') = E_\alpha\phi_\alpha(x,t')$$

where C_α are defined by

$$(4.7) \qquad C_\alpha = \langle\!\langle\phi_\alpha|\phi_0\rangle\!\rangle$$

and ϕ_0 is the initial t'-independent state. That is, for non-periodic field when the laser is turned on and after sometime it is turned off

$$(4.8) \qquad C_\alpha = \langle\varphi_{0,\alpha}|\phi_0\rangle$$

where $\varphi_{0,\alpha}$ is the $n = 0$ Fourier component of ϕ_α.

Even for the most narrow laser pulses used today the electromagnetic field carries out 10 to 15 oscillations. It seems that in most cases the adiabatic theorem holds and one may assume that at any given time t the electromagnetic field can be associated with an effective time-periodic Hamiltonian where the maximum field amplitude is varied adiabatically. That is

$$(4.9a) \qquad H(x,t') = H_0(x) + \hat{\mu}\epsilon(t)\cos wt'$$

where

$$(4.9b) \qquad \epsilon(t) = \epsilon_0 f(t)$$

and $f(t)$ describes the shape of the laser pulse. The adiabatically switching quasi-energy solutions are given by

$$(4.10) \qquad \phi_\alpha(x,t';t) = \phi_\alpha(x,t'+T;t)$$

and the corresponding quasi-energies (eigenvalues of the Floquet-like operator $\hat{\mathcal{H}}$) are varied in time, $E_\alpha(t)$. Therefore Eqs. (4.5, 4.6) can be replaced by

$$(4.11) \qquad \Psi(x,t) = \sum_\alpha C_\alpha(t)e^{-iE_\alpha(t)t/\hbar}\phi_\alpha(x,t'=t,t)$$

where

$$(4.12) \qquad \begin{aligned} C_\alpha &= \lang\!\langle \phi_\alpha(x,t';t)|\Psi(x,t=0)\rangle\!\rangle_{x,t'} = \\ &= \langle \varphi_{0,\alpha}(t)|\phi_0\rangle_x \end{aligned}$$

and $\varphi_{0,\alpha}(x)$ is the t'-Fourier component of $\phi_\alpha(x,t';t)$.

5. The complex scaled quasi-energy solutions. On the basis of the fundamental work of Balslev-Combes [7] and Simon [8] the narrow, broad and overlapping resonances were calculated for: autoionizing atomic and molecular systems; selective adsorbed atom on flat and corrugated surfaces; van der Waals complex; three atomic molecules. The resonances are the poles of the scattering matrix and are associated with the complex eigenvalues of the complex scaled Hamiltonian ($x \to x \exp(i\theta)$), which are θ-independent (providing that θ gets large enough value), $E_{res} = \epsilon_{res} - \frac{i}{2}\Gamma_{res}$. The complex scaled resonance states are square integrable functions regardless of their width (i.e. inverse lifetime.) The non-resonance scattering states are associated with the rotating continuum,

$$(5.1) \qquad E = (\epsilon - E_{th})e^{-2i\theta} + E_{th}$$

where E_{th} is the threshold energy.

When a molecule is exposed to an electromagnetic field, all bound states of the field-free Hamiltonian become resonances (i.e., they acquire a finite lifetime). The threshold energies are $\{\ldots, -2\hbar w, -\hbar w, 0, +\hbar w, +2\hbar w, \ldots\}$ when the dissociation limit of the field-free Hamiltonian is taken as zero. There are two different ways to calculate the complex quasi-energies. The straightforward procedure is by the diagonalization of the Floquet Hamiltonian [9, 10]

$$(5.2) \qquad \hat{\mathcal{H}}\phi_\alpha(x, t) = E_\alpha(x, t)$$

where

$$(5.3) \qquad x = (x' - x_0)e^{i\theta} + x_0$$

and x_0 is usually associated with the minimal value of the field-free molecular potential. This is the common approach first proposed by Chu and Reinhardt [9].

The other approach is to calculate the eigenvalues of the complex scaled evolution operator [11],

$$(5.4) \qquad e^{-i\hat{\mathcal{H}}T}\phi_\alpha(x, 0) = \lambda_\alpha\phi_\alpha(x, T)$$

where

$$(5.5) \qquad \lambda_\alpha = e^{-iE_\alpha/T/\hbar}$$

when the Floquet operator is unscaled λ_α get complex values which form a unit circle in the complex λ-plane. However, upon complex scaling only the threshold energies are mapped to a single point on the unit circle. In case of three atomic molecules ABC, for which three different dissociation limits, A+BC, AB+C and AC+B, can be obtained, all infinite numbers of threshold energies are mapped to three-isolated points on the limit circle in the λ-complex plane. The rotating continua are mapped to a spiral (or three spirals in the case of ABC molecules). The fact that a very small number of Fourier basis functions are required to construct the time evolution operator by the (t, t') method, makes us expect that the quasi-energies obtained by solving Eq. (5.4) will be much more accurate than the results obtained by direct diagonalization of the complex scaled Floquet Hamiltonian. Indeed the results presented in Fig. 2 in Ref. (12) for the model Hamiltonian which described the photodetachment reaction, $Cl^- \xrightarrow{\hbar w} Cl + e^-$, provide a remarkable numerical representation of this fact. When the quasi-energies were calculated by the (t, t') method (i.e. diagonalization of the time-evolution operator calculated by the (t, t') algorithm) converged results were obtained when the maximum field amplitude was increased up to 0.4 a.u. (atomic units.) When the quasi-energies were calculated by diagonalization of the Floquet Hamiltonian matrix using exactly the same basis set (for the x and t' coordinates) non converged results were obtained for $\epsilon_0 = 0.1$ and for $\epsilon_0 = 0.4$ most of the quasi-energies get out of the unit circle and could not provide even a qualitatively estimate for the values of the resonance and non-resonance quasi-energies.

6. Transition probabilities for half-collision processes by the time-independent scattering theory for time-dependent Hamiltonians. In half-collision processes we assume that the atomic or molecular systems are exposed to a cw laser. Due to the interaction of the atom or the molecule with the induced electromagnetic field, ionization or dissociation occurs. In the former case, the measurement of the kinetic energy of the photo-induced ionized electron is known as the ATI (above-threshold-ionization) spectroscopy. In the latter case the measurement of the relative kinetic energy of the dissociated particles due to the molecule/field interaction is known as ATD (above-threshold-dissociation) spectroscopy. In both cases spectra show series of isolated peaks separated by the photon energy $\hbar w$. The results are due to multi-photon absorption processes. All the peaks have the same width, and their relative heights are associated with the partial widths of the resonances which can be obtained by the combination of the (t, t') method and complex scaling as briefly described above in Section 4. Nonlinear dynamical effects are observed as the field intensity is increased. In contrast to what one might expect on the basis of perturbational analysis which holds for weak fields, the slow electron peak in the ATI/ATD spectra almost disappears and the peaks may become narrower rather than broader as one may expect.

The <u>initial</u> state which describes the fact that the system is suddenly exposed to the cw laser is given by

$$(6.1) \qquad \Psi(x, t', t = 0) = \psi_b(x)\delta(t') = \sum_n e^{iwnt'}\psi_b(x)$$

where $\psi_b(x)$ is an eigenfunction of the field-free Hamiltonian. Therefore, the Fourier components of the initial state are given by

$$(6.2) \qquad |\vec{\psi}_0\rangle = \begin{pmatrix} \vdots \\ \psi_b(x) \\ \psi_b(x) \\ \vdots \end{pmatrix}$$

The final state is given when the Hamiltonian is presented in the acceleration gauge

$$(6.3) \qquad \psi(x, t', t_f) = \sqrt{\frac{\mu}{\hbar k}}e^{ikx} \equiv \varphi_f(x)$$

where

(6.4)
$$E_f = \frac{(\hbar k)^2}{2\mu}$$

the E_f is the kinetic energy of the photoionized electron. Therefore, when Fourier basis functions are used as a basis set,

(6.5)
$$|\vec{\psi}_f\rangle = \begin{pmatrix} \vdots \\ 0 \\ \varphi_f(x) \\ 0 \\ \vdots \end{pmatrix}$$

the free electron propagation is given by

(6.6)
$$\vec{\psi}_f(t) = e^{-iE_f t/\hbar}\varphi_f(x) = e^{-i\hat{\mathcal{H}}_0 t/\hbar}\vec{\psi}_f(0)$$

whereas the propagation of the initial states is given by,

(6.7)
$$\vec{\psi}_0(t) = e^{-i\hat{\mathcal{H}}t/\hbar}\vec{\psi}_0(0) \ .$$

The transition probability from a bound state of the field-free Hamiltonian to a free electron with the kinetic energy E_f is obtained by calculating the overlap integral,

$$P(E_f) = |\lim_{t\to\infty} \langle\!\langle \overleftarrow{\vec{\psi}}_f(t)|\vec{\psi}_0(t)\rangle\!\rangle|^2$$

where $\langle\!\langle \ldots \rangle\!\rangle$ stands for the inner product in the generalized Hilbert space, as discussed by Howland. By using the fact that the limit as $t \to \infty$ can be analytically carried out, the closed-form expression for the state-to-state transition probability in a half-collision process is obtained [3]:

(6.8)
$$P(E_f) = |\langle\overleftarrow{\vec{\psi}}_0(0)|1 + G^+(E_f)V|\vec{\psi}_f(0)\rangle|^2$$

where

(6.9)
$$\hat{G}^+ V \equiv \hat{\mathcal{H}} - \hat{\mathcal{H}}_0$$

(6.10)
$$\hat{G}^+(E_f) = \lim_{\epsilon\to 0^+} \frac{1}{E_f - \hat{\mathcal{H}}(x, t') + i\epsilon}$$

When complex scaled Hamiltonians are used (see on the complex coordinate scattering theory in Ref. (13)) there is no need to define ϵ and

(6.11)
$$\hat{G}(E_f) = \frac{1}{E_f - \hat{\mathcal{H}}(x, t')}$$

where x stands for the complex scaled reaction coordinate. An important nontrivial result is the fact that only the $n = 0$ Fourier component of ψ_f (i.e. the $n = 0$ component in the vector $\vec{\psi}_f(0)$) is non-zero, whereas all the Fourier components of ψ_0 are identical and equal to the bound state of the field-free Hamiltonian.

7. Prediction of splitting in the ATD spectrum of heteronuclear diatomic molecules. When a diatomic molecule is exposed to a UV laser which couples the bound ground electronic state of the diatom with a repulsive excited electronic state, the Floquet-like operator is given by

$$(7.1) \quad \hat{\mathcal{H}} = \left(\hat{p}_{t'} + \frac{\hat{p}_x^2}{2m} \right) 1 + \left(\begin{array}{cc} V_1(x) & 0 \\ 0 & V_2(x) \end{array} \right) + \epsilon_0 x \left(\begin{array}{cc} \mu_{11} & \mu_{12} \\ \mu_{21} & \mu_{22} \end{array} \right) f(t)$$

where m is the reduced mass of the diatomic molecule; x is the deviation of the internuclear distance from the equilibrium value for the ground electronic state; $V_1(x)$ is the potential energy curve of the ground electronic state, and $V_2(x)$ is that for the excited electronic state; μ_{11} and μ_{22} are the permanent dipole moments in the two different electronic states, whereas $\mu_{12} = \mu_{21}^*$ are the dipole transition matrix elements. By carrying out gauge transformation as described in Ref. (14) the Floquet-like operator in the acceleration gauge is obtained,

$$(7.2) \quad \hat{\mathcal{H}} = \left(\hat{p}_{t'} + \frac{\hat{p}_x^2}{2m} \right) 1 + \left(\begin{array}{cc} \hat{V}_{11}(x, t') & \hat{V}_{12}(x, t') \\ \hat{V}_{21}(x, t') & \hat{V}_{22}(x, t') \end{array} \right) f(t)$$

where

$$(7.3) \qquad V_{11}(x, t') \to \frac{(\epsilon_0 \mu_+)^2}{4mw^2} \equiv E_{th}^+ \qquad \text{as} \quad x \to \infty$$

$$(7.4) \qquad V_{22}(x, t') \to \frac{(\epsilon_0 \mu_-)^2}{4mw^2} \equiv E_{th}^- \qquad \text{as} \quad x \to \infty$$

$$(7.5) \qquad V_{12}(x, t') = V_{21}^*(x, t') \to 0 \qquad \text{as} \quad x \to \infty$$

where $\mu_\pm$ are the two eigenvalues of the dipole moment matrix μ.

On the basis of the ATI/ATD spectra given in Ref. (15) it is clear that for heteronuclear diatomic molecules (such as HD$^+$ for example) the isolated peaks in the spectra are localized at

$$(7.6) \qquad n\hbar w - (E_{th}^\pm + D)$$

where D is the dissociation energy of the diatom at the ground electronic state. As discussed in Ofir E. Alon's thesis [16], a splitting in the ATD spectrum of HD^+ is expected and is associated with the two possible dissociation possibilities,

$$(7.7) \qquad H + D^+ \overset{\hbar w}{\leftarrow} HD^+ \overset{\hbar w}{\rightarrow} H^+ + D$$

We do hope that this prediction will bring about the motivation to carry out accurate ATD measurements of heteronuclear diatomic molecules.

8. Transition probabilities in full collision processes by the time-independent reactive scattering theory for time-dependent Hamiltonians. Let us consider a reaction involving three atoms,

$$A + BC \overset{\hbar w}{\rightarrow} AB + C$$

which is carried out in the presence of an electromagnetic field induced by a pulsed laser. A and BC are the reactants and are denoted by "R", whereas AB and C are the products and are denoted by "P". The nonreactive-process (when the field is off) is described by effective $H_0^{(R)}$ and $H_0^{(P)}$ Hamiltonians,

$$(8.1) \qquad \hat{H}_0^{(R)} \phi_R = E_R \phi_R$$

$$(8.2) \qquad H_0^{(P)} \phi_P = E_P \phi_P$$

where E_R and E_P are respectively the energies of the reactants and of the products. The transition from reactants to product require "absorbing" of energy E_f,

$$(8.3) \qquad E_f = E_R - E_P$$

by the reactants from the electromagnetic field. Therefore in the (t, t') "language"

$$(8.4) \qquad \hat{\mathcal{H}}_0^{(R)} \phi_R(x, 0) = E_R \phi_R(x, 0)$$

$$(8.5) \qquad \hat{\mathcal{H}}_0^{(P)} \phi_P(x, 0) = (E_P + E_f) \phi_P(x, t')$$

where

$$(8.6) \qquad \phi_P(x, t') = \phi_P(x, 0) e^{-iE_f t'/\hbar}$$

Let us assume that at a given time t the ABC complex interacts with the field such that at $t \to \infty$ the products will be obtained, but if $t \to -\infty$ the reactants would be re-obtained. That is,

$$(8.7) \qquad e^{-i\hat{\mathcal{H}}t/\hbar} \psi(x, 0) \to a_P e^{-i\mathcal{H}_0^{(P)} t'/\hbar} \phi_P(x, t')$$

as $(t' = t) \to \infty$ whereas,

$$(8.8) \qquad e^{-i\hat{\mathcal{H}}t/\hbar} \psi(x, 0) \to e^{-i\mathcal{H}_0^{(R)} t/\hbar} \phi_R(x, 0)$$

as $(t' = t) \to -\infty$. Consequently,

$$(8.9) \qquad \psi(x, 0) \to e^{-i\hat{\mathcal{H}}t/\hbar} e^{+i\hat{\mathcal{H}}_0^{(R)}t/\hbar} \phi_R(x, 0)$$

as $(t' = t) \to \infty$, and from Eq. (8.7) one immediately obtains that,

$$(8.10) \qquad a_P\phi_P = \lim_{(t'=t)\to\infty} e^{i\hat{\mathcal{H}}_0^{(P)}t/\hbar} e^{-2i\hat{\mathcal{H}}t/\hbar} e^{i\hat{\mathcal{H}}_0^{(R)}t/\hbar} \phi_R$$

The "reactant" to "product" transition probability is given by the absolute value of the coefficient a_P,

$$(8.11) \qquad P_{P\leftarrow R}(E_R) \equiv |a_P|^2$$

By taking the limit of $(t' = t) \to \infty$ one can get that,

$$P_{P\leftarrow R}(E_R) = \left| \langle \phi_P | \phi_R \rangle - 2\pi i \langle \phi_P | \hat{V}_P + \hat{V}_P \hat{G}^+ \left(\frac{E_R + E_P + E_f}{2} \right) \hat{V}_R | \phi_R \rangle \right|^2$$
$$(8.12)$$
where

$$(8.13) \qquad \hat{V}_P \equiv \hat{\mathcal{H}} - \hat{\mathcal{H}}_0^{(P)}$$

$$(8.14) \qquad \hat{V}_R = \hat{\mathcal{H}} - \hat{\mathcal{H}}_0^{(R)}$$

$$(8.15) \qquad \hat{G}^+(E) \equiv \lim_{\epsilon\to 0^+} (E - \hat{\mathcal{H}} + \epsilon)^{-1}$$

From the energy conservation law mentioned above, i.e. $E_R = E_P + E_f$, the final expression for the reactant to product transition probability is obtained

$$(8.16) \quad P_{P\leftarrow R}(E_R) = \left| \delta_{P,R} - 2\pi i \langle \phi_P | V_P + V_P G^+(E_R) V_P | \phi_R \rangle \right|^2$$

This is a generalization of the time-independent scattering theory for time-periodic Hamiltonians developed in Ref. (17) for not necessarily time-periodic Hamiltonians. This expression is closely related to the state-to-state field induced transition derived by Peskin and Miller [18].

9. Control of laser-induced dynamical processes.

Let us assume that the initial state is an eigenfunction or a wavepacket, ϕ_0, which is constructed of different populated eigenstates of the field-free Hamiltonian. By using given lasers we wish to bring the system to a state which is described by another wavepacket (or eigenstate) ϕ_f. The interaction of the atomic/molecular system with the electromagnetic field is given by

$$(9.1) \qquad F(x, t; \vec{\epsilon}_0, \vec{w}, \vec{\varphi}, \vec{\alpha}, \vec{\tau})$$

when $\vec{\varphi} = (\varphi_1, \varphi_2, \ldots)$ are phase parameters in a pulsed multi-colored laser with frequencies $(w_1, w_2, \ldots)$; $\vec{\alpha} = (\alpha_1, \alpha_2, \ldots)$ are the parameters which determine the shape of the laser pulses, and $\vec{\tau} = (\tau_1, \tau_2, \dots)$ are the time-delay parameters. $\vec{\epsilon}_0$ are the field intensities of the laser pulses. Of course one may include in F any other physical parameters which can be varied in the experiment. Note that $F \neq 0$ only when the field is on and therefore $F = 0$ for $t \leq 0$ and for $t \geq T$. Let us assume that the solution of the time-dependent Schrödinger equation which takes the system from $|\phi_0\rangle$ to $|\phi_f\rangle$ is known. Furthermore, let

$$(9.2) \qquad\qquad \psi(x, 0) = \phi_0$$

and

$$(9.3) \qquad\qquad \psi(x, t_f) = \phi_f$$

When t_f is larger than the time during which the electromagnetic field is on

$$(9.4) \qquad\qquad F(t = t_f) = 0 \ .$$

By solving the inverse problem one immediately obtains that,

$$(9.5) \qquad\qquad V(x, t) = \frac{i\hbar \frac{\partial}{\partial t}\psi(x, t) - \hat{H}_0(x)\psi(x, t)}{\psi(x, t)}$$

where $\hat{H}_0$ is the field-free Hamiltonian. To have an optimal control the field parameters, $(\vec{\epsilon}_0, \vec{w}, \vec{\varphi}, \vec{\alpha}, \vec{\tau})$, should be varied such that

$$(9.6) \qquad\qquad V(x, t) \cong F(x, t)$$

as much as possible. An optimal fit can be obtained by varying the field parameters to minimize

$$(9.7) \qquad\qquad J = \left| \langle\langle \psi | V(x, t) - F(x, t) | \psi \rangle\rangle \right| \ .$$

In the first step of the calculations we shall propagate the initial state to $t = t_f$. Using the initial guess parameters, $\tilde{\varepsilon}_0$, $\tilde{w}$, $\tilde{\varphi}$, $\tilde{\alpha}$, and $\tilde{\tau}$.

$$(9.8) \qquad\qquad \tilde{\psi}(x, t) = e^{-i\mathcal{H}(x, t')t/\hbar}\phi_0(x)\Big|_{t'=t}$$

where

$$(9.9) \qquad\qquad \mathcal{H}(x, t') = p_{t'} + \hat{H}_0(x) + F(x, t; \vec{\epsilon}_0, \vec{w}, \vec{\varphi}, \vec{\alpha}, \vec{\tau}) \ .$$

The result, $\tilde{\psi}(x, t = t_f)$, is not necessarily equal to the desired final state ϕ_f. The deviation from the desired result is measured by the overlapping integral between the two normalized functions:

$$(9.10) \qquad\qquad \Delta = S - 1$$

where

$$(9.11) \qquad S = \left| \langle \tilde{\psi}(t = t_f) | \phi_f \rangle \right|^2$$

In order to get a new guess for the parameter, $\tilde{\varepsilon}_0$, $\tilde{w}$, $\tilde{\varphi}$, $\tilde{\alpha}$, and $\tilde{\tau}$ we do not substitute $\psi = \tilde{\psi}$ into Eq. (9.5) but we substitute the following function,

$$(9.12) \quad \psi(x, t) = \phi_0(x) + \left(\frac{\tilde{\psi}(x, t) - \tilde{\psi}(x, 0)}{\tilde{\psi}(x, t_f) - \tilde{\psi}(x, 0)} \right) (\phi_f(x) - \phi_0(x))$$

$\psi(x, t)$ is a function that describes the transition of the system from the initial state ϕ_0 to the final one ϕ_f, as $t \to t_f$.

By substituting $\psi(x, t)$ given in Eq. (9.12) into Eq. (9.5), the new estimate of the field parameters can be obtained by minimizing J given in Eq. (9.7). By carrying out these iterative calculations, the deviation, Δ, from the final desired result as function of the field parameters is obtained (see Eqs. (9.12)-(9.13)). Of course, one should choose the field parameters for which Δ assumes a minimal value. To complete the description of this control procedure we should discuss the construction of the zero-iterative order $\psi(x, t)$. For the first iteration we may choose $\psi(x, t)$ to be adiabatically varied from ϕ_0 to ϕ_f. Namely,

$$(9.13) \qquad \psi(x, t) = \phi_0(x) + \sin^2(\Omega t)(\phi_f(x) - \phi_0(x))$$

where

$$(9.14) \qquad t_f = \frac{\pi}{2\Omega}$$

As Ω gets a smaller value the transition from ϕ_0 to ϕ_f is carried out more adiabatically.

An illustrated numerical example of this control procedure is currently under study [19].

REFERENCES

[1] F.H. Faisal, Theory of Multiphoton Processes (Plenum, New York, 1987).
[2] P. Brumer and M. Shapiro, Ann. Phys. Chem. 43, 257 (1992).
[3] U. Peskin and N. Moiseyev, J. Chem. Phys. 99, 4590 (1993).
[4] J.-S. Howland, Math. Ann. 207, 315 (1974); Indian University Mathematics Journal 28, 471 (1979).
[5] U. Peskin, R. Kosloff and N. Moiseyev, J. Chem. Phys. 100, 8849 (1994).
[6] R. Blumel, S. Fishman and U. Smilansky, J. Chem. Phys. 84, 2604 (1986).
[7] E. Balslev and J.J. Combes, Commun. Math. Phys. 22, 280 (1971).
[8] B. Simon, Ann. Math. 97, 247 (1973).
[9] S.I. Chu and W.P. Reinhardt, Phys. Rev. Lett. 39, 1195 (1997).
[10] N. Moiseyev and H.J. Korsch, Isr. J. Chem. 30, 107 (1990); N. Moiseyev, Isr. J. Chem. 31, 311 (1991).

[11] N. Moiseyev and H.J. Korsch, Phys. Rev. A 44, 7797 (1991); N. Ben-Tal, N. Moiseyev, C. Leforestier, and R. Kosloff, J. Chem. Phys. 94, 7311 (1991).

[12] U. Peskin, O. Alon and N. Moiseyev, J. Chem. Phys. 100, 7310 (1994).

[13] For time-independent Hamiltonians: E. Engdahl, N. Moiseyev and T. Maniv, J. Chem. Phys. 94, 1636 (1991); E. Engdahl, N. Maniv and N. Moiseyev, J. Chem. Phys. 94, 6330 (1991); U. Peskin and N. Moiseyev, J. Chem. Phys. 97, 2804 (1992); Int. J. Quantum Chem. 46, 343 (1993). For time-dependent Hamiltonians see: U. Peskin and N. Moiseyev, Phys. Rev. A. 49, 3712 (1994); Ref. (3); N. Moiseyev, Comments At. Mol. Phys. 31, 87 (1995).

[14] N. Moiseyev, O.E. Alon and V. Ryaboy, J. Phys. B (1995).

[15] N. Moiseyev, F. Bensch and H.J. Korsch, Phys. Rev. A 42, 4045 (1990); F. Bensch, H.J. Korsch and N. Moiseyev, J. Phys. B. 24, 1321 (1991); Phys. Rev. A 43, 5145 (1991).

[16] O. Alon, M.Sc. Thesis, Technion (Israel Institute of Technology, 1995).

[17] U. Peskin and N. Moiseyev, Phys. Rev. A 49, 3712 (1994).

[18] U. Peskin and W.H. Miller, J. Chem. Phys. 102, 4084 (1995).

[19] O. Alon, R. Kosloff and N. Moiseyev, in preparation.

BARRIER RESONANCES AND CHEMICAL REACTIVITY

RONALD S. FRIEDMAN* AND DONALD G. TRUHLAR†

Abstract. We survey the phenomenology of broad and overlapping resonances in the scattering theory of molecular collisions. We first discuss examples of resonances encountered in molecular collisions that are not described by the well known isolated narrow resonance formulae; nevertheless certain regularities are observed. We emphasize (i) the relationship between the total resonance width and the sum of the partial widths and (ii) the comparison of trapped-state resonances to barrier resonances, especially from the point of view of the change in background (direct) scattering over the width of the resonance. We then focus on quantal scattering by one-dimensional potential energy functions to provide further insight into the nature of barrier resonances, which are also called transition states. In studies of symmetric and unsymmetric potential functions, we show that reaction thresholds associated with barriers are associated with poles of the scattering matrix; i.e., chemical reaction thresholds are resonances. As the parameters of the potential function are varied, we follow the "trajectory" of the poles in the complex energy plane and examine the connection between barrier resonances and conventional resonances associated with wells between barriers. Resonances are further characterized by considering the relationship between the resonance width and the reactive delay time.

Key words. chemical reactions, poles of the scattering matrix, resonances, spectroscopy of the transition state

AMS(MOS) subject classifications. 81U10, 81U05, 81U20

1. Introduction. "Probably the most striking phenomenon in the whole range of scattering experiments is the resonance. Resonances are observed in atomic, nuclear and particle physics. In their simplest form they lead to sharp peaks in the total cross section as a function of energy ... There are many different theoretical approaches to the resonance phenomenon, all of them having in common that the sharp variation of the cross section ... is in some way related to the existence of a *nearly bound state* of the projectile-target system ..." [1].

Thus resonance states are metastable states, and such states have a much richer phenomenology than true bound states. The present contribution to the interdisciplinary workshop will focus on metastable states and resonance phenomena in chemically reactive molecular scattering processes that raise interesting and incompletely understood questions about correlating observed phenomena with the analytic structure of the scattering matrix. The chapter includes examples of specific case studies and relevant background. References are illustrative without attempting to give a complete survey of the literature.

---

* Department of Chemistry, Indiana University Purdue University Fort Wayne, Fort Wayne, IN 46805-1499.

† Department of Chemistry and Supercomputer Institute, University of Minnesota, Minneapolis, MN 55455-0431.

1.1. Trapped-state and barrier resonances. Metastable states in quantum mechanics are usually associated with quasibound states located in a potential well which may be between barriers or behind a barrier. The barrier separating the resonance state from the dissociative continuum may be in the potential energy or in an effective potential curve (potential energy plus centrifugal potential); in such cases the resonance is called a "shape resonance" or "single-particle resonance." Alternatively, the barrier may be associated with slow energy transfer from internal modes of one of the collision partners to the dissociative mode, i.e., to the collision coordinate; such resonances are variously known as core excited [1], internal-excitation [2], target excited [3], or Feshbach [4,5] resonances. The finite lifetime associated with the metastable state can manifest itself in a time delay for the scattering phenomenon (relative to the free particle motion in the absence of a potential) and in rapid variations in cross sections and transition probabilities. We refer to such metastable states as "trapped-state resonances."

However, metastability is not limited to states located "behind barriers". It is well known in *classical* mechanics that a potential *maximum* is associated with metastability at the energy of the maximum; passage over the barrier incurs a time delay due to the system slowing down *as it crosses* the barrier top. There should be, and in fact is, an analogue in *quantum* mechanics [6]. Quantum mechanically, barrier passage is also associated with a time delay, manifested as an increasing phase in elements of the scattering (S) matrix (relative to the background phase, i.e., the phase itself or show an increase or a less rapid decrease in the vicinity of the barrier) [3,7-9]. The time delays associated with barrier passage have been shown, as will be discussed below, to be associated with metastable states [10-12], which we will refer to as "barrier resonances."

Trapped-state and barrier resonances share one very important feature: they are both "resonances," which we define, using the most fundamental definition we know [1a,13-18], as observable, energy-localized manifestations of poles of the scattering matrix (S matrix) at complex (resonance) energies z_α given by

$$(1.1) \qquad z_\alpha = \varepsilon_\alpha - i\Gamma_\alpha/2$$

The index α (a collection of quantum numbers) labels the resonance, ε_α is the real part of the resonance energy and Γ is the (total) resonance width. That there is no distinction *in kind* between barrier and trapped-state resonances has been demonstrated in one-dimensional scattering studies [11,12] in which it has been shown that by continuously varying the form of the potential barrier function, a barrier resonance may be transformed into a trapped-state resonance and vice versa. In fact, in both the one-dimensional studies as well as three-dimensional, multiparticle reactive scattering work [19-21] we have found that a resonance need not be uniquely described as either a trapped-state or a barrier resonance but instead may partake of the characters of both. Nonetheless, although the classification of a pole of **S** as a barrier or trapped-state resonance is not always unambiguous, the distinction between the two types of resonances is useful, just as the subdivision [2,4] of trapped-state resonances into internal-excitation and shape resonances has proved useful in a wide variety of contexts. Trapped-state resonances in general tend to be associated with longer time delays and narrower widths than barrier resonances.

1.2. Isolated narrow resonances. The molecular collisions under consideration in this chapter are invariant under time reversal and, as a consequence, phases can be chosen so that the scattering matrix is symmetric [22], and we will do so. In the vicinity of an isolated resonance, the scattering matrix element $S_{n'n}$, where $n(n')$ denotes the set of quantum numbers for the initial (final) channel, can be written [1a,15,23-26]

$$(1.2) \qquad S_{n'n}(E) = S_{n'n}^b(E) - i\frac{\gamma_{an'}\gamma_{an}}{E - z_\alpha}$$

where z_α is given by eqn (1.1), $S_{n'n}^b(E)$ is the nonresonant background contribution to the scattering matrix element, E is the total energy, and $\gamma_{an'}$ is the partial width amplitude for channel n'. We immediately see that at the complex resonance energy z_α, there is a pole in the S matrix.

The total width Γ_α of eqn (1.1) is directly related to the rate constant k_α for the unimolecular decay of the resonance state [27,28]

$$(1.3) \qquad k_\alpha = \Gamma_\alpha/\hbar$$

where $\hbar$ is Planck's constant (h) divided by 2π. The rate constant itself is defined by

$$(1.4) \qquad dC_\alpha/dt = -k_\alpha C_\alpha$$

where C_α is the number density of state α. Defining the "lifetime" τ_α as the reciprocal of the rate constant yields

$$(1.5) \qquad \tau_\alpha = \hbar/\Gamma_\alpha$$

The "partial width" $\Gamma_{an'}$ for leaving the metastable resonance to go into final channel n' (or for entering it from channel n') is given by $|\gamma_{an'}|^2$. (It can also be defined using a golden-rule type of formula in terms of properly normalized resonance wave functions [29].)

For an isolated, narrow resonance (INR) one can show that [1a,15,23-29]

$$(1.6) \qquad \sum_n \Gamma_{an} = \Gamma_\alpha$$

where the sum extends over all open (i.e., energetically accessible) channels. The partial width Γ_{aj} controls the rate constant k_{aj} of unimolecular decay of the resonance α into a specific (initial or final) state j just as the total width Γ_α controls the total decay rate k_α. In particular [29],

$$(1.7) \qquad k_{\alpha j} = \Gamma_{\alpha j}/\hbar$$

Using eqn (1.2) and neglecting the background contribution $S^b_{n'n}$ and the interference between the background and resonant contributions, we see that the channel-to-channel transition probability $P_{n \to n'}$, to the extent that it is controlled by resonance α, is given by

$$(1.8) \qquad |S_{n'n}|^2 = \frac{\Gamma_\alpha^2}{(E - \varepsilon_\alpha)^2 + \Gamma_\alpha^2/4} P_{\alpha n} P_{\alpha n'}$$

where

$$(1.9) \qquad P_{\alpha j} = \Gamma_{\alpha j}/\Gamma_\alpha$$

We see that pure resonant scattering is statistical with branching ratios entirely governed by the partial widths. In other words, the decay of the resonance is independent of its mode of formation. This property obviously follows from the factorization of the residue in eqn (1.2).

Since the scattering matrix is unitary and symmetric, it can be diagonalized by a real, orthogonal matrix $\mathbf{B}$:

$$(1.10) \qquad S_{n'n} = \sum_m B_{n'm} e^{2i\Delta_m} B_{mn}$$

where Δ_m are called eigenphase shifts or the eigenphases. The eigenphase sum Δ is given by

$$(1.11) \qquad \Delta = \sum_m \Delta_m$$

For an isolated narrow resonance, the eigenphase sum follows the multi-channel Breit-Wigner formula [30]

$$(1.12) \qquad \Delta(E) = \Delta^b(E) + \arctan \frac{\Gamma_\alpha}{2(\varepsilon_\alpha - E)}$$

where Δ^b is the background contribution. If the background contribution is negligible, the eigenphase sum increases by exactly π as one transverses the real energy ε_α [31]. The eigenphase sum can be calculated by diagonalizing $\mathbf{S}$ and using eqn (1.11) or more simply by using

$$(1.13) \qquad \exp(2i\Delta) = \det \mathbf{S}$$

A resonance may also be characterized in terms of a channel-to-channel delay time $\Delta t_{nn'}$, which is the time difference between the injection of a pulse in initial channel n and the appearance of a pulse in final channel n', relative to the same delay time in the absence of a potential. In general, the delay time is defined by [3,11,12,32,33]

$$(1.14) \qquad \Delta t_{nn'} = \mathrm{Im}\left(\hbar(S_{n'n})^{-1}\frac{\mathrm{d}S_{n'n}}{\mathrm{d}E}\right)$$

or equivalently

$$(1.15) \qquad \Delta t_{nn'} = \hbar\frac{\mathrm{d}}{\mathrm{d}E}\arg S_{n'n}(E)$$

The scattered particle will experience a delay time at real energies in the proximity of ε_α which is expected to be large if the resonance width is small. In particular, if the resonance is an INR, then from eqns (1.2) and (1.14) and the assumption that $S_{n'n}^b$ is negligible, one can easily show that $\Delta t_{nn'}$ will have a maximum at $E = \varepsilon_\alpha$, at which

$$(1.16) \qquad \Delta t_{nn'} = 2\hbar/\Gamma_\alpha$$

Note that this result is independent of n and n' in this limit. The factor of two difference between eqns (1.5) and (1.16) is due to the fact that unimolecular decay may be considered a half collision, whereas eqn (1.16) has contributions from both the incoming and outgoing segments of the collision.

1.3. Overlapping and broad resonances. All of the above INR formulas are expected to hold very well for INRs except [34] where they are close to an energetic threshold. For example these formulas apply extremely well to the resonances observed in a quantum mechanical study of nonreactive atom-molecule scattering by Ashton et al. [35]. However, more generally, resonances encountered in molecular collisions are *not* well described by all of the INR formulas. In this section, we briefly review some literature on overlapping and broad resonances, and we present a few relevant examples from the literature in which partial widths of scattering resonances were determined.

Formal analysis of overlapping resonances is much more scarce than for isolated resonances. In paper III of his well known series on a unified theory of nuclear reactions, Feschbach [36] provided a general treatment of overlapping resonances in terms of complex effective potentials, but this analysis has been much less widely used than his analysis in paper II [5] of isolated resonances. The photoabsorption line shape in the case of *independent* overlapping resonances was considered by Shore [37]. The question of degenerate resonances and avoided resonance intersections has been treated recently by Mondragón and Hernández [38]. Desouter-Lecomte and coworkers [39] also discussed the intersection of two or more nearby resonance poles. In general, overlapping of resonances refers to their having real parts that are the same within approximately the sum of their widths, whereas an intersection would be where both their real and imaginary parts are equal.

Schwenke and Truhlar [27] characterized a total of 14 trapped-state resonances for several different collinear reactions

$$A + BC \rightarrow AB + C$$

treated as three bodies constrained to the z axis and interacting by an analytic potential energy surfaces. (After the center of mass motion has been removed from the collinear problem, two mathematical dimensions remain, which may, for example, be taken as the A-to-B and B-to-C distances.) For each resonance, z_α was evaluated and a full set of partial widths determined. In each case, the eigenphase sum Δ followed the INR formula (1.12) very well, and no special difficulties were encountered in performing fits to obtain ε_α and Γ_α. The background Δ^b was represented as a polynomial in energy. With the location of the scattering resonance determined, a fit of each scattering matrix element at real energies near ε_α to eqn (1.2) resulted in the evaluation of the residue. When we take account of the symmetry of S, there are $(N^2 + N)/2$ residues; but there are only N partial width amplitudes. Thus least square fitting of resonance parameters (as well as background) was used to extract best-fit values of the partial widths.

The ratio

$$(1.17) \qquad\qquad q_\alpha = \frac{\sum_n \Gamma_{\alpha n}}{\Gamma_\alpha}$$

was then computed to see how well INR formula (1.6) was followed. For all 14 resonances, the ratio was less than unity, and for 12 of them it was significantly less, falling in the range from 0.23 to 0.76; for the remaining two resonances, q_α was 0.99 and 0.98. The wider resonances showed the greatest deviation from unity. The conclusion reached in this study was that most of the reactive resonances are not narrow enough to be characterized as INRs.

The significance of q_α differing from unity has been discussed from several different points of view in the literature. Lane and Thomas [23] noted that partial widths smaller than the INR limits will result when the resonance scattering wavefunction has large amplitude in the region external to the Wigner R matrix [40] boundary. Weidenmüller, using analytic S matrix theory, showed that even when a resonance is isolated enough that eqn (1.12) is followed, q_α need not equal unity if the resonance is not narrow enough [25]. Weidenmüller [25] presented numerical results showing $q_\alpha \leq 1$; he showed that even when $q_\alpha \neq 1$, the sum of the eigenphases increases by π at a resonance and the resonant part of the S matrix still factorizes. Further analyses [24,29] have given explicit formal expressions for this ratio using correctly normalized resonance wavefunctions. Consistent with all these works is that q_α may deviate more significantly from 1 for the wider resonances. Of course when $q_\alpha \neq 1$, it would appear that the sum of the specific unimolecular rate constants does not equal the total unimolecular rate constant. The only satisfactory conclusion is that when $q_\alpha \neq 1$, the decay process is intrinsically nonexponential and the rate constants are not all well defined. Since the decay cannot be exponential at $t = 0$ [17], a short-lived resonance may have effectively decayed away before the exponential law has time to manifest itself. Examples of short-lived decaying states for which there is no time for the exponential law to manifest itself have been discussed by Beck and Nussenzweig [41]. The exponential decay law has been discussed by Desouter-Lecomte and Culot in the context of overlapping resonances; they concluded that in the statistical regime of highly overlapped resonances, the decay of a metastable system will be nonexponential, and they discusused the relation of recrossing effects (see below) to off-diagonal elements of the quantal width matrix [42].

Examination of nine published works from which q_α can be calculated showed that q_α was less than or equal to unity in eight [23,25,27,35,43-46] of them. The ninth reference [47] was concerned with shape resonances in the H_2 $X^1\Sigma_g^+$ system. In that paper, resonances were located (that is, ε_α and Γ_α were determined) by directly searching for the complex energy that gave rise to a pole in the scattering matrix. Then by solving the Schrödinger equation directly at complex energies z near z_α to determine $S_{n'n}(z)$, the partial widths were determined by numerically evaluating the contour integral around z_α

$$(1.18) \qquad \oint S_{n'n}(z)\mathrm{d}z = 2\pi\gamma_{\alpha n'}\gamma_{\alpha n}$$

This method of computing partial widths makes no assumptions about the background $S_{n'n}^b(E)$. For five of the twenty-two resonances characterized in Table 2 of Ref. 47, q_α differed from unity by more than 5%; the values of q_α in these cases were 0.85, 0.89, 1.06, 1.08 and 1.17. Thus this study gives examples where q_α exceeded unity. A mathematical analysis of the systematics of q_α and its dependence on the parameters of the scattering problem would be very useful.

2. Transition state resonances in reactive scattering. An important question posed by chemical dynamicists is whether the rates of chemical reactions are controlled by quantized dynamical bottlenecks. In other words, are there quantized structures in phase space that gate the flux of probability density from reactants to products, and, if so, is the gated flux quantized? The idea of a dynamical bottleneck is the central tenet in transition state theory [48-51], which postulates the existence of a structure in phase space - the "transition state" or "activated complex" - which controls chemical reactivity. In this section, we give an introduction to transition state theory and then summarize direct evidence which has been accumulated in accurate quantum dynamics calculations for a quantized spectrum of the transition state. We then discuss the implications of the identification of transition states as resonances.

2.1. Transition state theory. We focus on the bimolecular reaction between A and BC, representing the transition state as $ABC^{\ddagger}$

$$A + BC \rightarrow ABC^{\ddagger} \rightarrow AB + C$$

The transition state is constrained to a hypersurface in phase space dividing the reactants from products [49-51]. This dividing surface has one less degree of freedom than the phase space itself; the missing degree of freedom, which is orthogonal to the dividing surface, is called the reaction coordinate and denoted s. The temperature dependent rate constant $k(T)$ that appears in the rate expression

$$(2.1) \qquad -\frac{dC_A}{dt} = k(T)C_A C_{BC}$$

(where t denotes time) can be written, with the assumption that once the system gets to the transition state it inevitably proceeds to products, as

$$(2.2) \qquad k(T) \cong \nu C_{\neq}^{eq}.$$

where ν is the unimolecular rate constant for decay in the product direction of the transition state which has an equilibrium concentration $C_{\neq}^{eq}$. If we assume that $k(T)$ can be equated to the local-equilibrium phase-space flux in the product direction through the transition state hypersurface, then, in a classical world, one obtains [51]

$$(2.3) \qquad k(T) \cong \frac{k_B T}{h} K_{\text{classical}}^{\ddagger}$$

where k_B is Boltzmann's constant and $K_{\text{classical}}^{\ddagger}$ is a kind of equilibrium constant which relates the product $C_A C_{BC}$ reactant concentrations to $C_{\neq}^{eq}$ with the reaction coordinate removed in defining the hypersurface to which $C_{\neq}^{eq}$ refers. $K_{\text{classical}}^{\ddagger}$ is computed via an integral over the phase space of the hypersurface [51]. In the quantal world, in which there would not be a *continuum* of transition state structures along the dividing surface but rather a *discrete set* of states on the hypersurface, eqn (2.3) becomes

$$(2.4) \qquad k(T) \cong \frac{k_B T}{h} K_{\text{quantal}}^{\ddagger}$$

where $K_{quantal}^{\ddagger}$ is computed from the quantized partition functions (sums over discrete sets of levels) of A, BC, and the transition state hypersurface (again with motion along the reaction coordinate removed). Eqn (2.4), in effect, postulates the existence of a set of discrete levels associated with the transition state hypersurface orthogonal to the reaction coordinate. We will seek evidence in the accurate quantal calculations for these levels of the transition state; in other words, we seek to perform transition state spectroscopy, a term used in a series of papers by Polanyi and coworkers [52].

If the analysis that led to eqn (2.4) is repeated for a microcanonical ensemble (a system at fixed total energy rather than fixed temperature), the transition state theory rate constant $k^{TS}(E)$ at a given energy E can be written as [53-58]

$$(2.5) \qquad k^{TS}(E) = \frac{N^{TS}(E)}{h\rho^R(E)}$$

where $N^{TS}(E)$ is the number of energy levels of the activated complex with energies less than or equal to E, and $\rho^R(E)$ is the reactants' density of states per unit energy per unit volume. The quantity $N^{TS}(E)$ increases, obviously, by one at each of the energy levels of the transition state.

An exact quantum mechanical expression for the microcanonical rate constant $k(E)$ can be written as [19,54]

$$(2.6) \qquad k(E) = \frac{N(E)}{h\rho^R(E)}$$

where $N(E)$, which is called the cumulative reaction probability (CRP) [59], is a double sum over all the state-to-state reaction probabilities at an energy E:

$$(2.7) \qquad N(E) = \sum_{n}\sum_{n'} P_{n\to n'}(E)$$

For an atom-diatom reaction, $n(n')$ is an index representing the toal angular momentum J, its component M_J on an arbitrary space-fixed axis, and a set of initial (final) vibrational quantum number $v(v')$, rotational quantum number $j(j')$ and orbital angular momentum quantum number $l(l')$. The CRP can be written in terms of the scattering matrix elements $S_{n'n}$ as [60]

$$(2.8) \qquad N(E) = \sum_{n}\sum_{n'} |S_{n'n}|^2$$

Comparison of eqns (2.5) and (2.6) suggests, if the quantization of the transition state is taken literally, that the quantal CRP should increase in steps of one at each of the energy levels of the transition state. However, this analysis has neglected quantum mechanical tunneling at energies below the transition state energy level as well as nonclassical reflection above the energy level. In addition, it has been assumed up to this point that the transition state is a perfect dynamical bottleneck, i.e. systems crossing the activated complex hypersurface go right from reactants to products without

recrossing the dividing surface [51,57,61-64]. Generalizations of transition
state theory have been proposed to account for these effects [54,58,59,65].
For example, one way to account for the effects of recrossing, tunneling and
reflection, which we will use extensively in the next section, is to make the
replacement

$$(2.9) \qquad N^{TS}(E) \rightarrow \sum_\tau \kappa_\tau P_\tau(E)$$

in eqn (2.5), where τ is the index of a level of the transition state, κ_τ
is a transmission coefficient accounting for recrossing (more precisely, for
the quantal analog of classical trajectories that reach the transition state
hypersurface and then recross it - either before proceeding to products or
without ever proceeding to products [51,57,58]), and P_τ is a transmission
probability accounting for tunneling and reflection. Invoking substitution
(2.9), we see that the numerator of eqn (2.5) still contains a sum over energy
levels of the transition state but instead of increasing abruptly by a step of
unity at each new level of the activated complex, it will now increase more
gradually, by an amount given by $\kappa_\tau P_\tau(E)$. If we denote the transition
state energy level as E_τ, then for an ideal dynamical bottleneck, κ_τ is
unity, and $P_\tau(E)$ is expected to rise smoothly from zero at energies E well
below E_τ to unity at energies well above E_τ. Thus transition state theory
incorporating (2.9) predicts that the quantal CRP will increase in smooth
steps of height κ_τ at each new energy level of the transition state. We
have found clear evidence of such behavior in accurate quantal scattering
calculations, as described below.

2.2. Characterization of quantized transition states in chemical reactions. Most approximate theoretical treatments of chemical reaction rates involve the concept of a transition state, i.e. a dynamical
bottleneck. But, up until the time we began our studies, the spectrum of
this mathematical construct had never been directly observed, and one may
have questioned its physical significance. Thus we have sought to address
the question of whether one can observe the quantized nature of dynamical bottlenecks in chemical reaction dynamics. To seek an answer, we
have performed accurate three-dimensional quantal scattering calculations
for three-body systems interacting according to realistic potential energy
surfaces, and, via analysis of the cumulative reaction probability, we have
looked for evidence of quantized dynamical bottlenecks.

We can study the reaction dynamics for each value of the total angular
momentum J independently since it is conserved during the collision. Its
component M_J is also conserved and since results are independent of M_J,
we always set it to zero and forget about it. The J-specific contribution
to the CRP is denoted $N^J(E)$ and it is computed from the sum over J-
specific channel-to-channel transition probabilities $P^J_{n \rightarrow n'}(E)$ in a manner
analogous to eqn (2.7). The dynamical structure in the reaction rate constant is most clearly brought out by computing the energy derivative of

the cumulative reaction probability, which is called the density of reactive states. The J-specific density of reactive states is defined as

$$(2.10) \qquad \rho^J(E) = \frac{\mathrm{d}N^J(E)}{\mathrm{d}E}$$

We have looked for, and found, evidence of quantized transition states in a number of three-dimensional chemical reactions, including H+H$_2$ [19-21], O+H$_2$ [21,66-68], D+H$_2$ [69,70], F+H$_2$ [70,71], Cl+H$_2$ [70,72], Cl+HCl [73], I+HI [73], and I+DI [73]. In addition, the conclusions reached in these studies have been supported by a number of accurate calculations by other research groups on other atom-diatom systems including He+H$_2^+$ [74], Ne+H$_2^+$ [75], H+O$_2$ [76], and O+HCl [77]. Here we will focus on just two of the systems we have studied:

$$H + H_2 \rightarrow H_2 + H$$

$$O + H_2 \rightarrow OH + H$$

The first is the simplest and theoretically most-studied atom-diatom neutral reaction [78] and the first system for which we found evidence of global control of reactivity by a spectrum of quantized transition state levels. The second is also of interest because it is asymmetric (reactants and products are different chemical species), and this introduces the possibility of multiple bottleneck regions with transition states having different sets of energy levels on the reactant (O+H$_2$) side and on the product (OH+H) side.

For the above two reactions, we made the Born-Oppenheimer separation of electronic and nuclear motions [79], and we assume that the scattering event can be treated as the motion of three bodies (in full three-dimensional space) interacting through a single effective potential energy surface. The quantal results to be presented result from treating the two hydrogen atoms in the reactant H$_2$ molecule as distinguishable and giving results for one of the two possible sets of products, for example, A + H$_a$H$_b$ $\rightarrow$ AH$_a$ + H$_b$. Therefore, the factor of 2 for the products AH$_a$ + H$_b$ and AH$_b$ + H$_a$ is not included in the CRP nor the density of reactive states. The computational methods used to calculate accurate quantal CRPs and densities are briefly summarized in Appendix I.

2.2.1. H + H$_2$. The $J = 0$ cumulative reaction probability $N^0(E)$ was computed [19,20] from converged scattering matrix elements via eqn (2.8). The calculation employed the accurate double many-body expansion (DMBE) potential energy surface [80] for H$_3$. It was found that $N^0(E)$ is characterized by a series of smooth steps as predicted by transition state theory and discussed at the end of Section 2.1. Differentiation of a cubic spline fit to $N^0(E)$ yielded $\rho^0(E)$, the $J = 0$ density of reactive states; steps in N^0 were converted to peaks in ρ^0. The CRP reaches a value of 8.9 at 1.6 eV (we use energy units of eV; 1 eV/molecule = 96.48 kJ/mol). Thus if all transition states are perfect dynamical bottlenecks ($\kappa_\tau = 1$), transition state theory predicts nine energy levels of the quantized transition state at energies below 1.6 eV. From the analysis described in the following paragraphs, it was concluded that in fact ten energy levels contribute to the CRP and density of reactive states.

If an explicit functional form were available for the transmission probability $P_\tau(E)$, it would be desirable, based on eqn (2.9), to fit the accurate quantal CRP to

$$(2.11) \qquad N(E) = \sum_\tau \kappa_\tau P_\tau(E)$$

The fit would reflect the degree to which the quantal dynamics follow the predictions of transition state theory and also would allow us to extract parameters (such as κ_τ) characteristic of each transition state level τ.

To obtain a simple form for $P_\tau(E)$, we assume that there exists an effective potential energy barrier $V_\tau(s)$ for passage through the transition state region in level τ. We furthermore assume that the potential barrier is parabolic, having the form

$$(2.12) \qquad V_\tau(s) = E_\tau + \frac{1}{2}k_\tau s^2$$

where E_τ is the energy of the barrier maximum and k_τ is a negative force constant. Recall that the parameter E_τ also represents the energy of transition state level τ. Therefore, within this simple model, each level of the transition state is associated with a maximum in an effective potential barrier. In the case of a parabolic barrier, the quantum mechanical transmission probability P_τ is [81]

$$(2.13) \qquad P_\tau(E) = \frac{1}{1 + \exp[(E_\tau - E)/W_\tau]}$$

where W_τ is a width parameter given by

$$(2.14) \qquad W_\tau = \frac{\hbar|\omega_\tau|}{2\pi}$$

with ω_τ the imaginary frequency of the barrier

$$(2.15) \qquad \omega_\tau = \sqrt{k_\tau/\mu}$$

and μ the reduced mass. As anticipated, $P_\tau(E)$ rises smoothly from zero to one, accounting for tunneling and nonclassical reflection.

Since dynamical structure is brought out more clearly in the density of reactive states, it is advantageous to fit the accurate density as

$$(2.16) \qquad \rho(E) = \sum_\tau \left[\kappa_\tau \frac{dP_\tau(E)}{dE}\right]$$

Taking the energy derivative of $P_\tau(E)$ in eqn (2.13), we arrive at an expression useful for fitting the density of reactive states:

(2.17)
$$\rho(E) = \sum_\tau \kappa_\tau \rho_\tau(E)$$

where

(2.18)
$$\rho_\tau(E) = \frac{\exp[(E_\tau - E)/W_\tau]}{W_\tau(1 + \exp[(E_\tau - E)/W_\tau])^2}$$

The function $\rho_\tau(E)$ is a symmetric bell-shaped curve centered at E_τ; the wider the effective potential barrier $V_\tau(s)$ (the smaller the k_τ), the smaller the width parameter W_τ and the narrower the function $\rho_\tau(E)$. The accurate quantum scattering calculations [19] showed that the transition state theory prediction that the density of reactive states will be a sum of bell-shaped curves, each centered at some energy E_τ, appears to hold true for the density $\rho^0(E)$ for H+H$_2$. The strategy should now be clear; by fitting the accurate quantal density to a sum of terms $\kappa_\tau \rho_\tau(E)$, as given in eqn (2.17), we can determine the fitting parameters κ_τ, E_τ and W_τ for each transition state level τ.

A fit to $\rho^0(E)$ for H+H$_2$ using 10 terms in eqn (2.17) yields a curve which is nearly indistinguishable from the quantal result [19,21]. Integration of the fit to ρ^0 yields the fitted CRP curve which is indistinguishable from $N^0(E)$ to plotting accuracy.

The excellent agreement between the quantal and fitted densities of reactive states shows convincingly that the chemical reactivity is globally controlled by quantized transition states. All of the reactive flux up to an energy of 1.6 eV can be associated with energy levels of the transition state; there is no discernible background contribution. In addition, we can assign the quantized transition state structure to specific vibrational-rotational levels of the transition state, as discussed in the following paragraphs.

A quantized transition state of H$_3$ is a short-lived state (a resonance; see Section 2.3) with a set of linear-triatomic quantum numbers [82] $[v_1 v_2^K]$, where v_1 and v_2 are the stretch and bend quantum numbers, respectively, for modes orthogonal to the reaction coordinate and K is the vibrational angular momentum, the projection of J on the molecular axis. Whereas a stable triatomic molecule like CO$_2$ has a complete set of quantum numbers $(v_1 v_2^K v_3)$ with v_3 being the asymmetric stretch quantum number [82], the H$_3$ transition state appears to have one quantum number missing, corresponding to unbound motion along the reaction coordinate. The $[v_1 v_2^K]$ labelling therefore corresponds to the classical picture of the transition state being a hypersurface in phase space. We will see in Section 2.3 that in the quantum mechanical treatment of the transition state as a resonance, the 'missing' quantum number will reappear.

The lowest energy feature for H+ H$_2$ is readily assigned as $[00^0]$ since it is the overall reaction threshold and therefore corresponds to the lowest energy level of the transition state. The remainder of the assignments were made primarily by comparing the fitted E_τ values with features in semiclassical vibrationally adiabatic potential energy curves. The latter are defined by [65,83-87]

$$(2.19) \qquad V_a(v_1, v_2, K, J, s) = V_{\text{MEP}}(s) + \varepsilon_{\text{int}}(v_1, v_2, K, J, s)$$

where s is the distance along the reaction path, V_{MEP} is the Born-Oppenheimer potential energy along the reaction path, and ε_{int} is the vibrational-rotational energy of the stretch, bend and rotational motions excluding motion along the reaction coordinate. The stretching motion (with quantum number v_1) of the H_3 transition state correlates adiabatically with vibrational motion (with quantum number v) in the reactant H_2 [88]. The energies of the maxima in the vibrationally adiabatic curves are in good agreement with the fitted values of E_τ and allowed us to make the assignments. Furthermore, this good agreement strongly suggests that the reactive flux is "focused" [89] through dynamical bottlenecks that are locally vibrationally adiabatic. However, the quantal calculations [19,20] demonstrate that the overall chemical reaction is not *globally* adiabatic; e.g. many of the $P_{n \to n'}$ with $v \neq v'$ are significant in magnitude ($> 10^{-3}$). Thus reactive flux which is channelled through a particular $[v_1 v_2^K]$ level of the transition state need not originate from only those reactant states with $v = v_1$ but may originate from a wide set of reactant states; this point will be addressed in more detail in Section 2.3 below. One way to understand why the dynamics, although not globally adiabatic, is *locally* adiabatic at the transition state is to note that motion along the reaction coordinate is classically stopped at a barrier maximum. Thus the simplest criterion for vibrational adiabaticity, that vibrational motions normal to the reaction coordinate be fast compared to motion along the reaction coordinate, is satisfied locally.

With the assignments $[v_1 v_2^K]$ made, we can note several important trends in the fitted values of the transmission coefficient. We found that six of the first nine levels of the transition state are ideal dynamical bottlenecks with $\kappa_\tau \cong 1.0$. Deviations from unity are found for only a few of the levels, with the most significant breakdown of transition state theory being associated with a highly bend-excited level. We can also understand trends in the widths (W_τ values) of the quantized transition state features in terms of the widths of the vibrationally adiabatic potential curves; these kinds of trends have been discussed in detail in Refs. [11,12,20,21].

2.2.2. O + H$_2$. The accurate density of reactive states for $O(^3P) + H_2$, $J = 0$ was studied [66-68] using the Johnson-Winter-Schatz potential energy surface [90,91]. At energies below 1.3 eV, the density ρ^0 for this reaction looks strikingly similar to that of $H + H_2$ [19,21]. In fact, it turns out that the first six assignments $[v_1 v_2^K]$ are identical for the two systems [68]. Above 1.3 eV, however, the O + H$_2$ density exhibits more peaks due to a greater number of levels of the quantized transition state.

The density $\rho^0(E)$ for O+H$_2$ was fit using 17 terms in eqn (2.17) and the quantal and fitted densities are indistinguishable to plotting accuracy [68]. Many of the quantized transition states are nearly ideal dynamical bottlenecks; i.e. many κ_τ for values are close to unity. The excellent fit to $\rho^0(E)$ for O+H$_2$ demonstrates that quantized transition states globally control the reaction dynamics, as they did for H + H$_2$.

To assist in making assignments of the transition states, vibrationally adiabatic potential energy curves were computed using the same methods as for H + H$_2$ [68]. Unlike the curves for H+H$_2$, many of those for the O + H$_2$ reaction show two distinct transition states, one on the reactant side and one on the product side, each with its own set of energy levels. In such cases, it is interesting to see if bottlenecks in both regions influence the chemical reactivity. Each of the computed vibrationally adiabatic potential curves with $v_1 \geq 1$ exhibited several local maxima. Only the first and last local maxima of each curve can be associated with dynamical bottlenecks because only in the vicinity of these two maxima is the dynamics predicted, semiclassically, to be vibrationally adiabatic. If the chemical reaction were vibrationally adiabatic (i.e. $P_{n \to n'} = 0$ for all $v \neq v'$). only the higher of the two maxima, which appears at a reaction coordinate nearer to the reactants, would have an impact on the reactivity. We call these reactant-like bottlenecks *variational* transition states and designate them in the usual way, $[v_1 v_2^K]$. However, since the O + H$_2$ reaction has many nonnegligible $v \neq v'$ reaction probabilities, we know that the reaction is not globally vibrationally adiabatic and, therefore, it is possible that product-like dynamical bottlenecks influence the reactivity. We call these latter bottlenecks *supernumerary* transition states and denote them as $S[v_1 v_2^K]$.

The correspondence between the fitted values of E_τ and the global maxima in the 00^0 and 10^0 vibrationally adiabatic potential curves allowed [68] us to unambiguously assign the first and third features in the density curve to the $[00^0]$ and $[10^0]$ variational transition states. In addition, ρ^0 exhibits peaks at the energies of both the reactant-like and product-like maxima in the $v_1 = 2, v_2 = 0$ and $v_1 = 3, v_2 = 0$ vibrationally adiabatic curves. This allowed [68] us to identify two more variational transition states, $[20^0]$ and $[30^0]$, as well as two supernumerary transition states $S[20^0]$ and $S[30^0]$.

The remaining assignments were made largely [68,70] on the basis of analysis of channel-selected transition probabilities $P_n^J(E)$ defined as

$$(2.20) \qquad P_n^J(E) = \sum_{n'} P_{n \to n'}^J(E)$$

The quantity defined above gives the total reactive transition probability coming out from a specified initial (reactant) channel n by summing over all possible *final* channels n'. (By then summing $P_n^J(E)$ over all *initial* channels n, we retrieve the J-specific cumulative reaction probability.) We can then

define the corresponding density of channel-selected reaction probability
$\rho_n^J(E)$

$$(2.21) \qquad \rho_n^J(E) = \frac{\mathrm{d}P_n^J(E)}{\mathrm{d}E}$$

Similarly, we can make the analysis in terms of product channels n':

$$(2.22) \qquad P_{\to n'}^J(E) = \sum_n P_{n \to n'}^J(E)$$

$$(2.23) \qquad \rho_{\to n'}^J(E) = \frac{\mathrm{d}P_{\to n'}^J(E)}{\mathrm{d}E}$$

The quantities defined in eqns (2.22) and (2.23) provide useful information on the total reactive transition probability going into a specified final (product) channel n'.

The channel-selected densities (2.21) and (2.23) provide valuable information on how flux through a given level of the transition state couples with particular reactant and product states. In particular, when both supernumerary and variational transition states influence chemical reactivity, the former are primarily observed in $\rho_{\to n'}^J$ and the latter primarily in ρ_n^J. We can most easily understand this by recognizing that, from the principle of time reversal invariance, $P_{n \to n'}^J$ describes both forward $(n \to n')$ and reverse $(n' \to n)$ reactions. Therefore, whereas ρ_n^J describes reaction *out* of channel n for the *forward* reaction, $\rho_{\to n'}^J$ describes reaction *out* of channel n' for the *reverse* reaction. Thus, the quantity ρ_n^J will tend to be influenced by reactant-like (variational) transition states and $\rho_{\to n'}^J$ by product-like (supernumerary) transition states. By calculating channel-selected densities for O+H$_2$ and identifying maxima in the density curves as features due to variational and/or supernumerary transition states, we were able to assign most of the spectrum [68,70].

2.3. Interpretation of transition states as resonances. We have found strong evidence in quantal calculations, as discussed above, for the control of chemical reactivity by quantized transition states which can be thought of as being associated with the potential maxima of vibrationally adiabatic potential curves. In fits to the quantal density of reactive states, we have taken the energy levels of the transition states to be the potential energy maxima of effective parabolic barriers.

The parabolic barrier of eqn (2.12) has poles of the scattering matrix which are found at [10,21,92,93]

$$(2.24) \qquad E_\nu = E_\tau - i\hbar|\omega_\tau|(v + \frac{1}{2})$$

where $v = 0, 1, 2, \ldots$ labels the pole. Thus, there is a sequence, an infinite one in fact, of overlapping poles of the S matrix whose complex resonance energies all have a real part E_τ. The widths increase along the sequence as $2v+1$. This analysis and other studies of one-dimensional potential barriers in Section 3 show unambiguously that passage over a potential barrier is associated with poles in S, from which we can conclude that quantized transition states are reactive scattering resonances, which we refer to as barrier resonances. Note that we can associate the quantum number v which labels the poles for a given parabolic barrier with the 'missing' quantum number v_3 of the transition state corresponding to the 'missing' degree of freedom. (We point out here that a series of poles labelled by v is not unique to a parabolic barrier; for approximately parabolic barriers, as well, the labelling of members of a sequence of poles by a quantum number v emerges, as discussed in Section 3.)

However, there is an apparent paradox presented by the above discussion. Analyses of the poles of the scattering matrix show that even for a simple barrier (parabolic or approximately parabolic), there is more than one pole [10-12,92-94]. On the other hand, work on three-dimensional quantal dynamics such as that described in Section 2.2 shows that the cumulative reaction probability and density of reactive states are influenced by a small number of quantized transition states. In particular, the density can be fit extremely well using a model of scattering in which each effective parabolic potential barrier V_τ is associated with *one* level of the transition state at a real energy E_τ. This apparent paradox has more than one possible resolution. The simplest is to assume that only the poles close to the real axis have an appreciable effect on the dynamics. Then, since poles with $v \geq 1$ are three more times farther from the real axis than the corresponding poles with $v = 0$, one could consider only the $v = 0$ poles. Unfortunately, we know that the situation is not really quite so simple. Although the $v = 0$ poles provide a good quantitative account of the transition state energies and lifetimes [21], they do not account quantitatively for $P_\tau(E)$ [92]. Nevertheless when combined with a short-time correction, the sum over the $v = 0, 1, 2, \ldots$ states is rapidly convergent, giving reasonable results with 2 states and quite good results with $4 - 8$ states [92]. It is encouraging though that the initial state count in $N^{TS}(E)$ requires only the $v = 0$ states. One could imagine cases where the progressions in v are not readily assignable due to mixed character in the Siegert eigenfunctions [16] associated with the poles. Then a state might show mixed character such as 50% of $v = 0$ in the reaction coordinate with some set of quantum numbers τ for other degrees of freedom and 50% of some other combination, $(v'\tau')$, with $v' \neq 0$. Then it would be unclear whether to count this state in $N^{TS}(E)$. Such a situation would seem to indicate that one cannot identify a separable reaction coordinate, which is a necessity for applying transition state theory. Mathematical analysis clarifying the effect of various barrier poles on the rate constant would be very useful.

The $v = 0$ pole of the parabolic barrier has a width given by $\hbar|\omega_\tau|$, where we have used the general definition of the complex resonance energy given in eqn (1.1). We equate this resonance width with that of level τ of the quantized transition state characterized in the fits to $\rho(E)$. The resonance width is related to the collision lifetime $\Delta\tau$ by [95]

(2.25) $$\Delta\tau = 2\hbar/\Gamma$$

and upon substitution of Γ by $\hbar|\omega_\tau|$ and using eqn (2.14), one obtains

(2.26) $$\Delta\tau = \frac{2}{|\omega_\tau|} = \frac{\hbar}{\pi W_\tau}$$

From the W_τ values obtained in the fit to the density of reactive states for H + H_2 and given in Ref. 20, we calculated collision lifetimes via eqn (2.26). These lifetimes, labelled transition state resonance theory, are shown in Table 1 for the case of $J = 0$ discussed above and in addition for the case of $J = 1$ described elsewhere [20]. The transition state lifetimes range from 5 to 30 femtoseconds. They are compared in Table 1 to collision lifetimes computed [8] from accurate quantal scattering matrix elements without transition state theory or resonance theory. Since the W_τ values are obtained from fits to the energy dependence of reaction probabilities which do not depend on the phases of S matrix elements while the direct calculation [8,32,33] of collision lifetimes depends explicitly on phase information, there is no reason a priori why the two columns in Table 1 need agree. The fact that they do provides further evidence for the usefulness of treating transition states by resonance theory.

Furthermore, we can proceed to obtain spectroscopic constants for the quantized transition states, which are associated with poles of the scattering matrix at complex resonance energies $E_\tau - i\pi W_\tau$ (see above). By analogy to the procedure for getting spectroscopic constants for bound states [82], constants for the transition states can be obtained by fitting the resonance energies to the form

$$
\begin{aligned}
E_\tau(v_1, v_2) - i\pi W_\tau(v_1, v_2) \;=\; & E_0 + hc\omega_1(v_1 + 0.5) + hc\omega_2(v_2 + 1) \\
+ \;& x_{11}(v_1 + 0.5)^2 + x_{22}(v_2 + 1)^2 \\
+ \;& x_{12}(v_1 + 0.5)(v_2 + 1) + BJ(J + 1)
\end{aligned}
$$

(2.27)

where E_0 is a constant and $\omega_1, \omega_2, x_{11}, x_{12}, x_{22}$, and B are the spectroscopic fitting parameters, and c is the speed of light in a vacuum. Since we fit in eqn (2.27) both the energies E_τ and the widths W_τ of the quantized transition states, all of the fitting parameters, including E_0, are complex.

As a specific example of a spectroscopic fit we consider the O + H_2 quantized transition states discussed above. A least squares fit of the first six levels resulted in the following values (in wavenumber units of cm^{-1}) of the complex spectroscopic constants for the variational transition states:

$$E_0 = 3786 - 597i \qquad \omega_1 = 2241 + 507i \qquad \omega_2 = 737 - 307i$$
$$x_{12} = -45 - 89i \qquad x_{22} = -2 + 43i$$

TABLE 1

Lifetimes (fs) for H + H₂ quantized transition states.

J	E_τ (eV)	transition state resonance theory[a]	accurate[b]
0	0.65	10	11
	0.87	7	10
	0.98	28	28
	1.09	6	5
	1.19	10	8
1	0.65	10	11
	0.76	8	9
	0.88	7	10
	0.98	27	28
	0.99	6	8
	1.09	16	29
	1.10	10	5
	1.19	10	8
	1.21	7	6

[a]From eqn (2.26) and W_τ values given in Ref. 20
[b]From Ref. 8

We constrained the values of x_{11} and B to zero because none of the first six levels involve $v_1 > 1$ or $J > 0$.

The interpretation of transition states as reactive scattering resonances also enables us, through an analysis of channel-selected densities, to obtain information concerning the partial widths of these barrier resonances [20,21]. We considered channel-selected densities, and found that the major peaks in ρ_n^0 and $\rho_{\to n'}^0$ occur at or very near the values of E_τ determined earlier from the fit to the overall density of reactive states. We thus see that quantized levels of the transition state control state-selected as well as total chemical reactivity. For example, for H + H₂ with $J = 0$, reactive flux emanating from the $v = 0$, $j = 4$ channel passes predominantly through only the [02⁰] and [20⁰] levels of the transition state at their threshold energies. We find in general [20,21] that the asymptotic channels couple primarily to only a few transition state levels and, in particular, mainly to one or two consecutive bending (v_2) levels with stretch quantum number v_1 equal to channel vibrational quantum number v. Therefore, there is a large propensity for stretch adiabaticity ($v = v_1$) in the half collisions that take the system from reactants to the transition state.

To characterize quantitatively which asymptotic channels couple to which quantized levels of the transition state, we [20,21] fit $\rho_n^0(E)$ with a sum of lineshapes for scattering by parabolic potential energy barriers, just as we did for the total density in eqn (2.17):

$$(2.28) \qquad \rho_n(E) = \sum_\tau \kappa_{\tau n} \rho_\tau(E)$$

where $\rho_\tau(E)$ is given in eqn (2.18) with the values of E_τ and W_τ fixed at those values determined in the fit to the total density. The parameter $\kappa_{\tau n}$, which was determined using linear least squares, is a measure of the amount of reactive flux passing through transition state level τ at energy E_τ that originated from a specific state n. When all the values of $\kappa_{\tau n}$ for a particular level τ are summed, we retrieve the value of the transmission coefficient κ_τ of eqn (2.17). Since the ratio $\kappa_{\tau n}/\kappa_\tau$ represents how much of the total reactive flux passing through level τ of the transition state at energy E_τ originated from channel n, we can identify the partial width for entering the level τ at energy E_τ from incident channel n as (within our model of scattering off parabolic barriers)

$$(2.29) \qquad \Gamma_{\tau n} = \frac{\kappa_{\tau n}}{\kappa_\tau} \Gamma_\tau$$

with

$$(2.30) \qquad \Gamma_\tau = 2\pi W_\tau$$

Although eqn (2.29) would require, by definition, that the INR formula, (1.6), be strictly followed, which of course we don't expect to be the case for broad barrier resonances, eqn (2.29) is nonetheless useful because it provides a type of detailed picture of the reactive event that is available in no other way that does not involve analyzing wave functions.

By calculating the ratio of $\kappa_{\tau n}$ to κ_τ for each channel n and each level τ, we have determined which channels make significant partial width contributions to each level of the transition state [20,21]. Some results are presented in Table 2 for the low bend (v_2) states with $v_1 = 0, 1, 2$. M_{ad} and M_{nonad} are the number of initial channels which make stretch adiabatic ($v = v_1$) and stretch nonadiabatic ($v \neq v_1$) contributions, respectively, of at least 5% to the total reactive flux (i.e. $\kappa_{\tau n}/\kappa_\tau \geq 0.05$). The table's entries are percent contributions for those channels having a specified value of $|j - v_2|, v = v_1$; for example, 27% of the reactive flux passing through $[00^0]$ at its E_τ value originates from the channel with $|j - v_2| = 2$, that is, channel ($v = 0, j = 2$). When several initial channels with the same value of $|j - v_2|$ contribute, the number of channels is indicated in brackets, and the sum of their percent contributions is shown. We see from the table that the largest partial widths tend to be for those channels with $v = v_1$ and $|j - v_2| \leq 3$. Partial widths for channels with $v \neq v_1$ tend to become more important for high v_1.

If one pursues the identification of dynamical bottlenecks with resonances, the description of chemical reactivity ultimately becomes a quest

TABLE 2

State-selected contributions to the total reactive flux passing through thresholds for $H +$ H_2 with $J = 0$ and 1.

					$|j - v_2|$	for $v = v_1$		
threshold[a]	M_{ad}	M_{nonad}	0	1	2	3	4	> 4
$[00^0]$	4	0	22	41	27	9	0	0
$[01^1]$	4	0	15	35	32	15	0	0
$[02^0]$	5	0	0	21	42 [2][b]	23	6	0
$[10^0]$	3	0	33	51	20	0	0	0
$[11^1]$	3	0	24	45	26	0	0	0
$[12^0]$	5	1	7	34	37 [2]	10	0	0
$[20^0]$	2	6	13	17	0	0	0	0
$[21^1]$	3	2	28	41	13	0	0	0
$[22^0]$	4	3	9	31	24 [2]	0	0	0

[a] Results for even v_2 transition states from $J = 0$ quantal calculations. Results for $v_2 = 1$ transition states from $JP = 1+$ quantal calculations (P being the parity).

[b] For example, two initial channels with $v = v_1 = 0$, $|j - v_2| = |j - 2| = 2$ have contributions to the total reactive flux passing through $[02^0]$ at its threshold energy of greater than 5% and the sum of their percent contributions is 42.

to find, label, and characterize the set of resonances resulting from a many-body potential energy function. Resonances often provide simple explanations of complex observable phenomena, so we are motivated to develop a better understanding of resonances in many-body reactive systems.

The success in using a simplistic model of transmission through effective potential energy barriers to fit the total and channel-selected densities of reactive states and obtain resonance parameters for quantized transition states (as well as the good agreement between transition state energy levels and the maxima of semiclassical vibrationally adiabatic curves) in three-body, multichannel problems strongly suggests that detailed studies of scattering by one-dimensional potential barriers can provide further insight on barrier resonances. We have already seen how the analysis of the scattering matrix for the one-dimensional parabolic barrier proved useful in the interpretation of transition states as resonances. In the following section, we present results from other one-dimensional studies on both symmetric and unsymmetric potential energy barriers.

3. Barrier resonances in one-dimensional studies.

3.1. Barrier passage and poles of the S matrix. The most widely used model for a chemical reaction is passage over a potential energy barrier. The potential maximum is associated with the reaction threshold for

which we can provide insight via quantum mechanical studies of scattering by one-dimensional potential energy functions.

In two published studies [11,12] we have considered scattering in one dimension x off a potential barrier $V(x)$ with boundary conditions in the two limits $x \to \pm\infty$. In this two-channel (left-right) scattering problem, we let channel 1 denote the asymptotic region on the left ($x \to -\infty$) and channel 2 denote that on the right($x \to +\infty$). We solve the Schrödinger equation

$$(3.1) \qquad \frac{d^2\psi}{dx^2} + \frac{2\mu}{\hbar^2}[E - V(x)]\psi(x) = 0$$

where $\psi(x)$ is the wavefunction and determine the unitary scattering matrix S. From scattering matrix elements computed at real energies E, we can calculate transmission and reflection coefficients [96]

$$(3.2) \qquad P_{ij} = |S_{ji}|^2$$

as well as channel-to-channel delay times Δt_{ij} given by eqn (1.14). Furthermore, by directly solving the Schrödinger equation at complex energies [11,12,44,47], we search for poles of the S matrix; this allows us to find and characterize both narrow and broad resonances. The computational methods we used to solve eqn (3.1) are briefly described in Appendix II.

3.1.1. Symmetric potential functions. In the first study [11], we considered symmetric potential barrier functions of the form

$$(3.3) \qquad V(x) = V_0\left(\frac{e^{\beta x}}{(1 + e^{\beta x})^2} - \frac{\alpha}{2}\frac{e^{5\beta x}}{(1 + e^{5\beta x})^2}\right)$$

and the potential parameters were chosen to roughly resemble a one-dimensional model [97] of the H + H$_2$ reaction, in particular $\mu = \frac{2}{3}m_H = 1224.5m_e$, $V_0 = 2.177$ eV, and $\beta = 1a_0^{-1}$. When $\alpha = 0$, the potential is a simple Eckart barrier [98] and when this dimensionless parameter is greater than zero, the potential has twin maxima on either side of the $x = 0$ local minimum.

Reaction probabilities P_{12}, reactive delay times Δt_{12} and the locations of poles of the scattering matrix S were calculated for many values of α. A curve connecting the pole locations maps out their 'trajectory' in the complex energy plane as a function of α.

The various cases have been discussed at length in Ref. 11; here, we simply highlight the main features. For the pure Eckart potential, which corresponds to $\alpha = 0$, there is no structure in $V(x)$ other than the barrier maximum. In the vicinity of the potential maximum there is a smooth rise in the reaction probability from zero to unity and a maximum in the delay time of 33 fs. The real part ε_α of the resonance energy is close to the energy of the barrier maximum V_{max} and clearly demonstrates that passage over a potential barrier is associated with a pole of the S matrix.

For $\alpha = 0.17$ and 0.32, the real parts of the resonance energies are slightly above V_{max}. We also begin to see more structure for these cases in the reaction probability; P_{12} decreases slightly after its initial rise to unity and then returns to unity.

When $\alpha = 0.49$ there are two poles, one with ε_α below V_{max} and one above V_{max}. The dip in P_{12} after the threshold is more prominent and is beginning to be associated with a second feature in Δt_{12}. This shoulder in the delay time is clearly associated with the second very broad resonance coming from far off the real energy axis and 35 meV above V_{max}. For the four cases $\alpha = 0-0.49$, see that as the width Γ_α of the resonance decreases, the peak in the delay time increases.

The barrier resonance for $\alpha = 0 - 0.32$ is in the process of becoming a subthreshold trapped-state resonance for $\alpha = 0.49$ as the parameter α is varied. When $\alpha = 0.68$, the process is complete; there is a narrow trapped-state resonance 37 meV below V_{max} and a broad ($\Gamma_\alpha = 111$ meV) resonance 39 meV above V_{max} associated with the rise in the reaction probability from 0.2 to 1.0.

As the well in the potential function $V(x)$ gets deeper for $\alpha = 0.68 - 1.60$, the resonance with the smaller value of ε_α moves lower and lower in energy relative to V_{max} and is associated with an increasingly larger delay time Δt_{12}, consistent with its decreasing width [11]. It is obvious that what was once a barrier resonance is now clearly a trapped-state resonance, responsible for the sharp variation in the reaction probability from zero to unity and then almost back to zero. At the same time, the second very broad resonance first identified for $\alpha = 0.49$ is moving closer to the real energy axis and to the energy of the barrier maximum. For example, the broad resonance for $\alpha = 1.00$ has about the same width (and associated delay time) as the pole associated with the pure Eckart barrier. In addition there is an indication of a third resonance when $\alpha = 1.60$, as evidenced by a deep dip in P_{12} above V_{max} and by the hint of another feature in Δt_{12} at about 75 meV above the potential maximum. The story told with $\alpha = 0 - 0.49$ clearly repeats itself for $\alpha = 0.68 - 1.60$ [11].

In summary, this study showed that by varying the parameters of the potential function to change it from having a single maximum to having two maxima with a dip between them, one pole of the scattering matrix is transformed continuously from a barrier resonance (describing the threshold of a simple potential barrier) to a trapped-state resonance (describing a state trapped in the well between barriers). In the meantime, a second resonance comes into the picture above V_{max} and moves from a metastable state above threshold to a metastable state associated with the threshold, and it is in the process of turning into a second-trapped state when our picture stops. As α is further increased, the poles will move eventually onto the negative real energy axis and become bound states.

We discuss below the quantitative relation between the resonance width Γ and delay time Δt_{12}.

3.1.2. Unsymmetric potential functions. The potentials utilized in the study [11] just discussed were all symmetric functions. However, most chemical reactions do not possess this high degree of symmetry and, in addition, multiparticle three-dimensional quantal scattering calculations indicate that unsymmetric barriers can lead to new resonance phenomenon such as supernumerary transition states (see the $O + H_2$ discussion in Section 2.2.2). For these reasons, we were motivated to study one-dimensional quantal scattering by unsymmetric potential functions, with particular emphasis on potentials with two unequal local barrier maxima and their deformation to potentials having only one maximum. Such a study is described in detail in Ref. 12; we summarize its findings here.

In this second study [12], we considered potentials of the form

$$(3.4) \qquad V(x) = \frac{V_1 e^{\beta_1(x-x_1)}}{(1 + e^{\beta_1(x-x_1)})^2} + \frac{V_2 e^{\beta_2(x-x_2)}}{(1 + e^{\beta_2(x-x_2)})^2}$$

with reduced mass $\mu = 6526.3 m_e$. Parameters were chosen to roughly represent a one-dimensional model of the $O + D_2$ reaction. Two series of potential functions, all the form of eqn (3.4), were considered. In series 1, we focused on potential functions with two barrier maxima, taking $V_2 = 598.64$ meV, $\beta_1 = \beta_2 = 3a_0^{-1}$, $x_2 = -x_1 = 0.6a_0$, and V_1 variable. Nine different values of V_1 were considered and cases were labelled A-I. The potentials for representative cases A, C and I, with $V_1 = 598.64, 707.49$ and 870.75 meV, respectively, are shown in Figure 1. As we proceed from case A to case I, the lower local barrier maximum becomes less prominent.

In series 2, we continuously deformed the potentials so that eventually only one maximum appeared, rather than two. We took $V_1 = 870.75$ meV, $V_2 = 598.64$ meV, $x_2 = -x_1 = 0.6a_0$, and $\beta_1 = \beta_2 = \beta$ variable. Thirteen different values of β were considered and cases labelled I-U. (Case I is common to both series.) The potentials for representative cases K, N and T with $\beta = 2.80, 2.40$ and $1.80a_0^{-1}$, respectively, are shown in Figure 2.

Reaction probabilities P_{21} and reactive delay times Δt_{21} were computed. In addition, to further bring out structure in the reaction probability as a function of energy, we calculated the energy derivative of P_{21} by spline fitting the latter and analytically differentiating. Selected results are shown in Figures 1 and 2.

For all the potential cases, poles of the S matrix were located. The two resonances found in each case are characterized in the plots of the potential in Figures 1 and 2; the horizontal lines indicate values of ε_α and the number next to the horizontal line is the width Γ_α. The vertical lines in the other plots of Figures 1 and 2 indicate $\varepsilon_\alpha - V_{\max}$.

In many of the delay time curves, there are two peaks and in such cases, there is very good agreement between the two values of ε_α and the energies of the local maxima of Δt_{21} for that particular case. (For example, notice how the vertical arrows in Figures 1 and 2 coincide with peaks in

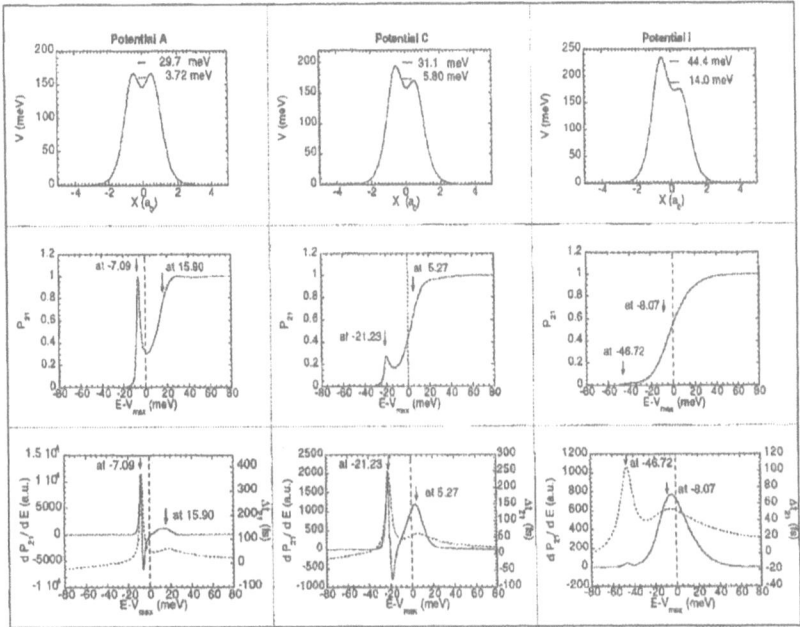

FIG. 1. *Potential energy functions (top row), reaction probabilities (middle row), and derivatives of reaction probabilities and reactive delay times (bottom row) for cases A, C, and I of Section 3.1.2. In the top row, horizontal lines indicate the real parts ε_α of the resonance energies, and the numbers next to the lines are the widths Γ_α in meV. In the middle and bottom rows the location of V_{max} is shown as a dashed vertical line, and the values of ε_α are given (with arrows) relative to V_{max} in each case. Derivatives of the reaction probabilities are solid curves with the scale at the left, and delay times are dashed curves with the scale at the right. (Modified from Ref. 12.)*

the reactive delay time.) Therefore, in such cases, we can associate each delay time peak with a resonance. In Figure 3 is plotted the value of the maximum in Δt_{21} versus the inverse of the width, $1/\Gamma$, of the closest resonance. We have also included results for $\alpha = 0 - 0.32$ and $0.68 - 1.60$ from Ref. 11 in addition to cases A, C, I, and K from Ref. 12. (Recall that for $\alpha = 0.49$ and cases N and T, two poles but only one maximum in the delay time were found.)

We now describe the main conclusions drawn from analyses of series 1 and 2, in turn.

For each of the potentials in series 1, two poles of the scattering matrix were located, and in general each resonance lies in the vicinity of a local potential barrier maximum (i.e. the value of ε_α is close to a local maximum in V.) The higher energy resonance (i.e. the one with larger ε_α) is always broader and is associated with a rise toward unity in the reaction probability and with a smaller peak in the reactive delay time. As the

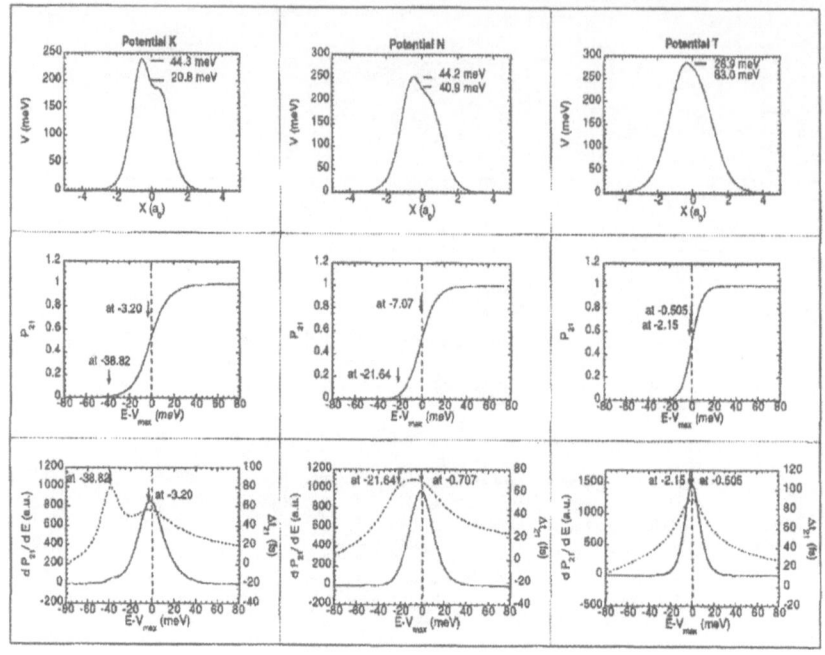

FIG. 2. *Same as Fig. 1 except cases K, N, and T. (Modified from Ref. 12.)*

resonance moves lower in energy relative to V_{max}, its width increases. The lower energy resonance, which lies in the vicinity of the lower local barrier maximum, has a smaller width and is associated with a larger peak in Δt_{21}. However, as we begin to deform the potential toward one having a single maximum, this resonance gets broader and its peak in the delay time gets smaller. In addition, its effect on the reaction probability is increasingly diminished; in case C, it is associated with a rise in P_{21} to only 0.27 and in case I, its effect on the reaction probability is really only discernible in the derivative curve dP_{21}/dE. Small rises in the reaction probability (to values significantly less than one) have also been observed in other model studies involving unsymmetric potentials [99,100].

For series 2, as we proceed from case I to T, the second potential maximum in $V(x)$ becomes a shoulder and then disappears, leaving only one maximum. Nonetheless, two poles of the scattering matrix are still located. Their values of ε_α approach one another and as a result the two peaks previously observed in Δt_{21} curves overlap more and more until they merge into one peak. While Γ of the higher energy resonance decreases, that of the lower energy resonance increases substantially; in case T, there is nearly a factor of 3 difference in widths.

Figure 3 clearly shows, for those potential cases in series 1 and 2 where

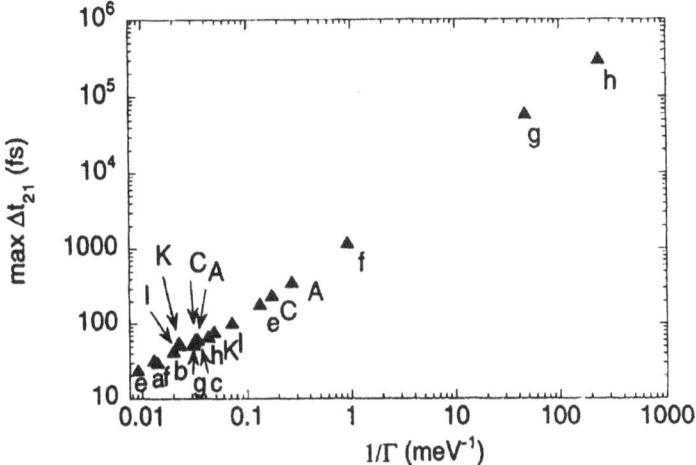

FIG. 3. *Local maximum in the reactive delay time* Δt_{21} *versus the inverse of the width of the resonance associated with the delay time feature. The cases shown are* $\alpha = 0(a)$, $0.17(b), 0.32(c), 0.68(e)$, $1.00(f)$, $1.45(g)$, *and* $1.60(h)$ *of Section 3.1.1 and* A, C, I, K *of Section 3.1.2.*

two peaks in Δt_{21} were observed, that the maximum delay time is inversely related to the width of the associated resonance. Cases N-T are not included in this figure because only one peak in Δt_{21} was observed in each case. However, even in these latter cases, there is an inverse relationship between the width of the *higher* energy resonance and the maximum delay time. (See Table 3 of Ref. 12.) While the width of the lower energy resonance increases, the higher energy resonance decreases in width, parallelling the increase in the maximum delay time. In addition there is an excellent correspondence between the value of ε_α of the higher energy resonance and the energy of maximum Δt_{21}. Thus, we conclude that when two resonances overlap significantly, it is the narrower (and higher energy) resonance which makes the dominant contribution to the delay time. We will revisit this point in Section 3.2 below.

The inverse relationship between the resonance width and the reactive delay time, clearly illustrated in Figure 3 for both symmetric and unsymmetric potential cases, can be elucidated in another manner. We have shown in Section 1.2 that for an isolated narrow resonance the reactive delay time will have a maximum at $E = \varepsilon_\alpha$, at which $\Delta t_{21}\Gamma_\alpha = 2\hbar$, eqn (1.16). This relationship is reminiscent of the time-energy uncertainty relation and we might expect that for broader resonances (such as barrier resonances) the product of the maximum reactive delay time and the resonance width will exceed $2\hbar$. In fact, we found that the very narrow resonances of potential (3.3) with $\alpha = 1.00 - 1.60$ are trapped-state resonances lying well below V_{max}, and for these, the product equals 2.0, as it should for an INR.

For cases in series 1, the product falls between 2.1 and 2.3 for the narrower resonances and between 2.8 and 3.6 for the broader ones; for cases in series 2, it falls between 2.3 and 3.1 for the narrower resonances and between 3.7 and 4.2 for the broader ones [12]. The results agree very well with our expectations for broad and narrow resonances.

3.2. Strings of poles and the cumulative reaction probability.
The reaction probability P_{21} in the model one-dimensional studies is the analogue of the cumulative reaction probability in the three-dimensional quantal studies and, likewise, the energy derivative dP_{21}/dE is analogous to the density of reactive states. From accurate three-dimensional quantal calculations, we have concluded that quantized levels of the transition state control the chemical reactivity of a number of chemical reactions. We have interpreted these quantized transition states as barrier resonances, responsible for steplike rises in the CRP. In the chemical reactions discussed in Section 2, we find no evidence for trapped-state resonances affecting the overall CRP or density of reactive states. In a similar manner, we have seen in our model one-dimensional studies that broad barrier resonances are also responsible for the rise from zero towards unity in the reaction probability. For a relatively structureless potential barrier, like the simple Eckart potential of Section 3.1.1, we see that the reaction probability rises smoothly from zero to one and remains at unity. We have identified this as a resonance phenomenon. However, we have discussed above (Section 2.3) that a parabolic or approximately parabolic (e.g. an Eckart) barrier has a series of poles of the scattering matrix. We now address the relation between a string of poles and the reaction probability.

For the parabolic barrier of eqn (2.12), Atabek et al. [10] analytically solved the Schrödinger equation and found a series of poles accumulating at an energy corresponding to the top of the barrier with complex resonance energies given by eqn (2.24). They proceeded to try to characterize these resonances numerically by invoking some procedures used for traditional (i.e. trapped-state) resonances, such as the analysis of stabilization graphs [101] and the search for a wave packet of maximum lifetime [102]. An identical result for the poles of the S matrix (Siegert eigenvalues [16]) has been obtained by Seideman and Miller [92] in a treatment of the parabolic barrier. Eqn (2.24) for the string of poles also results from a naive continuation of the analytic results for a harmonic oscillator. For the potential of eqn (2.12) with, in this case, k_τ a *positive* force constant, the bound state eigenvalues, which are poles of S along the real energy axis, are $E_\tau + \hbar\omega_\tau(\nu + 1/2)$, with ω_τ given by eqn (2.15). If we let k_τ now be *negative*, which transforms the potential from a harmonic oscillator to a parabolic barrier, we get poles given by eqn (2.24).

Analytical results for the poles associated with scattering by an Eckart potential barrier were presented by Ryaboy and Moiseyev [94]. For a simple, symmetric Eckart barrier of the form

$$(3.5) \qquad V(x) = \frac{\alpha e^{\gamma x}}{(1 + e^{\gamma x})^2}$$

where α and γ are real constants with dimensions of energy and inverse length, respectively, there are a series of Siegert resonance states at complex energies

$$(3.6) \qquad \bar{E}_v = \frac{\hbar^2 \gamma^2}{32\mu} \left[(1 + 2v) - i\sqrt{\frac{8\mu\alpha}{\hbar^2 \gamma^2} - 1} \right]^2$$

As for the parabolic barrier, the poles are labelled by a nonnegative integer v and the widths of the resonances vary as $(2v + 1)$. A more complicated analytical formula for the complex resonance energies arises for the nonsymmetric Eckart potential [94]. It should be noted that the nonsymmetric potentials of Ref. 94 differ from those studied in Ref. 12; in the case of Ref. 94, the potentials possess one maximum but are asymmetric in their asymptotic energies $[V(x = +\infty) \neq V(x = -\infty)]$. The series of poles for the symmetric and nonsymmetric Eckart potentials were also located numerically using the complex coordinate method [103-108]. The authors found that the complex scaled Siegert eigenfunctions are square integrable and localized within the region of the potential barrier and, in addition, that they may be characterized by their number of nodes v as for bound state wavefunctions.

In one relevant study [109], a string of poles of **S** was also located for the radial potential $V_0 r^2 \exp(-\alpha r)$ using the complex coordinate method. (This potential was also studied by Bain et al. [108].) The locations of the poles in the complex energy plane relative to the potential barrier height were analysed. Pole strings have also been observed in a number of other studies [110,111]. The manner in which the series of Siegert resonance states line up in the complex energy plane for some general types of potentials has been discussed by Meyer [111] and Seideman and Miller [92]. Recent work in the mathematical literature is also concerned with the number and distribution of scattering poles [112].

The variation we have observed [11,12] in the widths of the resonances agrees nicely with the analytical and numerical studies of model one-dimensional barriers discussed above. In particular, we find that as the unsymmetric potential of eqn (3.4) is deformed from one having two local potential maxima to having only a single maximum, the ratio of the widths of the two resonances located is nearly 3. (For example, for case T in Figure 2, the ratio is 2.87.) For a parabolic barrier or a symmetric Eckart barrier, the ratio of the two narrowest resonances ($v = 0, 1$) is exactly 3.

Since passage over potential barriers can give rise to a series of poles of the scattering matrix, the next question to address is the effect this string of poles has on the reaction probability. Ryaboy and Moiseyev [94] proved that the (cumulative) reaction probability for an Eckart potential barrier can be expressed entirely in terms of the Siegert poles in the S matrix, even though the energy dependence of P_{21} is structureless (it smoothly rises from zero to one). Likewise, Seideman and Miller [92] showed, using flux correlation functions [113,114], how the reaction probability for parabolic and Eckart barriers can be expressed in terms of Siegert eigenvalues. (In a related and earlier study, Atabek et al. [115] have shown that the scattering amplitude for single-channel scattering by a repulsive exponential potential can be accurately represented using a finite number of poles.) It is clear then that the rise in the reaction probability from 0 to 1 is due to a cumulative effect of the entire string of poles. Whereas an isolated narrow trapped-state resonance causes the reaction probability to rapidly vary from 0 to 1 and then back to 0, the reaction probability upon passage over a simple potential barrier (such as a simple Eckart potential) rises smoothly from 0 to 1 but does not return to 0 because of the cumulative effect of the string of broad and hence overlapping resonances.

However, it is important to reemphasize a point made earlier. The quantal three-dimensional studies described in Section 2 suggest that, when resonances are broad and overlapping, many features of the behavior of the cumulative reaction probability and density of reactive states can be understood in terms of the pole closest to the real energy axis (the $\nu = 0$ member of the pole sequence). We have discussed several examples of the utility of this suggestion in Section 2.3. In addition, both in our work on three-body reactions in the real three-dimensional world described in Section 2 and in our work on model one-dimensional unsymmetric potential functions described in Section 3.1.2, we concluded that when resonances overlap significantly, it is the narrowest one which makes the dominant contribution to the delay time. The nearest pole to the real energy axis is thus useful for understanding the dynamics in real time, an issue that is becoming increasingly important due to recent advances in femtosecond spectroscopy [116].

4. Concluding remarks. Analytical formulas characterizing isolated narrow resonances are well known. However, resonances that are encountered in molecular collisions usually are not well described by the INR formulas; in particular, we have given a number of examples from the literature where the sum of the partial widths does not equal the total resonance width.

A chemically important example of a type of resonance that is not expected to follow the isolated narrow resonance formulas is a quantized transition state resonance. These broad resonances have been shown, by analysis of the accurate quantum mechanical cumulative reaction probability and its energy derivative, to control chemical reactivity in a number of three-dimensional chemical reactions by acting as dynamical bottlenecks in phase space that gate the reactive flux from reactants to products. These transition state resonances, also called barrier resonances, do not differ in kind from conventional isolated narrow trapped-state resonances (i.e. both are poles of the scattering matrix); this has been demonstrated by one-dimensional quantal studies in which barrier resonances have been shown to transform continuously into trapped-state resonances by varying the potential energy function. In general, barrier resonances have larger widths (and are consequently associated with shorter delay times) than trapped-state resonances.

When the quantum mechanical description of a chemical reaction becomes more complete, it often becomes desirable or even necessary to use the language of resonance theory. The interpretation of transition states as resonances has been put to several practical uses. First, it allows us to compute the collision lifetime $\Delta\tau$ from the resonance width; $\Delta\tau$ values computed in this manner have been seen to compare favorably to those computed from the phase of accurate S matrix elements without resonance theory or transition state theory. Secondly, we can obtain complex spectroscopic constants for the quantized transition states by analogy to the well known procedure for obtaining (real) spectroscopic constants for bound states. Thirdly, by analysing channel-selected densities of reactive states, we can acquire information regarding the partial widths of these barrier resonances. This allows us to determine quantitatively which reactant states contribute to the reactive flux passing through a particular level of the transition state. In addition, the characterization of transition states as resonances has been used in a reformulation of variational transition state theory to compute anharmonic transition state energy levels [93].

The identification of transition states as reactive scattering resonances has been shown [117] to also result from application of a complex scaling transformation to the reaction coordinate of a chemical reaction. This analysis leads to a precise definition of the resonance width operator, whose expectation values are resonance lifetimes.

The levels of the transition state have been assigned stretch (v_1) and bend (v_2) quantum numbers for modes orthogonal to the reaction coordinate. The quantum number v_3, corresponding to the unbound motion along the reaction coordinate, appears to be missing in the $[v_1 v_2^K]$ assignments; however, when the transition state is treated properly as a quantum mechanical resonance, this quantum number reappears and is related to the quantum number ν used to label each resonance in the string of poles associated with passage over a one-dimensional potential barrier. We suggest that the "missing" quantum number in the transition state assignment be taken as $\nu = 0$, corresponding to the pole in the one-dimensional study string of poles that is nearest to the real energy axis. Our three-dimensional and one-dimensional studies suggest that it is the $\nu = 0$ pole that is most useful for understanding the chemical dynamics in real time.

Quantum mechanical scattering by one-dimensional potential energy functions has provided further insight into the nature of transition state, or barrier, resonances. In conclusion, we briefly point out how our one-dimensional model studies on unsymmetric potential energy functions can help us better understand the conclusions drawn from our multidimensional quantal studies of the asymmetric $O + H_2$ reaction. As described above, the three-dimensional quantal calculations suggest that reactant-like dynamical bottlenecks exert significant control on $O + H_2$ chemical reactivity; these bottlenecks are variational transition states associated with the global maxima of vibrationally adiabatic curves. In addition, there are product-like dynamical bottlenecks (supernumerary transition states) that also affect the cumulative reaction probability and density of reactive states. However, other product-like transition states do not appear to influence the overall CRP (such as product-like $[10^0]$.) We find in our model studies of one-dimensional unsymmetric potentials having two potential maxima that the resonance in the vicinity of the global potential maximum is associated with the rise in the reaction probability toward one; this resonance behaves like the variational transition state. The resonance in the vicinity of the lower local potential maximum sometimes appears to significantly influence P_{21} (as in case C of Section 3.1.2) like a supernumerary transition state and at other times appears to have very little influence (see case K in the same section). Although it is clear that comparison between model one-dimensional studies and quantal three-dimensional studies has its limitations, it is encouraging to see that some aspects of the three-dimensional results can be better understood in light of the one-dimensional studies. In the spirit of the present workshop we would like to suggest that the subject of broad overlapping resonances and especially barrier resonances provides a fertile area for further mathematical analysis that could be of great use in multiparticle scattering theory.

Acknowledgment: We gratefully acknowledge useful conversations with David Chatfield, Brian Kendrick, Claude Mahaux, David Schwenke, and Deborah Watson. This work was supported by the National Science Foundation, by the donors of The Petroleum Research Fund, administered by the American Chemical Society, by an Indiana University Purdue University Fort Wayne Summer Faculty Research Grant, and by a Purdue Research Foundation Summer Faculty Grant. Computational resources were provided by the Minnesota Supercomputer Institute and the Chemistry Department at IPFW.

Appendix I

Accurate quantum mechanical cumulative reaction probabilities and densities of reactive states for three-dimensional chemical reactions such as those discussed in Section 2.2 were obtained by performing converged quantum dynamics calculations. After removing the center-of-mass motion, this leaves 6 degrees of freedom. The scattering calculations were carried out by expanding the wavefunction in a multiple-arrangement basis set [118-122] and finding the basis set coefficients by linear algebraic methods utilizing a variational principle. Either the generalized Newton variational principle [15,123-126] or the outgoing wave variational principle [127-130] was used to obtain elements of the scattering matrix that are stationary with respect to small variations in the wavefunction. Details of the calculations are presented in previously published work [125,126,130-133]. We emphasize that the state-to-state reaction probabilities are very well converged with respect to numerical and basis set parameters; this is of crucial importance since densities of reactive states are obtained by numerical differentiation of CRPs with respect to energy, and the errors in the CRP must be small for the numerical derivatives to be smooth.

Appendix II

In this appendix we briefly describe the computational methods used in Refs. 11 and 12 for solution of the Schrödinger equation (3.1) describing two-channel, one-dimensional scattering. The scattering boundary conditions are

(II.1a) $$\psi \underset{x \to -\infty}{\sim} \frac{N_-}{\sqrt{k_-}}[C \exp(-ik_-x) + D \exp(ik_-x)]$$

(II.1b) $$\psi \underset{x \to +\infty}{\sim} \frac{N_+}{\sqrt{k_+}}[A \exp(-ik_+x) + B \exp(ik_+x)]$$

where $N_\pm$ is an arbitrary normalization factor, and asymptotic wave numbers $k_\pm$ are given by

(II.2) $$k_\pm = \sqrt{2\mu(E - V_{\pm\infty})}/\hbar$$

where $V_{\pm\infty}$ are the limits at $x = \pm\infty$ of $V(x)$. The S matrix relates the incoming wave coefficients A and D and the outgoing wave coefficients B and C:

(II.3) $$\begin{pmatrix} C \\ B \end{pmatrix} = \begin{pmatrix} S_{11} & S_{12} \\ S_{21} & S_{22} \end{pmatrix} \begin{pmatrix} D \\ A \end{pmatrix}$$

We used two different numerical methods to directly search for poles of S at complex energies. The first method [11,12] located most of the resonances studied but had difficulty characterizing very broad resonances. A second method [12] was used to locate the broadest resonances.

Before describing each method, we note that we considered potentials for which $V_{+\infty} = V_{-\infty} = 0$ and thus $k_+ = k_- = k$. We also set $N_\pm = \sqrt{k}$ for convenience.

In the first method, we transformed the second-order, complex differential equation (3.1) into four coupled first-order real differential equations which we then integrated directly. In particular, if we consider a wave incident only from $x > 0$, we can set $D = 0$ and thus, with $C = 1$ to normalize the solution, we know the solution of eqn (3.1) as $x \to -\infty$. This solution is then numerically propagated from a large negative value of x, $-a_1$, to a large positive value of x, a_2, using a variable stepsize Gear backward difference integrator [134] subroutine from the International Mathematical and Statistical Library (IMSL) [135]. From the solution at a_2, the coefficients A and B are computed and then S_{22} and S_{12} determined using eqn $II.3$. In a similar manner, we can compute the remaining two S matrix elements by again setting $D = 0$ and $C = 1$ but solving the Schrödinger equation with the potential $V(-x)$ rather than $V(x)$. The numerical parameters required for the numerical integration were varied until the scattering matrix elements were converged to better than 0.1%.

The second method we used to both locate the very broad resonances as well as check the results obtained from the first method is based on the technique of invariant embedding [136]. Equivalent to solving the second-order, linear eqn (3.1) subject to the above boundary conditions is to solve the following first-order, nonlinear initial value problem for the function $S(x)$ [137]:

(II.4a)
$$\frac{dS}{dx} = 2ikS(x) + \frac{1}{2ik}U(x)[1 + S(x)]^2$$

(II.4b)
$$S(-a_1) = 0$$

with $U(x) = 2\mu V(x)/\hbar^2$. Scattering matrix elements are then given by

(II.5a)
$$S_{22} = S(a_2) \exp(-2ika_2)$$

(II.5b)
$$S_{12} = \exp\left(\frac{1}{2ik} \int_{-a_1}^{a_2} dx\, U(x)[1 + S(x)]\right).$$

The first-order equation (II.4a) is transformed into two real, first-order coupled differential equations and the solution II.4b is propagated numerically from $-a_1$ to a_2 using the IMSL Gear integrator. S_{22} is computed via eqn II.5a and S_{12} is computed from a Simpson's rule [138] evaluation of the integral in II.5b.

Poles in the scattering matrix corresponding to resonances were located by numerically searching for zeroes in $1/S_{ji}$ in the complex energy plane. To accomplish this task, we used a modified IMSL subroutine which invoked Muller's method [138] for quadratic interpolation among three points to find the next estimate of the root. The search was typically stopped when the relative difference in two successive approximations to the root was smaller than 10^{-10}.

Transmission and reflection coefficients were computed at real energies from converged scattering matrix elements via eqn (3.2). Delay times Δt_{ij} defined by eqn (1.14) were also obtained by the method [33] of fitting scattering matrix elements at three consecutive real energies to

(II.6) $S_{ji}(E) = (c_1 + c_2 E + c_3 E^2) \times \exp[i(c_4 + c_5 E + c_6 E^2)]$

where c_n are real fitting parameters. The delay time at the central energy of the triad was then obtained by

(II.7) $\Delta t_{ij}(E) = \hbar(c_5 + 2c_6 E).$

REFERENCES

[1] J. R. Taylor, Scattering Theory: The Quantum Theory of Nonrelativistic Collisions, Krieger, Malabar, FL, 1983, p. 238, (a) p. 407.

[2] H.S. Taylor, G.V. Nazaroff, and A. Golebiewski, J. Chem. Phys. 45: 2872, 1966.

[3] B. C. Garrett, D. G. Truhlar, R. S. Grev, G. C. Schatz, and R. B. Walker, J. Phys. Chem. 85: 3806, 1981.

[4] J. Simons, ACS Symp. Ser. 263: 3, 1984.

[5] H. Feshbach, Ann. Phys. (N. Y.) 19: 287, 1962.

[6] N. Abusalbi, D. J. Kouri, M. Baer, and E. Pollak, J. Chem. Phys. 82: 4500, 1985.

[7] R. D. Levine and S.-f. Wu, Chem. Phys. Lett. 11: 557, 1971.

[8] S. A. Cuccaro, P. G. Hipes, and A. Kuppermann, Chem. Phys. Lett. 157: 440, 1989.

[9] Z. C. Kuruoglu and D. A. Micha, Int. J. Quantum Chem. Symp. 23: 103, 1989.

[10] O. Atabek, R. Lefebvre, M. Garcia Sucre, J. Gomez-Llorente, and H. Taylor, Int. J. Quantum Chem. 40: 211, 1991.

[11] R. S. Friedman and D. G. Truhlar, Chem. Phys. Lett. 183: 539, 1991.

[12] R. S. Friedman, V. D. Hullinger, and D. G. Truhlar, J. Phys. Chem. 99: 3184, 1995.

[13] A. Bohm, Quantum Mechanics: Foundations and Applications, 3rd ed., Springer-Verlag, New York, 1993, p. 452.

[14] Resonances, E. Brändas and N. Elander (eds.) [Lecture Notes in Physics, Vol. 325], Springer-Verlag, Berlin, 1989.

[15] R. G. Newton, Scattering Theory of Particles and Waves, 2nd ed., Springer-Verlag, New York, 1982, Section 11.2 and Chapter 16 for resonances as poles of the S matrix and Section 11.3 for variational principles.

[16] A. J. F. Siegert, Phys. Rev. 56: 750, 1939.

[17] B. Simon, Int. J. Quantum Chem. 14: 529, 1978.

[18] G. A. Hagedorn, Commun. Math. Phys. 65: 181, 1979.

[19] D. C. Chatfield, R. S. Friedman, D. G. Truhlar, B. C. Garrett, and D. W. Schwenke, J. Am. Chem. Soc. 113: 486, 1991.

[20] D. C. Chatfield, R. S. Friedman, D. G. Truhlar, and D. W. Schwenke, Faraday Discuss. Chem. Soc. 91: 289, 1991.

[21] D. C. Chatfield, R. S. Friedman, D. W. Schwenke, and D. G. Truhlar, J. Phys. Chem. 96: 2414, 1992.

[22] D. G. Truhlar, C. A. Mead, and M. A. Brandt, Adv. Chem. Phys. 33: 296, 1975.

[23] A. M. Lane and R. G. Thomas, Rev. Mod. Phys. 30: 257, 1958.

[24] J. Humblet and L. Rosenfeld, Nucl. Phys. 26: 529, 1961. J. Humblet and L. Rosenfeld, Nucl. Phys. 31: 544, 1962. J. Humblet, in Fundamentals in Nuclear Theory (Lectures Presented at and International Course, Trieste, 3 Oct.-16 Dec. 1966), A. de-Shalit and C. Villi, eds., International Atomic Energy Agency, Vienna, 1967, p. 369.

[25] H. A. Weidenmüller, Ann. Phys. (N.Y.) 28: 60, 1964; 29: 378, 1964. See also C. Mahaux and H. A. Weidenmüller, Shell-Model Approach to Nuclear Reactions, North-Holland, Amsterdam, 1969.

[26] P. A. Moldauer, Phys. Rev. 135B: 642, 1964.

[27] D. W. Schwenke and D. G. Truhlar, J. Chem. Phys. 87: 1095, 1987.
[28] A. Bohm, Quantum Mechanics: Foundations and Applications, 3rd ed., Springer-Verlag, New York, 1993, p. 549.
[29] D. K. Watson, Phys. Rev. A 34: 1016, 1986. D. K. Watson, J. Phys. B 19: 293, 1986.
[30] R. K. Nesbet, Adv. At. Mol. Phys. 13: 315, 1977. A. Hazi, Phys. Rev. A 19: 920, 1979.
[31] H.A. Weidenmüller, Phys. Lett. 24B: 441, 1967.
[32] F. T. Smith, Phys. Rev. 118: 349, 1960.
[33] M. Zhao, M. Mladenovic, D. G. Truhlar, D. W. Schwenke, O. Sharafeddin, Y. Sun, and D. J. Kouri, J. Chem. Phys. 91: 5302, 1989.
[34] R. H. Dalitz and G. Rajesekaran, Phys. Lett. 7: 373, 1963. R. J. Eden and J.R. Taylor, Phys. Rev. B 133: 1575, 1964. A. Herzenberg and D. Ton-That, J, Phys. B 8: 426, 1975.
[35] C. J. Ashton, M. S. Child, and J. M. Hutson, J. Chem. Phys. 78: 4025, 1982.
[36] H. Feshbach, Ann. Phys. (N. Y.) 43: 410, 1967.
[37] B. W. Shore, Phys. Rev. 171: 43, 1968.
[38] A. Mondragón and E. Hernández, J. Phys. A 26, 5595, 1993.
[39] M. Desouter-Lecomte, J. Lievin, and V. Brems, J. Chem. Phys. 103: 4524, 1995. M. Desouter-Lecomte and V. Jacques, J. Phys. B 28: 3225, 1995.
[40] E. P. Wigner and L. Eisenbud, Phys. Rev. 72: 29, 1947.
[41] G. Beck and H. M. Nussenzweig, Nuovo Cimento 16: 416, 1960.
[42] M. Desouter-Lecomte, and F. Culot, J. Chem. Phys. 98: 7819, 1993.
[43] B. C. Garrett, D.W. Schwenke, R.T. Skodje, D. Thirumulai, T.C. Thompson, and D. G. Truhlar, A.C.S. Symp. Ser. 263: 375, 1984.
[44] M. V. Basilevsky and V. M. Ryaboy, Int. J. Quantum Chem. 19: 611, 1981.
[45] M. V. Basilevsky and V. M. Ryaboy, Chem. Phys. 86: 67, 1984.
[46] C. W. McCurdy and T. N. Rescigno, Phys. Rev. A 20: 2346, 1979.
[47] D. W. Schwenke, Theor. Chim. Acta 74: 381, 1988.
[48] H. Eyring, J. Chem. Phys. 3: 107, 1935.
[49] E. Wigner, Trans. Faraday Soc. 34: 29, 1938.
[50] M. M. Kreevoy and D. G. Truhlar, Transition state theory, in Investigation of Rates and Mechanisms of Reactions (C. F. Bernasconi, ed.) [Techniques of Chemistry, 4th ed., A. Weissberger (ed.)], John Wiley & Sons, New York, 1986, Part I, pp. 13–95.
[51] S. C. Tucker and D. G. Truhlar, in New Theoretical Concepts for Understanding Organic Reactions (J. Bertrán and I. G. Csizmadia, Eds.), Kluwer, Dordrecht, 1989, pp. 291-346.
[52] J. C. Polanyi and R. J. Wolf, J. Chem. Phys. 75: 5951, 1981. H. R. Mayne, R. A. Poirier, and J. C. Polanyi, J. Chem. Phys. 80: 4025, 1984. H. R. Mayne, J. C. Polanyi, N. Sathyamurthy, and S. Raynor, J. Phys. Chem. 88: 4064, 1984. J. C. Polanyi, M. G. Prisant, and J. S. Wright, J. Phys. Chem. 91: 4727, 1987.
[53] R. A. Marcus and O. K. Rice, J. Phys. Colloid Chem. 55: 894, 1951. R. A. Marcus, J. Chem. Phys. 20: 359, 1952.
[54] R. A. Marcus, J. Chem. Phys. 45: 2138, 1966.
[55] H. M. Rosenstock, M. B. Wallenstein, A. L. Wahrhaftig, and H. Eyring, Proc. Natl. Acad. Sci. USA 38: 667, 1952.
[56] J. L. Magee, Proc. Natl. Acad. Sci. USA 38: 764, 1952.
[57] B. C. Garrett and D. G. Truhlar, J. Phys. Chem. 83: 1052, 1979.
[58] B. C. Garrett and D. G. Truhlar, J. Phys. Chem. 83: 1079, 1979; Errata: 84: 682, 1980; 87: 4553, 1983.
[59] W. H. Miller, J. Chem. Phys. 62: 1899, 1975.
[60] F. T. Smith, J. Chem. Phys. 36: 248, 1962.
[61] E. Wigner, J. Chem. Phys. 5: 720, 1937.
[62] J. Horiuti, Bull. Chem. Soc. Japan 13: 210, 1938.
[63] J. C. Keck, Advan. Chem. Phys. 13: 85, 1967.

[64] D. G. Truhlar and B. C. Garrett, Annu. Rev. Phys. Chem. 35: 159, 1984.

[65] D. G. Truhlar, A. D. Isaacson, and B. C. Garrett, in Theory of Chemical Reaction Dynamics (M. Baer, ed.), CRC Press, Boca Raton, FL, 1985, Vol. 4, pp. 65-137.

[66] K. Haug, D. W. Schwenke, D. G. Truhlar, Y. Zhang, J. Z. H. Zhang, and D. J. Kouri, J. Chem. Phys. 87: 1892, 1987.

[67] J. M. Bowman, Chem. Phys. Lett. 141: 545, 1987.

[68] D. C. Chatfield, R. S. Friedman, G. C. Lynch, D. G. Truhlar, and D. W. Schwenke, J. Chem. Phys. 98: 342, 1993.

[69] S. L. Mielke, G. C. Lynch, D. G. Truhlar, and D. W. Schwenke, J. Phys. Chem. 98: 8000, 1994.

[70] D. C. Chatfield, R. S. Friedman, S. L. Mielke, D. W. Schwenke, G. C. Lynch, T. C. Allison, and D. G. Truhlar, Computational Spectroscopy of the Transition State, in Dynamics of Molecules and Chemical Reactions, edited by R. E. Wyatt and J. Z. H. Zhang (Marcel Dekker, New York, 1996), pp. 323–386.

[71] J. D. Kress and E. F. Hayes, J. Chem. Phys. 97: 4881, 1992. G. C. Lynch, P. Halvick, M. Zhao, D. G. Truhlar, C.-h. Yu, D. J. Kouri, and D. W. Schwenke, J. Chem. Phys. 94: 7150, 1991.

[72] T. C. Allison, S. L. Mielke, D. W. Schwenke, G. C. Lynch, M. S. Gordon, and D. G. Truhlar, Die Photochemischen Bildung des Chlorwasserstoffs. Dynamics of Cl + H2 F HCl + H on a New Potential Energy Surface: The Photosynthesis of Hydrogen Chloride Revisited 100 Years after Max Bodenstein, in Gas-Phase Chemical Reaction Systems: Experiments and Models 100 Years after Max Bodenstein, edited by J. Wolfrum, H.-R. Volpp, R. Rannacher, J. Warnatz (Springer Series in Chemical Physics, Berlin, 1996), pp. 111–124.

[73] D. C. Chatfield, R S. Friedman, G. C. Lynch, and D. G. Truhlar, Faraday Discuss. Chem. Soc. 91: 398, 1991. D. C. Chatfield, R. S. Friedman, G. C. Lynch, and D. G. Truhlar, J. Phys. Chem. 96: 57, 1992. The cumulative reaction probabilities were presented in: G.C. Schatz, J. Chem. Phys. 90: 3582, 1989. G. C. Schatz, J. Chem. Phys. 90: 4847, 1989. G. C. Schatz, J. Chem. Soc. Faraday Trans. 86: 1729, 1990.

[74] Z. Darakjian, E. F. Hayes, G. A. Parker, E. A. Butcher, and J. D. Kress, J. Chem. Phys. 95: 2516, 1991. Erratum: 101: 9203, 1994. S. J. Klippenstein and J. D. Kress, J. Chem. Phys. 96: 8164, 1992.

[75] J. D. Kress, J. Chem. Phys. 95: 8673, 1991. J. D. Kress and S. J. Klippenstein, Chem. Phys. Lett. 195: 513, 1992. J. D. Kress, R. B. Walker, E. F. Hayes, and P. Pendergast, J. Chem. Phys. 100: 2728, 1994.

[76] R. T Pack, E. A. Butcher, and G. A. Parker, J. Chem. Phys. 99: 9310, 1993. R. T Pack, E. A. Butcher, and G. A. Parker, J. Chem. Phys. 102: 5998, 1995. C. Leforestier and W. H. Miller, J. Chem. Phys. 100: 733, 1994.

[77] M. J. Davis, H. Koizumi, G. C. Schatz, S. E. Bradforth, and D. M. Neumark, J. Chem. Phys. 101: 4708, 1994.

[78] D. G. Truhlar and R. E. Wyatt, Annu. Rev. Phys. Chem. 27: 1, 1976.

[79] C. A. Mead, in Mathematical Frontiers in Computational Chemical Physics (D. G. Truhlar, ed.), Springer-Verlag, New York, 1988, p. 1.

[80] A. J. C. Varandas, F. B. Brown, C. A. Mead, D. G. Truhlar, and N. C. Blais, J. Chem. Phys. 86: 6258, 1987.

[81] E. C. Kemble, The Fundamental Principles of Quantum Mechanics with Elementary Applications, Dover, New York, 1958, p. 109.

[82] G. Herzberg, Infrared and Raman Spectra of Polyatomic Molecules [Molecular Spectra and Molecular Structure, Vol. II], Van Nostrand Reinhold, New York, 1945, pp. 15, 75, 205.

[83] M. A. Eliason and J. O. Hirschfelder, J. Chem. Phys. 30: 1426, 1959.

[84] L. Hofacker, Z. Naturforsch. 18a: 607, 1963.

[85] D. G. Truhlar, J. Chem. Phys. 53: 2041, 1970.

[86] D. G. Truhlar and A. Kuppermann, J. Amer. Chem. Soc. 93: 1840, 1971.

[87] B. C. Garrett, D. G. Truhlar, R. S. Grev, and A. W. Magnuson, J. Phys. Chem. 84: 1730, 1980. Errata: 87: 4554, 1983.

[88] R. A. Marcus, J. Chem. Phys. 45: 4493, 1965.

[89] G. C. Lynch, P. Halvick, D. G. Truhlar, B. C. Garrett, D. W. Schwenke, and D. J. Kouri, Z. Naturforsch. 44a: 427, 1989.

[90] B. R. Johnson and N. W. Winter, J. Chem. Phys. 66: 4116, 1977.

[91] G. C. Schatz, J. Chem. Phys. 83: 5677, 1985.

[92] T. Seideman and W. H. Miller, J. Chem. Phys. 95: 1768, 1991.

[93] D. G. Truhlar and B. C. Garrett, J. Phys. Chem. 96: 6515, 1992.

[94] V. Ryaboy and N. Moiseyev, J. Chem. Phys. 98: 9618, 1993.

[95] A. Kuppermann, in Potential Energy Surfaces and Dynamics Calculations (D. G. Truhlar, ed.), Plenum, New York, 1981, p. 375.

[96] A. Messiah, Quantum Mechanics, Wiley, New York, 1966, p. 85.

[97] D. G. Truhlar and A. Kuppermann, J. Am. Chem. Soc. 93: 1840, 1971.

[98] C. Eckart, Phys. Rev. 35: 1303, 1930.

[99] V. A. Mandelshtam and H. S. Taylor, J. Chem. Phys. 99: 222, 1993.

[100] V. Ryaboy and R. Lefebvre, J. Chem. Phys. 99: 9547, 1993.

[101] A. U. Hazi and H. S. Taylor, Phys. Rev. A 1: 1109, 1970.

[102] D. G. Truhlar, Chem. Phys. Lett. 26: 377, 1974.

[103] J. Nuttall and H. L. Cohen, Phys. Rev. 188: 1542, 1969.

[104] E. Balslev and J. M. Combes, Commun. Math. Phys. 22: 280, 1971.

[105] B. Simon, Commun. Math. Phys. 27: 1, 1972; Ann. Math. 97: 247, 1973.

[106] B. R. Junker, Adv. At. Mol. Phys. 18: 207, 1982. B. R. Junker and C. L. Huang, Phys. Rev. A 18: 313, 1978.

[107] N. Moiseyev, Isr. J. Chem. 31: 211, 1991.

[108] R. A. Bain, J. N. Bardsley, B. R. Junker, and C. V. Sukumar, J. Phys. B. 7: 2189, 1974.

[109] O. Atabek and R. Lefebvre, Chem. Phys. Lett. 84: 233, 1981.

[110] See, for examples, N. Moiseyev, P. R. Certain, and F. Weinhold, Mol. Phys. 36: 1613, 1978. H. J. Korsch, H. Laurent, and R. Möhlenkamp, Mol. Phys 43: 1441, 1981; J. Phys. B 15: 1, 1982; J. Phys. B 15: L559, 1982. M. Rittby, N. Elander, and E. Brändas, Phys. Rev. A 24: 1635, 1981; Mol. Phys. 45: 553, 1982. H-D. Meyer and O. Walter, J. Phys. B 15: 3647, 1982.

[111] H-D. Meyer, J. Phys. B 16: 2265, 1983.

[112] J. Sjötrrand and M. Zworski, Commun. Partial Diff. Eq. 17: 1021, 1992. R. Froese and M. Zworski, Duke Math. J. 72: 275, 1993. J. Sjösrrand and M. Zworski, Commun. Partial Diff. Eq. 18: 847, 1993. J. Sjöstrand and M. Zworski, J. Funct. Anal. 123: 336, 1994. J. Sjöstrand, Int. J. Quantum Chem. 31: 733, 1987. M. Zworski, J. Funct. Anal. 73: 277, 1987; 89: 370, 1989. M. Zworski, Duke Math. J. 59: 311, 1989. J. Sjöstrand, Duke Math. J. 60: 1, 1990.

[113] T. Yamamoto, J. Chem. Phys. 33: 281, 1960.

[114] W. H. Miller, S. D. Schwartz, and J. W. Tromp, J. Chem. Phys. 79: 4889, 1983. T. Seideman and W. H. Miller, J. Chem. Phys. 96: 4412, 1992.

[115] O. Atabek, R. Lefebvre, and M. Jacon, J. Phys. B 15: 2689, 1982.

[116] A. H. Zewail and R. B. Bernstein, Chem. Eng. News 66: 24, Nov. 7, 1988; L. R. Khundkar and A. H. Zewail, Annu. Rev. Phys. Chem. 41: 15, 1990; A. H. Zewail, Faraday Discuss. Chem. Soc. 91: 207, 1991; A. H. Zewail, J. Phys. Chem. 97: 12427, 1993; A. Materny, J. L. Herek, P. Cong, and A. H. Zewail, J. Phys. Chem. 98: 3352, 1994.

[117] M. Zhao and S. A. Rice, J. Phys. Chem. 98: 3444, 1994.

[118] M. J. Seaton, Philos. Trans. Roy. Soc. London A 245: 469, 1953.

[119] H. S. W. Massey, Rev. Mod. Phys. 28: 199, 1956.

[120] D. A. Micha, Arkiv Fys. 30: 411, 1965.

[121] W. H. Miller, J. Chem. Phys. 50: 407, 1969.

[122] D. G. Truhlar, J. Abdallah, Jr., and R. L. Smith, Adv. Chem. Phys. 25: 211, 1974. D. G. Truhlar and J. Abdallah, Jr., Phys. Rev. A 9: 297, 1974.

[123] G. Staszewska and D. G. Truhlar, J. Chem. Phys. 86: 2793, 1987.

[124] D. W. Schwenke, K. Haug, D. G. Truhlar, Y. Sun, J. Z. H. Zhang, and D. J. Kouri, J. Phys. Chem. 91: 6080, 1987.

[125] D. W. Schwenke, K. Haug, M. Zhao, D. G. Truhlar, Y. Sun, J. Z. H. Zhang, and D. J. Kouri, J. Phys. Chem. 92: 3202, 1988.

[126] D. W. Schwenke, M. Mladenovic, M. Zhao, D. G. Truhlar, Y. Sun, and D. J. Kouri, in Supercomputer Algorithms for Reactivity Dynamics and Kinetics of Small Molecules (A. Laganà, ed.), Kluwer, Dordrecht, 1989, p. 131.

[127] L. Schlessinger, Phys. Rev. 167: 1411, 1968.

[128] Y. Sun, D. J. Kouri, D. G. Truhlar, and D. W. Schwenke, Phys. Rev. A 41: 4857, 1990.

[129] Y. Sun, D. J. Kouri, and D. G. Truhlar, Nucl. Phys. A508: 41c, 1990.

[130] D. W. Schwenke, S. L. Mielke, and D. G. Truhlar, Theor. Chim. Acta 79: 241, 1991.

[131] J. Z. H. Zhang, D. J. Kouri, K. Haug, D. W. Schwenke, Y. Shima, and D. G. Truhlar, J. Chem. Phys. 88: 2492, 1988.

[132] D. W. Schwenke and D. G. Truhlar, in Computing Methods in Applied Sciences and Engineering (R. Glowinski and A. Lichnewsky, eds.), SIAM, Philadelphia, 1990, p. 291. S. L. Mielke, D. G. Truhlar, and D. W. Schwenke, J. Chem. Phys. 95: 5930, 1991.

[133] G. J. Tawa, S. L. Mielke, D. G. Truhlar, and D. W. Schwenke, J. Chem. Phys. 100: 5751, 1994. G.J. Tawa, S.L. Mielke, D.G. Truhlar, and D.W. Schwenke, in Advances in Molecular Vibrations and Collision Dynamics, Vol. 2B (J.M. Bowman, ed.), JAI, Greenwich, CT, 1994, pp. 45–116.

[134] C.W. Gear, Numerical Initial Value Problems in Ordinary Differential Equations, Prentice-Hall, Englewood Cliffs, 1971.

[135] International Mathematical and Statistical Library, Version 2.0, IMSL Inc., Houston, September, 1991.

[136] R. Bellman, R. Kalaba, and M. Prestrud, Invariant Imbedding and Radiative Transfer in Slabs of Finite Thickness, American Elsevier, N.Y., 1963.

[137] M.E. Riley, Ph.D. Thesis, California Institute of Technology, Pasadena, CA, 1968.

[138] W.H. Press, B.P. Flannery, S.A. Teukolsky, and W.T. Vetterling, Numerical Recipes, Cambridge Univ. Press, Cambridge, 1986.

QUANTIZATION IN THE CONTINUUM - COMPLEX DILATED EXPANSIONS OF SCATTERING QUANTITIES

NILS ELANDER*

Abstract. Physical observables may be derived from the Green function, the Scattering matrix and the Titchmarsh Weyl m-function. Such objects possess, at least for Schrödinger-type problems, compact spectral representations encompassing a set of complex poles. These, so-called non-redundant poles, are identical to the eigenvalues of the associated Schrödinger equation. The above-mentioned spectral representations allow identification of the individual contributions from these non-redundant poles to observable physical spectral features in terms of their respective residues. A classification of these poles as resonant and background-building ones is suggested. Model potential studies indicate that the expansions of the S-matrix, the Green function as well as the associated Titchmarsh-Weyl m-function, converge rapidly, making the method attractive from both conceptional and computational viewpoints. This presentation includes applications of the above method to problems in atomic and molecular physics as well as a discussion of the search for general and accurate numerical methods utilizing, in particular, the exterior complex scaling formulation.

Key words. S-matrix, Green function, Titchmarsh Weyl m-function, spectral representation, resonant and background-building poles, uniform and exterior complex scaling

1. Introduction. Scientific work often has the goal of describing as many phenomena as possible with one grand unified theory. This was once the driving force behind the formulation of the theory of relativity which was able to join mechanics and electrodynamics.

Comparing bound state and scattering quantum mechanics we find that the different boundary conditions have demanded different treatments. The Complex dilation theory of Nuttall and coworkers[1], Balslev-Combes[2] Simon[3], van Winter[4] and others[5] allowed physicists and chemist the possibility to use the sometimes more powerful bound state methods to study scattering phenomena and, in this way, obtain a formulation of quantum mechanics which encompasses both bound states and scattering phenomena. This work has only been going on for about 25 years and has really only begun. Applications in physics and chemistry one first compute resonances which, within this framework, are well-defined eigenstates of some Hamiltonian, and then utilize these eigenstates and the associated eigenfunctions to derive observables like scattering cross sections, reaction rates etc.

In this contribution I will discuss some of the methods derived in the Stockholm - Uppsala collaboration and illustrate how they may be utilized.

2. Three routes to the cross section - the Green function, the Titchmarsh-Weyl $m-$function and the S-matrix. An experimental

* Atomic Physics, Stockholm University, Frescativägen 24, S-104 05 Stockholm, Sweden.

spectroscopist is interested in obtaining information on well-defined quantities such as energy differences between stable eigenstates of a system, lifetimes of the involved excited states, rates and intensities, the transitions between the states etc. In a scattering experiment one is interested in the outcome of the interaction between the incoming particles and a tergert. This is measured as, for instance, energy-dependent differential cross sections.

The interpretation of the experiment should relate the observed cross section to the details of the interactions and the structure of the incoming as well as the outgoing particles. Such an analysis reveals the existence of bound and short-lived (quasi-bound) states formed by the partners of the collision. These bound and quasi-bound states may be described in terms of the spectral density related to the process. This quantity also defines the potentials uniquely[6]. The fact that these states both have a well defined energy and in some sense a lifetime connects pure bound state theories with the classical non-resonant scattering formalism.

2.1. The Green function. Consider as an illustration the Schrödinger problem

$$(2.1) \qquad L_\ell u(r) = \lambda u(r) \quad \text{with} \quad L_\ell = -\frac{d^2}{dr^2} + \frac{\ell(\ell+1)}{r^2} + U_0 r$$

to which we can associate the Green function $G_\ell^+(r, r')$ through

$$(2.2) \qquad\qquad (\lambda - L_\ell)G_\ell^+(r, r') = \delta(r - r')$$

Since $G_\ell^+(r)$ for fixed r' is regular for both $r \to 0$ and $r \to \infty$ we have that $G_\ell^+(r)$ must be proportional to both the regular solution at origin $\psi_\ell(r)$ as well as the solution $\chi_\ell^+(r)$ which is regular at infinity. The proportionality constant is obtained by integration over the junction $r = r'$ and is found to be the reciprocal of the Wronskian between $\psi_\ell(r)$ and $\chi_\ell^+(r)$:

$$(2.3) \qquad\qquad G_\ell^+(r) = \psi_\ell(r_<)\chi_\ell^+(r_>)/W[\psi_\ell, \chi_\ell^+(r)].$$

2.2. The classical Titchmarsh-Weyl theory. is closely related to the Green function formalism. Weyl's[9] extension of the Sturm-Liouville theory for second order differential equations on a finite interval $[a, b]$ to the so called singular case which describes the open infinite interval $[0, +\infty)$ is described in textbooks[7,8].

Consider again the one-dimensional Schrödinger eq.(2.1) on a finite interval $[0, b]$ and let ϕ_ℓ and ψ_ℓ be two linearly independent solutions at a given eigenvalue λ which are defined through

$$(2.4) \qquad\qquad \begin{pmatrix} \phi_\ell & \psi_\ell \\ \phi_\ell' & \psi_\ell' \end{pmatrix}_{r=a} = \begin{pmatrix} \sin\alpha & \cos\alpha \\ -\cos\alpha & \sin\alpha \end{pmatrix}.$$

Here a is a point in $(0, \infty)$

Any solution to eq.(2.1) except ψ_ℓ can be written in a form

$$(2.5) \qquad \chi_\ell(r) = \phi_\ell(r) + \psi_\ell(r)m(E)$$

which defines Weyl's m−function. Now impose the following right bound-ary condition on $\chi_\ell(r)$ at $r = b > a$

$$(2.6) \qquad \cos \beta \chi_\ell(b) + \sin \beta \chi_\ell'(b) = 0$$

for some real angle $-\pi \le \beta \le \pi$. The fact that β is real implies

$$(2.7) \qquad \Im(\chi_\ell'/\chi_\ell)_{r=b} = 0$$

from which it can be shown that m in eq.(2.5) may be described by a circle in the complex plane with its center and radius given by

$$(2.8) \qquad C_b = -[\phi_\ell\psi_\ell](b)/[\psi_\ell\psi_\ell](b) \quad \text{and} \quad R_b = 1/[\psi_\ell\psi_\ell](b)$$

where

$$(2.9) \qquad [fg](t) = f(t)\bar{g}'(t) - f'(t)\bar{g}(t).$$

Taking the limit $b \to \infty$ we can now distinguish between two cases. Either there exists a finite, non-zero limiting radius R_b or the circle converges to a point. These two cases are known as the limit circle and the limit point cases[7,8,13]. Almost all applications in physics belong to the limit point case. Most often[13] both χ_ℓ and χ_ℓ' are square integrable on an interval $[a, \infty)$.

It can be shown that $m(E)$ is an analytic function whose imaginary part is the spectral density that will occur in the completeness relation

$$(2.10) \qquad \delta(r - r') = \int\limits_{-\infty}^{+\infty} \psi_\ell(\omega, r)\psi_\ell(\omega, r')d\rho(\omega),$$

where

$$(2.11) \qquad \rho(\omega_1) - \rho(\omega_2) = \lim_{\epsilon \to 0+} \frac{1}{\pi} \int\limits_{\omega_1}^{\omega_2} \Im\{m(\lambda + i\epsilon)\}d\lambda$$

Thus we obtain

$$(2.12) \qquad \Im(m(E)) = \pi\left(\frac{d\rho}{d\omega}\right)_{\omega=E}$$

From eq.(2.5) we find that the Titchmarsh-Weyl m−function can be written as

$$(2.13) \qquad m_{TW}(E) = \frac{W[\phi_\ell, \chi_\ell]}{W[\chi_\ell, \psi_\ell]}$$

$m_{TW}(E)$ can thus be expressed in the logarithmic derivative χ'_ℓ/χ_ℓ.

In scattering theory one commonly defines the wave number such that $k = \sqrt{E - U(\infty)}$ with a real $U(\infty)$. The asymptotic forms of the wave function solving the scattering Schrödinger problem have the properties

$$(2.14) \qquad \lim_{r \to 0}(2\ell + 1)!!(1/r^{\ell+1})\chi_\ell(r) = 1$$

$$(2.15) \qquad \lim_{r \to 0}\left(\chi_\ell(r) - \frac{|\mathcal{F}_\ell(k)|}{k^{\ell+1}}\sin(kr - \frac{\ell\pi}{2} + \delta_\ell(k))\right) = 0$$

where $f(k, r)$ is the Jost solution defined through

$$(2.16) \qquad \lim_{r \to \infty}\left[(\pm \imath)^\ell \exp(\pm \imath kr) - f_\ell(k, r)\right] = 1$$

and

$$(2.17) \qquad \mathcal{F}_\ell(k) = f_\ell(k, 0) = W[f_\ell, \psi_\ell]$$

Using the relation between m_ℓ and $\mathcal{F}_\ell(k)$ we obtain the Kodaira form[19,20] of the spectral density

$$(2.18) \qquad \Im(m_\ell(E)) = \pi\left(\frac{d\rho_\ell}{d\omega}\right) = \frac{\imath W[f^+, f^-]}{2W[f^+, \psi_\ell]^2} = \frac{k}{|\mathcal{F}_\ell(k)|^2}$$

The spectral density function has jumps at bound states ($E < 0$)

$$(2.19) \qquad |a_j^\ell|^2 = \left(\int_0^\infty |\psi_\ell(E_j^\ell, r)|dr\right)^{-2}$$

The spectral density can thus be shown to have the form

$$(2.20) \qquad \left(\frac{d\rho_\ell}{d\omega}\right)_{\omega=E} = \begin{cases} \dfrac{1}{\pi}\dfrac{E^{l+1/2}}{|f_\ell(E^{1/2})|^2} & E > 0 \\ \sum_j |a_j^\ell|^2\delta(E - E_j^\ell) & E < 0 \end{cases}.$$

This spectral density is also given by Newton[31]. We are from this able to construct a completeness relation

$$(2.21) \qquad \delta(r - r') = \int_{-\infty}^{+\infty} \psi_\ell(\omega, r)\psi_\ell(\omega, r')d\rho_\ell(\omega)$$

One can, by using the asymptotic form of the regular solution ψ_ℓ and utilizing the fact that

$$(2.22) \qquad \chi_\ell(kr) \propto f_\ell(kr) \to \exp(\imath kr)$$

show that the scattering amplitude may be expressed in terms of the Titchmarsh-Weyl m-function

$$(2.23) \quad \mathcal{F}_\ell(k) = \lim_{a \to \infty} \left\{ \frac{m^\ell_{(TW)}(k, a) - m^\ell_{(TW)0}(k, a)}{(1 - k^2)} \exp\left[-2\imath(ka - \frac{\ell\pi}{2}) \right] \right\}$$

Here $m^\ell_{(TW)0}(k, a)$ is the free-particle m-function.

Since the partial wave cross section may be written as

$$(2.24) \quad \sigma_\ell(k) = 4\pi(2\ell + 1)|\mathcal{F}_\ell(k)|^2$$

we thus have a representation of physical observables like the cross section in terms of the Titchmarsh-Weyl m-function.

2.3. A generalized spectral expansion of the Green function.

Following eq.(2.2) we[10,11,12] use the resolvent formulation of the Green function and write

$$G^+_\ell(\lambda; r, r') = \qquad \lim_{\epsilon \to 0} \int_{-\infty}^{+\infty} \frac{d\hat{\tau}(\omega)}{\lambda + \imath\epsilon - \omega}$$

$$= \lim_{\epsilon \to 0} \left\{ \sum_j \frac{\text{Res } G^+_\ell(\lambda_j; r, r')}{\lambda + \imath\epsilon - \lambda_j} + \int_0^{+\infty} \frac{[d\hat{\tau}(\omega)/d\omega]d\omega}{\lambda + \imath\epsilon - \omega} \right\} \lambda \neq \lambda_j$$

If we extract the imaginary part of $G^+_\ell(\lambda; r, r')$ by the distribution formula

$$(2.25) \quad \lim_{\epsilon \to 0} \frac{1}{\lambda + \imath\epsilon - \omega} = \mathcal{P}\frac{1}{\lambda - \omega} - \imath\pi\delta(\lambda - \omega)$$

we find

$$(2.26) \quad \Im(G^+_\ell(\lambda; r, r')) = -\pi \left(\frac{d\hat{\tau}(\omega)}{d\omega} \right)_{\omega=\lambda}$$

Using $\chi^+_\ell = \phi_\ell + \psi_\ell m^+$ and $W[\phi_\ell \psi_\ell]$ we find that

$$(2.27) \quad \Im(G^+_\ell(\lambda; r, r')) = -\psi_\ell(r)\psi_\ell(r')\Im(m^+)$$

This makes it possible to rewrite eq.(2.25) as

$$(2.28) \quad G^+_\ell(\lambda; r, r') = \lim_{\epsilon \to 0} \int_{-\infty}^{+\infty} \frac{\psi_\ell(\omega, r)\psi_\ell(\omega, r')d\rho(\omega)}{\lambda + \imath\epsilon - \omega}$$

where now

$$(2.29) \quad \frac{d\rho(\omega)}{d\omega} = \begin{cases} \sum_j \dfrac{\delta(\omega - \lambda_j)}{\langle\psi_\ell(\lambda_j)|\psi_\ell(\lambda_j)\rangle} & \text{for } \omega < 0 \\ \frac{1}{\pi}\Im\{m^+(\omega)\} & \text{for } \omega > 0 \end{cases}$$

By letting the resolvent operator work on eq.(2.28) it can be shown that the spectral density in eq.(2.29) and (2.21) are the same. The Green function above is thus uniquely defined from the Titchmarsh-Weyl m-function. We now want to extend the Green function analytically. It can be shown that the m-function is of Nevanlinna type and we will use that.

2.3.1. On the Nevanlinna representation of the m-function.
A Cauchy analytic function $f(z)$ is said to be of Nevanlinna type[32,33] if it maps the upper (lower) complex half-plane onto itself. The Nevanlinna functions posses a uniquely defined spectral function $\sigma(\omega)$ such that

$$(2.30) \qquad f(z) = \int_{-\infty}^{+\infty} \frac{d\sigma(\omega)}{\omega - z}$$

where

$$(2.31) \qquad \frac{d\sigma(\omega)}{d\omega} = \begin{cases} \sum_j \text{Res } f(z_j)\delta(\omega - z_j) & \text{for } \omega < 0 \\ \frac{1}{\pi}\Im\{f(\omega + i0)\} & \text{for } \omega > 0 \end{cases}$$

Now let us for the sake of simplicity limit ourselves to $\ell = 0$ and consider $f(\lambda) = m(\lambda) - m_{\text{free}}$ where the latter may be shown to be $m_{\text{free}} = i\sqrt{\lambda}$.

Hence we have

$$(2.32) \qquad m(\lambda) - i\sqrt{\lambda} = \int_{-\infty}^{+\infty} \frac{d\sigma(\omega)}{\omega - z}$$

with

$$(2.33) \qquad d\sigma(\omega) = d\sigma(\omega) - d\sigma_{\text{free}}(\omega)$$

2.4. A complex dilated Titchmarsh-Weyl theory.
may be obtained by complex dilation of the radial coordinate. We first briefly consider the corresponding Schrödinger problem.

2.4.1. Analytical continuation of the Schrödinger problem.
is one of the themes of this workshop. It is a technique to analytically continue a self-adjoint Hamiltonian operator to a family of non-self-adjoint operators $\{H_\eta\}$[30]. Most of the work done in this area is based on what we call uniform scaling of the independent variable r

$$(2.34) \qquad r \to \eta r; \qquad \eta = \exp i\theta, \quad \theta < \theta_{\text{crit}}$$

Here θ_{crit} is the critical angle related to the properties of the potential. If there are resonances in the physical system described by H, then H_η will have them as complex eigenvalues. The family $\{H_\eta\}$ is constructed such that the point spectra $\sigma_p(H_\eta)$ have the property

$$(2.35) \qquad \sigma_p(H_{\eta_1}) \subset \sigma_p(H_{\eta_2}) \qquad 0 < arg(\eta_1) \leq arg(\eta_2) < \theta_{\text{crit}}$$

If the potential has singularities in the first quadrant of the complex coordinate plane or if it is only available as a tabulated point array it is not practical to use uniform scaling. Simon introduced, for this purpose, the exterior complex scaling[18]

In Ref. [18] η was chosen to be

$$(2.36) \qquad \eta(r) = \begin{cases} 1 & r < R \\ \eta & r \geq R \end{cases}.$$

It is important to notice that for $R > 0$ the domains $D(H_\eta)$ depend on η in the form[15]

$$(2.37) \qquad \Psi \in D(H_\eta): \quad \begin{cases} \Psi(R+0) &= \eta^{1/2}\Psi(R-0) \\ \Psi'(R+0) &= \eta^{3/2}\Psi'(R-0) \end{cases}.$$

Conditions (2.37) mean that H_η is only defined on functions that are discontinuous at $r = R$, and also the derivatives must be discontinuous. A definition of the derivative of a δ-function of a discontinuous function is needed. This can be achieved as the values and derivatives at $x = 0$ were replaced by the mean values of the left and right limits to 0, which is implied by the definition

$$(2.38) \quad f(x)\delta'(x) = \frac{f(-0) + f(+0)}{2}\delta'(x) - \frac{f'(-0) + f'(+0)}{2}\delta(x).$$

While δ' does not have an interpretation as an operator in the Hilbert space, the operator, which is formally defined by

$$(2.39) \qquad A\Psi := \quad \left[-d^2/dr^2 + c\delta'(r-R) \right] \Psi = \chi,$$

can be given meaning by demanding that χ contain no δ-like singularities. It was shown[15] that

$$(2.40) \qquad A_\eta = \eta^{-1}\left[-\frac{d^2}{dr^2} + 2\frac{1 - \sqrt{\eta}}{1 + \sqrt{\eta}}\delta'(r-R) \right]\eta^{-1},$$

The present version of exterior complex dilation was initiated by our work on the finite elements[29,43]. The formal results below are discussed in terms of uniform scaling but hold equally well when using exterior scaling.

2.4.2. Application to Titchmarsh Weyl theory. The dilated one-dimensional Schrödinger equation reads

$$(2.41) \qquad \left(-\frac{d^2}{dr^2} + \frac{\ell(\ell+1)}{r^2} + \eta^2 U_0(\eta r) - \eta^2\lambda \right)\chi_{\ell,\eta}(r) = 0$$

Noting that the complex dilation implies that

$$(2.42) \qquad \chi'_{\eta,\ell}(r) = \frac{d\chi_\ell(\eta r)}{d(\eta r)}$$

and similarly for $\phi'_{\eta,\ell}(r)$ and $\psi'_{\eta,\ell}(r)$. The boundary condition (2.4) now takes the form

(2.43)
$$\begin{pmatrix} \phi_{\ell,\eta} & \psi_{\ell,\eta} \\ \phi'_{\ell,\eta} & \psi'_{\ell,\eta} \end{pmatrix}_{r=a} = \begin{pmatrix} \sin\alpha & \cos\alpha \\ -\eta\cos\alpha & \eta\sin\alpha \end{pmatrix}$$

Provided the energy parameter λ in eq.(2.41) is within the rotated sector it can be shown that the dilated Weyl circle will still converge to a point.

The resonance poles of the m–function in the fourth quadrant of the complex energy plane are obtainable through the analytic continuation of eq.(2.13). Consider again the Schrödinger problem (eq.(2.41)). Rewriting it in terms of the logarithmic derivative $Z_{\ell,\eta} = u'_{\ell\eta}(u_{\ell\eta})^{-1}$ we get the complex dilated Riccati equation

(2.44)
$$Z'_{\ell,\eta} = 2\eta^2(U_0(\eta r) - \lambda) - Z^2_{\ell,\eta}$$

The Titchmarsh-Weyl m–function is then evaluated as

(2.45)
$$m_{(TW),\ell,\eta}(\lambda) = \frac{Z^J_{\ell,\eta} Z^R_{\ell,\eta}}{Z^R_{\ell,\eta} - Z^J_{\ell,\eta}}$$

where $Z^J_{\ell,\eta}$ and $Z^R_{\ell,\eta}$ are the logarithmic derivatives of the Jost and regular solutions, respectively.

2.4.3. Deformation of the integration contour in the Nevanlinna representation of the m–function.

Above we defined $m^+_\eta(\lambda)$ and $m^-_\eta(\lambda)$ to have analytic continuation onto a higher order Riemann sheet. If $m^+(\lambda)$ and $m^-(\lambda)$ are non real for real λ–parameters they differ only by the signs of their imaginary parts. Thus, for real λ–parameters we have

(2.46)
$$\Im(m^+(\lambda)) = (m^+(\lambda) - m^-(\lambda))/2\imath$$

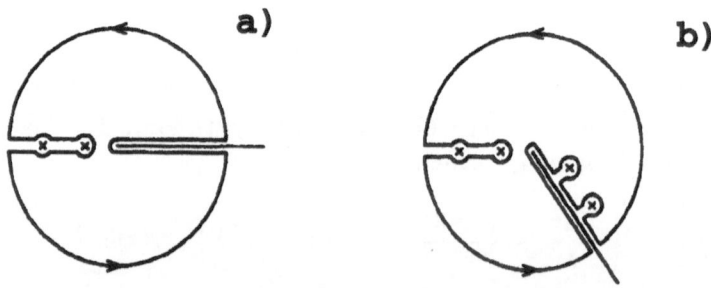

FIG. 2.1. *Integration contour for the Cauchy representation displaying the deformation around bound states and (a) the unrotated cut along the real axis and (b) also the resonance poles and the rotated cut. (Modified from ref)*

We can now simplify our further analysis by defining a generalized imaginary part of $m^+(\lambda)$, from now on denoted $\Im\text{mg}$, as the left hand side of eq.(2.46).

This enables us to write a generalized Nevanlinna representation as

$$(2.47)\; m^+(\lambda) - \imath\sqrt{\lambda} = \sum_j \frac{\text{Res } m^+(\lambda)}{\lambda_j + \lambda} + \int_C \frac{\Im\text{mg }(m^+(\omega) - \imath\sqrt{\omega})d\omega}{\omega - \lambda}$$

where the contour is that of figure 2.1. We are now ready to apply this formalism to a spectral representation of the Green function with deformed contour.

2.4.4. A spectral representation of the Green function with deformed contour in the complex energy plane. From eq.(2.29) it follows that

$$(2.48)\qquad \text{Res } (m^+(\lambda)) = -\langle\psi(\bar{\lambda}_j)|\psi(\lambda_j)\rangle^{-1}$$

where the scalar product has the form

$$(2.49)\qquad \langle\psi(\bar{\lambda}_j)|\psi(\lambda_j)\rangle = \int_{\eta R^+} \psi(\bar{\lambda}_j, r')\psi(\lambda_j, r')dr'$$

We thus arrive at the expansion

$$G^+(\lambda; r, r') = \sum_j \frac{\psi(\lambda_j, r)\psi(\lambda_j, r')\text{Res }(-m^+(\lambda_j))}{\lambda - \lambda_j}$$

$$+ \int_C \frac{\psi(\omega, r)\psi(\omega, r')\Im\text{mg }(m^+(\omega))d\omega}{\pi(\lambda - \omega)} \qquad \lambda \neq \lambda_j$$

By applying the operator $\lambda - \hat{L}(r)$ on (2.50) we find

$$\delta(r - r') = \sum_j \psi(\lambda_j, r)\psi(\lambda_j, r')\text{Res }(-m^+(\lambda_j))$$

$$(2.50)\qquad + \int_C \psi(\omega, r)\psi(\omega, r')\Im\text{mg }(m^+(\omega))d\rho_\eta(\omega)$$

where

$$(2.51)\qquad \frac{d\rho_\eta(\omega)}{d\omega} = \frac{1}{\pi}\Im\text{mg }(m^+(\omega))$$

2.5. The partial wave S-matrix may be defined in terms of the Jost functions. as

$$(2.52)\qquad S_\ell(k) = \frac{\mathcal{F}_\ell^-(k)}{\mathcal{F}_\ell^+(k)}.$$

It is thus well motivated to search for an analytic continuation of the S-matrix by studying the properties of the Jost functions[16,17].

2.5.1. The Jost functions. may be defined in terms of the potential V and the Riccati-Bessel function $u_\ell(k, r)$

$$(2.53) \qquad \mathcal{F}_\ell^\pm = 1 + \frac{(-1)^\ell}{k} \int_0^\infty u_\ell(k, r) V(r) f_\ell^\pm(k, r) dr$$

or equivalently through the Wronskian

$$(2.54) \qquad \mathcal{F}_\ell^\pm(k) = (-k)^\ell W[f_\ell^\pm(k, r), \psi_\ell(k, r)]$$

The former definition implies that

$$(2.55) \qquad \lim_{|k| \to \infty} \mathcal{F}^+(k) = 1$$

2.5.2. The analytic continuation of the partial wave S-matrix. Consider again the Schrödinger-type operator

$$(2.56) \qquad H = -\frac{1}{2} \frac{d^2}{dr^2} + U(r)$$

which fulfills Green's formula

$$(2.57) \qquad \int_0^b \{\phi^*(r) H\psi(r) - [H\psi(r)]^* \psi(r)\} dr = W[\psi, \phi^*]|_0^b$$

Using the generalized inner product

$$\langle f|g \rangle_\eta = \int_0^{b\eta} f^*(\eta^* r) g(\eta r)$$

$$(2.58) \qquad = \int_0^{b\eta} \{(\eta^*)^{\frac{1}{2}} f(\eta^* r)\}^* \eta^{\frac{1}{2}} g(\eta r) dr = \langle f(\eta^*)|g(\eta) \rangle$$

we are able to show that the complex dilated Schrödinger equation fulfills an analytically continued relation

$$\int_0^{b\eta} [\{\phi(\eta^* r)\}^* H(\eta)\psi(\eta r) - \{H(\eta^*)\psi(\eta^* r)\}^* \phi(\eta r)] d\eta r =$$

$$\int_0^b \{\phi(\eta^*)\}^* H(\eta)\psi(\eta) - \{H(\eta^*)\phi(\eta^*)\}^* \psi(\eta) dr = \int_0^b \eta^{-2} \frac{d}{dr} W[\psi(\eta), \phi^*(\eta^*)] dr =$$

$$\eta^{-1} W[\psi(\eta), \phi^*(\eta^*)]|_0^b$$

Through these simple arguments we have thus obtained the complex dilated Jost functions which may be used to construct an analytically continued partial wave S-matrix.

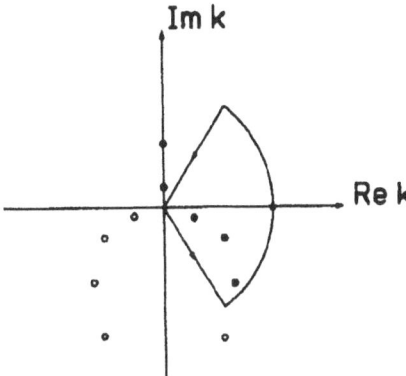

FIG. 2.2. *Integration contour for evaluation of the integral part of the partial wave S-matrix expansion in the text.(Modified from ref)*

Note that since $\mathcal{F}_\ell^+(k)$ and $\mathcal{F}_\ell^-(k)$ are well-defined in the upper and the lower complex k−planes respectively, different scaling parameters η should be used in order to continue $S_\ell(k)$ into a sector of the complex k−plane.

Using the fact that the analytically continued partial wave S-matrix now is a meromorphic function we can by choosing some arbitrary point a study

(2.59) $$D_\ell(k) = S_\ell(k) - S_\ell(a)$$

and through Cauchy's integral theorem immediately obtain

(2.60) $$S_\ell(k) = S_\ell(a) + \sum_{i=1}^{N} \text{Res}[S_\ell(k_i)]\left(\frac{1}{k - k_i} + \frac{1}{k_i - a}\right) + \frac{k - a}{2\pi\imath}\int_C \frac{S_\ell(z)}{(z - k)(z - a)}dz$$

Note that in order to evaluate the partial wave S-matrix at any k we need to know the poles of $S_\ell(k)$, $S_\ell(a)$ as well as the contour integral. The computational usefulness of eq.(2.60) must then be that, at least, the contour integral is easy to approximate or even neglect.

3. Generalization to multi-channel problems.

3.1. Spectral densities and Green functions.

The single channel methods presented above are readily generalized to the multi-channel case [56,58,57,59,29,61] The Schrödinger equation (2.1) becomes

(3.1) $$\left(-\frac{1}{2}\frac{d^2}{dr^2}\mathbf{1} + \mathbf{V}(r) - E\mathbf{1}\right)\mathbf{\Psi} = 0$$

where $\mathbf{\Psi}$ is now a column vector describing the different channels and $\mathbf{V}$ is a symmetric square matrix describing the interaction between zero order states. Two types of logarithmic derivatives can be determined

$$(3.2) \qquad\qquad \mathbf{R} = \ \mathbf{\Psi}'\mathbf{\Psi}^{-1}$$

$$(3.3) \qquad\qquad \tilde{\mathbf{R}} = \ \tilde{\mathbf{\Psi}}^{-1}\tilde{\mathbf{\Psi}}'$$

from which matrix Riccati equations can be written

$$(3.4) \qquad\qquad \mathbf{R}' = (\mathbf{V} - \mathbf{1}\lambda) + \mathbf{R}^2$$

$$(3.5) \qquad\qquad \tilde{\mathbf{R}}' = (\mathbf{V} - \mathbf{1}\lambda) + \tilde{\mathbf{R}}^2$$

The single channel regular and irregular solutions ψ and ϕ as well as the Jost solution $f(k,r)$ are generalized to solution matrices $\mathbf{\Psi}$, $\mathbf{\Phi}$ and $\mathbf{F}$ where each component fulfills

$$(3.6) \qquad \psi_k^{(i)} = \delta_{ki}\frac{r^{\ell+1}}{(2\ell+1)!!} \quad \text{for small } r$$

$$(3.7) \qquad \phi_k^{(i)} = \delta_{ki}\frac{(2\ell-1)!!}{a^\ell}$$

$$(3.8) \qquad f_k^{(i)\pm} = \lim_{r\to\infty}\delta_{ki}\exp(\pm\imath k_i r) \ ; \ k_i = \lim_{r\to\infty}\sqrt{\lambda - V_{ii}(r)}$$

respectively. These definitions imply that

$$(3.9) \qquad\qquad \mathbf{W}[\mathbf{\Phi},\mathbf{\Psi}] = 1$$

A Titchmarsh-Weyl $\mathbf{M}$−matrix is defined through

$$(3.10) \qquad\qquad \chi = \mathbf{\Phi} + \mathbf{\Psi M} = \mathbf{Fd}$$

and it can be proven that

$$(3.11) \qquad\qquad \mathbf{M} = -\{\mathbf{W}[\mathbf{F},\mathbf{\Psi}]\}^{-1}\mathbf{W}[\mathbf{F},\mathbf{\Phi}]$$

$$(3.12) \qquad\qquad = (\mathbf{R_J\Psi} - \mathbf{\Psi}')^{-1}(\mathbf{\Psi}' - \mathbf{R_J\Psi}) = \tilde{\mathbf{M}}$$

and

$$(3.13) \qquad\qquad \mathbf{d} = -(\mathbf{W}[\mathbf{\Psi},\mathbf{F}])^{-1}.$$

An associated Green function can further be defined as

$$(3.14) \quad \mathbf{G}(r,r') = \begin{cases} \mathbf{\Psi}(kr)(\tilde{\mathbf{W}}[\mathbf{\Psi},\mathbf{F}])^{-1}\tilde{\mathbf{F}}(kr') & r < r' \\ \mathbf{F}(kr)(\tilde{\mathbf{W}}[\mathbf{\Psi},\mathbf{F}])^{-1}\tilde{\mathbf{\Psi}}(kr') & r > r' \end{cases}$$

where $k^2 = \lambda$

A matrix spectral density

$$(3.15) \qquad\qquad \pi\left(\frac{d\boldsymbol{\rho}}{d\omega}\right)_{\omega=E} = \mathbf{M}_i$$

which also may be written as

$$(3.16) \qquad \left(\frac{d\boldsymbol{\rho}}{d\omega}\right)_{\omega=E} = \frac{1}{\pi}\tilde{\mathbf{W}}[\boldsymbol{\Psi}, \mathbf{F}^+]\mathbf{k}\mathbf{W}\{\boldsymbol{\Psi}, \mathbf{F}^-\} \qquad E > 0$$

can be derived. A resolution of the identity matrix

$$(3.17) \qquad \mathbf{1} = \int\limits_{-\infty}^{+\infty} |\boldsymbol{\Psi}(\mathbf{k})\rangle d\boldsymbol{\rho}(\omega)\langle\boldsymbol{\Psi}(\mathbf{k})| = \int\limits_{-\infty}^{+\infty} \boldsymbol{\Psi}(\mathbf{k}) d\boldsymbol{\rho}(\omega)\tilde{\boldsymbol{\Psi}}(\mathbf{k})$$

also exists.

3.2. A Multi-channel scattering theory. may as in the single channel case be based on the multi-channel Titchmarsh-Weyl theory. Here we only note that the Jost function matrix takes the form

$$(3.18) \qquad \mathbf{J}^+ = -\mathbf{W}\{\boldsymbol{\Psi}, \mathbf{F}^+\} = (\mathbf{d}^+)^{-1}.$$

and that the partial wave S-matrix now may be expressed in $\mathbf{J}^\pm$ as

$$(3.19) \qquad \mathbf{S} = \left(\mathbf{W}[\boldsymbol{\Psi}, \mathbf{F}^+]\right)^{-1}\mathbf{W}[\boldsymbol{\Psi}, \mathbf{F}^-] = (\mathbf{J}^+)^{-1}\mathbf{J}^-$$

For computational purposes the form

$$(3.20) \qquad \mathbf{S} = (\mathbf{F}^+)^{-1}(\mathbf{R}_{\mathbf{J}}^+ - \mathbf{R}_{\mathbf{R}})^{-1}(\mathbf{R}_{\mathbf{J}}^- - \mathbf{R}_{\mathbf{R}})\mathbf{F}^-$$

is attractive.

Most of eqs.(3.2)- (3.20) are able to transform into their complex dilated analogues.

4. Computational methods for resonances using complex dilation. The applications that we have studied have been based essentially on five methods. The single-channel quantities discussed above were studied numerically on the real axis as well as in the complex plane using uniform as well as exterior scaling.

The search for the resonance poles was first performed by using complex valued versions of the Riccati equation based on Johnson's algorithm [25,26,27]. Wave function information was obtained from the renormalized Numerov method also described by Johnson[26,27] The pole search was then based on the Newton-Raphson method. This implies that of close or or degenerate eigenvalues only the "dominant" one in the Newton-Raphson meaning would be found. Still it is our experience that these direct numerical methods are superior to global basis set methods since they are able to determine complex poles accurately even far away from the real axis. The finite element (FEM) and the B-spline methods are bases on a local trial function basis. We have in particular used FEM and found it superior to the previously tested methods. It was not only found to have the same flexibility as the direct recursion formula related methods but also able to find close even degenerate eigenvalues. Since it seems that we have accumulated considerable experience in accurately determining complex poles far out in the complex plane, our techniques may be of interest to the reader.

4.1. Recursive direct numerical techniques : Numerov-Cowel and the Johnson Riccati technique.

4.1.1. The Johnson Riccati technique.
The fact that a scattering solution is a combination of exponentially growing and decreasing functions makes them hard to treat numerically. The logarithmic derivative of such a combination of function is however a nice, smooth linear function and although the Riccati equation is not linear it is a stable and reliable alternative. The multi-channel radial Schrödinger equation may then be written as

$$(4.1) \quad \left\{ \mathbf{I} \cdot \frac{d^2}{dr^2} + \mathbf{Q}(\eta r) \right\} \Psi(\eta r) = 0 \quad \text{where} \quad \mathbf{Q}(\eta r) = 2\mu\eta^2 \{ \mathbf{I} \cdot \lambda - \mathbf{V}(\eta r) \},$$

for a complex energy $\lambda = E + i\epsilon = E - i\Gamma/2$. Substitution of the logarithmic derivative $\mathbf{Z} = \boldsymbol{\Psi}' \boldsymbol{\Psi}^{-1}$ this expression into the radial Schrödinger equation (4.1), leads to the well known nonlinear Riccati equation which on the real axis reads

$$(4.2) \quad \mathbf{Z}'(r) + \mathbf{Q}(r) + \mathbf{Z}^2(r) = 0,$$

for the logarithmic derivative. As recently mentioned, Johnson[25,26,27] has developed an algorithm for the numerical integration of Riccati equations of the type above. One then starts by putting

$$(4.3) \quad \mathbf{Y}_n = \mathbf{I} + h\mathbf{Z}_n,$$

where $\mathbf{I}$ denotes the identity matrix and where h is the spacing between the $N+1$ grid points $r_0, r_1, ..., r_N$ and $\mathbf{Y}_n \equiv \mathbf{Y}(r_n)$. This transformation leads to the following recurrence relation for $\mathbf{Y}_n$

$$(4.4) \quad \mathbf{Y}_n = (2 \cdot \mathbf{I} - \mathbf{U}_n) - \mathbf{Y}_{n-1}^{-1},$$

where the integration weight factor matrices $\mathbf{U}_n$ are the same as in the Simpson integration rule, i.e.

$$(4.5) \quad \mathbf{U}_n = \begin{cases} \frac{h^2}{3}\mathbf{Q}(r_n) & n = 0, \text{N} \\[2mm] \frac{2h^2}{3}\mathbf{Q}(r_n) & n = 2,4,...,\text{N-2} \\[2mm] 8 \cdot \mathbf{I} - 8[\mathbf{I} + \frac{h^2}{6}\mathbf{Q}(r_n)]^{-1} & n = 1,3,...,\text{N-1} \end{cases}$$

where $n = 0$ correspontds to the starting point.

4.1.2. The renormalized Numerov method. The ordinary Numerov method is an efficient algorithm that can be used to obtain numerical solutions of the radial Schrödinger equation. The basic formula is the three term recurrence relation

$$(4.6) \quad [\mathbf{I} - \mathbf{T}_{n+1}]\Psi_{n+1} - [2\cdot\mathbf{I} + 10\mathbf{T}_n]\Psi_n + [\mathbf{I} - \mathbf{T}_{n-1}]\Psi_{n-1} = 0,$$

where $\Psi_n \equiv \Psi(r_n)$ and $\mathbf{T}_n = -(h^2/12)\mathbf{Q}(r_n)$. Here h is the spacing between the grid points and $\mathbf{Q}(r)$ is defined by eq.(4.1) above. The renormalized Numerov algorithm is derived from eq.(4.6) by making the following transformations

$$\begin{aligned} \mathbf{W}_n &= (\mathbf{I} - \mathbf{T}_n), \\ \mathbf{F}_n &= \mathbf{W}_n\Psi_n, \\ \mathbf{U}_n &= 12\mathbf{W}_n^{-1} - 10\cdot\mathbf{I}, \end{aligned}$$

yielding the recurrence relation

$$(4.7) \qquad\qquad \mathbf{F}_{n+1} - \mathbf{U}_n\mathbf{F}_n + \mathbf{F}_{n-1} = 0.$$

With $\mathbf{R}_n = \mathbf{F}_{n+1}\mathbf{F}_n^{-1}$ we obtain $\mathbf{R}_n = \mathbf{U}_n - \mathbf{R}_{n-1}^{-1}$.

4.1.3. Calculation of Resonance Poles. In order to obtain the complex poles of the S-matrix or the matrix representation of the associated classical Green's function discussed above, we require that

$$(4.8) \qquad\qquad \det(\mathbf{Z_R}(\lambda) - \mathbf{Z_J}(\lambda)) = 0,$$

should be satisfied for the resonant energy vector $\lambda = \mathbf{E} + i\epsilon = \mathbf{E} - i\frac{\Gamma}{2}$. $\mathbf{Z_R}$ and $\mathbf{Z_J}$ are the logarithmic derivatives of the regular and Jost solutions, respectively. In our programs we have used a Newton-Raphson technique in an iterative procedure to fulfill the condition on the determinant above. The situation is somewhat complicated, compared to the bound state situation, since we have introduced additional degrees of freedom in making the energy as well as $\mathbf{Z_R}$ and $\mathbf{Z_J}$ complex[37].

4.1.4. A defect corrected Numerov method. Both the log-derivative method and the Renormalized Numerov methods discussed above are fourth order methods in the sense that the local estimated error is $\mathcal{O}^4(h)$ where h is the local discretization step length. This can be improved by constructing a two-cycle method[62]. The fourth order solution is constructed in the first step. It is then used in the second cycle to approximate higher order terms that were neglected in the first cycle. We write the Numerov integration formula like

$$(4.9) \quad \frac{y(z_*) - 2y(z_0) + y(z_1)}{h^2} = \tfrac{1}{12}\left\{ f(z_*, y(z_*)) + 10f(z_0, y(z_0)) + f(z_{-1}, y(z_{-1})) \right\} + \mathcal{O}(h^4)$$

This formula can be improved in such a way that the residue r can be computed to higher orders by using the solution obtained in the fist cycle. The expression is

$$r_n = \frac{1}{h^2}\left\{\delta^2 y_n - h^2(1 + 1/12\delta^2)f_n + a_3\delta^6 y_n + a_4\delta^8 y_n + ...\right.$$

(4.10)
$$\left. -h^2(c_2\delta^4 f_n + c_3\delta^6 f_n + ..)\right\}$$

The parameters a_ν and c_ν are algebraically derived in order to minimize r_n. The result is that this defect corrected method has a local error of $\mathcal{O}(h^8)$. The recursion formulas have to be modified if exterior scaling is used[62] to determine complex Schrödinger eigenvalues with highly accuracy and more important accurate eigenfunctions.

4.2. Basis set methods.

4.2.1. The finite element method - general. The idea behind the finite elements method to divide the configuration space in a finite number of line or volume elements. In each of the elements one uses a local basis which is so constructed that only a few of the basis functions are different from zero on the boundary of the element. These local basis sets are tailored to describe the behavior of the system of interest in the element. Originally the methods were mainly applied in mechanical engineering, but during the last decade they began to be used in physics and chemistry [38,39,40,41,42]. The idea behind this approach is to have a method which is independent of the dimensionality.

4.2.2. One-dimensional problems. The discussion below will first be directed to one-dimensional coupled channel Hamiltonians for the form

$$(4.11) \qquad H\Psi := \sum_\beta [\frac{\delta_{\alpha\beta}}{2m}(-\frac{d^2}{dr^2} + \frac{\ell(\ell+1)}{r^2}) + V_{\alpha\beta}(r)]\Psi_\beta(r).$$

In the most general way, finite elements basis functions are defined as functions identical to 0 outside a given interval

$$(4.12) \qquad f_{im}(r) \equiv 0 \quad r \notin [r_{i-1}, r_i], \quad i = 1, \ldots, N,$$

on the real axis. The approximate eigenfunctions of H are then assumed to be of the form

$$(4.13) \qquad \tilde{\Psi}_c(r) = \sum_{im} c_{im} f_{im}(r)$$

The $\tilde{\Psi}_c$ will in general be discontinuous at the element boundaries r_i. At the least, continuity is required to be able to define the integrals $\langle \tilde{\Psi}_c^* | (-\Delta)_{\theta, R_0}$

$\tilde{\Psi}_c\rangle$. This subjects the c_{im} to a linear constraint, which is usually implemented as follows: First construct f_{im} such that

(4.14) $$f_{i,1}(r_{i-1}) = 1 ; \quad f_{i,2}(r_i) = 1$$
$$f_{im}(r_{i-1}) = f_{im}(r_i) = 0, \qquad \text{otherwise.}$$

and

(4.15) $$\tilde{\Psi}_c(r) = \sum_{im} c_{im} f_{im}(r).$$

With such f_{im} the limits of $\tilde{\Psi}_c$ from the left and from the right to a given interval boundary defined as $c_{i,2}$ and $c_{i+1,1}$ respectively. The continuity condition $\tilde{\Psi}_c(r_i - 0) = \tilde{\Psi}_c(r_i + 0)$ becomes a simple constraint on the variational parameters c_{im}

(4.16) $$c_{i+1,1} = c_{i,2}.$$

To implement the discontinuity required by Eq. (2.37) the Ith element boundary is chosen to coincide with the exterior scaling radius $r_I = R_0$ and Eq. (4.16) is replaced by

(4.17) $$c_{I+1,1} = e^{(n/2)\theta} c_{I,2}.$$

Now we proceed to construct a set of f_{im}, $m = 1, \ldots, p + 1$ with the boundary values of eq. (4.14) as a linear combination of a set of differentiable functions.

The locality of the finite elements method manifests itself in the fact that only matrix elements between f_{im} from the same interval $[r_{i-1}, r_i]$ are nonzero. Therefore the complex dilated matrices $\tilde{H}_\eta$ and $\tilde{S}_\eta$ are sparse matrices with a total number of $\sim N \times (p + 1)^2$ non-vanishing elements. The computation times for the solution of the full eigenvalue problem are proportional to $N^3 \times (p + 1)^3$.

A further important consequence is that the matrix $\tilde{S}_\eta$ is well-behaved, while global basis set methods in general suffer from ill-conditioning for large basis sets.

From the sub-matrices $H^{(i)}_{mm'} := \langle f^*_{im} | H_\eta f_{im'} \rangle$ and $S^{(i)}_{mm'} := \langle f^*_{im} | f_{im'} \rangle$ the global matrices $\tilde{H}_\eta$ and $\tilde{S}_\eta$ are built. The only connection between the individual sub-matrices is established by the constraints (4.16) and (4.17), which link variational coefficients belonging to different sub-matrices. After properly ordering the coefficients $c_{i,m}$ one obtains the following block-diagonal form of $\tilde{H}_\eta$:

$$\tilde{H}_\eta\, c =$$

$$
\begin{pmatrix}
\ddots & \vdots & 0 & \cdots & & \\
 & H^{(i-1)}_{p+1,2} & 0 & 0 & 0 & \\
H^{(i-1)}_{2,p+1} & H^{(i-1)}_{2,2} + H^{(i)}_{1,1} & \cdots & H^{(i)}_{1,2} & 0 & \\
0 & H^{(i)}_{3,1} & \cdots & H^{(i)}_{3,2} & 0 & \\
\vdots & \vdots & \ddots & \vdots & 0 & \\
0 & H^{(i)}_{p+1,1} & \cdots & H^{(i)}_{p+1,2} & 0 & \\
0 & H^{(i)}_{2,1} & \cdots & H^{(i)}_{2,2} + H^{(i+1)}_{1,1} & \cdots \\
0 & 0 & 0 & \vdots & \ddots
\end{pmatrix}
\begin{pmatrix}
\vdots \\
c_{i-1,p+1} \\
c_{i,1} = c_{i-1,2} \\
c_{i,3} \\
\vdots \\
c_{i,p+1} \\
c_{i,2} = c_{i+1,1} \\
\vdots
\end{pmatrix}
$$

and analogously for $\tilde{S}_\eta$. To implement the condition (4.17) at the element boundary $r_I = R_0$, one further needs to multiply the elements of $H^{(I+1)}_{mm'}$ and $S^{(I+1)}_{mm'}$ by $e^{n\theta/2}$ if only one of the indices is 1, and by $e^{n\theta}$ if both indices are 1.

Convergence of the approximate eigenvalues is the criterion for the choice of the number of intervals N and polynomial degree p. Generally finite elements methods with polynomials variationally approximate eigenvalues like

$$(4.18) \qquad |\tilde{E}_k - E| \approx a_k N^{-2p},$$

where the constant a_k is related to the pth derivative of the exact eigenfunction. Provided this eigenfunction is known to be a smooth function inside each element, one will therefore always try to choose p, within reasonable limits, to be as large as possible.

4.2.3. A three-dimensional extension. While a 1 D implementation is easily made accurate and fast the application of FEM to 3 D problems is, in practice, quite a bit more complicated. Not only does the number of elements and thereby the entire basis grow quickly but even worse is that fact that the grid topology is more involved. While 1 D problems only have one type of boundaries between two adjacent elements a volume element in a 3 D rectangular implementation used[43] possesses of the order of 27 neighbors which furthermore can not be sorted in a linearly connected sequence. The boundaries will now be surfaces, edges or nodes. The Hamiltonian and the overlap matrices H and S will now still be banded but the bandwidth will increase considerably when compared to a typical 1 D case.

The implementation is further such that T-shaped connections between surfaces and edges may occur. A 3-D topology which has T-shaped connections as this one is also complicated by the fact that the surface side of one element is to be connected to several edges and nodes and surfaces of other elements. A flexible variation of element size over the grid space is then possible. This is important since it is meaningless to have a dense grid where the variation of the solution functions is small while it on the method on the other hand gives more accurate solutions is the grid is dense where the solution functions vary rapidly.

A rectangular grid makes it possible to define the elementary 3-dimensional functions $f_{i;lmn}(x,y,z)$ as products of 1-dimensionally elementary functions

$$f_{i;lmn}(x,y,z) = f_{il}^x(x)f_{im}^y(y)f_{in}^z(z),$$

where $f_{il}^t(t), t = x, y, z$ are 1-dimensional elementary functions on the intervals in the respective directions.

The continuity of the wave function across the element boundaries here in the 3-D case can be realized by rewriting the wave function

(4.19)
$$\tilde{\Psi} = \sum_{im} c_{im} f_{im} = \sum_i d_{i0} g_i^b + \sum_{k>2,i} d_{ik} g_{ik}^e$$

in internal $g_{ik}^{(e)} = f_{ik}, k > 2$ which are zero at any boundary and element boundary functions $g_i^{(b)} = f_{i2} + f_{i+1,1}$ which connects an element to its neighbors.

The final search of the eigenvalues was achieved by making use of the block structure of the still banded generalized eigenvalue matrix problem.

5. Model potential studies and applications.
Since one of the aims of this contribution was to illustrate the structure of the pole-strings found in numericalcalculations, I have here mainly limited myseld to show application of several methods to a few test cases only.

5.1. Single channel models.

5.1.1. Poles-strings - the $W_0 r^2 e^{-r}$ potential.
The $W(r) = W_0 r^2 e^{-r}$ potential[45] has by construction no bound states. Several, if not an infinite number of, eigenenergies with both positive and negative real parts are found. A real-valued threshold above which no real part of an eigenenergy exits. Following Jensen [44] one can show that a boundary for the eigenvalues exist and that no eigenvalue can be found outside this boundary in the complex energy plane. The pole-strings for different angular momenta ℓ expressed as those of

(5.1)
$$V_J(x) = V_0(x) + \frac{\ell(\ell+1)}{x^2},$$

are found to follow each other closely in the complex energy plane. The absolute values of the imaginary parts of the eigenenergies grow along the string and with ℓ. It is thus only the first few eigenstates which possibly will appear as peaks in an elastic scattering against this potential. We have tentatively called these primary or structure building eigenstates whereas the rest are described as secondary or background building ones.

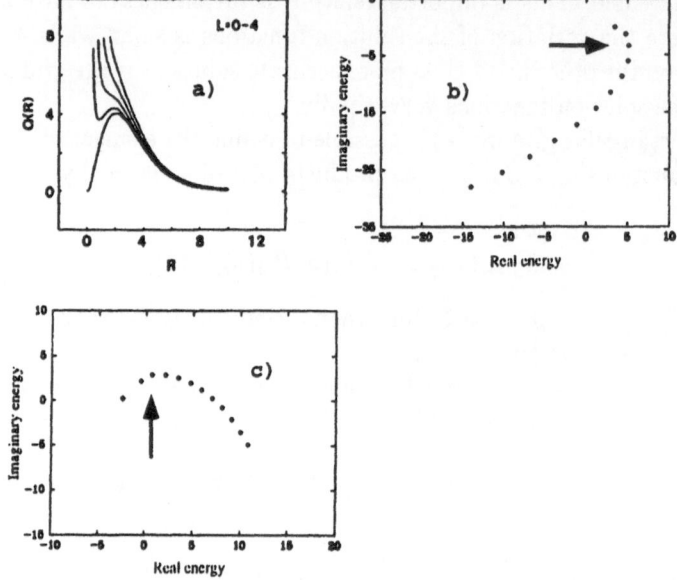

FIG. 5.1. *The potential by Bain et al.a) seems to have infinitely many poles and by construction no bound states. The different angular momenta ℓ correspond to close-lying and almost parallel trajectories b). Here $\ell = 0 - 4$ are displayed. The m−function residues (here only for $\ell = 0$) also fall on a string. The arrows in (b) and (c) point to the pole closest to the maximum of the real part of the complex eigenenergy. (Modified from ref)*

5.1.2. The Moiseyev, Weinhold and Certain and the Morse potential. The spectrum associated with the potential of the Moiseyev[46] which is described as

$$(5.2) \qquad V(x; s, J) = \left(\frac{1}{2}x^2 - J\right) \exp(-Sx^2) + J$$

seems likewise to have an infinite number of poles. The Bain[45] and the Moiseyev[46] potentials differ on the other hand in the sense that the critical angles are $\pi/2$ and $\pi/4$ respectively. For the Moiseyev case the real part of the poles does however not approach minus infinity but seems to oscillate between two thresholds with increasing "quantum number".

The Morse potential, well known from molecular physics

$$(5.3) \qquad V(x; r_e, \beta, T_e, D_e) = T_e - D_e \left\{ 1 - \exp(-\beta[x - r_e]) \right\}^2.$$

seems on the other hand only to possess a finite number of stable complex poles.[14]. The string of stable poles ends shortly after the maximum real energy threshold is reached as illustrated in figure 5.3. All three examples

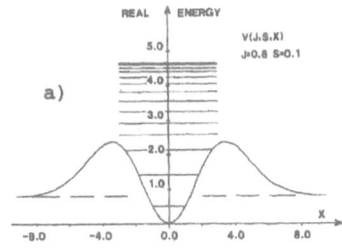

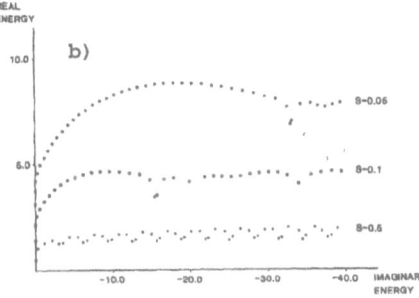

FIG. 5.2. *The potential described by Moiseyev, Weinhold and Certain and its first 44 poles. Note that the poles of this potential oscillate between two thresholds. (Modified from ref)*

shown do have eigenenergy strings with a real maximum energy threshold, critical angles and close lying strings associated to different angular momenta (ℓ). The spacing between two adjacent poles in the complex plane seems to be constant or possibly only slowly varying which makes it easy to interpolate and extrapolate missing or higher eigenenergies not yet calculated.

5.1.3. The Green function - m-function expansion and the emptying of the spectra and the primary and secondary poles.

Following the formalism of eq.(2.41 - 2.49) we calculated the m-function and its residues associated with the potential $W(r)$. The results are displayed in figure 5.1c. Not only do the poles follow a trajectory in the complex plane but this example illustrates that the residues have the same property of forming a smooth string with almost equally spaced points. The residue of m for a bound state is a negative quantity. If we follow the pole-string we note that $\Re\{\text{Res}\,(m^+)\}$ is still negative but that Res (m^+) now has a non-zero imaginary part. When the resonance pole trajectory reaches its maximum real energy we note that the real part of Res (m^+) changes sign from $-$ to $+$. Is this a general property for similar potentials or is it an artifact of our model? If generalizable, we suggest that this sign change may define the distinction between primary, structure building

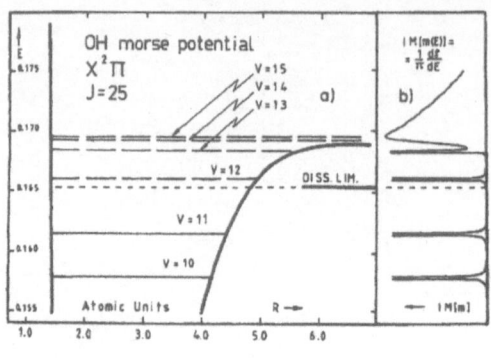

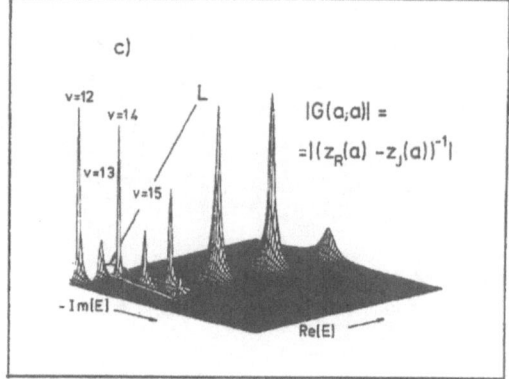

FIG. 5.3. *a) A Morse type potential and its Titchmarsh-Weyl m-function b) The absolute value of the diagonal part ($r = r'$) of the Green function $G(r, r', E)$ corresponding to the Morse potential. Note how the higher poles build the spectral density bump in a). (Modified from ref)*

eigenenergies and background building ones discussed above.

The expansion of the m−function and, based on this, the Green function found in eq.(2.47) and (2.50) can be used to compute spectral properties which makes it possible for us to numerically investigate these expansions on rays in the complex energy plane. For a free particle ($V(x) = 0$) with the kinetic energy ω the generalized imaginary part of the m−function (eq.(2.46) is $-\Im\mathrm{mg}\,(m^+_{\mathrm{free}}) = -(1/\pi)\sqrt{\omega}$. The effective spectral density described in the integral representation (2.47) has this value inside the eigenvalue string and we are led to a deflation based approximation

$$(5.4) \quad m_{\mathrm{defl}}(\lambda) = \imath\sqrt{\lambda} + \sum_j \frac{\mathrm{Res}\,[m^+(\lambda_j)]}{\lambda - \lambda_j} - \frac{2}{\pi}\sqrt{\Omega} + \frac{\sqrt{\lambda}}{\pi}\ln\left[\frac{\sqrt{\lambda} + \sqrt{\Omega}}{[\sqrt{\lambda} - \sqrt{\Omega}]}\right]$$

There are essentially three parts in this equation. The first is the free-particle background function that describes the contributions from the

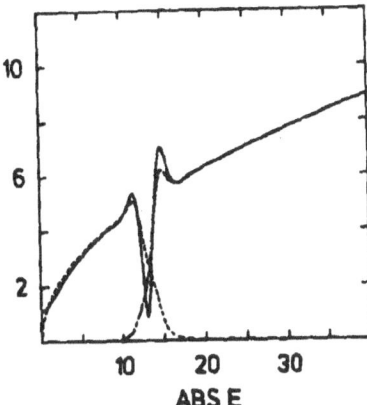

FIG. 5.4. *The absolute value of the generalized imaginary part of the* m^+*-function (solid line), the absolute value of* $\Im mg\ (m^+)$ *(semi-dashed line) and the absolute value of* $\Im mg\ (m^+ - m_{free})$ *(dashed line) as a function of the absolute energy. (Modified from ref)*

bound states and the uncovered resonances while the remainder includes the pole-background interaction. Ω describes the transition point on the rotated continuum branch cut and is found somewhere between the last uncovered pole and the first still covered one on the pole-string.

Is this deflation property now again something that is an artifact of our model or is it something more general?

5.2. The partial wave S-matrix expansion.

5.2.1. The $W_0 r^2 e^{-r}$ **potential** . The poles and the residues of the partial wave S-matrix expansion[16,17] using eq.(2.60) was computed. It was of special interest to evaluate the integral part of eq.(2.60) on the complex dilated branch cut. If $S_\ell(k)$ is small or at least has a smooth variation on the integration path it is easily approximated and little computational effort is needed to evaluate it. If, on the other hand, it has a rapid variation eq.(2.60) has only formal interest since (1) we have not gained any understanding of the structure of the cross section associated with the S-matrix or (2) gained any computational speed in its evaluation. First consider the full partial wave S-matrix on the real axis. The results are displayed in figure 5.5a. The real (as well as the imaginary part) of $S_\ell(k)$ is quickly varying with k. Then consider the analytic continuation of the same function along the a ray in the complex k-plane as in figure 5.5b. It has a much smoother behavior. Only at the point in the complex plane where the eigenvalue trajectory crosses the rotated branch-cut do we find that the dilated partial wave S-matrix clearly deviates from zero. This shows that at least the resonant structure is contained in the sum over the residues in

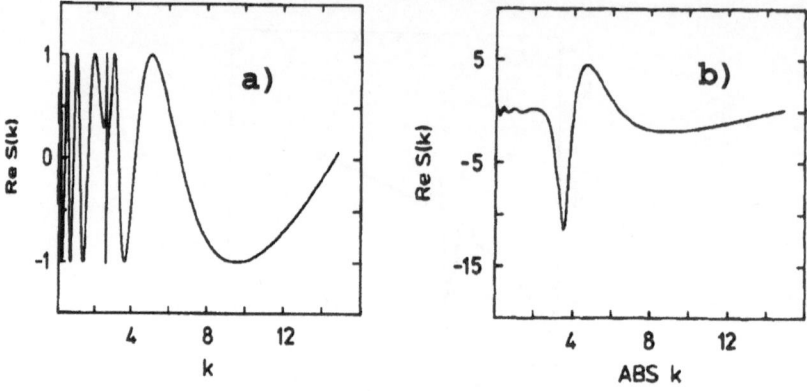

FIG. 5.5. *The (a) real the non-dilated partial wave S-matrix $\ell = 0$. as a function of k. The real (b) part of the same but complex dilated S-matrix along the complex ray $arg(k) = -0.2$. (Modified from ref)*

eq.(2.60). In fact further analysis of this case shows that the main part of the cross section can be attributed to uncovered eigenstates if a sufficient but usually very small number of them have been taken into account.

5.2.2. Shape resonances in negative alkaline earth ions. have now been studied for some time[47]. We made an early contribution in which the alkaline earth atom was described by its MSXα potential to which a parameterized polarization potential and various model exchange potentials were added to describe the single channel potential for the resonant or loosely bound electron. The integral and constant terms in eq.(2.60) are collected to $\mathcal{G}_\ell^{RD}(k)$, which was assumed to vary only slowly, was obtained by doing a full Riccati-Bessel type calculation[25,26,27] at a few points (about 10) and then fit the result minus the resonant contributions to a third order spline. We thus partitioned the now approximate partial wave S-matrix as

$$S_\ell^{RD}(k) = S_\ell^R(k) + \mathcal{G}^{RD}(k)$$

$$(5.5) \qquad = \sum_{j=1}^N \text{Res }(S_\ell(k)) \left[\frac{1}{k - k_j} + \frac{1}{k_j - 0} \right] + \mathcal{G}_\ell^{RD}(k)$$

Using this form we are not only able to decrease the computer time for evaluating a scattering cross section, here by a factor of about 30, but also to identify and understand the resonant structures.

5.3. Applications to multi-channel generalizations. for computation of complex eigenvalues are in principle straight forward [52,56,58,57,59,29,61,63,62] but computationally an order of magnitude more complicated. We have only been able to obtain good numerical accuracy and stability (relative calculated uncertainty of less 10^{-12}) using the

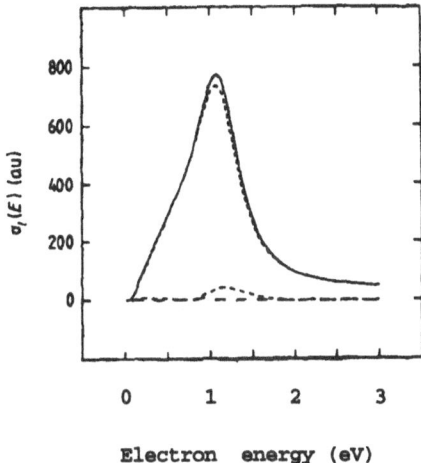

FIG. 5.6. *Exterior complex dilated partial wave S-matrix expansion using eq.(5.5) (Modified from ref)*

finite element method.

5.3.1. Photo fragmentation and predissociation of diatomic molecules.

Vibrational fragmentation and association phenomena in diatomic molecules are the most obvious application for the one-dimensional multi-channel case. The electronic molecular Schrödinger problem is solved with the nuclear coordinates as parameters in the adiabatic Born-Oppenheimer picture[64]. These adiabatic solutions are non-crossing potential energy curves (PEC) for the nuclear problem where the interaction terms between different electronic states are neglected. The neglected interaction terms may be calculated using perturbation theory on the electronic wave functions. Different electronic states represented by different PEC often come close at a number of nuclear geometries causing strong mixing of two or more electronic states. These types of problems are, to our opinion, best treated by transforming the, in principle infinite, set of adiabatic potential energy curves, including the nuclear geometry dependent, interaction terms to so called diabatic PEC where the diagonal terms now may cross each other. This is illustrated in figure 5.7 by our work on the four lowest $^2\Sigma^+$ states in the CaH-radical[61]. Note how the structure of the rovibrational spectrum changes where two or more diagonal potentials come close. The detailed analysis illustrates how the electronically excited states interfere with each other as well as with the ground state vibrational continuum. The width of a level as well as its local rotational frequency is shown to be a clear sign of the dominating character of the level.

The detailed variation is most easily studied in a simple two-channel model like the model for the two lowest $^2\Sigma^+$ electronic states below. Here it is more clear how the real part of an eigenvalue-string follows essentially

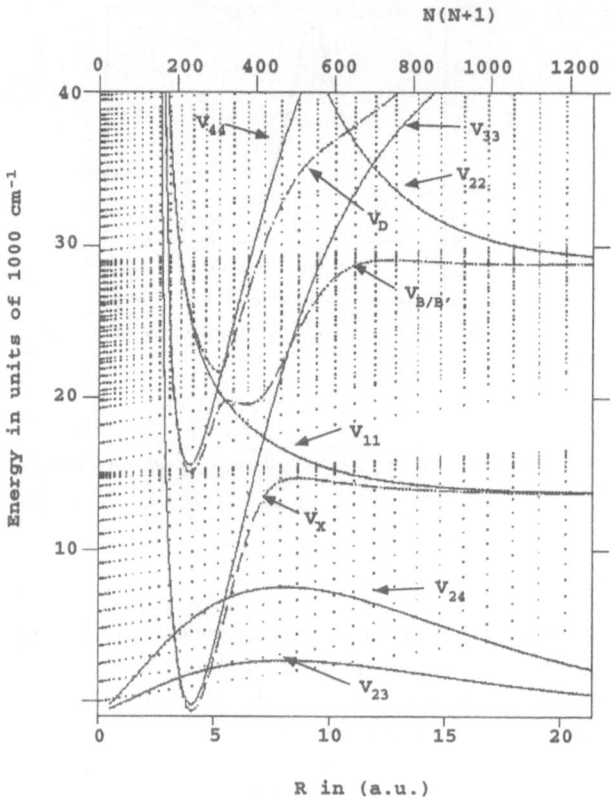

FIG. 5.7. *Adiabatic(dashed) and non-adiabatic potential energy curves (full drawn) for the four lowest $^2\Sigma^+$ states in the CaH-radical and the corresponding rovibrational eigenstates.*

the same kind of pattern as found for single channel cases. The widths do however oscillate regularly as an effect of the interference with the ground state continuum[65,66] In this two-channel study of predissociations of the MgH radical we find that the complex part of the energy is more sensitive to changes in the interaction potential (Figure 5.8). Korsch[56] has derived the resonances spectrum related to the same MgH $^2\Sigma^+$−potentials using semi-classical methods and obtained a similar variation of the poles with the strength of the Gaussian shaped interaction potential. The difference between our[57] complex Riccati study and the results of Korsch is that he found that some vibrational levels split in two with increasing interaction strength. We have so far not confirmed these interesting and somewhat remarkable results by Korsch *et al.*

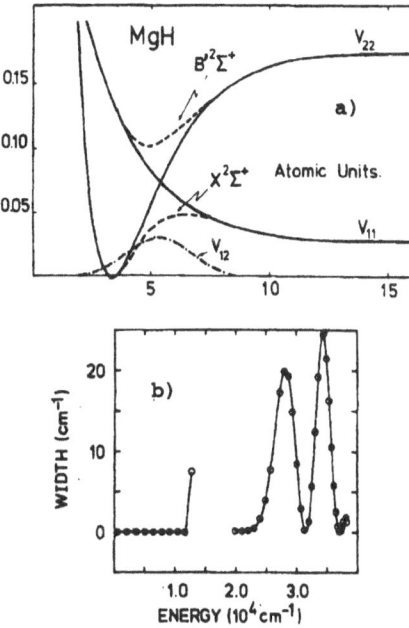

FIG. 5.8. (a) *Adiabatic (dashed) and non-adiabatic(full) potential energy curves and the corresponding* b) *pole string for the two channel models of the MgH radical. (Modified from ref)*

5.3.2. Eigenvalue expansions - Orbiting in ion atom collisions.

Orbiting resonances in a charge exchange collision of N^{3+} and H almost at threshold energy were studied in a two channel model[36,37]. Direct scattering calculations of the S and K-matrices using a Riccati technique [25,26,27] give a cross section rich in resonant structures. Figure 5.10 illustrates a small part of this scattering spectrum. Note that while the eigenvalues marked with arrows do correspond to sharp peaks, this is not true for the ones marked with stars. This illustrated the fact that the scattering contribution from an eigenvalue is given both in the form of its position and its residue. Application generalizations of the expansion theories above are in principle straight forward except for the fact that more than one complex rotated branch cut will appear. A contour integration like in figure 5.9 will then have to follow the first cut out to infinity and then back to the real threshold along the second cut on its "backside" and out again on the "uncovered" side. Engdahl[34] used a basis set founded method in which he approximated the contributions to the partial wave S-matrix from the branch-cut by those of the unphysical eigenstates that follow the rotated cuts.(see figure 5.11) His results indicate that the contributions from these eigenstates is small except possibly close to the points where an eigenvalue string crosses the cut. The fact that a resonant energy level in a multi-channel case may be associated with more than one thresh-

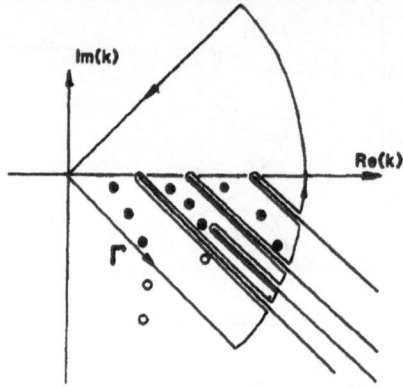

FIG. 5.9. *Suggested integration contour for an S-matrix expansion of a multi-channel problem. Note how the contour has to come back along the dilated cuts. (Modified from ref)*

old energy $U(\infty)$ corresponding to different channels introduces another complication[35].

5.3.3. A non-local potential problem - dynamic effects in small atomic systems.

Dynamic effects like photodetachment, excitation ionization and dieletronic recombination involve multiply excited states of atoms and molecules. The electron-electron interaction is here of the same order of magnitude or in rare cases even larger than the electron nuclear interaction. A thorough treatment of electron correlation effects in these decaying systems is thus needed in order to have an appropriate theoretical treatment of these effects. Lindroth[48] has applied a complex dilated many-body formalism to the method by Salomonson and Öster[50] which is based on the use of a discrete numerical basis. Thereby she is able to theoretically describe dielectronic recombination (DR) involving $n = 2$ and 3 levels in He. The same method was later applied in very accurate work on the photo detachment of H^- and Li^-[47]. The excellent agreement between recent He DR-experiments[49] and the cited many-body calculation is likely due to the fact that in an accurate many-body calculation combined with the complex dilation method one is able to take the electron correlation of the bound and auto-ionizing orbitals into account on an equal footing. In the photo-detachment study Lindroth[47] used the spectral decomposition of the photo-absorption cross section of Rescigno and McKoy[51]

$$(5.6) \quad \sigma(\omega) = \frac{e^2}{4\pi\varepsilon_0} \frac{4\pi}{3} \frac{\omega}{c} \Im\left(\sum_n \frac{< \Psi_0 \sum_j r_j \eta \mid \Psi_n >< \Psi_n \mid \sum_j r_j \eta \mid \Psi_0 >}{E_n - E_0 - \hbar\omega}\right)$$

Following Lindroth each Ψ represents here a correlated many-particle state with a complex energy, $E_n = E^r - i\Gamma/2$. The sum over final states, Ψ_n, is

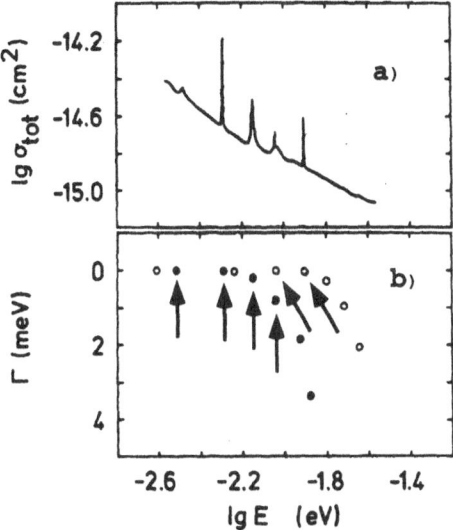

FIG. 5.10. *(a)Part of the very low energy cross section of the process $N^{3+}(1s^2 2s^2)^1 S + H(1s) \rightarrow N^{2+*}((1s^2 2s^2 3s)^2 S + H^+$. (b) Positions of the associated complex resonance poles. (Modified from ref)*

taken over all optically allowed states, continuum states as well as doubly excited states. Eq. 5.6 accounts thus for direct photo detachment as well as auto detachment preceded by photo excitation to a doubly excited state. The initial ground state is denoted by Ψ_0. The usual expression which requires summation over discrete states, Ψ_n, and integration over continuum states is here replaced by a summation over the discretized two electron spectrum. Due to the use of complex rotation, this sum can be carried out directly without any special considerations close to the poles in the energy denominator. Eq. 5.6 is obtained with only a few approximations; a discretized continuum is used, the treatment is non-relativistic and the photon - ion interaction is considered in the lowest order within the dipole approximation.

Eq.(5.6) is based on the analytic continuation of the polarizability obtained as an expanded resolvent describing an atom or molecule in the presence of an electric field. It is in that sense closely related to the formal Green function or S-matrix expansions discussed above. The success of this photo detachment modeling is thus a result of the fact that the electron correlation was properly treated and that the resolvent expansion of the dipole polarizability contains sufficiently many poles and is quickly convergent.

5.3.4. 3-D problems - Laser atom interactions. New effects appear when the interaction between an electron in an atom and an applied laser field become of the same order of magnitude as the electron nucleus

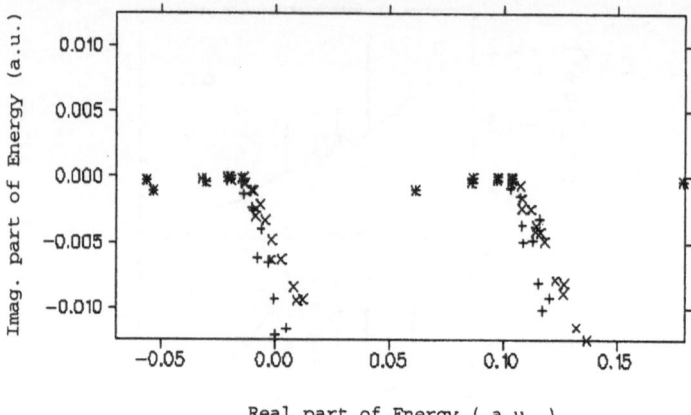

FIG. 5.11. *Eigenvalue spectrum corresponding to the Kramers–Henneberger Floque Hamiltonians below. The true resonant eigenvalues are stable with respect to the dilation angle while the unstable ones correspond to the discretized dilated continua. (Modified from ref)*

interaction. The ionization of Rydberg atoms in strong electro-magnetic fields has recently, in contradiction to what naively may be believed, been found to decrease with increasing field intensity. Using the 3D finite element method recently developed we studied the hydrogen atom in an intense laser field[28]. The system is described by a time dependent Schrödinger equation with a time–periodic potential. In the Kramers–Henneberger frame of reference[67] this equation takes the form

$$(5.7) \qquad \left[-\frac{1}{2}\Delta + \frac{1}{|\vec{r} + \vec{\alpha}_0 \cos \omega t|} \right] \Psi(\vec{r}, t) = i\frac{d}{dt}\Psi(\vec{r}, t)$$

with the laser frequency ω and a field strength $E_0 = \alpha_0 \omega^2$.

A fully numerical solution of this equation using the Floquet ansatz[68] for the wave function

$$(5.8) \qquad \Psi(\vec{r}, t) = e^{-iEt} \sum_{n=-\infty}^{\infty} e^{-in\omega t}\Psi_n(\vec{r}).$$

gives an eigenvalue equation for E

$$(5.9) \qquad \left[-\frac{1}{2}\Delta + V_0 - (E + n\omega) \right] \Psi_{E,n} = - \sum_{\substack{m=-\infty \\ m \neq n}}^{\infty} V_{n-m}\Psi_{E,m}$$

where

$$(5.10) \qquad V_n(\alpha_0, \vec{r}) = \frac{1}{2\pi} \int_{-\pi}^{\pi} \frac{e^{in\phi}}{|\vec{r} - \vec{\alpha}_0 \cos \phi|} d\phi$$

A mathematical proof that the complex dilation method is applicable to this kind of problems was given by Graffi and Yajima[69].

Figure 5.11 illustrates the discretized complex dilated spectrum obtained. The ionization rate is determined from the complex part of the true, stable eigenvalues.

The inverse process of laser-induced ionization - laser induced radiative recombination - was previously studied by Faisal[70]. We the 3D finite element code for the numerical evaluation. The Hamiltonian describing the laser - atom system can be written in terms of eigenstates of the atomic Hamiltonian without the laser and free photons. The coupling matrix element between a system initially consisting of a free electron and and a Rydberg atom and the bound final recombined atom is described within the length gauge of the dipole approximation. The time evolution generated by the Hamiltonian gives us the recombination gain of a final state $|i>$ after a given time. With the electrons in a continuum state $|\vec{k}>$ at $t = 0$, for an single ion and an electron density n_e it is given by

$$(5.11) \qquad P_{i\vec{k}}(t) = n_e| < i|e^{-\imath tH}|\vec{k} > |^2.$$

It is shown how this evolution operator may be expanded in terms of the eigenvalues of an associated effective Hamiltonian as

$$(5.12) \qquad < i|e^{-\imath tH}|\vec{k} >= \sum_{j,l} \frac{c_{il}c_{jl}}{W_l - E_k} V_{i\vec{k}} \left(e^{-\imath tW_l} - e^{-\imath tE_k} \right)$$

where W_l are the complex eigenvalues of the above mentioned effective Hamiltonian and c_{il} are the corresponding eigenvectors.

Thus, also in this case we find how complex eigenvalue or pole expansions are useful in a description of physical phenomena.

6. Conclusion. In this paper I have tried to show the importance of the resonance poles as they appear in analytically continued scattering theory. I have briefly showed how quantities like the Titchmarsh-Weyl $m-$function, the Green function and the S-matrix are analytically continued using the complex dilation technique. It is shown both formally and briefly numerically how the poles of the analytically continues quantities are obtained . It is also briefly shown how spectral expansions of the analytically continued Titchmarsh-Weyl $m-$function, the Green function and the S-matrix as well as the evolution operator were studied and used to express observable quantities like cross sections, rates etc. One can in this way relate an observed peak in a cross section to a number of poles in the associated Titchmarsh-Weyl $m-$function, the Green function or the S-matrix. By calculating the corresponding residues one is also able to quantify the respective contributions.

The poles by themselves are found to appear on pole strings in the complex plane. At most the first few poles in a string are found to contribute to observable structures in cross sections etc. We have classified the rest as background building poles. This may be understood from the fact that the residues that associate with a pole sting also seem to appear in strings in the complex energy or k-plane. The contribution from a pole to a scattering quantity at a real energy is given in terms of the residue divided by the difference of between the associated complex eigenenergy and the particular real energy. It is often is found to be smaller for poles far out in the complex plane than from those close to the real axis. This is not only due to the simple fact that the absolute value of the denominator is larger but also due to the fact that the absolute value of the numerator i.e. the residue is smaller. Based on the study of the potential of Moiseyev *et al.*[46] we have suggested a partition of the poles as structure building ones and background building ones related to the sign change of the real part of the Titchmarsh-Weyl m-function from plus to minus. Does this numerically based suggestion make sense from a more mathematical point of view ?

Our experience, confirmed by others, is that it is not enough to derive a quantity just from the first poles in a string. Rather one has to include a number of them. Numerical comparisons between conventionally computed cross sections derived from the evaluation of for instance the S-matrix on the real axis and the here advocated spectral expansion methods show that only a few poles have to be included in the expansion. The structures in a cross section are this sense to be associated with the complex eigenstates of the Schrödinger problem describing the process. The structures are in that sense a sign of quantization in the continuum.

7. Aknowledgements. The results presented here were mainly obtained within the Stockholm - Uppsala collaboration. I would here like to take the opportunity to thank Dr. Erkki Brändas for his freindship during our common long time efforts. I would also like to thank the four students Magnus Rittby, Peter Krylsted, Erik Engdahl and Pavel Kurasov, who have presented their theses in connection with this work. Dr Armin Scrinzi spent more than two years with me and was mainly reponsible for the FEM work presented here. I am also grateful to Dr. Bernard Piraux for introducing me into the mysteries of laser-atom interation. All of you have helpt me to understand and develop this subject and I am very grateful to you.

I am also grateful to Barry Simon and Donald Truhlar and the IMA staff for the organization of the scientifically and socially nice workshop where I had the opportunity to present this paper.

REFERENCES

[1] G. DOOLEN, M. HIDALGO, J. NUTTALL AND R. STAGAT, *in Atomic Physics, S.J. Smith and G.K. Walters,Eds.*, (Plenum, New York, 1973), Vol. 3, p. 257.
G. DOOLEN, J. NUTTALL AND R.W. STAGAT, *Phys. Rev.A*, Vol. 10, 1612 (1974).
G. DOOLEN, J. NUTTALL AND C.J. WHERRY, Phys. Rev. Lettr., Vol. 40, 313

(1978).

J. NUTTALL AND H.L. COHEN, Phys. Rev., Vol. 188, 1542 (1969).

[2] F. AGUILAR AND J.M. COMBES, Commun. Math. Phys., Vol. 22, 269 (1971).

E. BALSLEV AND J.M. COMBES, Commun. Math. Phys., Vol. 22, 280 (1971).

E. BALSLEV, Commun. Math. Phys., Vol. 52, 127 (1977).

[3] B. SIMON, Commun. Math. Phys., Vol. 27, 1 (1972).

B. SINOM, Ann. Math., Vol. 97, 247 (1973).

[4] C. VAN WINTER, J. Math. Anal. Appl. Vol. 47,633 (1974).

-ibid. Vol. 48, 368 (1974).

ibid, Vol. 49,88 (1975).

[5] WHOLE ISSUE OF INTERNAT. J. QUANTUM CHEM., VOL. 14, Complex Scaling in the Spectral Theory of the Hamiltonian, 1978.

[6] K. CHARDAN AND P.C. SABATIER, Inverse Problems in Scattering Theory, (Springer Verlag, New York, 1977).

[7] E.A. CODDINGTO AND N. LEVINSON, Theory of Ordinary Differential Equations, (McGraw-Hill, New York, 1955).

[8] E.C. TITCHMARSH, Eigenfunction Expansion Associated with Second-Order Differential Equations, (Oxford Univ. Press 1946).

[9] H. WEYL, Math. Annl., Vol. 68, 220 (1910).

[10] E. BRÄNDAS, M. RITTBY AND N. ELANDER, J. Math. Phys., Vol. 26,2648 (1985).

[11] E. ENGDAHL, E. BRÄNDAS, M. RITTBY AND N. ELANDER , J. Math. Phys., Vol. 27, 2629 (1986).

[12] E. ENGDAHL, E. BRÄNDAS, M. RITTBY AND N. ELANDER, Phys. Rev.A, Vol. 37, 3777 (1988).

[13] M. HEHENBERGER, B. LASKOWSKI, AND E. BRÄDAS, J. Chem. Phys., Vol. 65, 4559 (1976).

[14] M. RITTBY, N.ELANDER AND E. BRÄNDAS, Internat. J. Quantum Chem Symp., Vol. 23, 865 (1983).

[15] P.B. KURASOV, A. SCINZI AND N.ELANDER, Phys. Rev. A, Vol. 49, 5095 (1994).

[16] M. RITTBY, N. ELANDER AND E. BRÄNDAS, in : Resonances, E. Brändas and N. Elander Ed., Lecture Notes in Physics, Vol. 325, (Springer, Berlin, 1989), p. 129.

[17] P. KRYLSTEDT, C. CARLSUND AN N. ELANDER, J. Phys. B, Vol 22, 1051 (1989).

[18] B.SIMON,Physics Letters, Vol. 71A,211 (1979).

[19] M. HEHENBERGER, H.V. MCINTOSH AND E. BRÄNDAS, J. Chem. Phys., Vol. 65, 4559 (1976).

[20] K. KODAIRAAm. J. Math., Vol. 71, 921 (1949).

[21] B. SIMON, Int. J. Quant. Chem., Vol. 14, 529 (1978);

WM. REINHARDT, Ann. Rev. Phys. Chem., Vol. 33, 223 (1982).

[22] M.REED AND B.SIMON, Methods of Modern Mathematical Physics, (Academic Press, New York, 1975).

[23] N.LIPKIN, N.MOISEYEV, AND E.BRÄNDAS, Phys.Rev.A, Vol.40, 549 (1989).

[24] P.B.KURASOV, PH.D. THESIS, Dept. Math., Stockholm University, Sweden, 1993.

[25] B.R. JOHNSON,J. Comp. Phys., Vol. 13, 445 (1973).

[26] B.R. JOHNSON, J. Chem. Phys., Vol. 67, 4086 (1977).

[27] B.R. JOHNSON, J. Chem. Phys., Vol. 13, 4678 (1978).

[28] A.SCRINZI, N.ELANDER, AND B. PIRAUX, Phys. Rev.A, 48, R2527, (1993).

[29] A.SCRINZI AND N.ELANDER, J. Chem.Phys., Vol. 98, 3866 (1993).

[30] M. REED AND B. SIMON Methods of Modern Mathematical Physics, (Academic, New York, 1982), Vol.IV, p.183 ff.

[31] R. NEWTON, Scattering theory of particles and waves, (Springer Verlag Heidelberg, New York, 1982).

[32] N.I. AKHIEZER, The classical moment problem, (Oliver and Boyd, Edingburgh, 1965).

[33] W.N. EVERITT AND C. BENNERWITZ, in " Tribute to Åke Pleijel", Proceedings of the Conference 1979, (Dept of Math. Uppsala University, 1980).

[34] E. ENGDAHL, *Private communication*.

[35] U. PESKIN, N. MOISEYEV AND R. LEFEBVRE, *J. Chem. Phys.*, Vol. 92, 2902 (1990). N. MOISEYEV AND U. PESKIN, *Phys. Rev. A*, Vol. 42, 255 (1990).

[36] A. BARANY, E. BRÄNDAS, N. ELANDER, M. RITTBY, *Physica Scripta*, Vol. T3, 233 (1983), RITTBY, N. ELANDER, E. BRÄNDAS, AND A. BARANY, *J. Phys. B* Vol. 17, L667 (1984).

[37] M. RITTBY, *Thesis, Department of Physics University of Stockholm*, 1985.

[38] C. BOTTCHER, *Adv. At. Mol. Phys.*, Vol. 20 (1985) 241.

[39] J. LINDERBERG , *in Finite Elements in Physics*, P. Gruber, Ed. (North Holland, Amsterdam, 1987), p. 209.

[40] D. HEINEMANN, B. FRICKE, AND D. KOLB *Phys. Rev. A* Vol. 38, 4994 (1988).

[41] J. OLSEN, P. JORGENSEN, AND J. SIMONS, CHEM. PHYS. LETT., Vol. 169, 463 (1990).

[42] D. SUNDHOLM AND J. OLSEN, *Chem. Phys. Lett.*, Vol. 171,53 (1990).

[43] A. SCRINZI, *Computer Physics Commun.*, Vol. 86, 67 (1995).

[44] , A. JENSEN*Private Communication*.

[45] R.A. BAIN, J.N. BARDSLEY, B.R. JUNKER, C.V. SUKUMAR, *J. Phys. B* , Vol. 7, 2189 (1974).

[46] N. MOISEYEV, P.R. CERTAIN AND F. WEINHOLD, *Molec. Phys.*, Vol. 36, 1613 (1978).

[47] E. LINDROTH, *Phys. Rev. A* , Vol. 52, 2737 (1995).

[48] E. LINDROTH, *Phys. Rev. A*, Vol. 49, 4473 (1994).

[49] D.R. DEWITT ET AL., *J. Phys. B*, Vol. 28, L147 (1995).

[50] S. SALOMONSON AND P. ÖSTER , *Phys. Rev. A*, Vol. 40, 5559 (1989).

[51] T.N. RESCIGNO AND V. MCKOY, *Phys. Rev. A*, Vol. 12, 522 (1975).

[52] J. TURNER AND C.W.MCCURDY, *Chem. Phys.*, Vol. 71, 127 (1982).

[53] O. ATABEC, R. LEFEBVRE AND A. REQUENA, *Molec. Phys.*, Vol. 40, 1107 (1980).

[54] O. ATABEC AND R. LEFEBVRE, *Chem. Phys. Lett.*, Vol. 84, 233 (1982).

[55] H.J. KORSCH AND H. LAURENT, *J. Phys. B*, Vol. 14, 4213 (1982). H.J. KORSCH, H. LAURENT AND R. MÖHLENKAMP, *J. Phys. B*, Vol. 15, 1 (1982). H.J. KORSCH, H. LAURENT AND R. MÖHLENKAMP, *Molec. Phys.* , Vol. 43, 1441 (1981).

[56] H.J. KORSCH, *in : Resonances, E. Brändas and N. Elander Ed.*, Lecture Notes in Physics, Vol. 325, (Springer, Berlin, 1989). p. 253.

[57] E. BRÄNDAS, M. RITTBY AND N. ELANDER, *in : Resonances, E. Brändas and N. Elander Ed.* Lecture Notes in Physics, Vol. 325, (Springer, Berlin, 1989). p. 345.

[58] R. LEFEBVRE , *in : Resonances, E. Brändas and N. Elander Ed.* Lecture Notes in Physics Vol. 325, (Springer, Berlin, 1989), p. 313.

[59] M.A. NATIELLO AND A.R. ENGELMANN, *in : Resonances, E. Brändas and N. Elander Ed.*, Lecture Notes in Physics Vol. 325, (Springer, Berlin, 1989). p. 329.

[60] A. SCRINZI, N. ELANDER AND A. WOLF, *Zeitschrift für Physik* Vol. D34,185 (1995).

[61] A. NUNEZ, C. CARLSUND-LEVIN, A. SCRINZI, N. ELANDER AND H. MARTIN , *An Exterior Complex Rotated Close-coupling Description of Predissociation in Diatomic Molecules applied to a model of the four lowest $^2\Sigma^+$ states in CaH*, Manuscript.

[62] C. CARLSUND-LEVIN, B. LINDBERG AND N.ELANDER, *An exterior Complex dilated Cowell (Numerov) integration method, Part 1. - Single Channel Problems* Prel. Manuscript.
-ibid carlsunddefc*An exterior Complex dilated Cowell (Numerov) integration method, Part 2. - Applications to a set of coupled Schrödinger type Problems*, Prel. Manuscript.

[63] CHU , This Volume.

[64] R.T. PACK AND J.O. HIRSCHFELDER*J.Chem.Phys.*, Vol. 52, 521 (1970). R.T. PACK AND J.O. HIRSCHFELDER, *J.Chem.Phys*, Vol 49,4009 (1970).

[65] M.S. CHILDS AND R. LEFEBVRE, *Chem. Phys. Lettr.*, Vol. 55, 213 (1978).

[66] J.M. ROBBE, *Thesis, L'Universite des Science et Tech. de Lille* (1978).

[67] W.C. HENNEBERGER, *Phys. Rev. Lett.*, Vol. 21, 838 (1968).

[68] M. Gavrila and J. Kaminski, *Phys.Rev.Lett.*, Vol. 52, 614 (1984).

[69] S. GRAFFI AND K. YAJIMA, *Commun. Math. Phys.*, Vol. 80, 277 (1983).

[70] F.H.M.FAISAL, A. LAMI, AND N.K.RAHMAN, *J. Phys. B*, Vol. 14, L569 (1981).
   A.LAMI, N.K.RAHMAN, AND F.H.M.FAISAL, *Phys. Rev. A*, Vol. 30, 2433 (1984).

[71] F.H.M.FAISAL, A. LAMI, AND N.K.RAHMAN *J. Phys. B*, Vol. 14, L569 (1981).

MICROSCOPIC ATOMIC AND NUCLEAR MEAN FIELDS

CLAUDE MAHAUX*

Abstract. The single-particle properties of the electron-atom and nucleon-nucleus systems can be accurately reproduced with the help of a one-body "mean field." It is shown that many one-body potentials exist which all reproduce the single-particle elastic scattering wave function. Among these potentials, the self-energy plays a privileged role, because it also yields information on the single-particle properties of the bound states.

Key words. Atomic and nuclear mean fields. Optical-model potential.

1. Introduction. Accurate fits to the electron-atom and nucleon-nucleus elastic scattering cross sections can be obtained by replacing the projectile-target interaction by a complex one-body "optical-model potential." [1-5] In phenomenological analyses, this potential is assumed to be local,

$$(1.1) \qquad M(r; E) = V(r; E) + iW(E),$$

where E denotes the kinetic energy of the projectile. Good fits to the experimental data are obtained with potentials whose radial shape and energy dependence are simple and vary smoothly with the number of particles in the target. When these properties are required, fits to the experimental data accurately determine the phenomenological optical-model potential. The latter is thus often considered as being an "observable."

Besides expressions of the optical-model potential which have been found in the framework of approximation schemes (e.g. the eikonal approximation [6] or the multiple scattering expansion [7, 8]) two "exact" microscopic expressions have been derived. One is due to Feshbach [9-11], and the other to Bell and Squires [12, 13] and Namiki [14]. These two expressions are different. In particular, they do not have the same analytical properties in the complex energy plane. It is thus necessary to discuss which one of these expressions is most closely related to the phenomenological optical-model potential. This problem became even more acute after the realization that, actually, an infinite number of potentials exist which all yield exactly the same elastic scattering wave function [15]. These topics are surveyed in the present paper, which is mainly based upon recent work done in collaboration with Franco Capuzzi [16, 17].

2. Generalized optical-model Hamiltonians.

2.1. Elastic single-particle wave function. We assume that the target particles have spin 1/2 and mass m, and are all identical to the incident projectile. In the nuclear case, the difference between protons and neutrons can be handled by means of an isospin quantum number. In

* Institut de Physique B5, Université de Liège, Sart Tilman, 4000 Liège 1, Belgium.

the atomic case, the effect of the central nucleus is taken into account by adding a central Coulomb field to the expressions derived below. We shall omit reference to the spin and isospin degrees of freedom. The target is the ground eigenstate of the A-particle Hamiltonian,

$$(2.1) \qquad H^{(A)} \Psi_0^{(A)}(\xi_A) = \mathcal{E}_0^{(A)} \Psi_0^{(A)}(\xi_A),$$

where ξ_A generically denotes the coordinates $\mathbf{r}_1, ..., \mathbf{r}_A$. The combined (projectile + target) system is an eigenstate of the (A+1)-particle Hamiltonian,

$$(2.2) \qquad H^{(A+1)} \Psi_\mathbf{k}^{(A+1)}(\xi_{A+1}) = (E + \mathcal{E}_0^{(A)}) \Psi_\mathbf{k}^{(A+1)}(\xi_{A+1}),$$

where $\xi_{A+1} = \mathbf{r}_1, ..., \mathbf{r}_A, \mathbf{r}_{A+1}$. The index $\mathbf{k}$ refers to the asymptotic boundary condition that the incident projectile has momentum $\mathbf{k}$ and energy

$$(2.3) \qquad E = E_k = k^2/2m.$$

Since we deal with identical particles, it is convenient to use the second quantization formalism. The Lippmann-Schwinger equation can be written as [6],

$$(2.4) \qquad \left| \Psi_\mathbf{k}^{(A+1)} \right\rangle = \frac{i\eta}{E_k - (H^{(A+1)} - \mathcal{E}_0^{(A)}) + i\eta} a_\mathbf{k}^\dagger \left| \Psi_0^{(A)} \right\rangle.$$

Here and below, the limit $\eta \to +0$ is implicit. The "elastic single-particle wave function" is defined as the projection of $\left| \Psi_\mathbf{k}^{(A+1)} \right\rangle$ upon $\left| \Psi_0^A \right\rangle$,

$$(2.5) \qquad \chi_\mathbf{k}(\mathbf{r}_{A+1}) = \sqrt{A+1} (\Psi_0^{(A)}(\xi_A) \left| \Psi_\mathbf{k}^{(A+1)}(\xi_{A+1}) \right\rangle,$$

where $(\,|\,)$ denotes an integration over the coordinates ξ_A. Equivalently,

$$(2.6) \qquad \chi_\mathbf{k}(\mathbf{r}) = \left\langle \Psi_0^{(A)} \left| a(\mathbf{r}) \right| \Psi_\mathbf{k}^{(A+1)} \right\rangle.$$

By inserting (2.4) in (2.6), one obtains

$$(2.7) \qquad \chi_\mathbf{k}(\mathbf{r}) = \left\langle \Psi_0^{(A)} \left| a(\mathbf{r}) \frac{i\eta}{E_k - (H^{(A+1)} - \mathcal{E}_0^{(A)}) + i\eta} a_\mathbf{k}^\dagger \right| \Psi_0^{(A)} \right\rangle.$$

The asymptotic behaviour of $\chi_\mathbf{k}(\mathbf{r})$ at large distance yields the elastic scattering phase shift, and thus the elastic scattering cross section. The value of $\chi_\mathbf{k}(\mathbf{r})$ at finite distance is also of interest, in particular for the description of nonelastic processes in the framework of the distorted wave Born approximation [5, 11].

2.2. Bound overlap functions. Besides scattering eigenstates, the $(A+1)$-particle Hamiltonian also has bound eigenstates,

$$(2.8) \qquad H^{(A+1)}\Psi_\lambda^{(A+1)}(\xi_{A+1}) = (E_\lambda^{(+)} + \mathcal{E}_0^{(A)})\Psi_\lambda^{(A+1)}(\xi_{A+1}),$$

$$(2.9) \qquad \left\langle \Psi_\lambda^{(A+1)} \middle| \Psi_{\lambda'}^{(A+1)} \right\rangle = \delta_{\lambda\lambda'}.$$

The "overlap functions" are defined as

$$(2.10) \qquad \chi_\lambda^{(+)}(\mathbf{r}) = \left\langle \Psi_0^{(A)} \middle| a(\mathbf{r}) \middle| \Psi_\lambda^{(A+1)} \right\rangle.$$

They play an important role in the description of one-nucleon transfer processes [18]. In nuclear physics, the square of their norm is called the "spectroscopic factor." In atomic physics, it is related to the "spectroscopic intensity."

2.3. Requirements. The "generalized" optical-model Hamiltonian is required to have the elastic scattering wave function as eigenstate,

$$(2.11) \qquad h(E)|\chi_\mathbf{k}\rangle = E|\chi_\mathbf{k}\rangle.$$

In general, $h(E)$ is nonlocal in the spatial coordinates, so that a more explicit form of (2.11) is,

$$(2.12) \qquad \int d\mathbf{r}' h(\mathbf{r}, \mathbf{r}'; E)\chi_\mathbf{k}(\mathbf{r}') = E\chi_\mathbf{k}(\mathbf{r}).$$

In practice, the "generalized" optical-model Hamiltonian is also required to have the bound overlap functions $\left|\chi_\lambda^{(+)}\right\rangle$ as eigenstates,

$$(2.13) \qquad h(E_\lambda^{(+)})\left|\chi_\lambda^{(+)}\right\rangle = E_\lambda^{(+)}\left|\chi_\lambda^{(+)}\right\rangle.$$

Note that, in this equation, the Hamiltonian is evaluated at the eigenvalue $E_\lambda^{(+)}$. It is in that sense that we shall use the expression "eigenvalue" below. In nuclear physics, $E_\lambda^{(+)}$ is called the "single-particle energy." In atomic physics, it is called the "electron attachment energy" and is closely related to the electron "affinity."

The word "generalized" refers to the fact that $h(E)$ yields the exact value of the transition matrix at the energy E [9]. The optical-model Hamiltonian proper only yields the energy average of the transition matrix. It can be identified with

$$(2.14) \qquad \tilde{h}(E) = h(E + i\Delta),$$

where Δ measures the size of the averaging interval [11, 19-21]. In the following, we shall drop the word "generalized," and call $h(E)$ "the optical-model Hamiltonian," without further ado. Correspondingly, we shall identify the "optical-model potential" with the difference,

$$(2.15) \qquad M(E) = h(E) - \tau.$$

where τ is the one-body kinetic energy operator.

3. Particle-type generalized optical-model Hamiltonians.

3.1. Definition of the particle Hamiltonian. The "particle" one-body Green's function $G^{(p)}(E)$ is defined as,

$$(3.1) \quad \langle \mathbf{r} | G^{(p)}(E + i\eta) | \mathbf{k} \rangle = \left\langle \Psi_0^{(A)} \left| a_{\mathbf{r}} \frac{1}{E - (H^{(A+1)} + \mathcal{E}_0^{(A)}) + i\eta} a_{\mathbf{k}}^\dagger \right| \Psi_0^{(A)} \right\rangle.$$

The comparison between (2.7) and (3.1) yields,

$$(3.2) \qquad \langle \mathbf{r} | \chi_{\mathbf{k}} \rangle = i\eta \left\langle \mathbf{r} \left| G^{(p)}(E + i\eta) \right| \mathbf{k} \right\rangle.$$

The "particle" Hamiltonian $h^{(p)}(E)$ is the operator of which $G^{(p)}(E + i\eta)$ is the resolvent [16, 17],

$$(3.3) \qquad \left[E - h^{(p)}(E) \right] G^{(p)}(E + i\eta) = 1.$$

$$(3.4) \qquad h^{(p)}(E) = E - \left[G^{(p)}(E + i\eta) \right]^{-1}.$$

The "particle" potential is the difference

$$(3.5) \qquad M^{(p)}(E) = h^{(p)}(E) - \tau.$$

3.2. Properties of the particle Hamiltonian. From (3.1) and (3.3), one readily verifies that $h^{(p)}(E)$ is an optical-model Hamiltonian, i.e. that it fulfills the requirements (2.11) and (2.13),

$$(3.6) \qquad h^{(p)}(E_k) | \chi_{\mathbf{k}} \rangle = E_k | \chi_{\mathbf{k}} \rangle,$$

$$(3.7) \qquad h^{(p)}(E_\lambda^{(+)}) \left| \chi_\lambda^{(+)} \right\rangle = E_\lambda^{(+)} \left| \chi_\lambda^{(+)} \right\rangle.$$

Its main properties are the following:

(a) It is symmetric in the coordinate space representation,

$$(3.8) \qquad h(\mathbf{r}, \mathbf{r}'; E) = h(\mathbf{r}', \mathbf{r}; E).$$

This property is important because, if it did not hold, one would not know how to construct a local potential which would be approximately "equivalent" to $M^{(p)}(\mathbf{r}, \mathbf{r}'; E)$ [22]. Therefore, one would not be able to compare $M^{(p)}(\mathbf{r}, \mathbf{r}'; E)$ to empirical optical-model potentials.

(b) It diverges at large energy,

$$(3.9) \qquad h^{(p)}(E) \sim -\frac{K}{1-K}E,$$

where K denotes the one-body density matrix related to the target,

$$(3.10) \qquad K_{\mathbf{rr}'} = \left\langle \Psi_0^{(A)} \left| a_{\mathbf{r}'}^\dagger a_{\mathbf{r}} \right| \Psi_0^{(A)} \right\rangle.$$

The divergence (3.9) is at variance with the fact that, at large energy, the empirical optical-model Hamiltonians approach the kinetic energy plus, possibly, a finite "static" potential.

(c) The eigenvalues of $h^{(p)}(E)$ fulfill the inequality,

$$(3.11) \qquad E_\lambda^{(+)} \geq E_F^{(+)},$$

where $E_F^{(+)}$ is the "Fermi energy,"

$$(3.12) \qquad E_F^{(+)} = \mathcal{E}_0^{(A+1)} - \mathcal{E}_0^{(A)}.$$

The inequality (3.11) is a direct consequence of (3.1) and (3.3). It implies that $h^{(p)}(E)$ is quite different from phenomenological optical-model Hamiltonians. Indeed, the latter have eigenvalues smaller than $E_F^{(+)}$. For instance, $E_F^{(+)} \approx -10 MeV$ in the nuclear case, and empirical nuclear optical-model Hamiltonians have eigenvalues $E_\lambda^{(+)}$ as small as $-40 MeV$. In the atomic case, $h^{(p)}(E)$ may have eigenvalues smaller than $E_F^{(+)}$ because of the contribution of the Coulomb potential created by the atomic nucleus. However, these "deeply-bound" eigenvalues have no physical meaning.

3.3. Definition of Feshbach's optical-model Hamiltonian. Let us multiply (3.6) by $(1 - N)$, where N is an arbitrary energy-independent operator. The resulting equation reads,

$$(3.13) \qquad h_N^{(p)}(E_k) |\chi_{\mathbf{k}} \rangle = E_k |\chi_{\mathbf{k}} \rangle,$$

where,

$$(3.14) \qquad h_N^{(p)}(E) = (1 - N)h^{(p)}(E) + EN.$$

Likewise, (3.7) gives

$$(3.15) \qquad h_N^{(p)}(E_\lambda^{(+)}) \left| \chi_\lambda^{(+)} \rangle = E_\lambda^{(+)} \left| \chi_\lambda^{(+)} \rangle.$$

Therefore, $h_N^{(p)}(E)$ is also an optical-model Hamiltonian, for any value of the operator N. The operator obtained by setting $N = K$ reads,

$$(3.16) \qquad h_K^{(p)}(E) = (1 - K)h^{(p)}(E) + EK.$$

It can be shown [16] that $h_K^{(p)}(E)$ is identical to the optical-model Hamiltonian which had been introduced by Feshbach in the framework of his projection operator theory [10]. The latter is outlined in Section 6.

3.4. Properties of Feshbach's optical-model Hamiltonian. (a) In contrast to $h^{(p)}(E)$, Feshbach's Hamiltonian $h_K^{(p)}(E)$ is not symmetric in the coordinate space representation, because of the asymmetric appearance of one factor $(1 - K)$ on the right-hand side of (3.16). This hinders the construction of a local Hamiltonian which would be "equivalent" to Feshbach's Hamiltonian. (b) According to (3.9), $h_K^{(p)}(E)$ is finite at large energy. However, its high-energy limit differs from the kinetic energy operator [16]. (c) As indicated by (3.15), the discrete eigenvalues of $h_K^{(p)}(E)$ are identical to those of $h^{(p)}(E)$. This is a consequence of the relation

$$(3.17) \qquad \frac{1}{E - h_K^{(p)}(E)} = (1 - K)^{-1}\frac{1}{E - h^{(p)}(E)}.$$

This implies that $h_K^{(p)}(E)$ has no eigenvalue smaller than $E_F^{(+)}$. Therefore, it is also very different from empirical optical-model Hamiltonians.

4. Self-energy.

4.1. Hole Green's function. The factor η which appears on the right-hand side of (3.2) indicates that, in that relation, one can replace $G^{(p)}(E + i\eta)$ by

$$(4.1) \qquad \tilde{G}(E + i\eta) = G^{(p)}(E + i\eta) + C(E),$$

where $C(E)$ denotes any operator which is regular in a domain containing the positive real axis. For instance, one can set $C(E)$ equal to the so-called "hole" Green's function,

$$(4.2) \quad \langle \mathbf{r}|G^{(h)}(E + i\eta)|\mathbf{k}\rangle = \left\langle \Psi_0^{(A)} \left| a_\mathbf{r}^\dagger \frac{1}{E + (H^{(A-1)} - \mathcal{E}_0^{(A)}) + i\eta} a_\mathbf{k} \right| \Psi_0^{(A)} \right\rangle,$$

where $H^{(A-1)}$ is the $(A - 1)$-particle Hamiltonian.

4.2. Retarded Green's function. The "retarded" one-body Green's function is the sum of the "particle" and "hole" Green's functions,

$$(4.3) \qquad G(E + i\eta) = G^{(p)}(E + i\eta) + G^{(h)}(E + i\eta).$$

The corresponding "self-energy" Hamiltonian $h(E)$ is defined by; see (3.3) and (3.5),

$$(4.4) \qquad [E - h(E)]G(E + i\eta) = 1.$$

4.3. Eigenstates and eigenvalues. Since $G^{(h)}(E + i\eta)$ is analytic for $E > E_F^{(+)}$, the one-body Hamiltonian $h(E)$ defined by (4.4) fulfills the requirements (2.11) and (2.13). In addition, it can readily be checked that

$$(4.5) \qquad\qquad h(E_\mu^{(-)}) \left| \chi_\mu^{(-)} \right\rangle = E_\mu^{(-)} \left| \chi_\mu^{(-)} \right\rangle ,$$

where $\left| \chi_\mu^{(-)} \right\rangle$ is the "overlap function" associated with a bound eigenstate of the $(A-1)$-particle Hamiltonian,

$$(4.6) \qquad\qquad \chi_\mu^{(-)}(\mathbf{r}) = \left\langle \Psi_\mu^{(A-1)} \left| a(\mathbf{r}) \right| \Psi_0^{(A)} \right\rangle ,$$

$$(4.7) \qquad H^{(A-1)} \Psi_\mu^{(A-1)}(\xi_{A-1}) = \left(\mathcal{E}_0^{(A)} - E_\mu^{(-)} \right) \Psi_\mu^{(A-1)}(\xi_{A-1}),$$

$$(4.8) \qquad\qquad \left\langle \Psi_\mu^{(A-1)} \left| \Psi_{\mu'}^{(A-1)} \right. \right\rangle = \delta_{\mu\mu'}.$$

4.4. Properties of the self-energy. The "self-energy" is the difference between $h(E)$ and the kinetic energy operator,

$$(4.9) \qquad\qquad M(E) = h(E) - \tau.$$

It has the following main properties: (a) The self-energy is symmetric in the coordinate space representation,

$$(4.10) \qquad\qquad M(\mathbf{r}, \mathbf{r}'; E) = M(\mathbf{r}', \mathbf{r}; E).$$

Therefore, a local potential equivalent to $M(E)$ can constructed, provided that the spatial nonlocality of $M(E)$ is sufficiently simple. (b) The self-energy remains finite at large energy,

$$(4.11) \qquad\qquad M(E) \sim V^{(s)},$$

where the "static part" of the self-energy has the simple form,

$$(4.12) \qquad\qquad V_{\alpha\alpha'}^{(s)} = \sum_{\beta\beta'} \langle \alpha\beta |v| \alpha'\beta' - \beta'\alpha' \rangle K_{\beta\beta'}.$$

Here, v denotes the two-body interaction, and $\{\alpha\}$ is an arbitrary single-particle basis. Expression (4.12) can be viewed as an extension of the Hartree-Fock potential. It is routinely evaluated in self-consistent calculations of the ionization spectra and electron affinities of molecules [23]. (c) The self-energy Hamiltonian has eigenvalues smaller than $E_F^{(+)}$, namely at the energies $E_\mu^{(-)}$ associated with the bound states of the $(A-1)$-particle system; see (4.7). In the atomic case, these are called the "ionization energies;" they are measured in $(e, 2e)$ reactions, for instance [24]. In the nuclear case, they are called the "separation energies"; they are observed by means of $(e, e'p)$ "knockout" reactions [25]. (d) The intensity of the excitation of the bound states $\left| \Psi_\mu^{(A-1)} \right\rangle$ by means of $(e, 2e)$ or $(e, e'p)$ reactions on the target $\left| \Psi_0^A \right\rangle$ is measured by the "spectroscopic factor." The latter is the square of the norm of the overlap function :

$$(4.13) \qquad\qquad S_\mu^{(-)} = \left\langle \chi_\mu^{(-)} \middle| \chi_\mu^{(-)} \right\rangle .$$

It is given by [16],

$$(4.14) \qquad S_\mu^{(-)} = 1 + \left[\left\langle \chi_\mu^{(-)} \middle| \frac{d}{dE} M(E) \middle| \chi_\mu^{(-)} \right\rangle \right]_{E=E_\mu^{(-)}} .$$

This formula also applies to the spectroscopic factor associated with $\left| \Psi_\lambda^{(A+1)} \right\rangle$, which can be measured by means of one-nucleon transfer reactions on $\left| \Psi_0^A \right\rangle$

5. Dispersion relations.

5.1. Particle Hamiltonian. The "particle" Hamiltonian $h^{(p)}(E)$ is the limit of an operator $h^{(p)}(z)$ which is analytic in the upper-half on the complex z-plane. Its real and imaginary parts are thus connected by a dispersion relation. The latter reads,

$$Re\, h^{(p)}(E) = -\frac{K}{1-K} E + \frac{1}{1-K} C^{(p)} \frac{1}{1-K} + \frac{P}{\pi} \int_{E_F^{(+)}}^\infty dE' \frac{Im\, h^{(p)}(E')}{E' - E},$$
$$(5.1)$$

where P denotes a principal value, and the energy-independent operator $C^{(p)}$ is given by,

$$(5.2) \qquad\qquad C_{\alpha\alpha'}^{(p)} = \left\langle \Psi_0^{(A)} \middle| [a_\alpha, H]\, a_{\alpha'}^\dagger \middle| \Psi_0^{(A)} \right\rangle .$$

Approximate forms of (5.1) have been used in nuclear [26] and in atomic [27] physics. Three features should be noted. (a) The first term on the right-hand side of (5.1) diverges at large energy, in keeping with (3.9). (b) The energy-independent contribution to $h^{(p)}(E)$, i.e. the second term on the right-hand side of (5.1), is not related to the kinetic energy operator and to the Hartree-Fock potential. (c) In (5.1), the dispersion integral extends from $E_F^{(+)}$ to ∞ . Actually, $Im\, h^{(p)}(E)$ vanishes for E smaller than the lowest nonelastic threshold $T_{in}^{(+)} = \mathcal{E}_1^{(A)} - \mathcal{E}_0^{(A)}$, except at isolated energies. Therefore, the integral in (5.1) actually corresponds to a sum over pole contributions for $E' < T_{in}^{(+)}$ and to an integral running from $T_{in}^{(+)} > 0$ to ∞ [17].

5.2. Self-energy. The self-energy $M(E)$ is the limit of an operator $M(z)$ which is analytic in the upper-half of the complex z-plane. Its real and imaginary parts are thus connected by a dispersion relation,

$$(5.3) \qquad Re\, M(E) = V^{(s)} + \frac{P}{\pi} \int_{-\infty}^{\infty} dE' \frac{Im\, M(E')}{E' - E}.$$

The following differences with (5.1) are noticeable. (a) The dispersion relation (5.3) relates the real and imaginary parts of a potential operator, while (5.1) connects the real and imaginary parts of a Hamiltonian. This reflects the fact that the kinetic energy operator appears naturally in the expression of the self-energy Hamiltonian, but not in $h^{(p)}(E)$. (b) In contrast to (5.1), the right-hand side of (5.3) contains no linearly energy-dependent contribution. The static contribution $V^{(s)}$ is given by (4.12); it has a straightforward physical interpretation. (c) In (5.3), the dispersion integral extends from $-\infty$ to $+\infty$. This is due to the fact that the retarded Green's function contains the "hole" Green's function. The latter has a branch cut which extends from $-\infty$ up to the energy

$$(5.4) \qquad T_0^{(-)} = \mathcal{E}_0^{(A)} - \mathcal{E}_0^{(A-2)}$$

associated with the threshold for particle emission from the $(A-1)$-particle system. In the nuclear case, $T_0^{(-)} \approx -20 MeV$. The self-energy in general has isolated poles in the energy domains

$$(5.5) \qquad T_0^{(-)} < E < E_F^{(-)} = \mathcal{E}_0^{(A)} - \mathcal{E}_0^{(A-1)},$$

and

$$(5.6) \qquad E_F^{(+)} < E < T_{in}^{(+)}.$$

We recall that the physical mean field is obtained after having performed an energy average. It is given by; see (2.14),

$$(5.7) \qquad \tilde{M}(E) = M(E + i\Delta).$$

This quantity is free from pole singularities. The properties of $M(E)$ strongly suggest that $\langle \tilde{M}(E) \rangle$ is the quantity that should be identified with the mean field. Since $M(z)$ is analytic for $z > -\Delta$, the quantity $\langle \tilde{M}(E) \rangle$ also fulfills the dispersion relation (5.3),

$$(5.8) \qquad Re\tilde{M}(E) = V^{(s)} + \frac{P}{\pi} \int_{-\infty}^{\infty} dE' \frac{Im\tilde{M}(E')}{E' - E}.$$

Calculations performed in nuclear matter indicate that the spatial nonlocality of $M(E)$ is simple [28]. Therefore, the dispersion relation (5.8)can be adapted to yield a relation between the real and imaginary parts of the empirical mean field. This enabled one to extrapolate the mean field from positive energies (where many experimental data are available) towards negative energies, i.e. to construct the shell-model potential by extrapolating the optical-model potential [29]. A similar work has not yet been performed in the atomic case, possibly because only little information is available on the size of $Im\tilde{M}(E)$ at negative energies.

6. Projection operator approach. In his pioneering work on the optical-model potential, Feshbach developed a projection operator approach. The latter proved most useful for many purposes, in particular for the description of resonance phenomena [9, 10]. It consists in writing each vector of the Hilbert space as a sum of two orthogonal components, with the help of two orthogonal projection operators P and Q ,

$$(6.1) \qquad PQ = 0 \ , \ P^2 = P \ , \ Q^2 = Q,$$

$$(6.2) \qquad P + Q = 1.$$

For instance; see (2.4),

$$(6.3) \qquad \left|\Psi_k^{(A+1)}\right\rangle = P\left|\Psi_k^{(A+1)}\right\rangle + Q\left|\Psi_k^{(A+1)}\right\rangle .$$

By substituting this expression in the wave equation

$$(6.4) \qquad H\left|\Psi_k^{(A+1)}\right\rangle = E\left|\Psi_k^{(A+1)}\right\rangle ,$$

one readily obtains the coupled equations

$$(6.5) \qquad (E + \mathcal{E}_0^{(A)} - PHP)P\left|\Psi_k^{(A+1)}\right\rangle = PHQ\left|\Psi_k^{(A+1)}\right\rangle ,$$

$$(6.6) \qquad (E + \mathcal{E}_0^{(A)} - QHQ)Q\left|\Psi_k^{(A+1)}\right\rangle = QHP\left|\Psi_k^{(A+1)}\right\rangle .$$

Provided that the projection operators are constructed in such a way that $Q\left|\Psi_k^{(A+1)}\right\rangle$ contains only outgoing waves at large distance, (6.6) yields

$$(6.7) \qquad Q\left|\Psi_k^{(A+1)}\right\rangle = (E + \mathcal{E}_0^{(A)} - QHQ + i\eta)^{-1}QHP\left|\Psi_k^{(A+1)}\right\rangle .$$

By substituting (6.7) in (6.5), one finds,

$$(6.8) \qquad \left[E + \mathcal{E}_0^{(A)} - H_{eff}^{(A+1)}(E)\right] P\left|\Psi_k^{(A+1)}\right\rangle = 0 ,$$

where we introduced the many-body "effective" Hamiltonian,

$$(6.9) \qquad H_{eff}^{(A+1)}(E) = PHP + PHQ\frac{1}{E + \mathcal{E}_0^{(A)} - QHQ + i\eta}QHP.$$

In the present context, the main problem consists in finding an expression for P which would be such that $Q \left| \Psi_{\mathbf{k}}^{(A+1)} \right\rangle$ only has components in nonelastic channels. This is achieved by requiring that,

$$(6.10) \qquad P\Psi_{\mathbf{k}}^{(A+1)}(\xi_A) = \frac{1}{\sqrt{A+1}} \mathcal{A}\{\varphi_{\mathbf{k}}(\mathbf{r}_{A+1})\Psi_0^{(A)}(\mathbf{r}_1, ..., \mathbf{r}_A)\},$$

$$(6.11) \qquad (\Psi_0^{(A)}(\mathbf{r}_1, ..., \mathbf{r}_A) \mid Q\Psi_E^{(+)}(\mathbf{r}_1, ..., \mathbf{r}_A, \mathbf{r}_{A+1}) >= 0 \text{ for all } \mathbf{r}_{A+1},$$

where $\mathcal{A}$ denotes the antisymmetrization operator, and where we used the same notation as in (2.5). It turns out that (6.10) and (6.11) are sufficient to determine the projection operator P and the relationship between $\varphi_{\mathbf{k}}(\mathbf{r})$ and $\chi_k(\mathbf{r})$,

$$(6.12) \qquad P = \int d\mathbf{r} d\mathbf{r}' a_{\mathbf{r}}^{\dagger} |\Psi_0^A\rangle \left\langle \Psi_0^{(A)}|a_{\mathbf{r}'},\right.$$

$$(6.13) \qquad |\chi_{\mathbf{k}}\rangle = (1 - K)|\varphi_{\mathbf{k}}\rangle .$$

By introducing (6.12) in (6.8) and then projecting on $\left\langle \Psi_0^{(A)} \right|$, one obtains the one-body wave equation (3.6).

Since we only dealt with vectors of the Hilbert space of $(A+1)$-particles, it is not surprising that Feshbach's optical-model Hamiltonian yields no information on the $(A-1)$-particle system, in contrast to the self-energy Hamiltonian. In order to derive the latter from the projection operator approach, it is necessary to introduce a projection operator P which acts in an "extended" Hilbert space, spanned by antisymmetrized vectors of $(A+1)$ and $(A-1)$ particles. It has recently been shown [17] that one should take

$$(6.14) \qquad P = \int d\mathbf{r}(a_{\mathbf{r}}^{\dagger} + a_{\mathbf{r}}) \left| \Psi_0^{(A)} \right\rangle \left\langle \Psi_0^{(A)} \right| (a_{\mathbf{r}}^{\dagger} + a_{\mathbf{r}}).$$

7. Discussion. Two theories of the microscopic optical-model potential had been developed independently by Feshbach [10] and by Bell and Squires [12]. Feshbach's expression is adopted in many semi-phenomenological descriptions and theoretical discussions [1, 4, 5, 11], because its interpretation is simpler when the effects of antisymmetrization are neglected. However, that simplicity is lost whenever antisymmetrization effects are taken into account. That complication is due to the appearance of the one-body density matrix K, as illustrated by (3.9). Furthermore, Feshbach's optical-model Hamiltonian has no bound eigenstate below the Fermi surface. This property exhibits that it bears no close relationship with the empirical optical-model Hamiltonians, at least at low energy. This should be ascribed to the fact that the spatial nonlocality of Feshbach's potential is very complicated at low energy [15, 16].

In contrast, the optical-model Hamiltonian derived by Bell and Squires, i.e. the self-energy Hamiltonian, has bound eigenstates below the Fermi energy; the corresponding eigenvalues are the "separation" or "ionization" energies [23, 30, 31]. Accordingly, it is the self-energy that is usually identified to the mean field at negative energies.

We showed that an infinite number of Hamiltonians can be constructed which are all equivalent to the self-energy Hamiltonian; see (4.1). However, it has been argued that the self-energy Hamiltonian is the only one which fulfills a number of physical requirements [16].

In the present survey, we focussed on definitions and properties. We did not discuss methods of calculations. It is worth noticing that various computational approaches have been developed to evaluate the self-energy [30-32]. In particular, perturbation theory can be used, possibly after prior summation of subseries as in Brueckner's approach [33]. In contrast, perturbation theory cannot be applied to the calculation of Feshbach's Hamiltonian, because the "particle" Green's function has no inverse in the independent particle approximation [16].

REFERENCES

[1] I.E. McCarthy *Microscopic Optical Potentials*, edited by H.V. von Geramb (Springer Verlag, Berlin, 1979), p. 447, and references contained therein.

[2] G. Staszewska, D.W. Schwenke, and D.G. Truhlar, Intern. J. Quantum, Chem. 17, 163 (1983).

[3] G. Staszewska, D.W. Schwenke, and D.G. Truhlar, J., Chem. Phys. 81, 335 (1984).

[4] P.E. Hodgson, *Nuclear Reactions and Nuclear Structure*, (Clarendon Press, Oxford, 1971), and references contained therein.

[5] G.R. Satchler, *Direct Nuclear Reactions*, (Clarendon Press, Oxford, 1983), and references contained therein.

[6] C.J. Joachain, *Quantum Collision Theory*, (North-Holland Publ. Comp., Amsterdam, 1975), vol. 2.

[7] M.L. Goldberger and K.M. Watson, *Collision Theory*, (John Wiley and Sons New York, 1964), and references contained therein.

[8] A.K. Kerman, M. McManus, and R. Thaler, Ann. Phys. (N.Y.) 8, 551 (1959).

[9] H. Feshbach, Ann. Phys. (N.Y.) 5, 357 (1958).

[10] H. Feshbach, Ann. Phys. (N.Y.) 19, 287 (1962).

[11] H. Feshbach, *Theoretical Nuclear Physics: Nuclear Reactions*, (John Wiley & Sons, New York, 1992).

[12] J.S. Bell and E.J. Squires, Phys. Rev. Lett. 3, 96 (1959).

[13] J.S. Bell, *Lectures on the Many-Body Problem*, edited by E.R. Caianiello (Academic Press, New York, 1962), p. 91.

[14] M. Namiki, Prog. Theor. Phys., 23, 629 (1960).

[15] C. Mahaux and R. Sartor, Nucl. Phys., A530, 303 (1991).

[16] F. Capuzzi and C. Mahaux, Ann. Phys. (N.Y.), 239, 57 (1995).

[17] F. Capuzzi and C. Mahaux, Ann. Phys. (N.Y.), 245, 147 (1996).

[18] J.M. Bang, F.G. Gareev, W.T. Pinkston, and J.S. Vaagen, Physics Reports 125, 253 (1985), and references contained therein.

[19] H. FESHBACH, A.K. KERMAN, AND R. LEMMER, Ann. Phys. (N.Y.) **41**, 230 (1967).

[20] R. LIPPERHEIDE, Z. Phys. **202**, 58 (1967).

[21] C. MAHAUX AND H.A. WEIDENMÜLLER, *Shell Model Approach to Nuclear Reactions*, (North-Holland, Amsterdam, 1969), and references contained therein.

[22] F.G. PEREY AND B. BUCK, Nucl. Phys. **32**, 353 (1962).

[23] H.-G. WEICKERT AND L.S. CEDERBAUM, Few-Body Systems **2**, 33 (1987), and references contained therein.

[24] I.E. MCCARTHY AND E. WEIGOLD, Rep. Progr. Phys. **54**, 789 (1991).

[25] A.E.L. DIEPERINK, AND P.K.A. DE WITT HUBERTS, Ann. Rev. Nucl. Part. Sci. **40**, 239 (1990).

[26] G. PASSATORE, *Nuclear Optical Model Potential*, edited by S. Boffi and G. Passatore (Springer Verlag, Berlin, 1976) p. 177, and references contained therein.

[27] D. THIRUMALAI, G. STASZEWSKA, AND D.G. TRUHLAR, *Comments on Atomic and Molecular Physics*, **20**(1987), 217, and references contained therein.

[28] M. BALDO, I. BOMBACI, G. GIANSIRACUSA, U. LOMBARDO, C. MAHAUX, AND R. SARTOR, Nucl. Phys. **A545**, 741 (1992), and references contained therein.

[29] C. MAHAUX AND R. SARTOR, Adv. Nucl. Phys. **20**, 1 (1991), and references contained therein.

[30] A. TARANTELLI AND L.S. CEDERBAUM, Phys. Rev. **A49** 3407 (1994), and references contained therein.

[31] C. MAHAUX, P.F. BORTIGNON, R.A. BROGLIA AND C.H. DASSO, Physics Reports **120C**, 1 (1985), and references contained therein.

[32] N. VINH MAU, *Microscopic Optical Potentials*, edited by H.V. von Geramb (Springer Verlag, Berlin, 1979), p. 40, and references contained therein.

[33] J. HÜFNER AND C. MAHAUX, Ann. Phys. (N.Y.) **73**, 525 (1972).

THE PAULI PRINCIPLE IN MULTI-CLUSTER STATES OF NUCLEI

JENS BANG*

Abstract. Some ways of taking the Pauli principle into account in scattering of composite particles and in cluster formation in nuclei are reviewed. The emphasis is on the limits of applicability of these approaches. A useful method is the replacement of the full antisymmetrization by a principle of exclusion or blocking, using the existence of strongly bound shells in nuclei

1. Mean field methods. The Pauli principle is often taken into account in nuclear physics by use of mean field methods. Since the single particle wave functions in such a mean field are orthogonal, it is trivial to construct antisymmetric wave functions in the form of Slater determinants and to go further, introducing occupation number formalism etc. The mean field may either be introduced empirically, as in shell model or optical model, or from the fundamental interactions, in a self consistent way, as in the Hartree-Fock approximation [1]. This leads to a non-local mean field with exchange terms. *If* a single Slater determinant approximation is used, the main Pauli problem seems to be how to include the exchange terms in the calculations. The simple problem of scattering of nucleons on nuclei in this way becomes a generalization of the bound state problem, at least when the contributions of scattering states to the self-consistent field are neglected. Here, however, the exchange terms play a smaller role than the direct ones. They may even be simulated by local terms and together with the direct ones (which lead to a folding potential) give a local optical potential.

Let us look at the diagonal term of the exchange potential in a scattering state, $\psi_{i'}$

$$\sum_{i<i_F} \int \int d\mathbf{r}_1 d\mathbf{r}_2 \psi_i^*(\mathbf{r}_1)\psi_i(\mathbf{r}_2)V(\mathbf{r}_1,\mathbf{r}_2)\psi_{i'}^*(\mathbf{r}_2)\psi_{i'}(\mathbf{r}_1)$$

Here $V(\mathbf{r}_1,\mathbf{r}_2)$ is an interaction between two nucleons depending on their coordinates $\mathbf{r}_1,\mathbf{r}_2$. We assume here and in the following that we have single particle states of the (self consistent) Hamiltonian, with consecutive energies, denoted by $i = 1, 2, .., i_F$ so that the lowest state of the nucleus is a Slater determinant corresponding to all the lowest single particle levels up to i_F (the Fermi level with internal wave number k_F) are occupied. If the sum were extended over a complete set of states it would just give

$$\sum \psi_i^*(\mathbf{r}_1)\psi_i(\mathbf{r}_2) = \delta(\mathbf{r}_1,\mathbf{r}_2)$$

* The Niels Bohr Institute, Blegdamsvej 17, 2100 Kobenhavn Ø Copenhagen, Denmark.

and for $V = V(|\mathbf{r_1} - \mathbf{r_2}|)$ the integral would be

$$\int V(0)|\psi_{i'}(\mathbf{r})|^2 d\mathbf{r} = V(0)$$

In the case of the limited sum, $i < i_F$, we still have a similarity to a δ-function; only we have, instead of the width being 0, a width (extension in $|\mathbf{r_1} - \mathbf{r_2}|$) of the order of $1/2k_F$ and it is valid only inside the nuclear radius. The width is of the order of or smaller than the range of the nucleon-interaction, in which case the approximation $V = V(0)$ is not so bad. This means that at least the diagonal matrix elements of the exchange force may be approximated by those of a state independent potential i.e. contribute to the mean field we were talking about.

We shall here mainly discuss low-energy phenomena. Note that here the mean field methods often have a certain self-consistency coming from the fact that residual interactions, although they are relatively strong, are limited in their possibilities of creating correlations, just by the Pauli principle. This is why short range residual forces are more inclined to create superconductivity-like pairing states than local clusters of two-nucleon or α-type.

There also exist, though, low-energy phenomena where a clustering is important and a Hartree-Fock-like description is impossible. Such phenomena are mainly of two types: Those we mainly create ourselves like collisions between composite particles, e.g. α scattering on nuclei, and those made by nature, like α-radioactivity. The latter case is more complicated, since the α-particle must somehow be *formed*. This happens mainly in the surface region where the density is low and the Pauli principle therefor less efficient in preventing the formation of α-particles. We shall argue that scattering of composite particles at low energies and their existence in losely bound (and resonant) states of nuclei are closely related phenomena. This may be seen from the fact that the S-matrix of low-energy scattering is dominated by poles of the Green functions, and that the asymptotic forms of the losely bound (resonant) states are given by the same poles. For composite particles, the Green functions are found by solution of the Faddeev-Jakubovsky equations [2], and so must be the asymptotic forms, understood as the forms of the wave functions in the relevant coordinates in regions where interactions are negligible. We shall here not go in details of these solutions but only note that they can generally not be written in the product form which leads to the Slater determinants, but lend themselves easily to the Pauli methods suggested below.

2. Elastic scattering of simple and composite particles. Methods for taking the Pauli principle into account in elastic scattering were first suggested by Feshbach [3] for the case of scattering of one nucleon, on a nucleus and later generalized by Saito [4], to the scattering of two composite particles α, β, which are, in the incoming channel, in an internal wave function, ϕ_0, which is a product of their ground states. The projection operator on these ground states is called $P = 1 - Q$ (Q is also a projection operator). During the scattering the two systems will be subject to a potential which includes all interactions between one particle in α and one in β and which will not only induce elastic scattering but also admix components of $Q\Psi$ to an initial $P\Psi$. From the Schrödinger equation for the total system

$$H\Psi = E\Psi$$

one obtains, eliminating $Q\Psi$, a similar equation for $P\Psi$ with the effective Hamiltonian

$$H_f = PHP + PHQ\frac{1}{E - QHQ}QHP$$

An equation for the wave function of the relative motion of the two clusters $u(\mathbf{R})$ is now

$$< \mathbf{R}_0\phi_0|H_f - E|\tilde{A}u\phi_0 >= 0$$

Here, and in the following, $\tilde{A}$ is an antisymmetrizing operator and the bracket notation includes integration over all the internal variables of the ϕ of the bra. Performing these integrations the energy term appears with a factor $u - Ku$ where the kernel K contains a sum of terms corresponding to all permutations of particles between α and β necessary for construction of the totally antisymmetric wave function from the product of internally antisymmetric wave functions.

$$K(\mathbf{R}_0, \mathbf{R}_0') = \sum_{j=1}^{n}(-1)^{j+1}c_j < \mathbf{R}_0, \phi_0|\mathbf{R}_0', \phi_j >$$

where j is the number of permuted particles, c_j the corresponding number of permutations, n the number of particles of the smallest cluster, ϕ_j the same function as ϕ_0 but of the variables after permutation, $\mathbf{R}_0'$ is the integration variable of the kernel K (but for each j the integration variables must be transformed to the corresponding internal variables and $\mathbf{R}_j$) Now eigenfunctions of K with eigenvalue 1, i.e. satisfying

$$u_\nu = Ku_\nu$$

are easily seen also to satisfy

$$\tilde{A}[u_\nu(\mathbf{R})\phi_0(1...N)] = 0$$

This means that these u_ν states are Pauli forbidden. If, in the mentioned case of Feshbach, the nuclear wave function, like in the Hartree-Fock approximation, is given by a single Slater determinant of orthogonal normalized single particle wave functions

$$\phi_{0,H-F} = \tilde{A}\Pi_{i,n}^{N-1}w_n(\mathbf{r_i})$$

we have

$$K(\mathbf{R}, \mathbf{R}') = \sum_{n:\text{occupied}} w_n^*(bfR)w_n(\mathbf{R}')$$

so here $u_\nu = w_n$ and the eigenvalues 1 correspond to forbidden states. As a much less trivial example Saito gives $\alpha - \alpha$ scattering where the occupied internal orbits with radial and angular momentum quantum numbers n,l all have 00; here the forbidden states of the relative $\alpha - \alpha$ motion are, in a similar notation, 00,10 and 02. Although the calculations are a little complicated, the result is very simple. The first is, e.g., necessarily forbidden, because it puts all 8 particles in a state with no nodes whereas only 4 can be there. In this case the role of eigenstates of K with eigenvalues different from 1 can be proven to be very small.

The apparent simplicity of the H_f formalism, also due to Feshbach [3], is treacherous since the second term in reality is extremely complicated and must *always* be approximated in some way. Here a large part of the Q-space generally consists of break-up channels, as already seen in the example of the α particle which, like another important projectile, the deuteron, has no bound, excited states. Neglect of break-up channels may be a good approximation for low energy scattering of α particles, but in general this problem represents a severe limitation on the Saito method.

3. Inclusion of break-up channels. Other methods, common for bound state and scattering, were constructed, based in a different way on the tightly bound character of closed shell nuclei, of which the α-particle is just the best example. These methods are so to say complementary to Saito's in the sense that break-up in some channels are explicitly taken into account even when virtual, whereas elastic scattering is playing a less fundamental role. Let us think of a closed shell and some other nucleons which may, to a good approximation, be described as bound in the next, open, shells or impinging as a more losely bound cluster as e.g. a deuteron. As far as residual interactions between the closed shell core and the outer nucleons are not so strong as to give rise to large amplitudes of core excitation with high energies (compared to shell distances) such systems look very much like few-body systems with one heavy particle [2], the closed shell, and some light ones. Here we have, compared to ordinary few-body systems, two problems: Internal degrees of freedom of the heavy particle may be excited; so, looking apart from its break-up, we may, like in scattering of two heavy nuclei, introduce coupled channels, one for each excited state. The other problem is that although, in agreement with our introduction, a single particle outside a closed shell will automatically be in states which are orthogonal to the occupied ones of the closed shell of the samefield,interactions between several outer particles must in general lead to admixtures of the occupied states, thus violating the antisymmetry. From this way of expressing the problem, it is already hinted that instead of writing a fully antisymmetrized wave function for all particles, one may look at an approximation where we just keep some occupied states, those of the closed shells, and forbid the other particles w.f. to have admixtures of these occupied ones. [5]

4. The blocking approximation. How good is such a **blocking** approximation? I shall throw some light on this question by looking at the case of an even number of particles interacting with what is usually called a pairing force. I go over to the occupation number formalism [1], which is just another way of describing the Slater determinants mentioned above. We can e.g. describe the Hartree-Fock ground state, where all the lowest-energy single particle orbits are occupied, by operating with corresponding creation operators a_n^+ on some vacuum state $|0>$:

$$\psi_{0,H-F} >= \Pi_{n:\text{occupied}}\, a_n^+ |0>$$

We also introduce annihilation operators a_n, and some anticommutation rules between our operators, given in standard textbooks, e.g. [1]. In this representation we write a Hamiltonian

$$H = \sum_\nu^\Omega \epsilon_\nu(a_\nu^+ a_\nu + a_{\nu'}^+ a_{\nu'}) - G\sum_{\nu,\mu} B_\nu^+ B_\mu$$

$$B_\nu = a_\nu a_{\nu'}$$

We see that the single particle states occur in Ω two-fold degenerate pairs with energies $\epsilon_\nu = \epsilon_{\nu'}$ (e.g. time reversed states). Further there is an interaction with a coupling constant G, with equal matrix elements between all paired ν, ν' states. This is just the Bardeen, Cooper,Schrieffer Hamiltonian [6] but instead of their approximate solutions we shall use the exact ones of Richardson [7] (unnormalized). We shall neglect pair breaking, which leads to trivial complications, and only look at completely paired states among which the ground state Φ_0 is found:

$$\Phi_{\text{Pairing}} >= \Pi_\lambda^N \left(\sum_\nu^\Omega \frac{B_\nu^+}{2\epsilon_\nu - z_\lambda} \right) |0>$$

where the z_λ are solutions of the coupled equations

$$\sum_\nu^\Omega \frac{1}{2\epsilon_\nu - z_\lambda} - \sum_{\lambda' \neq \lambda} \frac{2}{z_{\lambda'} - z_\lambda} = G^{-1}$$

and the total energy is

$$E = \sum_\lambda^N z_\lambda$$

We see that for $G \rightarrow 0$ the solutions must behave like

$$z_\lambda \rightarrow 2\epsilon_\nu, z_{\lambda'} \rightarrow 2\epsilon_\mu, \nu \neq \mu$$

thus satisfying the Pauli principle (whereas $z_\lambda = z_{\lambda'}$ is prevented by the last term, the coupling term, on the left side of the coupled equations). The interaction is realistic in the sense that the main term in the neutron-neutron interaction admixes the same states. To get quantitatively realistic results one must, however, in general use a model space with a finite Ω and then adjust the coupling constant G according to this number; for fixed G the energy will diverge when Ω goes to infinity.

Let us look at the simplest non-trivial case with three levels, $2\epsilon_\nu = E_1, 0, E_2$, and two pairs. The coupled equations are here

$$\frac{1}{E_1 - z} - \frac{1}{z} + \frac{1}{E_2 - z} - \frac{2}{z' - z} = G^{-1}$$

$$\frac{1}{E_1 - z'} - \frac{1}{z'} + \frac{1}{E_2 - z'} - \frac{2}{z - z'} = G^{-1}$$

Let us first look at the opposite of the blocking approximation, an approach of neglecting the Pauli principle completely. We see again that the coupling terms represent the antisymmetry, if they were not there we would have a boson-like approximation for the pairs, i.e. only one equation

$$\frac{1}{E_1 - z} - \frac{1}{z} + \frac{1}{E_2 - z} = G^{-1}$$

This has two types of solutions for $G \rightarrow 0$:a) The two factors of Ψ_{Pairing} are identical, say $z = z' \rightarrow E_1$. This condensate is obviously in violent contradiction with the Pauli principle. b) The two factors contain different roots, say $z \rightarrow 0, z' \rightarrow E_1$.

We shall here follow a philosophy in agreement with what was said above about the relation between the model space and the coupling constant G: For different approximations we look at similar z, z' values and adjust the G values. If the necessary adjustment of G from its exact value is small, we may say that the approximation is good. Calling the coupling constant of the above boson-like approximation G_2 we have

$$G^{-1} - G_2^{-1} = \pm \frac{2}{z' - z} \rightarrow \pm \frac{2}{E_1}$$

for small G, G_2. So even in this case the approximation may be inaccurate by an arbitrary amount. Now, for larger G values, we may have z-crossings as, e.g.,

$$z = -z' \sim 0; G^{-1} \sim E_1^{-1} + E_2^{-1}$$

whereas the boson-like equation never leads to crossing. It is clear that in the neighborhood of the crossing points, the difference $G^{-1} - G_2^{-1}$ may take arbitrarily large values.

Now let us in stead look at blocking of the lowest state.

$$\frac{1}{-z} + \frac{1}{E_2 - z} = G_1^{-1}$$

This gives a second-order equation for determination of z

$$-(E_2 - 2z)G_1 + (E_2 - z)z = 0$$

with the solutions

$$z = +E_2/2 - G_1 \pm ((E_2/2 - G_1)^2 + G_1 E_2)^{1/2}$$

We now use these z values in one of the coupled equations, say, the first, whereas z' from the blocked case then must be E_1. In accordance with what was said above, we let the two z values correspond to coupling constants G_+, G_-, respectively. We then get

$$G_\pm^{-1} = G_1^{-1} - (E_1 - E_2/2 + G_1 \mp S)^{-1}$$

where the square root is

$$S = ((E_2/2)^2 + G_1^2)^{1/2}$$

With the usual relation between level density and pairing coupling constant we get $G_+/G_1 \sim 0.9$. G_b/G_1 is a little smaller ($\sim .85$) corresponding to the obvious fact, that the blocking approximation is less exact the nearer one is to the Fermi level. For the more appropriate case that E_1 corresponds to a deeply bound closed shell, $-E_1 >> G_1$, we get $G_\pm \sim G_1$. In other words, looking at the case of a fixed coupling constant, we see that the blocking approximation is good when the blocked states are deeply bound states.

5. **Blocking and few-body problems.** We have here got some support for the blocking method. Let us now look for how this method is carried out in practical calculations with realistic interactions that allow for both bound state and scattering solutions. We shall here again introduce a projection operator on Pauli allowed states

$$P\Psi = \Psi_P$$

Let us now solve

$$P(H - E)\Psi_P = 0$$

(in the allowed space Ψ_p satisfies the Schrödinger equation) This may also be expressed

$$(H - E)\Psi_P = \sum_{ij..} \sum_n c_n(ij..)\Phi_n^{ij..}$$

where $\Phi_n^{ij..}$ is a complete set in Q space, which may be written in terms of the forbidden internal states,$\phi(i,j,..)$ of clusters

$$\Phi_n^{ij..} = \phi(i,j..)f_n^{ij..}$$

$f_n^{ij..}$ a complete set of ortonormalized functions of the remaining Jakobian variables. For each $\Phi_n^{i,j..}$ we may solve the inhomogenous equation with the appropriate boundary conditions

$$(H - E)\Psi(i,j,..,n) = \Phi_n^{ij..}$$

Then we may write

$$\Psi = \sum c_n(i,j,..)\Psi(i,j,..,n)$$

where the c's are given by the equations $< \Phi_n^{ij..}|\Psi >= 0$ We may also write

$$Q_{ij..} = |\phi(ij..) >< \phi(ij..)| = \sum_n |\Phi_n^{ij..} >< \Phi_n^{ij..}|$$

Then

$$\left(H - E + \lambda \sum Q_{ij..}\right)\Psi = \sum_{i,j,..,n} c_n(i,j,..)\Phi_n^{ij..}$$

$$c_n(i,j,..) = c_n(i,j,..\lambda) = \lambda < \Phi_n^{ij..}|\Psi >$$

So, for finite c's and $\lambda \to \infty$ we must have

$$< \Phi_n^{ij..}|\Psi >\to 0$$

For bound state problems, where it is obvious that finite c's exist, this latter variant is very efficient; with continuum problems the first may be better. It should be noted that it is easily shown that a set of $f_n^{ij..}$ can be chosen which goes sufficiently fast to 0 for the relevant Jakobian coordinates going to infinity. It should also be noted that both formulations fit into the methods of multi cluster scattering, since the Pauli terms, although non-local, are limited in space in a similar way as the potentials, and therefore lead to the same type of connected kernel structure as these.

For the three-body system of an alpha-particle and two nucleons, ^6He we have, e.g., the Faddeev equations

$$(E - T - \tilde{V}_i)\psi_i = \tilde{V}_i \sum_{j \neq i} \psi_j$$

$$\tilde{V}_{n\alpha} = V_{n\alpha} + \lambda|s, \mathbf{r}_{n\alpha} >< s, \mathbf{r}_{n\alpha}|$$

$$\lambda \to \infty$$

It is interesting to note that if we here take $V_{n\alpha}$ as the optical potential which reproduces the scattering, we are automatically forced to approximate the forbidden bound state with the $l = n = 0$ state of the *same* potential. This state may not be so good an approximation for the α - wave function, but for the present purpose it is sufficient and consistent.

When the wave function is written as a finite expansion in a complete set of functions,

$$\Psi = \sum_{j=1}^{J} d_j \psi_j$$

an alternative to $\lambda \to \infty$ is to write the Schrödinger equation (using one index for i,j,..n)

$$\sum_{j'=1}^{J} (H_{jj'} - E\delta_{j,j'})d_{j'} + \sum_{k=1}^{K} \gamma_{jk}c_k = 0, j = 1, ..J$$

$$\sum_{j'=1}^{J} \gamma_{kj'}^{+}d_{j'} = 0, k = 1, ..K$$

So we have J+K equations for J+K coefficients and may determine E as an eigenvalue. The γs represent overlaps between basis functions (index j) and forbidden ones (index k) The method is similar to Saito's; after elimination of the d's we get the Hermitian eigenvalue equation

$$M_{kk'}c_{k'} = 0; M = \gamma_{kj}^{+}(H - E)^{-1}\gamma_{jk}$$

In the scattering case we may fix d_1 to describe the incoming wave, thus getting J+K-1 equations for J+K-1 coefficients.

6. Low energy states. Let me finally look at a particular problem related to very low energy scattering or very losely bound states of complex particles, halo systems. The point is the following: let us look at two potentials, one with a $n = 1, l = 0$ single particle state (resonance) and one with a $n = 0, l = 0$ state at the same energy. Further (due, e.g., to l-dependence of the potential, which is that of a nucleus which is at least as heavy as an α particle) there is e.g., a dominating $n = 0, l = 1$ state (resonance), which also has a small energy. Let the potentials be square wells: $-V_0$ for $r < R, 0$ for $r > R$. We introduce

$$\frac{2m|E|}{\hbar^2} = \kappa^2; \frac{2m(V_0 - |E|)}{\hbar^2} = k^2$$

For $l = 0$ the w.f. in the well is $Aj_0(kr)$, outside the well $Bh_0(i\kappa r)$. Boundary condition (eigenvalue equation):

$$kR\cot(kR) = -\kappa R$$

κ small gives $kR = \pi(n + 1/2)$ which also gives $A^2 = B^2$, normalization: $A^2(R/2 + 1/(2\kappa)) = 1$ for the state with no node. For sufficiently small κ we have $A^2 = 2\kappa$. The same will be the case for the state with one node, and in both cases we see that the contribution from the inner part of the wave function is negligible. The interaction between two outer particles in $n = 0, l = 1$ states will lead to admixtures which due to the near-identity between the two outer wave functions must be practically identical. In the second case with one node for small r-values, there must, however, also be an admixture of the deeper bound 0s state. That is a spurious contribution, if this, as in the α particle, is occupied. We see, that in **this** case the replacement of the original deep potential with a more shallow one, which has only the bound state with no nodes, to some extent takes care of the Pauli principle. The example is important but quite special and the replacement of the Pauli principle by a change of the local potential should in no way be generalized.

REFERENCES

[1] A. BOHR AND B. MOTTELSON, *Nuclear Structure*, Benjamin, Reading, 1969.
[2] O.A. YAKUBOVSKIJ, Yadernaya Fizika 5 (1967), 1312.
[3] H. FESHBACH, Ann. of Phys.19 (1962), 287.
[4] S. SAITO, *Interactions between clusters and Pauli principle*, Prog. Theor. Phys. **41** (1969), 705.
[5] J. BANG, J.J. BENAYOUN, C. GIGNOUX AND I.J. THOMPSON, *Scattering and break-up of deuterons on alpha-particles in a realistic three-body model*, Nucl. Phys A **405** (1983), 126, and references therein.
[6] J. BARDEEN, L.N. COOPER AND J.R. SCHRIEFFER, Phys. Rev. **106** (1957), 162.
[7] R.W. RICHARDSON AND N. SHERMAN, Nucl.Phys. **52** (1964), 221

NONPERTURBATIVE APPROACHES TO ATOMIC AND MOLECULAR MULTIPHOTON (HALF-COLLISION) PROCESSES IN INTENSE LASER FIELDS

SHIH-I CHU*

Abstract. In this article, we describe some recent development of generalized Floquet formalisms and computational methods for nonperturbative treatments of atomic and molecular multiphoton (half-collision) processes in intense and superintense laser fields. We start with a brief review of the conventional Floquet matrix techniques, applicable to multiphoton bound-bound transitions in finite-level systems in periodic fields, and their limitations. Several generalized Floquet formalisms, beyond the Floquet theorem, are then introduced for the treatment of more complicated systems, such as the many-mode Floquet theory for multi-frequency laser excitation with non-periodic Hamiltonians, the non-Hermitian Floquet formalism for bound-free and free- free multiphoton ionization and dissociation etc. Finally we describe several recent case studies of strong-field processes using the generalized Floquet techniques: intensity-dependent ionization potential and threshold shift, a.c. Stark shifts of Rydberg states in strong fields, above-threshold multiphoton detachment of H^- in one-color, two-color, and pulsed laser fields, stabilization and ionization suppression of negative ions in superintense high-frequency laser fields, and laser- induced chemical bond softening and hardening and molecular stabilization, etc.

1. Introduction. The recent advancement in short-pulse high-power laser technology has generated considerable interest in the study of the structure and dynamics of atoms and molecules exposed to intense and superintense laser fields. This has led to the discovery of a number of novel strong- field multiphoton and nonlinear optical phenomena, such as multiphoton and above threshold ionization (MPI/ATI) of atoms, giant A.C. Stark shifts in Rydberg states, very high-order multiple harmonic generation, multiphoton and above-threshold dissociation (MPD/ATD) of small molecules, atomic stabilization in superstrong high-frequency fields, and laser-induced chemical bond softening and hardening, etc. [1]. In the presence of intense laser fields, atomic and molecular structures are strongly distorted, resulting in energy level shifting, broadening, and mixing. In such cases, perturbation theory often becomes invalid, and *nonperturbative* formalisms and methods are required for the description of high-order multiphoton processes.

Multiphoton processes may be formally treated in a fully quantum mechanical or semiclassical fashion. In the former approach, both the system (atoms/molecules) and the electromagnetic fields are treated quantum mechanically, whereas in the semiclassical approach, the system is described by the time-dependent Schrödinger equation and the field satisfies the classical Maxwell equations. The fully quantum approach is generally difficult to apply as it involves an infinite set of time-independent coupled second-

* Department of Chemistry, University of Kansas and Kansas Institute for Theoretical and Computational Science, 2010 Malott, Lawrence, Kansas 66045-0046.

order differential equations subject to boundary conditions. So far the fully quantum approach has been used for the treatment of two- level systems only [2]. In contrast, the semiclassical approach leads to a set of time-dependent coupled first-order differential equations which are initial value problems and computational more tractable. Furthermore, it is well known that the semiclassical theory gives rise to results that are equivalent to those obtained from the fully quantized theory in strong fields [3]. Another advantage of the semiclassical approach is that when the Hamiltonian is periodically in time as is often the case for monochromatic laser fields with sinusoidal time-dependence, the Floquet theorem [4,5,6] is applicable, leading to significant simplification in both theoretical and computational studies of multiphoton processes. However, the conventional Floquet methods are not without limitations. In this article we describe some *generalized* Floquet formalisms, [6] beyond the Floquet theorem, and methods recently developed at the University of Kansas for the studies of atomic and molecular half-collision processes in the presence of intense and superintense laser fields. No attempt is made to cover the whole field.

2. Floquet theorem and time-independent Floquet Hamiltonian method in periodic fields. When the perturbing laser field is sufficiently strong, it is useful to introduce the notion of *quasienergy* which can be considered to be a (time-averaged) characteristic energy of the combined system (namely, atoms/molecules) and electromagnetic (EM) fields together. The description of the response of atoms and molecules to monochromatic (or polychromatic) laser fields is greatly facilitated by the use of the Floquet (or many-mode Floquet [6]) theorem. We start the discussion from the conventional Floquet theorem for monochromatic fields.

2.1. The Floquet theorem. The solutions of linear differential equations with periodic coefficients were first considered by Floquet and Poincaré bout ta century ago [4]. The Floquet theorem was later used by Autler and Townes [7] to obtain wave functions for the two-level system in terms of infinite continued fractions. Application of Floquet theory to quantum systems began to grow only after the mid 1960's. In particular, Shirley [3] reformulated the time-dependent problem of the interaction of a two-level quantum system with a strong oscillating classical field as an equivalent time-independent infinite-dimensional Floquet matrix.

Consider the general features of the wave functions of a quantum system driven by a periodic external field with frequency ω and period $\tau(= 2\pi/\omega)$. The Schrödinger equation for the system may be written as $(\hbar = 1)$

$$(2.1) \qquad\qquad \hat{H}(\vec{r}, t)\Psi(\vec{r}, t) = 0,$$

where

$$(2.2) \qquad\qquad \hat{H}(\vec{r}, t) \equiv H(\vec{r}, t) - i\partial/\partial t.$$

$H(\vec{r}, t)$ is the total Hamiltonian given by

$$(2.3) \qquad H(\vec{r}, t) = H_0(\vec{r}) + V(\vec{r}, t),$$

and $V(\vec{r}, t)$ is the periodic perturbation due to the interaction between the system and the monochromatic field,

$$(2.4) \qquad V(\vec{r}, t + \tau) = V(\vec{r}, t).$$

The unperturbed Hamiltonian $H_0(\vec{r})$ has a complete orthonormal set of eigenfunctions:

$$(2.5) \qquad H_0(\vec{r})|\alpha(\vec{r})\rangle = E_\alpha^0|\alpha(\vec{r})\rangle, \quad \langle \beta(\vec{r})|\alpha(\vec{r})\rangle = \delta_{\beta\alpha}.$$

The wavefunction Ψ, called the quasienergy state (QES), can be written, according to the Floquet theorem, in the following form

$$(2.6) \qquad \Psi(\vec{r}, t) = \exp(-i\epsilon t)\Phi(\vec{r}, t),$$

where $\Phi(\vec{r}, t)$ is periodic in time, i.e.,

$$(2.7) \qquad \Phi(\vec{r}, t + \tau) = \Phi(\vec{r}, t),$$

and ϵ is a real parameter, which is unique up to multiples of $2\pi n/\tau$, called the *Floquet characteristic exponent or the quasienergy*. The term *quasienergy* reflects the formal analogy of the states, Eq. (2.6), with the Bloch eigenstates in a solid with the *quasimomentum* $\vec{k}$.

Substituting Eq. (2.6) into Eq. (2.1), we obtain an eigenvalue equation for the quasienergy,

$$(2.8) \qquad \hat{H}(\vec{r}, t)\Phi_\gamma(\vec{r}, t) = \epsilon_\gamma \Phi_\gamma(\vec{r}, t),$$

subject to the periodicity condition (2.7).

For the Hermitian operator $\hat{H}(\vec{r}, t)$, one can introduce the composite Hilbert space $R \oplus T$. The spatial part is spanned by square-integrable (L^2) functions in the configuration space, with the inner product defined by

$$(2.9) \qquad \langle \alpha(\vec{r})|\beta(\vec{r})\rangle \equiv \int \alpha^*(\vec{r})\beta(\vec{r})d\vec{r} = \delta_{\alpha\beta}.$$

The temporal part is spanned by the complete orthonormal set of functions $\{\exp(im\omega t)\}$, where $m = 0, \pm 1, \pm 2, \ldots$ is the Fourier index, and

$$(2.10) \qquad \frac{1}{\tau}\int_0^\tau \exp[i(n - m)\omega t]dt = \delta_{nm}.$$

The eigenvectors of $\hat{H}$ satisfy the orthonormality condition

$$(2.11) \qquad \langle\langle \Phi_\gamma|\Phi_{\gamma'}\rangle\rangle \equiv \frac{1}{\tau}\int_0^\tau dt \langle \Phi_\gamma(\vec{r}, t)|\Phi_{\gamma'}(\vec{r}, t)\rangle = \delta_{\gamma\gamma'},$$

and form a complete set in $R \oplus T$:

$$(2.12) \qquad \sum_\gamma |\Phi_\gamma\rangle\langle\Phi_\gamma| = I.$$

2.2. General properties of quasienergy states. The quasienergy states (QES) play a similar role in studying periodically time-dependent quantum systems as the bound states do for the time-independent Hamiltonian. The QES with different quasienergies ϵ_γ are mutually orthogonal for each moment of time and form a complete orthonormal set, as indicated in Eqs. (2.11)–(2.12).

The quasienergy eigenvalue equation (2.8) has the form of the "time-independent" Schrödinger equation in the composite Hilbert space $R \oplus T$. It can be readily shown that all the general quantum-mechanical theorems for the time-independent Schrödinger equation, such as the variational principle, the Hellman-Feynman, hypervirial, Wigner-Neumann and other theorems, can be extended also to the QES in periodic fields. Thus, for example, the variational form of Eq. (2.8) can be written as

$$(2.13) \qquad \delta\epsilon[\Phi] = 0, \quad \epsilon[\Phi] \equiv \langle\langle\Phi|\hat{H}|\Phi\rangle\rangle \,/\, \langle\langle\Phi|\Phi\rangle\rangle.$$

While the energy of the system is not a conserved quantity for an explicitly time-dependent Hamiltonian, it is possible to determine the "mean energy" $\overline{H}_\epsilon$ of the system in the QES $\Psi_\epsilon(\vec{r}, t)$:

$$
\begin{aligned}
\overline{H}_\epsilon &= \frac{1}{\tau} \int_0^\tau \langle\Psi_\epsilon(\vec{r}, t)|\hat{H}(\vec{r}, t)|\Psi)_\epsilon(\vec{r}, t)\rangle dt \\
&= \epsilon + \langle\langle\Phi_\epsilon(\vec{r}, t)|i\partial/\partial t|\Phi_\epsilon(\vec{r}, t)\rangle\rangle.
\end{aligned}
$$
$$(2.14)$$

Using the Hellman-Feynman theorem, it can be shown that

$$(2.15) \qquad \overline{H}_\epsilon = \epsilon - \omega\partial\epsilon/\partial\omega.$$

Other properties of quasienergy states are reviewed in refs. 6(a)-6(c).

2.3. Time-independent Floquet Hamiltonian method for multiphoton bound-bound transitions. Exact analytical solution of the time-dependent Schrödinger equation with temporally periodic Hamiltonian is possible only in exceptional cases. Thus it is in general necessary to develop approximate methods for the treatment of multiphoton excitation (MPE) of atoms and molecules. There are two general approaches for dealing with periodically time-dependent Hamiltonian systems. The first is direct numerical solution of the time-dependent Schrödinger equation from $t = 0$ to $t = T$ (one optical cycle) to obtain the time-evolution operator $\hat{U}(0, T)$, followed by an eigen-analysis of $\hat{U}(0, T)$ [8]. The second approach is the Floquet Hamiltonian method in which the time-dependent

Hamiltonian is transformed into an equivalent time- independent infinite-dimensional matrix eigenvalue problem [3,6]. In this section, we discuss the essence of time-independent Floquet Hamiltonian method.

The quasienergy state (QES) function $\Psi_\alpha(\vec{r}, t)$, Eq. (2.6), can be expanded in a Fourier series,

$$(2.16) \qquad \Psi_\alpha(\vec{r}, t) = \exp(-i\epsilon_\alpha t) \sum_{n=-\infty}^{\infty} A_\alpha^{(n)}(\vec{r}) \exp(-in\omega t).$$

Thus a QES can be expressed as a superposition of stationary states with energies equal to $(\epsilon_\alpha + n\omega)$. The functions $A_\alpha^{(n)}(\vec{r})$ of (2.16) can be further expanded in terms of the orthonormal set of unperturbed eigenfunctions of $H_0(\vec{r})$, namely, $\{|\beta(\vec{r})\rangle\}$,

$$(2.17) \qquad A_\alpha^{(n)}(\vec{r}) = \sum_\beta \Phi_{\alpha\beta}^{(n)} |\beta(\vec{r})\rangle.$$

Substituting Eqs. (2.16) and (2.17) into Eq. (2.1), we obtain the following (time-independent) system of equations

$$(2.18) \qquad \sum_n \sum_\beta [\langle\alpha|\hat{H}^{(m-n)}|\beta\rangle - (\epsilon_\alpha + m\omega)\delta_{mn}\delta_{\alpha\beta}]\Phi_{\alpha\beta}^{(n)} = 0,$$

where

$$(2.19) \qquad \hat{H}^{[q]}(\vec{r}) \equiv (\omega/2\pi) \int_0^{2\pi/\omega} H(\vec{r}, t) \exp(iq\omega t) dt.$$

Note that for a linearly polarized monochromatic field, $V \sim \cos\omega t$, only the matrix elements $\langle\alpha|\hat{H}^{[q]}|\beta\rangle$ with $q = 0, \pm 1$ are nonvanishing. The system of equations (2.18) is similar to the system for a constant (i.e. time-independent) perturbation.

It is convenient to introduce the *Floquet-state* nomenclature $|\alpha n\rangle \equiv |\alpha\rangle \otimes |n\rangle$, where α is the system index, and $|n\rangle$ are the Fourier vectors $(n = 0, \pm 1, \pm 2, ...)$ such that $\langle t|n\rangle = \exp(in\omega t)$. The system of equations (2.18) can be recast into the form of a matrix eigenvalue equation:

$$(2.20) \qquad \sum_\gamma \sum_k \langle\alpha n|\hat{H}_F|\gamma k\rangle \Phi_{\gamma\beta}^{(k)} = \epsilon_\beta \Phi_{\alpha\beta}^{(n)},$$

where $\hat{H}_F$ is the *time-independent Floquet Hamiltonian* whose matrix elements are defined by

$$(2.21) \qquad \langle\alpha n|\hat{H}_F|\beta m\rangle = \hat{H}_{\alpha\beta}^{[n-m]} + n\omega\delta_{\alpha\beta}\delta_{nm}.$$

It follows from Eq. (2.20) that the quasienergies are eigenvalues of the secular equation

$$(2.22) \qquad \det|\hat{H}_F - \epsilon I| = 0.$$

As an example, consider the multiphoton excitation (MPE) of the vibrational-rotational states of a diatomic molecule with dipole moment $\vec{\mu}(\vec{r})$ by a monochromatic field with amplitude $\vec{E}_0$, frequency ω, and phase ϕ, respectively. In the electric dipole approximation, the interaction potential energy between the quantum system and the classical field is given by

$$(2.23) \qquad V(\vec{r}, t) = -\vec{\mu}(\vec{r}) \cdot \vec{E}_0 \cos(\omega t + \phi).$$

The Floquet matrix $\hat{H}_F$ possesses a block tridiagonal form as shown in Figure 1. The determination of the vibrational-rotational quasienergies $\epsilon_{\alpha n}$ and quasienergy eigenstates $|\epsilon_{\alpha n}\rangle$ thus reduces to the solution of a time-independent Floquet matrix eigen-problem. Figure 1 shows that $\hat{H}_F$ has a periodic structure with only the number of ω's in the diagonal elements varying form block to block. This structure endows the quasienergy eigenvalues and eigenvectors of $\hat{H}_F$ with the following general periodic properties:

$$(2.24) \qquad \epsilon_{\alpha n} = \epsilon_{\alpha 0} + n\omega,$$
$$(2.25) \qquad \langle \alpha, n+p | \epsilon_{\beta, m+p} \rangle = \langle \alpha, n | \epsilon_{\beta m} \rangle.$$

Consider now the transition probability from an initial quantum state $|\alpha\rangle$ to a final quantum state $|\beta\rangle$. The time-evolution operator $\hat{U}(t, t_0)$, in matrix form, can be expressed as

$$(2.26) \qquad \begin{aligned} \hat{U}_{\beta\alpha}(t, t_0) &\equiv \langle \beta | \hat{U}(t, t_0) | \alpha \rangle \\ &= \sum_n \langle \beta n | \exp[-i\hat{H}_F(t - t_0)] | \alpha 0 \rangle \exp(in\omega t). \end{aligned}$$

Eq. (2.26) shows that $\hat{U}_{\beta\alpha}(t, t_0)$ can be interpreted as the amplitude that a system initially in the Floquet State $|\alpha 0\rangle$ at time t_0 evolve to the Floquet State $|\beta n\rangle$, by time t according to the time-independent Floquet Hamiltonian $\hat{H}_F$, summed over n with weighing factors $\exp(in\omega t)$. The latter interpretation enables one to solve problems involving Hamiltonians periodic in time by methods applicable to time-independent Hamiltonians. The transition probability going from the initial quantum state $|\alpha\rangle$, and a coherent field state to the final quantum state $|\beta\rangle$, summed over all final field states, can now be written as,

$$P_{\alpha \to \beta}(t, t_0) = |\hat{U}_{\beta\alpha}(t, t_0)|^2$$

$$= \sum_k \sum_m \langle \beta k | \exp[-i\hat{H}_F(t - t_0)] | \alpha 0 \rangle \exp(im\omega t_0)$$

$$\langle \alpha m | \exp[-i\hat{H}_F(t - t_0)] | \beta k \rangle.$$

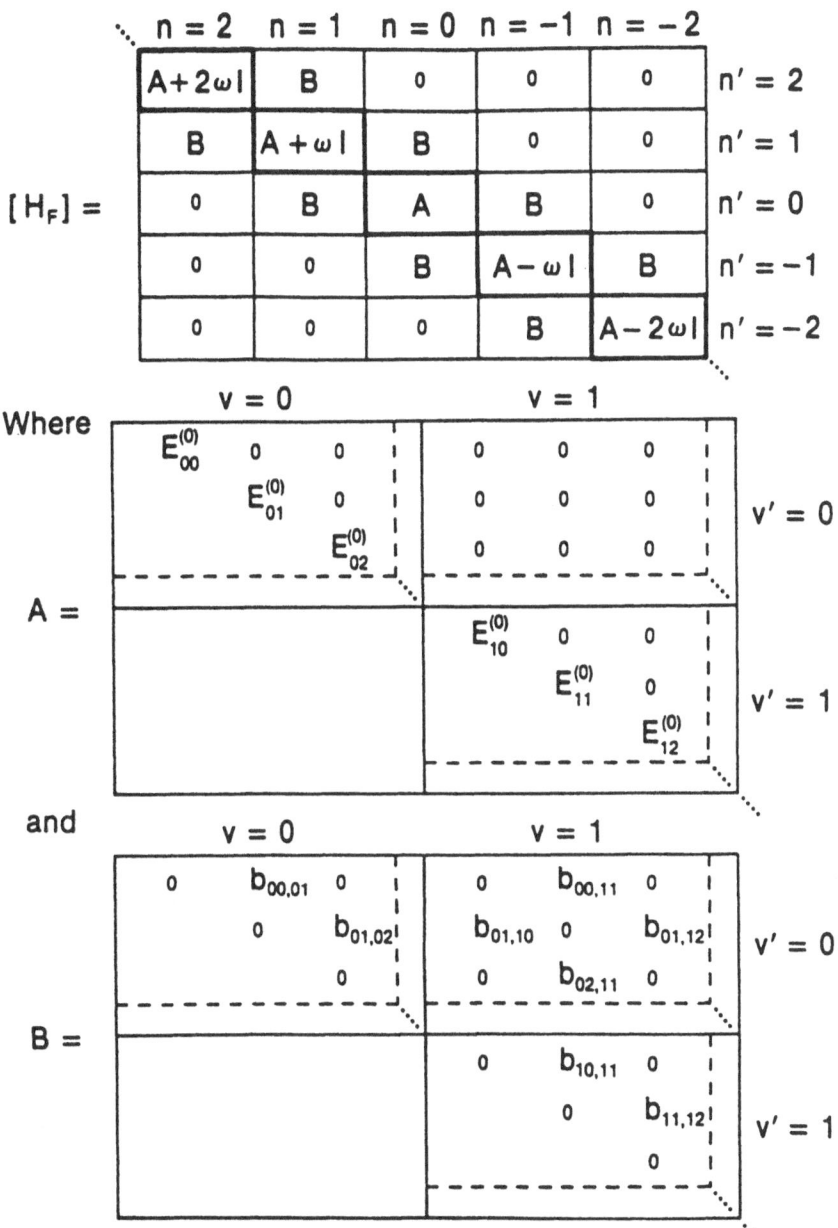

FIG. 1. *Structure of the time-independent Floquet Hamiltonian $\hat{H}_F$ in the Floquet state basis $(|vj, n\rangle)$. Hamiltonian is composed of the diagonal Floquet blocks of type A and the off-diagonal blocks of type B. $E_{vj}^{(0)}$ are unperturbed vibrational-rotational energies and $b_{vj,v'j'}$, are electric dipole coupling matrix elements.*

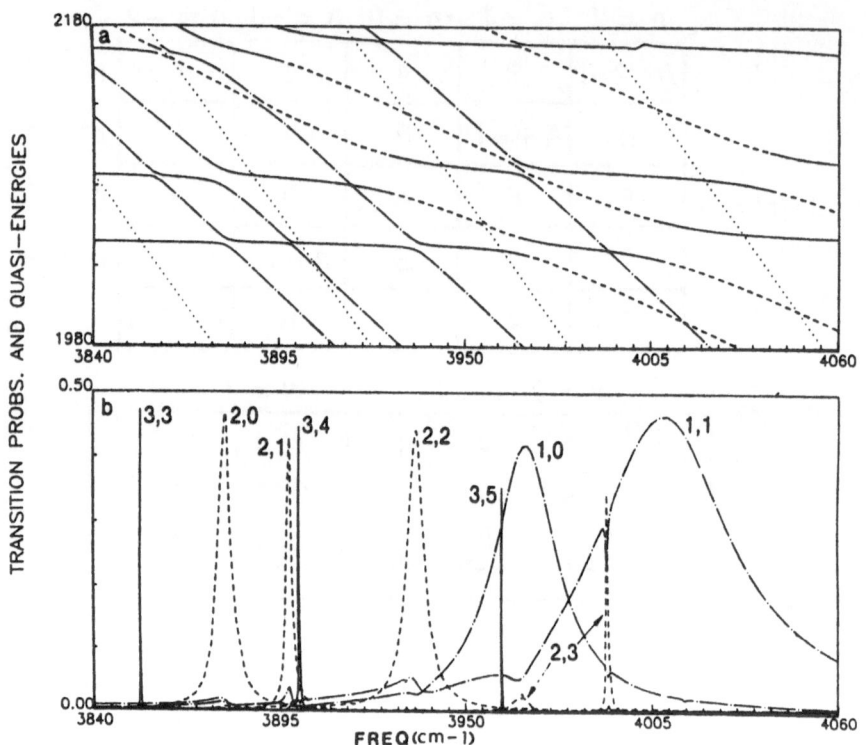

FIG. 2. (a) Quasi-energy plots and (b) time-averaged multiphoton excitation transition probabilities $\overline{P}_{00 \to vj}$ for HF molecule subject to both the laser ($E_{ac} = 1.0 TW/cm^2$) and the dc electric ($E_{dc} = 10^{-4}$ a.u.) fields simultaneously. Dot-dash lines: one-photon peaks. Dashed lines: two-photon peaks. Solid lines: three-photon peaks. Nonlinear effects such as power broadening, dynamical Stark shift, Autler-Townes multiplet splitting, hole burning, and S-hump behaviors are observed and can be correlated with avoided crossing pattern of the quasi-energy levels.

The quantity of experimental interest, however, is the transition probability averaged over initial times t_0 (or equivalently averaged over the initial phases of the field seen by the system), keeping the elapsed time $t - t_0$ fixed. This yields

$$(2.27) \qquad P_{\alpha \to \beta}(t - t_0) = \sum_k |\langle \beta k| \exp[-i\hat{H}_F(t - t_0)]|\alpha 0\rangle|^2.$$

Finally, averaging over $t - t_0$, one obtains the long-time average transition probability

$$(2.28) \qquad \overline{P}_{\alpha \to \beta} = \sum_k \sum_{\gamma \ell} |\langle \beta k|\epsilon_{\gamma \ell}\rangle \langle \epsilon_{\gamma \ell}|\alpha 0\rangle|^2.$$

Much information can be obtained from the plot of quasienergy eigen-values of the Floquet Hamiltonian. The essential feature of the quasienergy plot is illustrated in Figure 2 for the case of HF molecule subject to both laser and dc electric fields [9]. Nonlinear optical effects such as power broadening, a.c. Stark shift, Autler-Townes multiplet splitting, [7] hole burning etc., are observed and may be correlated with the avoided crossing patterns in the quasienergy diagram (Figure 2). Many of the salient features of the spectral line shapes may be qualitatively understood in terms of an analytical three- or four-level model. The addition of a dc electric field spoils the restriction of the rotational dipole selection rule, strongly enhances the multiphoton excitation probabilities and results in a much richer spectrum, in accord with the experimental observations.

3. Beyond the Floquet Theorem: Generalized Floquet formalisms and methods. The conventional Floquet methods described in the last section provide a powerful nonperturbative technique for the treatment of bound-bound multiphoton transitions for simple systems. However, there are several fundamental limitations of the methods for more complicated systems. The major limitations are listed below:

(i) For complex systems such as large molecules or Rydberg atoms involving a large number of energy levels and photons, the dimension of the Floquet matrix can become prohibitively large.

(ii) Conventional Floquet methods with Hermitian Hamiltonians can treat only bound-bound multiphoton excitation (MPE) transitions, but not bound-free and free-free multiphoton ionization (MPI) or multiphoton dissociation (MPD) processes.

(iii) For "multi-color" (multi-frequency) laser excitations in which the laser frequencies $\omega_i (i = 1, 2, ...)$ are *incommensurate*, the Hamiltonian becomes *nonperiodic* in time, and the Floquet theorem is simply *not* valid.

(iv) Conventional Floquet methods (based on the Schrödinger equation)*cannot* deal with nonlinear optical processes with (radiative and collisional) *relaxations*.

(v) For laser fields with arbitrary pulse shape, the Hamiltonians are *nonperiodic* in time and the Floquet theorem is again *not* valid.

These are the common problems encountered by the conventional Floquet techniques. To overcome these difficulties, several generalizations of the Floquet theorem and methods have been developed in the past decade [6]. Listed below are some of these *generalized Floquet formalisms* and associated computational techniques. For a more complete list, see review 6(c).

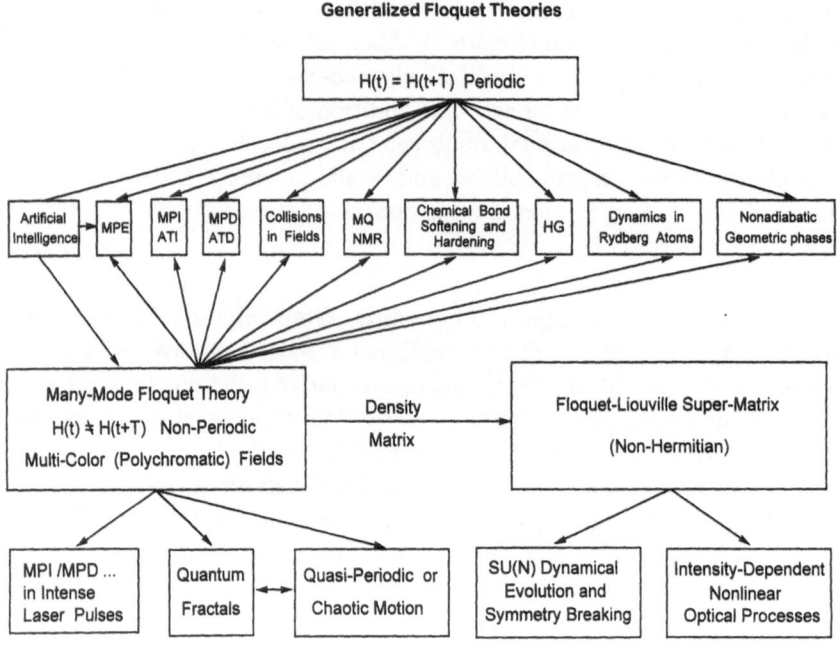

Fig. 3. *Scope of recent generalized Floquet theoretical developments and their applications.*

(i) *Non-Hermitian* Floquet matrix formalism and *complex* quasi-energy methods for nonperturbative treatment of bound-free and free-free MPI/ATI of atoms [10,11] and MPD/ATD [12] of molecules.

(ii) *Many-mode Floquet* theorem (MMFT) [13] for the treatment of *multi-color* (multi-frequency) laser excitations, where the Hamiltonian is polychromatic (non-periodic) in time.

(iii) *Most probable path approach* (MPPA) [14] using *artificial intelligence* (AI) algorithms for selecting the most important multiphoton excitation Floquet-state pathways, allowing many orders of magnitude reduction in the size of Floquet matrix yet maintaining good accuracy.

(iv) *Floquet-Liouville supermatrix* (FLSM) formalism [15] (i.e., non-Hermitian MMFT formulation of density matrix) for nonperturbative treatment of intensity- and frequency-dependent nonlinear optical processes allowing for relaxation mechanisms.

(v) *Stationary* treatments of multiphoton processes in intense (arbitrarily shaped) laser pulses [16].

Figure 3 shows an outline of recent generalized Floquet theoretical developments and their applications pursued at the University of Kansas. Some of these works have been described in previous review articles [6]. Because of space limitation, we shall confine our discussion in this section to only two of these generalized Floquet developments, namely, the many-mode Floquet theorem and the non-Hermitian Floquet matrix method.

3.1. Many-mode Floquet theorem. One of the major limitations of conventional Floquet techniques described in Section II is that they are applicable only to monochromatic (i.e. one-laser-field) problem. However, many recent experiments involve the use of more than one laser field, namely, multi-color or multi-frequency excitation (with frequencies ω_i incommensurate). In such cases, the Hamiltonians are no longer periodic in time and the Floquet theorem is simply not valid. Such limitation has been removed with the development of the so-called *many-mode Floquet theorem* (MMFT) [6,13]. The theorem allows the exact transformation of any *polychromatic* time-dependent Schrödinger equation into an equivalent time-*independent* infinite-dimensional generalized Floquet eigenvalue problem.

Without loss of generality, let us consider the interaction of an arbitrary N-level system with two monochromatic radiation fields. In the electric dipole approximation, the Schrödinger equation can be written as ($\hbar = 1$)

$$(3.1) \qquad i\partial\Psi(\vec{r},t)/\partial t = \hat{H}(\vec{r},t)\Psi(\vec{r},t)$$

where the Hamiltonian $\hat{H}(\vec{r},t)$ is *bichromatic* in time,

$$(3.2) \qquad \hat{H}(\vec{r},t) = H_0(\vec{r}) - \vec{\mu}(\vec{r}) \cdot [\vec{E}_1(t) + \vec{E}_2(t)].$$

H_0 and μ are respectively the unperturbed Hamiltonian and the dipole moment of the system, $\vec{E}_1$ and $\vec{E}_2$ classical fields given by

$$(3.3) \qquad \vec{E}_i(t) = Re[E_i\hat{\zeta}_i e^{-i\omega_i t}],$$

where Re signifies "the real part of," and E_i , $\hat{\zeta}$ and ω_i are respectively the electric field amplitude, the polarization vector, and the frequency associated with the ith field. The MMFT theorem states that the exact solution of the time-dependent Schrödinger equation, with the Hamiltonian Eq. (3.2), has the following explicit form:

$$(3.4) \qquad \Psi(\vec{r},t) = \exp(-i\epsilon t)\Phi(\vec{r},t),$$

where ϵ is the generalized (two-mode) quasienergy, and $\Phi(t)$ is *bichromatic* in time. Thus the concept of quasienergy is preserved even when the Hamiltonian is no longer periodic in time.

$$[\mathbf{H_F}] = \begin{bmatrix} A + 2\omega_2 I & B & 0 & 0 & 0 \\ B & A + \omega_2 I & B & 0 & 0 \\ 0 & B & A & B & 0 \\ 0 & 0 & B & A - \omega_2 I & B \\ 0 & 0 & 0 & B & A - 2\omega_2 I \end{bmatrix}$$

WHERE

$$A = \begin{bmatrix} C + 2\omega_1 I & X & 0 & 0 & 0 \\ X & C + \omega_1 I & X & 0 & 0 \\ 0 & X & C & X & 0 \\ 0 & 0 & X & C - \omega_1 I & X \\ 0 & 0 & 0 & X & C - 2\omega_1 I \end{bmatrix}$$

$$C = \begin{bmatrix} E_\alpha & 0 & 0 \\ & E_\beta & 0 \\ & & E_\gamma \end{bmatrix} \qquad X = \begin{bmatrix} 0 & V_{\alpha\beta}^{(1)} & V_{\alpha\gamma}^{(1)} \\ & 0 & V_{\beta\gamma}^{(1)} \\ & & 0 \end{bmatrix}$$

$$B = \begin{bmatrix} \begin{array}{ccc} 0 & V_{\alpha\beta}^{(2)} & V_{\alpha\gamma}^{(2)} \\ & 0 & V_{\beta\gamma}^{(2)} \\ & & 0 \end{array} & 0 \\ 0 & \begin{array}{ccc} 0 & V_{\alpha\beta}^{(2)} & V_{\alpha\gamma}^{(2)} \\ & 0 & V_{\beta\gamma}^{(2)} \\ & & 0 \end{array} \end{bmatrix}$$

FIG. 4. *Floquet Hamiltonian for two linearly polarized radiation field problems constructed in a symmetric pattern. Here ω_1 and ω_2 are the two radiation frequencies, $E_\alpha, E_\beta, \ldots$ are unperturbed energies of H_0, and $V_{\alpha\beta}^{(i)} (i = 1, 2)$ is the electric dipole coupling matrix element between unperturbed states $|\alpha>$ and $|\beta>$ by the ith field. Note that the diagonal block A possesses an identical Floquet structure to that of one laser problem. Figure 4 can be generalized to the N-field problem.*

The many-mode Floquet theory allows the *exact* transformation of the bichromatic time-dependent problem, Eq. (3.1), into an equivalent *time-independent* infinite-dimensional matrix eigenvalue problem [13]

$$(3.5) \quad \sum_{\gamma_2} \sum_{k_1} \sum_{k_2} \langle \gamma_1 n_1 n_2 | \hat{H}_F | \gamma_2 k_1 k_2 \rangle \langle \gamma_2 k_1 k_2 | \lambda \rangle = \lambda \langle \gamma_1 n_1 n_2 | \lambda \rangle,$$

where

$$(3.6) \quad \begin{aligned} &\langle \gamma_1 n_1 n_2 | \hat{H}_F | \gamma_2 k_1 k_2 \rangle \\ &= H_{\gamma_1 \gamma_2}^{[n_1 - k_1, n_2 - k_2]} + (n_1 \omega_1 + n_2 \omega_2) \delta_{\gamma_1 \gamma_2} \delta_{n_1 k_1} \delta_{n_2 k_2}, \end{aligned}$$

with

$$(3.7) \quad H_{\gamma_1 \gamma_2}^{[n_1, n_2]} = E_{\gamma_1} \delta_{\gamma_1 \gamma_2} \delta_{n_1 0} \delta_{n_2 0} + \sum_{i=1}^{2} V_{\gamma_1 \gamma_2}^{(i)} (\delta_{n_i, 1} + \delta_{n_i, -1}),$$

$$(3.8) \quad V_{\gamma_1 \gamma_2}^{(i)} = -\frac{1}{2} E_i \langle \gamma_1 | \vec{\mu} \cdot \vec{\zeta_i} | \gamma_2 \rangle.$$

Here $\hat{H}_F$ is the (time-independent) two-mode Floquet Hamiltonian defined in terms of the *generalized* Floquet basis state $|\gamma n_1 n_2\rangle = |\gamma > \otimes | n_1 > \otimes | n_2 \rangle$, with γ being an atomic (or molecular) state of H_0, and the integer index $n_i (= 0, \pm 1, \pm 2, ...)$ a Fourier component of the ith field. Figure 4 depicts the structure of the two-mode Floquet Hamiltonian for the linear polarization case. The components are ordered in such a way that γ runs over unperturbed states (denoted by Greek letters) of H_0 before each change in n_1 and n_1, in turn, runs over before n_2. The quasienergy eigenvalues $\{\lambda_{\gamma n_1 n_2}\}$ and their corresponding eigenvectors $\{|\lambda_{\gamma n_1 n_2}\rangle\}$ of $\hat{H}_F$ have the following useful *bichromatic* forms, namely,

$$(3.9) \quad \lambda_{\gamma n_1 n_2} = \lambda_{\gamma 00} + n_1 \omega_1 + n_2 \omega_2,$$

and

$$(3.10) \quad \langle \gamma_1, n_1 + q_1, n_2 + q_2 | \lambda_{\gamma_2, n_1 + q_1, n_2 + q_2} \rangle = \langle \gamma_1 n_1 n_2 | \lambda_{\gamma_2 n_1 n_2} \rangle.$$

The time evolution operator $\hat{U}(t, t_0)$ can be expressed in the following matrix form

$$\hat{U}_{\beta \alpha}(t, t_0) \equiv \langle \beta | \hat{U}(t, t_0) | \alpha \rangle$$

$$(3.11) \quad = \sum_{n_1 = -\infty}^{\infty} \sum_{n_2 = -\infty}^{\infty} \langle \beta n_1 n_2 | \exp[-i \hat{H}_F (t - t_0)] | \alpha 00 \rangle \exp[i(n_1 \omega_1 + n_2 \omega_2)t].$$

The transition probability averaged over the initial time t_0 while keeping the elapsed time $t - t_0$ fixed is given by

$$(3.12) \quad P_{\alpha \to \beta}(t - t_0) = \sum_{k_1 k_2} |\langle \beta k_1 k_2 | \exp[-\hat{H}_F (t - t_0)] | \alpha 00 \rangle|^2.$$

Performing the long time average over $t - t_0$ gives the time averaged transition probability

$$(3.13) \qquad \overline{P}_{\alpha \to \beta} = \sum_{k_1 k_2} \sum_{\gamma \ell_1 \ell_2} |\langle \beta k_1 k_2 | \lambda_{\gamma \ell_1 \ell_2} \rangle \langle \lambda_{\gamma \ell_1 \ell_2} | \alpha 00 \rangle|^2.$$

Application of many-mode Floquet theory to various multiphoton processes can be found in the review articles [6].

3.2. Non-Hermitian Floquet Hamiltonian methods for multiphoton ionization and dissociation.

The time-independent Floquet matrix methods described in previous sections involving *Hermitian* Floquet Hamiltonians provide nonperturbative *ab initio* techniques for the treatment of *bound-bound* multiphoton transitions. These methods cannot, however, be applied directly to *bound-free* and *free-free* transitions such as multiphoton ionization (MPI) of atoms or multiphoton dissociation (MPD) of molecules. One of the major generalizations of the Floquet theory is the development of *non-Hermitian* Floquet matrix methods, [10,12] invoking the use of the method of *complex scaling transformation* [17] and L^2-continuum discretization [18] techniques, for MPI/MPD processes. In this section we discuss the essence of *non-Hermitian Floquet Hamiltonian* method for periodic perturbation. Extension to *multi-color* MPI/MPD processes can be similarly made by means of the MMFT.

Applying the complex scaling transformation, $\vec{r} \to \vec{r} e^{i\alpha}$, to the Schrödinger equation, we obtain from Eqs. (2.1)-(2.2),

$$(3.14) \qquad i\partial \Psi(\vec{r} e^{i\alpha}, t)/\partial t = H(\vec{r} e^{i\alpha}, t)\Psi(\vec{r} e^{i\alpha}, t),$$

where $H(\vec{r} e^{i\alpha}, t)$ is now a *non-Hermitian* periodic Hamiltonian. The QES wavefunction $\Psi(\vec{r} e^{i\alpha}, t)$, can be written, according to the Floquet theorem,

$$(3.15) \qquad \Psi(\vec{r} e^{i\alpha}, t) = \exp(-i\epsilon t)\Phi(\vec{r} e^{i\alpha}, t),$$

where ϵ is the *complex* quasienergy, and the periodic function $\Phi(\vec{r} e^{i\alpha}, t) = \Phi(\vec{r} e^{i\alpha}, t + \tau)$ satisfies the eigenvalue equation

$$(3.16) \qquad \hat{H}(\vec{r} e^{i\alpha}, t)\Phi(\vec{r} e^{i\alpha}, t) = \epsilon \Phi(\vec{r} e^{i\alpha}, t),$$

where $\hat{H}$ is given in Eq. (2.3). Following the procedure described in Sec. 2.3, Eq. (3.16) can be converted into a matrix eigenvalue equation, except that $\hat{H}_F(\vec{r} e^{i\alpha}) \equiv \hat{H}_F(\alpha)$ is now an analytically continued, time-independent *non-Hermitian* Floquet Hamiltonian. The complex scaling transformation distorts the continuous spectrum away from the real axis, exposing the a.c. Stark resonances (i.e. complex quasienergy states) in appropriate higher Riemann sheets, and also allowing use of finite variational expansions employing L^2 basis function chosen from a complete discrete basis [10,11]. The use of a complete L^2 basis obviates the necessity of explicit introduction of exact atomic or molecular bound and continuum states, thus reducing all computations to those involving finite-dimensional non-Hermitian matrices. The use of complex coordinates not only allows direct calculation of eigenvalue parameters associated with complex quasienergy states, but completely avoids numerical problems arising from strong coupling between overlapping atomic or molecular continua. The real part of the complex quasienergy (E_R) provides the a.c. Stark shifted energy, whereas the imaginary part ($\Gamma/2$) determines the total MPI or MPD rate (width). In the remaining of this section, we discuss the implementation of the non-Hermitian Floquet Hamiltonian method.

3.2.1. Implementation and solution of the non-Hermitian Floquet Hamiltonian: Complex-scaling generalized pseudospectral technique.

The complex quasienergy eigenvalues and eigenfunctions of the non-Hermitian Floquet Hamiltonian $\hat{H}_F(\alpha)$ are normally solved by means of the L^2 basis set expansion-variational method [6,10,11,12]. Recently we have introduced an alternative procedure, the *complex-scaling generalized pseudospectral* (CSGPS) method, [19,20] for accurate and efficient solution of the resonance state problems, including the complex quasienergy states. The method does not require the computation of potential matrix elements (which is usually the most time-consuming part of atomic and molecular structure calculations using the conventional basis set expansion-variational method), is simple to implement, and provides the values of the wavefunctions directly at the space grid points. As has been shown elsewhere [19,20], the generalized pseudospectral methods are far more efficient and accurate than the finite difference method and computationally more efficient and advantageous than the basis set expansion-variational method. The generalized pseudospectral method is a natural extension of the Fourier-grid Hamiltonian (FGH) method [21] and its generalizations [22,23] recently developed for the studies of atomic and molecular bound and resonances. The FGH methods employ Fourier series and require the mesh points to be equally spaced. The generalized pseudospectral methods employ orthogonal polynomials (such as Legendre or Chebyshev polynomial) and allow for uneven mesh spacing. It has been shown recently that the generalized FGH methods work well for potentials without singularity, such as the Morse potential for chemical bond etc. However, for problems involving singularity and/or long range potentials (such as the Coulomb potential), the generalized pseudospectral method with appropriate mapping is the more effective approach [19,20].

4. Recent applications of generalized Floquet methods to atomic and molecular multiphoton processes in intense laser fields.

The generalized Floquet formalisms and methods have been applied to a number of atomic and molecular multiphoton processes in intense and superintense laser fields. Examples of recent studies include

(i) Multiphoton and above-threshold ionization (MPI/ATI) of atoms [6,11,24];

(ii) A.C. Stark shifts of Rydberg states and intensity-dependent ionization potential [6,25];

(iii) Multiphoton and above-threshold dissociation (MPD/ATD) of diatomic molecules [26];

(iv) Multiphoton detachment of H^- [20,27,28,29];

(v) Stabilization and ionization suppression of neutral atoms [30] and negative ions [31] in superintense high-frequency laser fields;

(vi) Laser-induced chemical bond-softening and hardening and molecular stabilization [32,33];

(vii) Two-color phase control of MPI/ATI of atoms [28], MPD/ATD of molecules [33], and multiple high-order harmonic generation [34];

(viii) Generalized geometric phases in multiphoton and nonlinear optical processes [35];

(ix) Laser-assisted ion-atom charge exchange reactions [36];

(x) Quantum fractal behavior of quasienergy states in polychromatic fields [37].

In this section, we confine our discussion to several recent case studies and illustrate some of the novel high-intensity phenomena uncovered.

4.1. Intensity-dependent ionization potential and threshold shift [6,25] . In the presence of strong laser fields, all the bound states are coupled to the continuum and become complex quasienergy resonance states possessing complex eigenvalues $(E_R, -\Gamma/2)$. The real parts of the complex quasienergies, E_R, provide the a.c. Stark shifts of the energy levels, whereas the imaginary parts (Γ) give rise to total multiphoton ionization (MPI) widths (rates). These complex quasienergies can be determined by the non-Hermitian Floquet matrix formalism described in the last section.

The ionization potential in intense fields may be defined as

$$(4.1) \qquad \epsilon_{th}(F) = \bar{\epsilon}_{osc} + |E_R(F)|,$$

where F is the (peak) field strength, $E_R(F)(< 0)$ is the field-dependent perturbed ground-state energy obtained from the complex quasienergy calculation, and

$$(4.2) \qquad \bar{\epsilon}_{osc} = e^2 F^2/4m\omega^2,$$

is the average *quiver* kinetic energy (also known as the *pondermotive potential*) [38] picked up by an electron of mass m and charge e driven sinusoidally by the field. Since in the limit of high quantum numbers, a Rydberg electron becomes a free electron, the continuum threshold is shifted up by the amount equal to $\bar{\epsilon}_{osc}$. Electrons traversing a laser beam scatter elastically from regions of high light intensity by the pondermotive potential. Thus an electron with energy $\epsilon_{el}(F)$ less than $\bar{\epsilon}_{osc}$ cannot escape from the Coulomb potential and is trapped. From Eq. (4.1), we can define the *threshold shift* as

(4.3) $$\Delta\epsilon_{th}(F) = \epsilon_{th}(F) - \epsilon_{th}(F = 0),$$

where $\epsilon_{th}(F = 0)$ is the field-free ionization threshold.

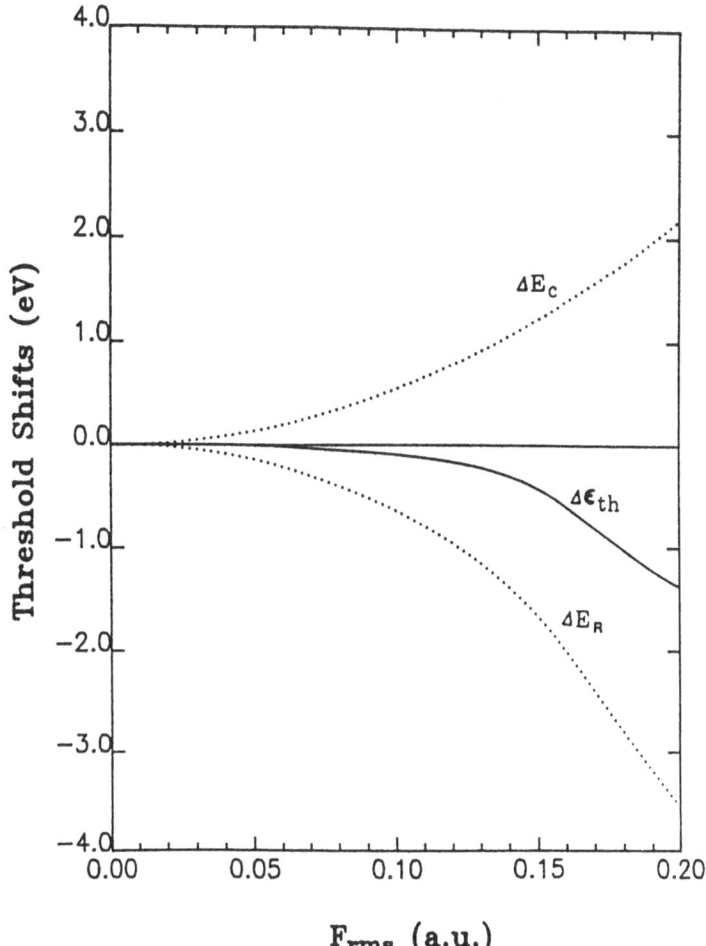

FIG. 5. *Intensity-dependent threshold shift* $\Delta\epsilon_{th}(F) = \Delta E_c + \Delta E_R$ *for* $\omega = 0.5$ *a.u.* ($N_m = 1$). *Here* $\Delta E_R = E_R(0) - E_R(F)$ *is the ac Stark shift of the ground state of atomic H, while* $\Delta E_C = e^2 F_{rms}^2/2m\omega^2$ *is the continuum threshold upshift.*

The total energy of the emitted electron in the field can be written as

(4.4) $$\epsilon_{el}(F) = N\hbar\omega + E_R(F) = e^2F^2/4m\omega^2 + P_T^2/2m,$$

where $N = (N_m + S)$ is the total number of photons absorbed by the electron near the atom. Since a free electron cannot absorb or emit photons after leaving the Coulomb field, the electron has an energy ϵ_{el} which is the same in the laser field as it is at the detector. Thus the pondermotive potential acts to alter the kinetic energy $(P_T^2/2m)$ of the electron from its value outside the laser field to a lower value inside the laser.

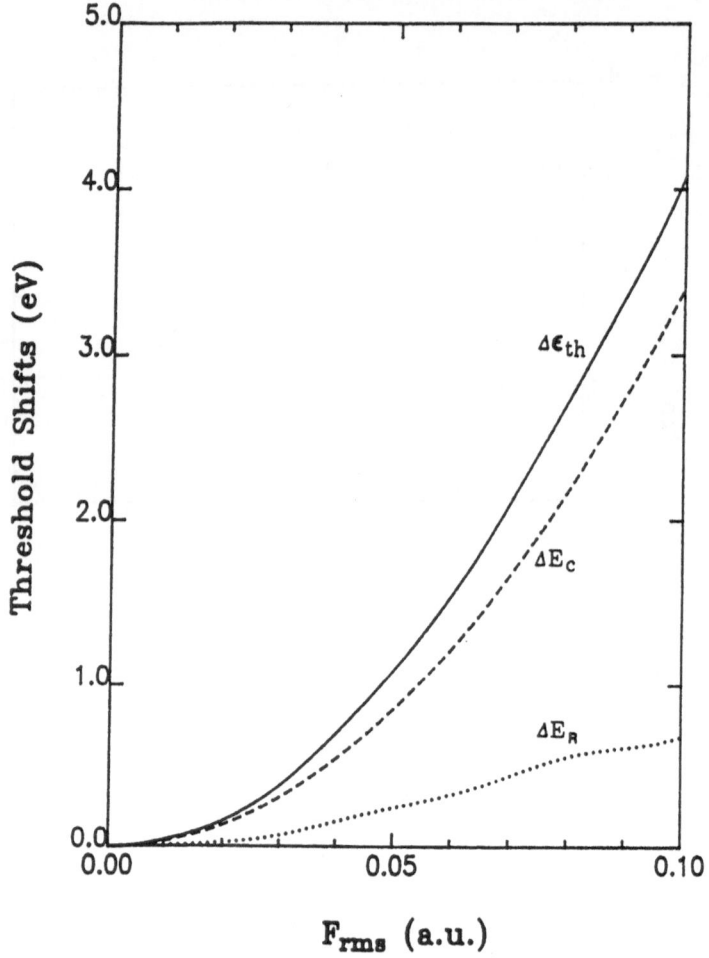

FIG. 6. *Intensity-dependent threshold shifts of atomic H for $\omega = 0.2$ a.u. ($N_m = 3$).*

Figures 5 and 6 show typical examples of intensity and frequency dependent threshold shifts of atomic hydrogen in intense laser fields, where $\Delta E_R = E_R(F = 0) - E_R(F)$ is the ac Stark shift, $\Delta E_C = e^2 F^2/4m\omega^2$ is the continuum threshold upshift due to pondermotive potential, and $\Delta\epsilon_{th}(F)$ is the net threshold shift defined by Eq. (47) and is equivalent to the sum of ΔE_R and ΔE_C. Figure 5 shows the threshold shifts typical to the one-photon ($N_m = 1$) dominant process ($\omega \geq 0.5 a.u.$) while Figure 6 shows the typical phenomena for multiphoton ($N_m = 3$ in this case) dominant process ($\omega < 0.5a.u.$). Note the marked difference between the two cases. For $N_m = 1$ (Figure 5), both the ground state ($E_R(F) > E_R(0)$) and the continua are upshifted with the ac Stark shift $|\Delta E_R|$ being greater than ΔE_C. The resulting net threshold shift $\Delta\epsilon_{th}(F)$ becomes more negative as the field strength increases. Hence the ionization potential *decreases* with increasing F. On the other hand, for $N_m \geq 2$ such as the case $\omega = 0.2$ a.u. shown in Figure 6, the ground state energy shifted downward ($E_R(F) < E_R(0)$) while ΔE_C shifts the continuum threshold upward. The result is a large positive net threshold shift and the ionization potential *increases* rapidly with increasing field strength F. As a general rule, the pondermotive potential ΔE_C becomes more and more important than the ac Stark shift $|\Delta E_R|$ as N_m increases or ω decreases. The consequence is that the ionization potential increases rather rapidly with both F and N_m. The disappearance of the lowest energy electrons in the MPI/ATI experiments of atoms [1], for example, can be attributed to this threshold shift effect.

4.2. A. C. Stark shifts of Rydberg states in strong fields. Giant ac Stark shifts of high-lying atomic states in strong fields have been observed recently [39]. While perturbation calculation can be performed to arbitrary excited states, it is valid only for very weak fields. The behavior of ac Stark Shift of Rydberg atoms in strong fields is a largely unexplored area of research. In this section, we discuss some recent theoretical results obtained from a generalized Floquet technique recently developed, using the Sturmian basis [25]. The method allows nonperturbative treatment of a.c. Stark shifts of arbitrary excited states. Figure 7 shows the a.c. Stark shifts of $n = 12$ atomic states of atomic H for $\ell = 0, 1, 2, 3$. Several essential energy-shift behaviors of excited states are noted: (a) All the excited levels shown are shifted upward and closely follow the shift caused by the pondermotive potential $\bar{\epsilon}_{osc}$ (shown by dotted curves) in the weaker-field region. This effective potential has its origin in the A^2 term (where A is the vector potential) and has been shown (in Sec. 4.1) to be equal to the average quiver kinetic energy picked up by an electron of mass m and charge e driven sinusoidally by the fields. The results lend further support to the view that all Rydberg states and the continuum are upshifted by the same amount, described by $\bar{\epsilon}_{osc}$. However, this description appears valid only in the weak-field or non-resonant transition regime where no strong mixings exist among atomic states. (b) Above some critical field strengths (F_c), the atomic energy levels (for a given n but different ℓ) split, and significant deviation from the A^2 curve occurs. The critical field strength F_c depends on n and decreases rather rapidly as n increases. One should therefore use the A^2 pondermotive shift law with caution in the interpretation of energy-level shifts in high-intensity MPI/ATI experiments. (c) For $F > F_c$, strong mixings exist among nearby atomic states, and the level identities usually cannot be discerned.

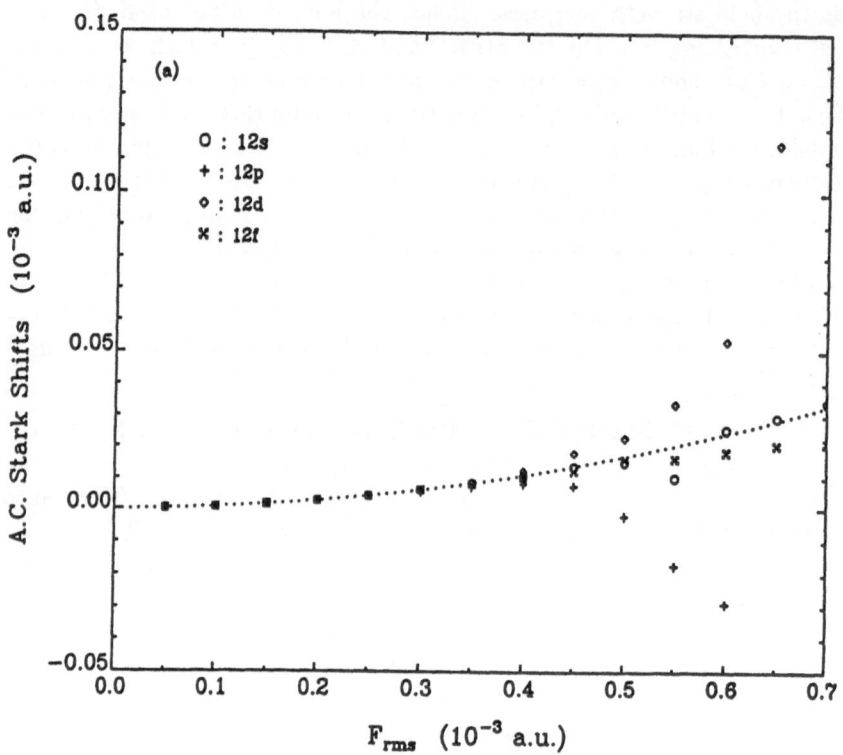

FIG. 7. *Field-dependent A.C. Stark shifts of $n = 12$, $\ell = 0, 1, 2, 3$ states of atomic hydrogen. The dotted curve is the pondermotive potential quadratic shift.*

Figure 8 shows the intensity-dependent energy-level shift pattern for highly excited states ($n = 49, 50, 51, 52$) [25]. (For clarity, only the even-parity (s, d, g) states are shown. The level-shift pattern is similar to that shown in Figure 7, except that the critical field strengths F_c are now considerably lower. For $F > F_c$, a large departure from the A^2 shift occurs, and strong inter-n mixings take place. This behavior is expected to prevail for all Rydberg levels.

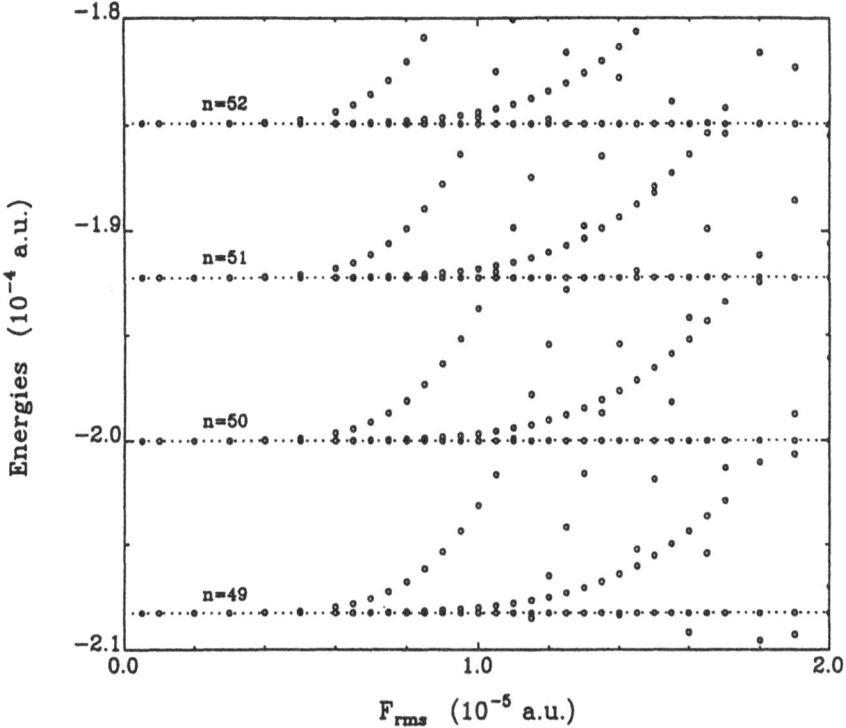

FIG. 8. *Field-dependent energy-level shift pattern for even-parity* (s, d, g) *states of Rydberg hydrogen atoms* $(n = 49 - 52)$. *Pondermotive potential shifts are shown in dotted lines.*

4.3. Multiphoton detachment of H^-. The study of multiphoton detachment of H^- is a subject of much current interest, stimulated mainly by the recent experimental work at LAMPF in Los Alamos [40,41]. The experimental setup uses the relativistic Doppler effect, allowing the continuous tuning of the laser frequency (in the atom frame) over a wide range of photon energies. For moderately strong laser intensities (10^9 to $10^{11} W/cm^2$) used in the experiments, doubly excited states are far above the ionization threshold of the ground state and can thus be safely ignored.

In a recent study, we have constructed an accurate one-electron model potential [42] which reproduces precisely the known H^- detachment energy and the low-energy $e - H(1s)$ elastic-scattering phase shifts. The model potential so constructed produces essentially exactly the one-photon detachment cross sections [43] obtained from ab initio two-electron correlated

calculations, demonstrating its accuracy. We then use this model potential
to pursue the following studies.

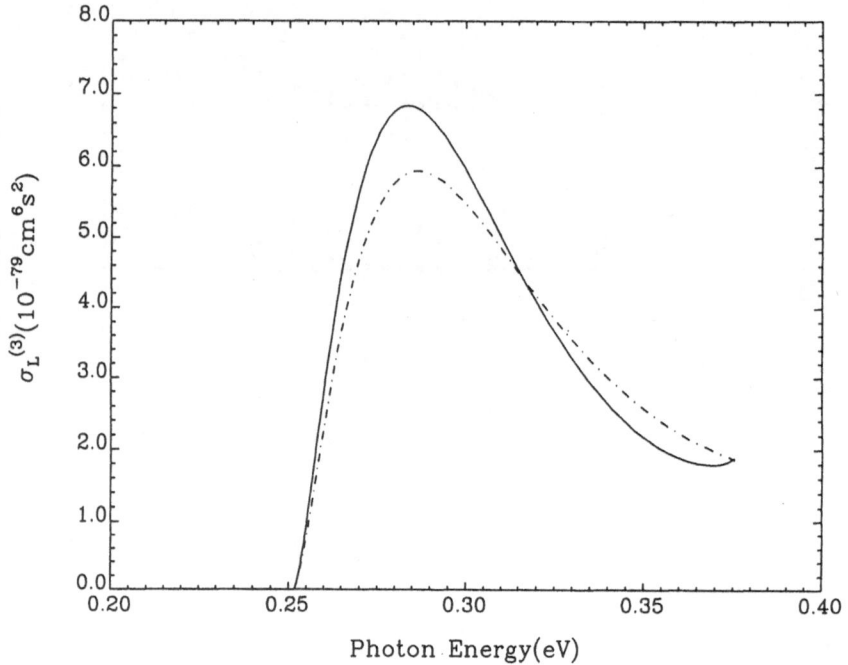

FIG. 9. *Generalized three-photon deetachment cross sections* $\sigma_L^{(3)}$. *Solid curve: present perturbative calculation. Dashed-dotted curve: zero-range plane-wave (ZRPW) approximation.*

4.3.1. Perturbative calculations. We first performed high-order
perturbative calculations of multiphoton detachment of H^-. Generalized
n-photon detachment cross sections are evaluated based on the lowest non-
vanishing order perturbation theory [42]. An accurate and efficient numerical procedure for the solution of the associated set of inhomogeneous differential equations was developed. Two- to eight-photon detachment cross
sections were determined. (Figures 9–10 show the examples of 3– and

8–photon results). Comparison with several existing ab initio two-electron calculations was made. It appears that the present study provides the first consistent results with the accuracy of high-order ($n > 3$) photon detachment cross sections comparable to that of lower orders ($n = 2, 3$). Intensity-averaged multiphoton detachment rates for linearly polarized light are calculated and compared with recent experimental measurements [41]. The overall agreement is generally rather good except nearby each n-photon threshold where the perturbation theory tends to predict sharper transition than that of experimental results (Figure 11). This suggests *nonperturbative* treatment of the multiphoton processes is necessary.

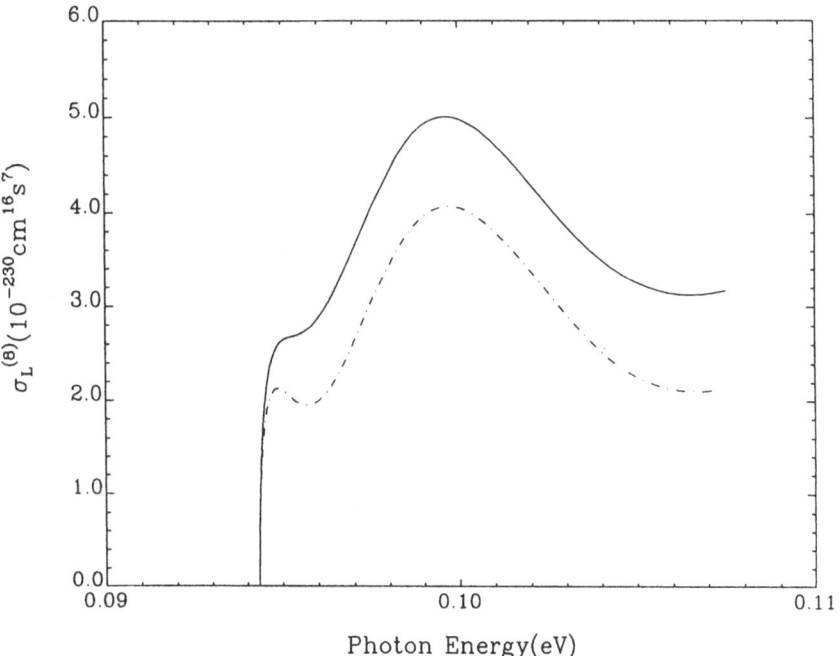

FIG. 10. *Generalized eight-photon detachment cross sections* $\sigma_L^{(8)}$. *Solid curve: present perturbative calculation. Dashed-dotted curve: ZRPW results.*

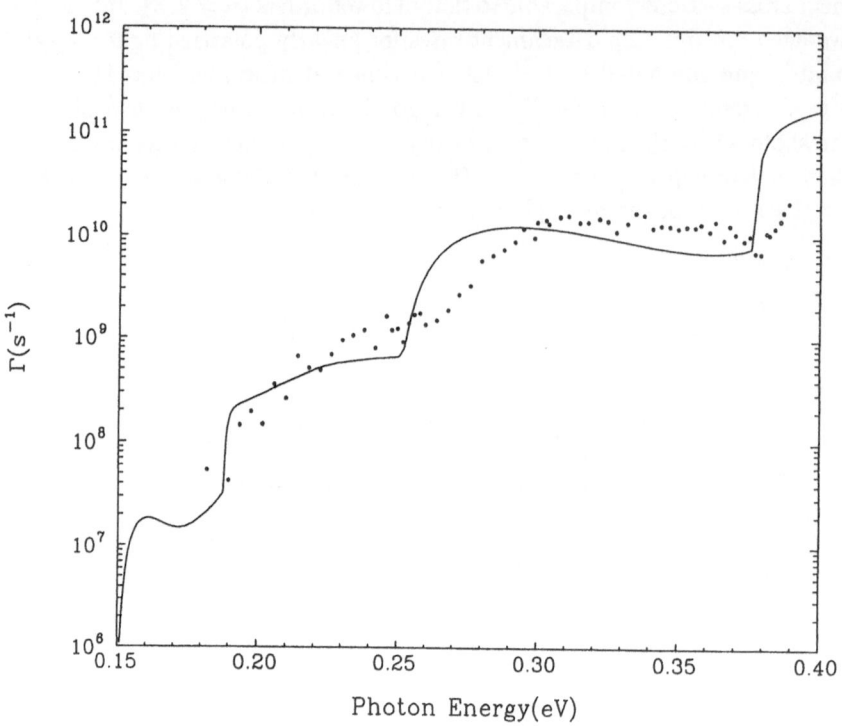

FIG. 11. *Comparison of experimental and calculated intensity-averaged multiphoton detachment rates of H^- for linearly polarized Gaussian pulses with peak laboratory-frame intensity of 4 GW/cm^2. The theoretical results are obtained from perturbative calculations.*

4.3.2. Nonperturbative study of intensity-dependent multiphoton detachment rates and threshold behavior. Since H^- has a small binding energy, the intensity range employed in the Los Alamos experiments requires theoretical treatment beyond the perturbation theory. We thus perform nonperturbative study and consider the intensity-dependent multi-photon detachment rates and threshold behavior [20]. The complex-scaling generalized pseudospectral (CSGPS) method is used for

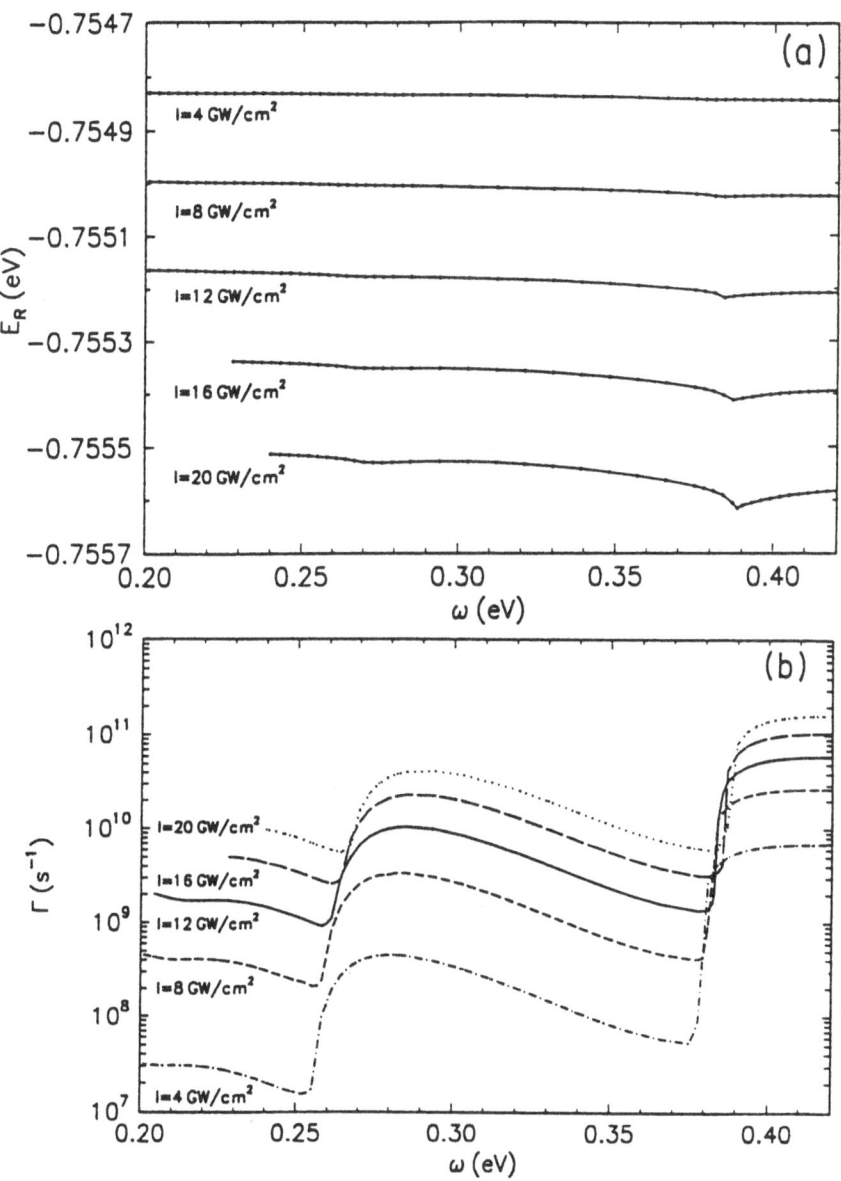

FIG. 12. *The frequency- and intensity-dependent complex quasienergies* $(E_R, -\Gamma/2)$ *of* H^- *for* $I = 4, 8, 12, 16,$ *and* $20 GW/cm^2$ *and* $\omega = 0.20 - 0.42 eV$: *(a)* E_R *(real energies), showing the ac Stark shifts of* H^- *ground state, and (b)* Γ *(imaginary energies), showing the multiphoton detachment rates.*

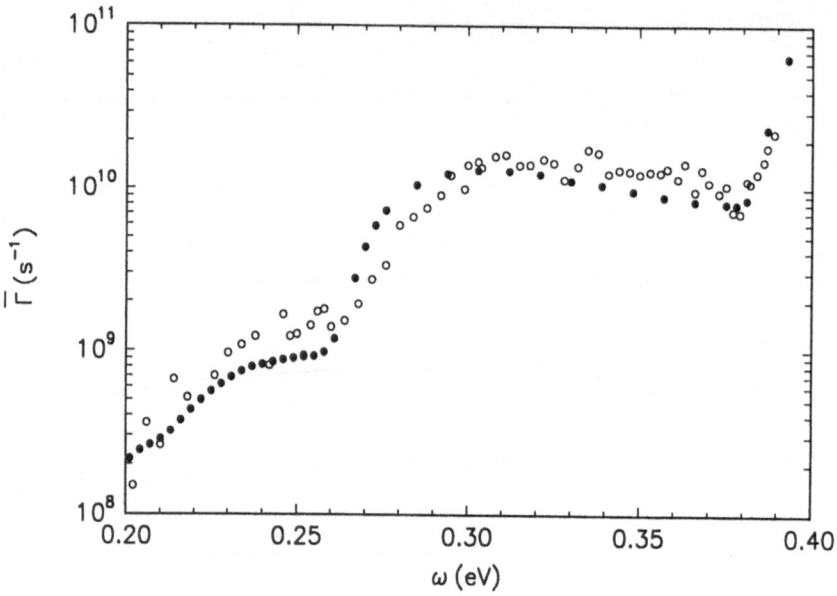

FIG. 13. *Comparison of intensity-averaged multiphoton detachment rates of H^- for the case of laser intensity (in the laboratory frame) = $4GW/cm^2$. Theoretical prediction ..., Los Alamos experimental data o o oo.*

the discretization and accurate treatment of the time-independent non-Hermitian Floquet Hamiltonian $\hat{H}_F$. We obtained detailed nonperturbative results of the intensity- and frequency-dependent complex quasi-energies $(E_R, -\Gamma/2)$, the complex eigenvalues of H_F, providing directly the a.c. Stark shifts and multiphoton detachment rates (widths) of H^-. The laser intensity considered ranges from 1 to $40GW/cm^2$ and the laser frequency covers 0.20 to 0.42 eV (in the c.m. or atom frame). Figure 12 shows typical examples of the results. Finally to compare with the experimental results, we perform a simulation of the intensity-averaged multiphoton detachment rates by considering the experimental conditions of the laser and H^- beams. The results (without any free parameters) are in very good agreement with the recent Los Alamos experimental data [41] (which contain an error bar of a factor of five) both in the absolute magnitude and in the threshold behavior [see Figure 13].

4.3.3. Above-threshold multiphoton detachment in one-color laser fields: Angular distributions and partial widths.

In the presence of strong fields, multiphoton above-threshold ionization (ATI), [44] [namely, electrons can absorb more photons than necessary to get ionized], can occur. It is essential to consider the angular distributions and partial rates associated with the ATI processes. We have recently developed a general theory [27] for the nonperturbative treatment of the angular distribution and partial rates in monochromatic fields through a formal analysis of the integral form of the Schrödinger equation. Further, we present a procedure for the "back rotation" of the total complex scaled quasienergy wavefunctions to the real axis, allowing the evaluation of the partial rates and angular distribution in an accurate and more straightforward manner. The total complex scaled quasienergy wavefunctions are obtained by the non-Hermitian Floquet formulation and the complex-scaling generalized pseudospectral discretization technique mentioned in the last section. Alternatively, a nonperturbative *complex-scaling adiabatic* theory [27] can be formulated for more efficient determination of the complex quasienergy wavefunctions. The adiabatic Floquet theory is valid when the laser frequency is smaller than the binding energy which is the case for multiphoton detachment of H^- in Los Alamos experiments.

The method is applied to the first nonperturbative study of multiphoton ATI of H^- ions [27]. We perform detailed study of the angular distribution and partial rates for ATI of H^- in moderate strong laser intensities (10^{10} to $10^{11} W/cm^2$) at $10.6\mu m$, involving the absorption of 20-40 photons. Figures 14(a)–14(c) show the ATI spectra for three different laser intensities and Figure 15 shows an example of the angular distribution from different electron peaks.

4.3.4. Multiphoton and above-threshold ionization in two-color laser fields.

The works described above are for one-color fields. Recently there are also considerable interest in the study of ATI processes in two-color fields. We have recently presented a general theory of electron energy and angular distributions and partial rates in two-color [28] (two-frequency) fields for both commensurate and incommensurate frequencies based on the extension of the many-mode Floquet theorem [13] and the integral form of the Schrödinger equation. The theory is applied to the study of multiphoton above-threshold detachment of H^- for the fundamental frequency of CO_2 laser (wavelength $10.6\mu m$) and its third harmonic [28]. The calculations are performed for the fundamental field intensity $10^{10} W/cm^2$ and harmonic field intensities 10^9 and $10^8 W/cm^2$. The results show the following novel features. First, the total and partial rates for the two-color detachment, when the harmonic field is relatively strong (such as 10 times smaller than that of fundamental field), are generally much larger than

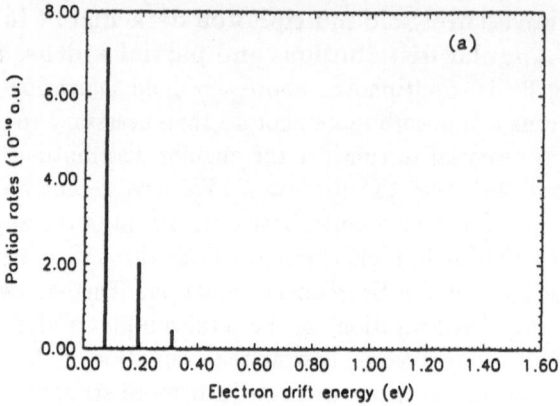

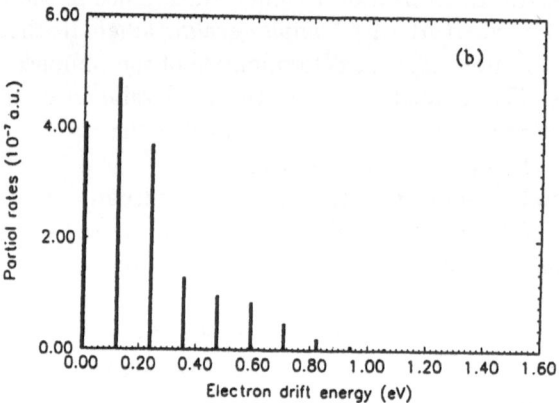

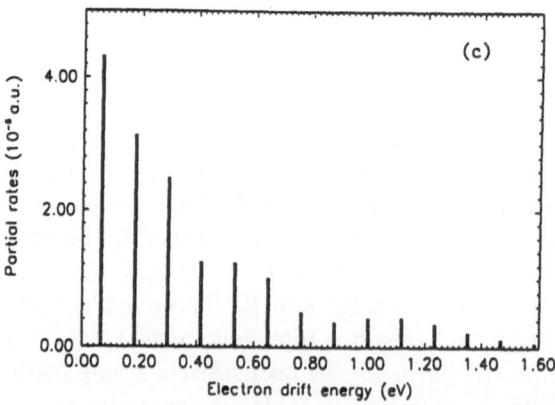

Fig. 14. *(a) Electron energy distribution after multiphoton above-threshold detachment of H^- by 10.6μm, 10^{10} W/cm^2 laser field. The heights of the bars correspond to the partial rates after absorption of n photons, starting with $n_{min} = 8$. (b) The same as Figure 14(a) for the intensity $5 \times 10^{10} W/cm^2$; $n_{min} = 11$. (c) The same as Figure 14(a) for the intensity $10^{11} W/cm^2$; $n_{min} = 16$.*

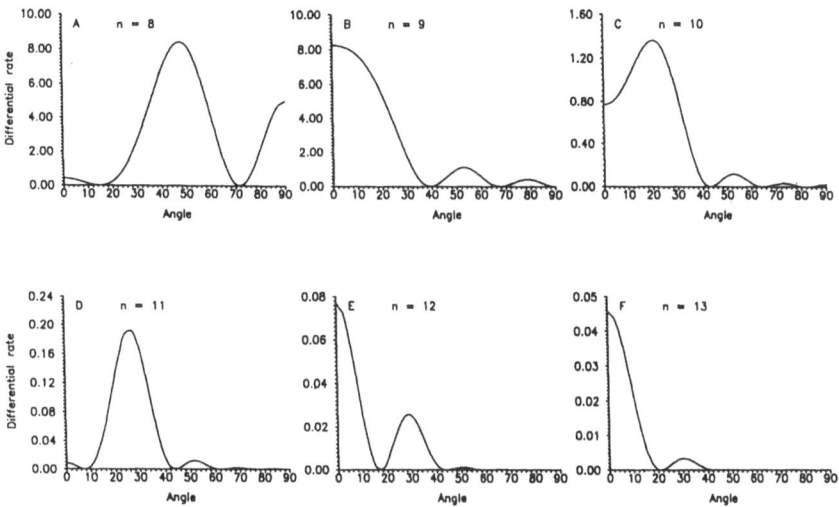

FIG. 15. *Electron angular distributions* $(1/\sin\theta)$ $(d\Gamma_n/d\theta)$ *(in units of* 10^{-10} *a.u.) for the first six above-threshold peaks in the energy spectrum, for the laser intensity* 10^{10} *W/cm*^{2} *and the wavelength* 10.6μm. *Graphs A-F correspond to the number of absorbed photon* $n = 8 - 13$, *as indicated.*

the rates for the one-color detachment by the fundamental or harmonic field alone. However, the opposite situation is also possible if the harmonic field is weaker (say, 100 times smaller than that of fundamental field intensity), and relative phase δ is close to π. Second, the total and partial rates manifest a strong dependence on the relative phase between the two fields. The total rate is the largest for the phase $\delta = 0$, and the smallest for $\delta = \pi$. Such a dependence on the relative phase is also valid for the first few ATI peaks. However, for the subsequent ATI peaks, the picture is quite different. The energy spectrum for the case of $\delta = \pi$ is broader and the peak heights decrease more slowly compared to the case of $\delta = 0$. The strong phase dependence is also manifested in the angular distributions of the ejected electrons. Figures 16 and 17 show some typical results of the ATI energy and angular distributions in two-color laser fields [28].

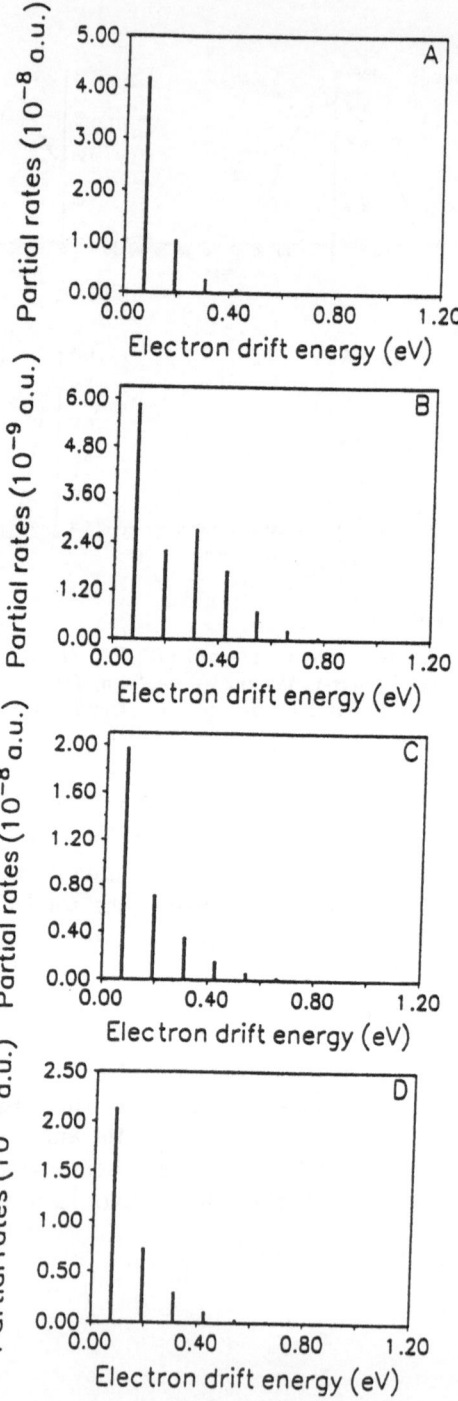

FIG. 16. *Electron energy distrubtions after multiphoton above-threshold detachment from H^- by 10.6μm radiation and its third harmonic with the intensities $I_L = 10^{10}$ and $I_H = 10^9$ W/cm^2, respectively. The heights of bars correspond to the partial rates after absorption of n fundamental frequency photons, starting with $n_{min} = 8$. Phase shift (A) $\delta = 0$, (B) $\delta = \pi$, (C) $\delta = \pi/2$, (D) $\delta = -\pi/2$.*

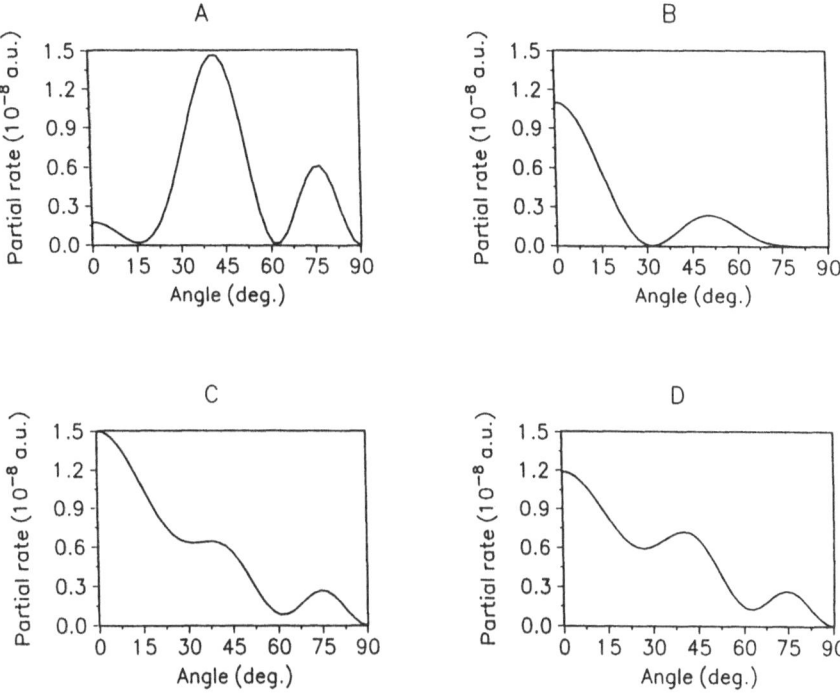

FIG. 17. *Electron angular distributions* $(1/\sin\theta)$ $(d\Gamma/d\theta)$ *for the nine-photon peak in the energy spectrum. Graphs A-D correspond to detachment of* H^- *by* $10.6\mu m$ *radiation and its third harmonic with the intensities* $I_L = 10^{10}$ *and* $I_H = 10^9$ W/cm^2 *and phase shift (A)* $\delta = 0$, *(B)* $\delta = \pi$, *(C)* $\delta = \pi/2$, *(D)* $\delta = -\pi/2$.

4.3.5. A new adiabatic approach for multiphoton detachment by intense laser pulses.

Recent advances in short-pulse high-power laser technology have generated considerable interest in the study of MPI/ATI of atoms and negative ions in intense laser *pulsed* fields [44]. To facilitate such a study, we have recently presented an integral equation formulation for the nonperturbative treatment of MPI/ATI processes [29]. A new *adiabatic* approach is introduced [29]. It is based on the concept of *adiabatic* quasienergy states and allows to predict the oscillatory behavior of the electron energy spectra. It is shown that for the laser pulse with slowly varying envelope $F(t)$, the electron energy distribution can be expressed via the detachment transition amplitudes in the *monochromatic* fields: [29]

$$\frac{dP(\vec{k})}{dE d\Omega} = \frac{k}{(2\pi)^3} |\sum_n \int_{-\infty}^{\infty} dt \exp[iE(\vec{k})t + i(2\omega)^{-2} \int_{-\infty}^{\infty} F^2(\tau)d\tau$$

$$(4.5) \qquad\qquad -i \int_{-\infty}^{t} \Delta\epsilon(\tau)d\tau - i(E^{(0)} + n\omega)t]A_n(\vec{k}, t)|^2,$$

where $\vec{k}$ is the electron momentum, $E^{(0)}$ is the unperturbed energy of the initial state and $\Delta\epsilon(t)$ is a complex value including the instantaneous a.c. Stark shift and (minus) halfwidth of the state. The monochromatic amplitudes $A_n(\vec{k})(t)$ depend on time via the field envelope $F(t)$.

The adiabatic approach is valid provided that (i) the laser frequency is smaller than the binding energy; (ii) the laser pulse shape is smooth and contains at least tens of optical cycles of the laser frequency; and (iii) only one adiabatic quasienergy is involved. Such conditions are well satisfied for the case of negative ions. We apply the theory to the study of the pulse shape effects on the electron energy and angular distributions for the multiphoton above-threshold detachment of H^- at $10.6\mu m$ [29].

Figure 18 shows a typical example of the angle-integrated electron energy spectrum dP/dE for the peak intensity $5 \times 10^{10} W/cm^2$ (Gaussian pulse for three different pulse lengths T: (A) $T = 5ps$; (B) $T = 10ps$; (C) $T = 20ps$). Also shown in Figure 18(D) is the monochromatic-field results with the same intensity. The oscillatory satellite structures to the main peaks can be attributed to the interference of the electrons detached on the rising and falling edges of the pulse. Simple analytical formulas describing these subpeak structures can be constructed [29].

4.4. Stabilization of negative ions in superintense high-frequency laser fields. Under normal conditions, one would expect that atoms will become less stable and the ionization rates increase with increasing laser intensity. However, in the case of high-frequency excitation (i.e. laser frequency larger than the binding energy), it has been theoretically predicted that atoms can become more stabilized at higher fields, contrary to the normal expectation. This subject of the stabilization or ionization suppression of neutral atoms in high-frequency superintense laser fields has attracted considerable recent attention [29]. However, stabilization of ground-state neutral atoms (such as H and rare gas atoms) cannot be realized under current experimental capabilities.

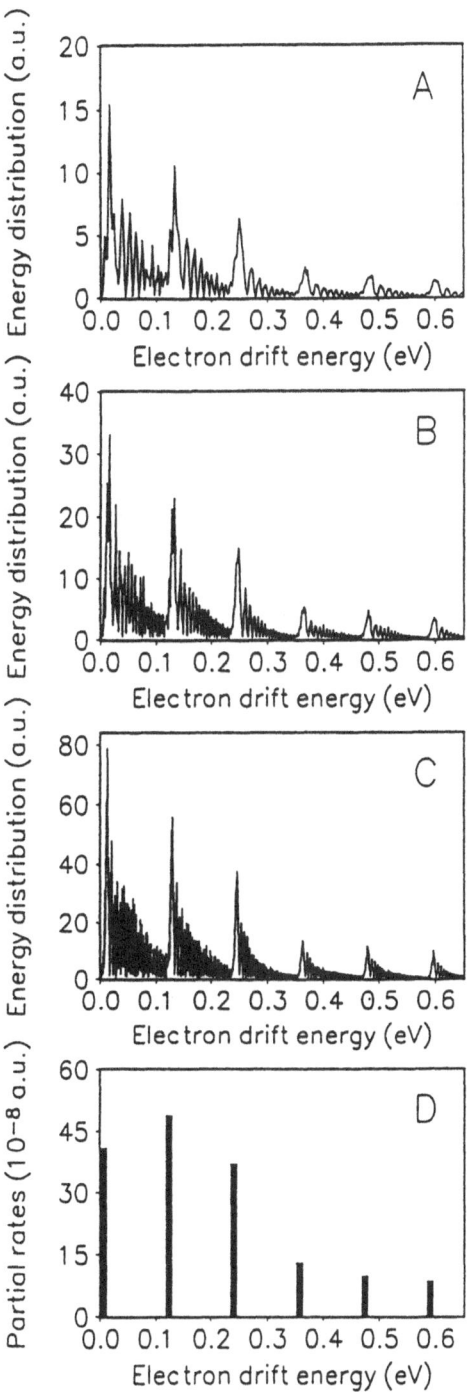

FIG. 18. *Angle-integrated electron energy spectra dp/DE for multiphoton above-threshold detachment of H^- in a Gaussian pulse with peak intensity $5 \times 10^{10} W/cm^2$, and pulse length T. (a) T = 5ps, (B) T = 10ps, (C) T = 20ps, (D) monochromatic results.*

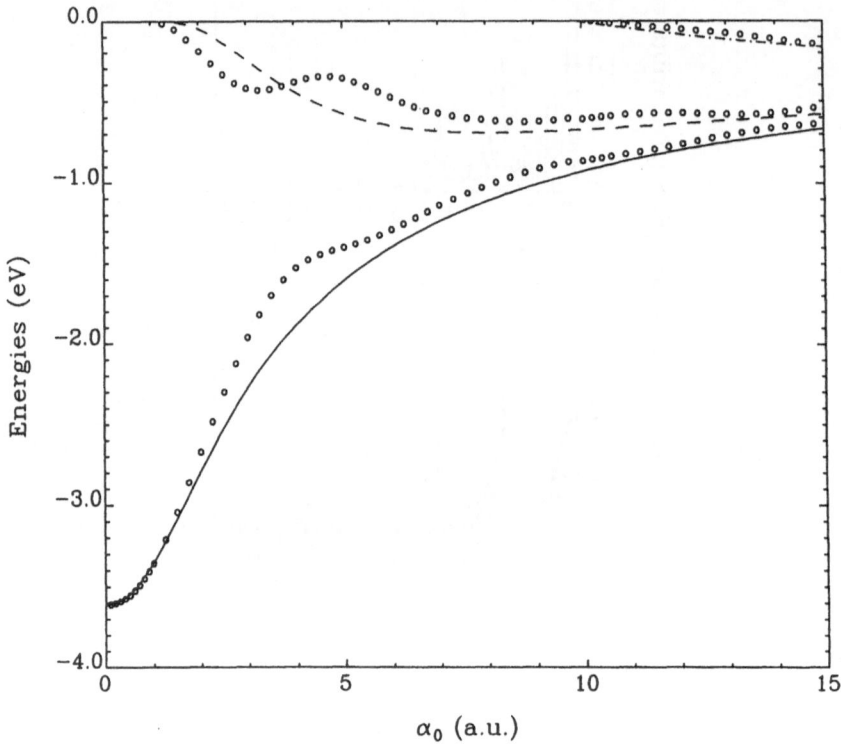

FIG. 19. *Energies of the three field-deformed bound states of the model $C\ell^-$ ion as a function of α_0. The solid, dashed and dash-dotted lines correspond to the energies of these states when high-order couplings $V_n(n \neq 0)$ are neglected whereas the circle lines are the corresponding exact coupled-channel Floquet results (the real parts of the complex quasienergies).*

Recently we have considered the feasibility of observing the stabilization of negative ions in superintense laser fields. Negative ions have smaller binding energies than those of neutral atoms, and the high-frequency and high-intensity conditions required for stabilization [30] may be met by the state-of-the-art of excimer laser technologies. We developed a complex-scaling Fourier-grid Hamiltonian (CSFGH) technique in *momentum* representation [31] for the discretization and solution of the non-Hermitian Floquet Hamiltonian. The method is applied to the nonperturbative treatment of the complex quasienergies $(E_R, -\Gamma/2)$ for a model $C\ell^-$ ion in an ArF excimer-laser field [31]. A variety of laser intensities are considered covering from perturbative to very intense laser fields and the stability of the ion is investigated under realistic experimental conditions. We report some novel new phenomena. First note that $C\ell^-$ has only one bound state in the absence of laser fields. In the presence of strong fields, however, new bound or quasi-bound states may occur. Figure 19 shows the energies of the lowest three bound states, supported by the field-deformed potential curves, versus α_0, where $\alpha_0 = eE_0/m\omega^2$. [$\alpha_0$ is a measure of the strength of the laser-atom coupling. For $\alpha_0 = 10.0$ a.u., for example, it corresponds to the laser intensity of roughly $10^{16}W/cm^2$ for laser frequency fixed at 193 nm]. Figure 19 shows that two extra bound states can be formed in the presence of strong fields. The laser modifies the dressed potential and produces extra new bound states. These three bound states are in fact not bound states but quasi-bound states and possess complex quasi-energies, the imaginary parts of their energies, $\Gamma/2$, are related to their multiphoton detachment rates or lifetimes.

Shown in Figure 20(a)-20(c) are the photodetachment rates (Γ) of the model negative $C\ell^-$ ion at 193 nm for a range of laser intensities (α_0) from the weak to the very intense regime. Figure 20(a) corresponds to the detachment rate of the ground state and Figures 20(b) and 20(c) to that of the first and of the second new bound states. Striking oscillatory structures of the photodetachment rates as a function of field strength are found, revealing that the stabilization phenomenon may not necessarily be a monotonic behavior. Physical insight into this interesting feature is explored in the Kramers-Henneberger frame, and the feasibility for the experimental observation of the oscillatory behavior of the negative ions is discussed in ref. 31.

4.5. Multiphoton and above threshold dissociation of molecules: Laser-induced molecular stabilization and chemical bond softening and hardening. It has long been known that multiphoton dissociation (MPD) of polyatomic molecules is an efficient process [45,46] and can occur in relatively weak infrared laser fields. In contrast, MPD of small molecules, particularly diatomic molecules, is a very slow and inefficient process, due to the low density and anharmonicity of vibrational states. Indeed, MPD of diatomic molecules from the *ground* vibrational states of diatomic molecules has never been observed experimentally until 1986. The only exception, as far as diatomic molecules are concerned, is the experimental observation of two-photon dissociation from *highly excited* vibrational states of HD^+ [47]. Theoretical studies [48] have shown that MPD from the weaker-bound high vibrational levels exhibiting large amplitude vibration is usually far more efficient than from those (tighter-bound) low-lying levels, when the laser intensity is weak.

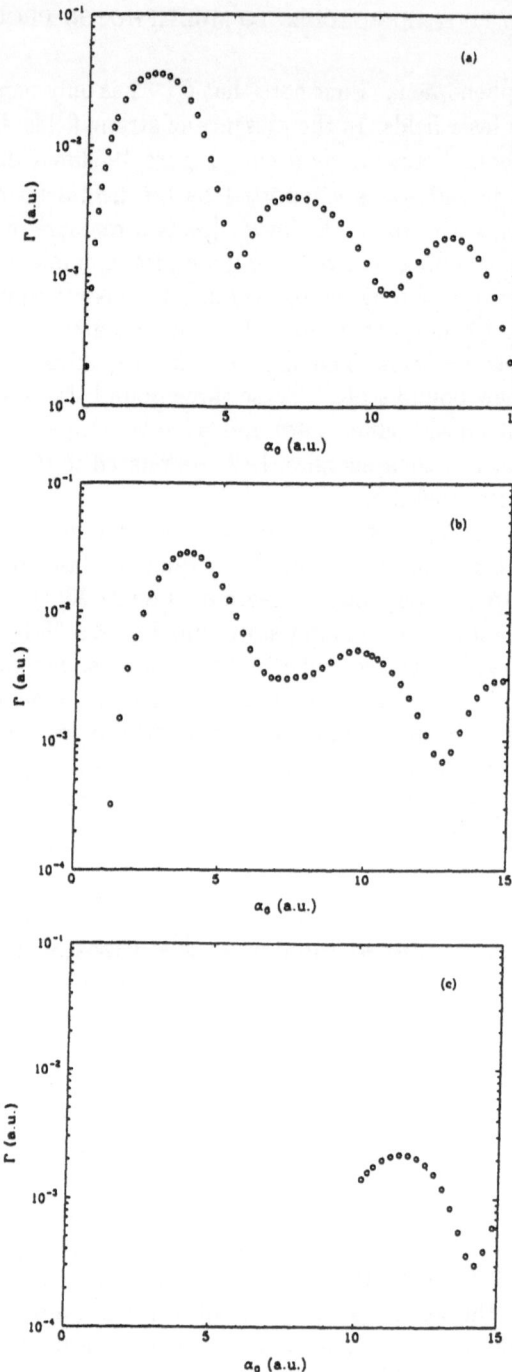

FIG. 20. *Photodetachment rates (widths Γ) obtained from the CSFGH method. The system being studied is a model $C\ell^-$ ion subject to a ArF laser field with $\gamma = 193nm$ (6.424 eV). Here $\alpha_0 = eE_0/m\omega^2$ (E_0 is the electric field amplitude) is related to the laser field intensity. For example, $\alpha = 7.0$ a.u. corresponds to 5.33×10^{15} W/cm^2 for the present case. The essential feature here is that both the ground state (Figure 20(a)) and the (field-induced) first and second excited states (Figures 20(b)–(c)) show oscillatory behaviors.*

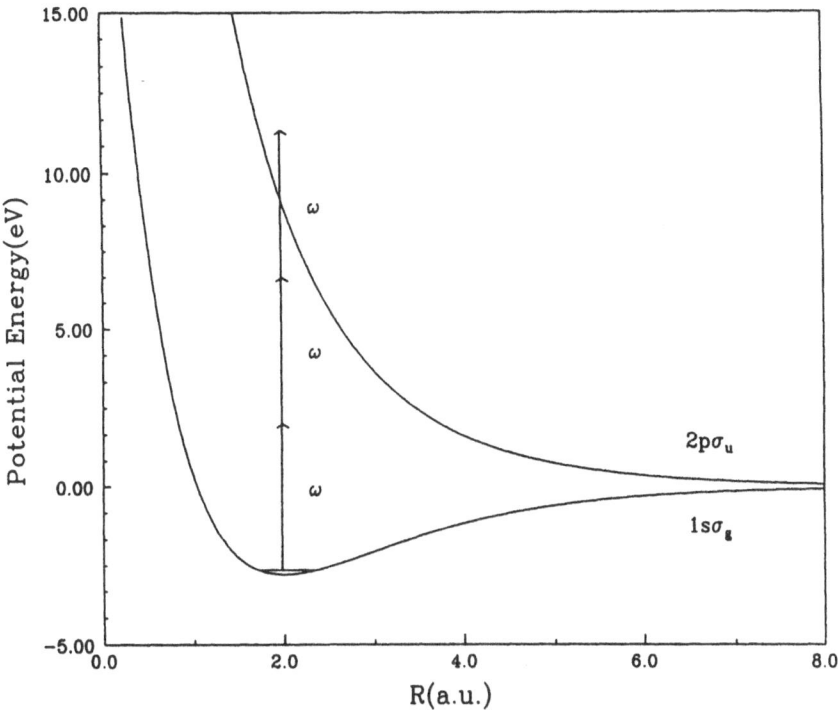

FIG. 21. *Potential-energy curves for the ground* $(1s\sigma_g)$ *and first excited* $(2p\sigma_u)$ *states of* H_2^+ *as a function of internuclear separation R. Also displayed is a schematic diagram showing the absorption of one, two, and three photons of wavelength 2660 Å. from the ground vibrational level of the* $(1s\sigma_g)$ *state.*

Stimulated by the discoveries of a number of novel nonlinear phenomena in the response of atoms to strong laser fields, there is now a growing new interest in the study of nonlinear multiphoton dissociation (MPD) dynamics of diatomic molecules, [26,32,33,49-52] particularly molecular hydrogen. In this section we discuss laser-induced molecular structure deformation and novel high-intensity phenomena recently uncovered. As will be shown below, the presence of additional interatomic degrees of freedom in molecules enriches greatly the problem of the nonlinear interaction of molecules with intense laser fields.

Consider the response of the H_2^+ molecules to intense monochromatic laser fields. Figure 21 shows the potential energy curves of the ground $(1s\sigma_g)$ and first excited $(2p\sigma_u)$ states of H_2^+ as a function of internuclear separation R. Also displayed is a schematic diagram showing the multiphoton and above threshold dissociation (MPD/ATD) process from the ground vibrational level of the $1s\sigma_g$ state. In the presence of external electromagnetic fields, all the vibrational levels of the H_2^+ molecules in the ground ($1s\sigma_g$) electronic state are coupled to the dissociative continuum of the upper (repulsive) electronic state $2p\sigma_u$, and thus become (shifted and broadened) *vibrational quasi-energy* (VQE) resonances. Each VQE resonance possesses an intensity- and frequency-dependent complex energy eigenvalue ($E_R, -\Gamma/2$), the real part of which is related to the ac Stark shift and the imaginary part (width) provides the total MPD/ATD rate. Figure 22 shows the dressed-state (electronic-field potential energy curves) picture (solid lines: diabatic curves; dotted lines: adiabatic curves). Each curve corresponds to $U_i(R)+n\hbar\omega$, where $U_i(R)$ are the electronic potential energy for $1s\sigma_g$ or $2p\sigma_u$ states, and $n = 0, -1, -2, -3$ are the Fourier photon indices. Formally, the photodissociation or multiphoton dissociation between a bound and a repulsive electronic states is a half-collision process and can be regarded as a (diabatic) curve-crossing or an (adiabatic) avoided-crossing predissociation problem.

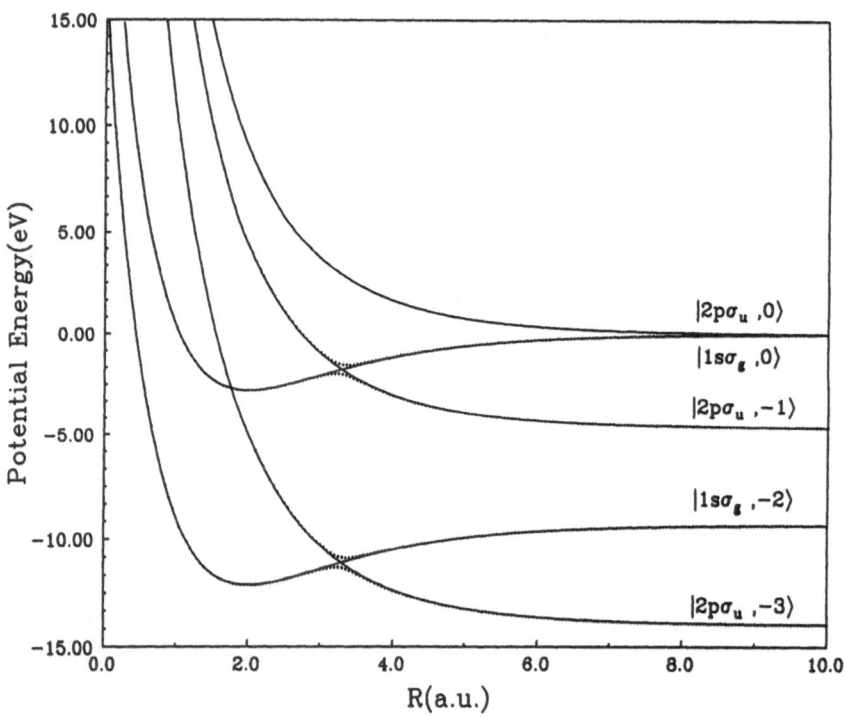

FIG. 22. *Electronic-field potential-energy curves of the two electronic states of H_2^+ dressed by $n = 0, -1, -2, -3$ photons of wavelength 2660 Å. Solid lines: diabatic curves. Dotted lines: adiabatic curves.*

In the following we describe some novel laser-induced VQE structure predicted from a new nonperturbative procedure based on the extension of complex quasi-vibrational energy formalism [12] and the complex-scaling Fourier-grid Hamiltonian (CSFGH) method [26]. The procedure is simple to implement and allows accurate (exponentially convergent w.r.t. the number of grid points) and efficient determination of complex VQEs of both low lying and highly excited resonance states *without* the need of using L^2 basis set expansion.

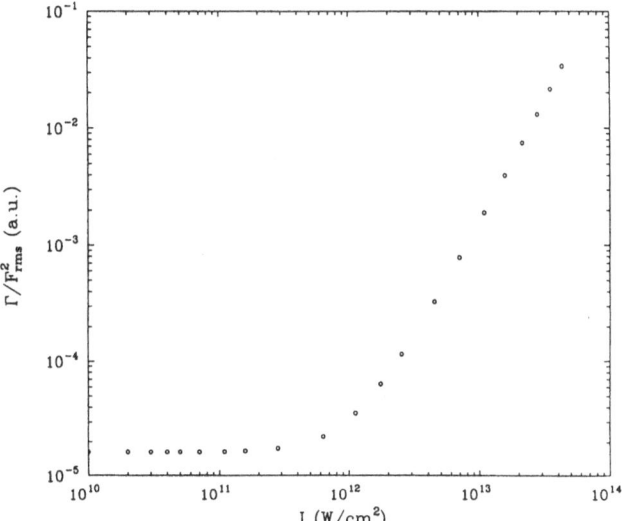

FIG. 23. *Reduced widths* (Γ/F_{rms}^2) *vs intensity I, for the ground vibrational level* $(v = 0)$ *of the* $H_2^+(1s\sigma_g)$ *state at* $\lambda = 2660$ *Å. At weaker fields, the photodissociation is a dominant one-photon process. Above some critical field intensity* $(\sim 10^{12}W/cm^2)$, *MPD/ATD become significant and dominant in the photodissociation process.*

Figure 23 shows the intensity-dependent MPD half-widths ($\Gamma/2$), obtained from the CSFGH method, as a function of the laser intensity I for the *ground* vibrational level ($v = 0$) of the $H_2^+(1s\sigma_g)$ electronic state. At weaker fields, Γ/F_{rms}^2 (proportional to Γ/I) is seen to be independent of the laser intensity I, and the photodissociation is dominantly a one-photon process. Above some critical field intensity ($I \simeq 10^{12}W/cm^2$), ATD sets in, and the process becomes highly nonlinear [26].

We now discuss vibrational quasienergy resonance structure in intense laser fields as well as laser-induced molecular *stabilization* and chemical bond *softening* and *hardening* phenomena [32(a)-(b)]. Figures 24–26 show the real (E_R) and imaginary ($\Gamma/2$) parts of the complex vibrational quasienergies at 775 nm for laser intensity at $I = 10^{11}W/cm^2$ (weaker field case), $5 \times 10^{12}W/cm^2$ (medium strong field case), and $5 \times 10^{13}W/cm^2$ (strong field case), respectively. In Figures 24(a)–26(a), the field-modified adiabatic potentials are displayed and labeled by Floquet-state basis index $|g(\text{or } u)$, $n >$ in the asymptotic (R) region. The horizontal line segments at the left side column represent the converged (real parts of the) energies of VQE resonances. The line segments at the right side column(s) denote the energy positions of bound vibrational levels or shape resonances supported by the corresponding adiabatic potential well(s). The imaginary (half-) widths ($\Gamma/2$) of VQE resonances are shown in Figures 24(b)–24(b) in ascending order (label by v') according to the magnitude of their corresponding E_R's displayed in Figures 24(a)–24(b).

Figures 24(a)–24(b) show the expected weak-field behavior. At this intensity ($I = 10^{11}W/cm^2$), the (perturbed) VQE resonance positions shown in Figure 24(a) are very close to those of (field-free) vibrational states supported by the ground $1s\sigma_g$ potential curve. The behavior of the photodissociation widths (Figure 24(b)) of these resonance states is also expected: low-lying states generally have smaller widths and longer (photodissociation) lifetimes than those of the high-lying levels. In fact, the photodissociation rates of high-lying resonances ($v' \geq 8$) can be nine to ten orders of magnitude larger than the photodissociation rate of the tightly bound ground vibrational level.

The situation becomes more dedicated when the laser intensity increases. For example, at the medium strong intensity $I = 5 \times 10^{12}W/cm^2$, the gap of one-photon avoided crossing ($R \simeq 5$ a.u.) already becomes sufficiently large and the structure of VQE resonances significantly distorted. In fact, the VQE resonances now break into two groups (Figure 25(a)): lower-lying resonance group ($v' = 0 - 10$) and higher-lying resonance group ($v' \geq 11$), widely separated in energy. The photodissociation widths (Figure 25(b)) exhibit distinct behaviors for these two separated groups. The widths of the upper-group resonances ($v' \geq 11$) are consistently smaller than those of the higher members of the lower-lying group resonances (e.g. $v' = 7 - 10$). As compared with the weaker field case (Figure 25(b)), all the VQEs in the lower-lying group are now broadened substantially, i.e. molecules become more unstable in stronger fields, a phenomenon known as "bond softening" [51]. What is more striking here is the unexpected behavior of the upper group resonances. A comparison of Figures 24(b) and 25(b) reveals that the photodissociation rates of these high-lying VQE resonances actually *decrease* with *increasing* laser intensity! That is, molecules become more *stable* at *stronger* fields, a novel phenomenon which may be termed as "chemical bond *hardening*" [32]. These bond hardened states arise from the trapping of molecular vibrational wave functions at longer ($R \simeq 5$ a.u.) internuclear separation by the (one-photon) adiabatic potential well. These trapped states are in fact *not* bound states but slowly leaking *quasi-bound* resonance states due to the *nonadiabatic* couplings to other Floquet-state channels. As the laser intensity increases, the one-photon gap becomes larger, leading to weaker nonadiabatic couplings and therefore smaller photodissociation rates (widths).

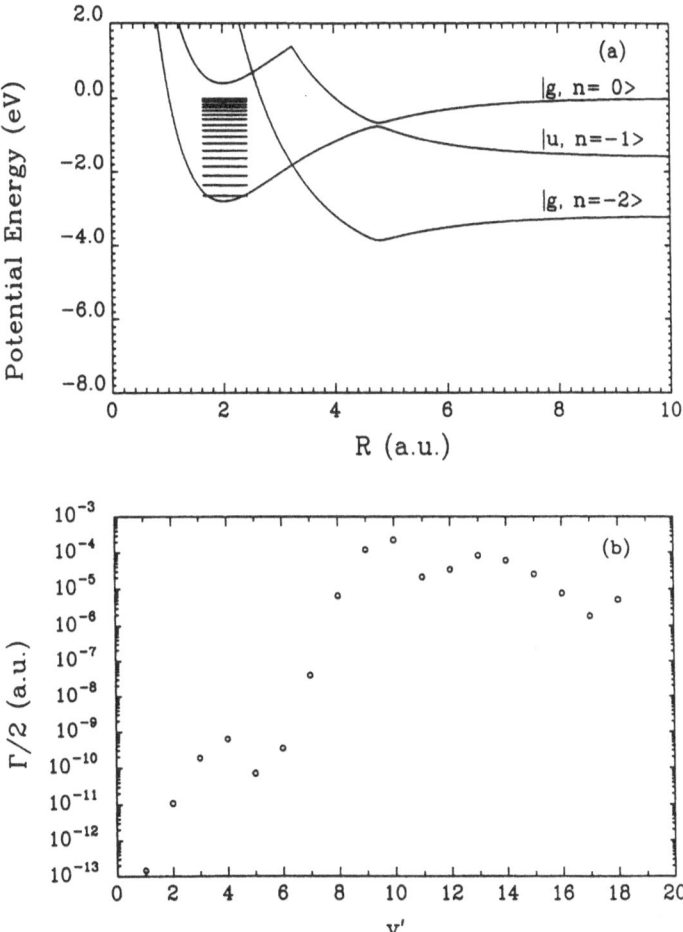

FIG. 24. *(a) Vibrational quasi-energy level structure and dressed adiabatic potentials of H_2^+, (b) Photodissociation (half-) widths ($\Gamma/2$) of vibrational quasi-energy states (labeled by v') of H_2^+, at $\lambda = 775$ nm and $I = 10^{11} W/cm^2$ (weaker field case).*

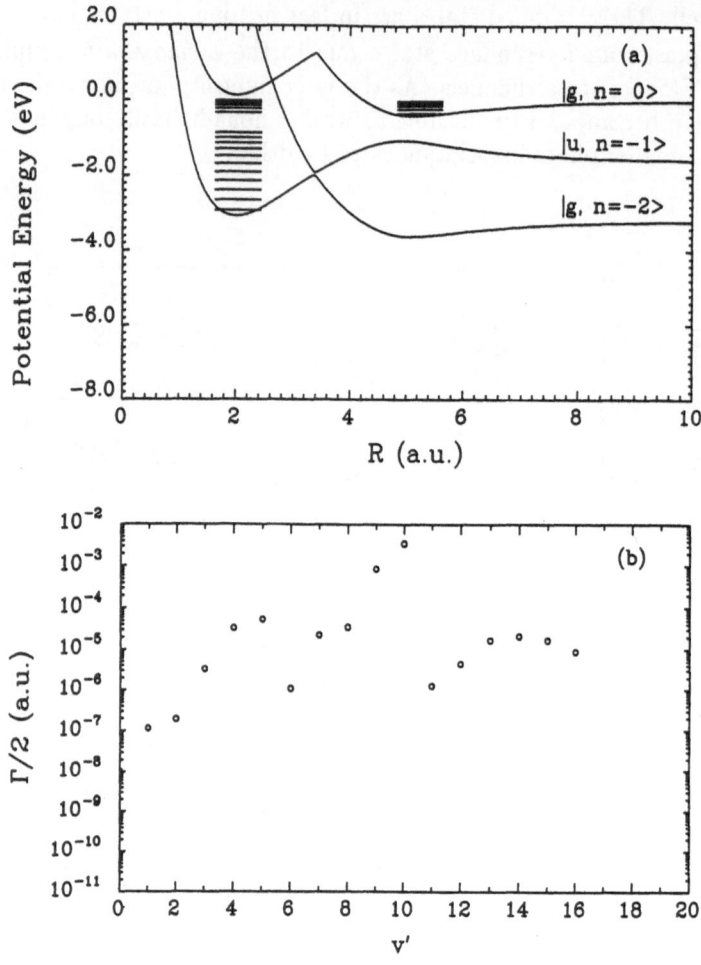

FIG. 25. *Same as Figure 24 except for* $I = 5 \times 10^{12} W/cm^2$ *(medium strong field case).*

As the laser intensity further increases to strong field regime, multi-photon avoided crossings now play significant role and the resonance structure undergoes dramatic changes. Figure 26 shows a stronger field case at $I = 5 \times 10^{13} W/cm^2$, in which the VQE resonances now break into several different groups. The topmost resonance-group states ($v' = 16 - 19$) are well separated in energy from the lower-lying groups. These highest-lying resonances exhibit the following distinct features: (a) They are supported by the (now very shallow) adiabatic potential well near the one-photon avoided cross region ($R_e \simeq 6.4 a.u.$). (b) Their photodissociation rates are extremely small, smaller than that of any lower-lying group resonance states. Molecules associated with these VQE states are therefore very stable against photodissociation even their binding energies are very small, a full manifestation of the bond hardening phenomenon. The phenomenon of laser-induced *stabilization* of molecules has been recently observed experimentally [53].

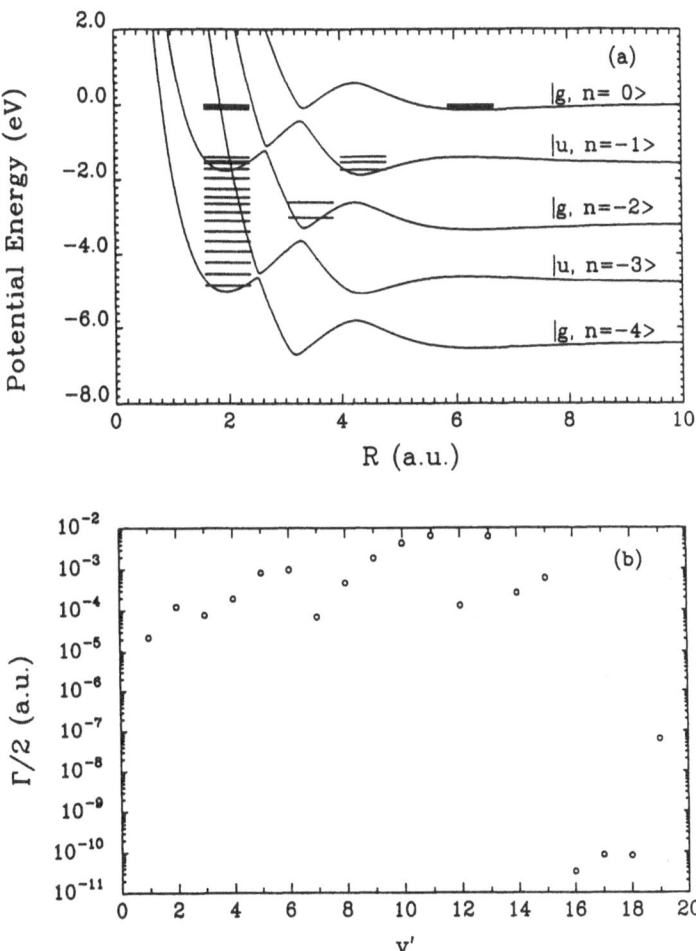

FIG. 26. *Same as Figure 24 except for* $I = 5 \times 10^{13} W/cm^2$ *(strong field case).*

Acknowledgments. This work was supported by the Division of Chemical Sciences, Office of Basic Energy Sciences of the U.S. Department of Energy.

REFERENCES

[1] For recent reviews of strong-field researches, see, for example, *Atoms in Intense Laser Fields*, Adv. At. Mol. Opt. Phys. Sup. 1, edited by M. Gavrila (Academic Press 1992).

[2] See, for example, P.K. Aravind and J.O. Hirschfelder, J. Phys. Chem. 88 (1984) 4788.

[3] J.H. Shirley, Phys. Rev. 138 (1965) B979.

[4] G. Floquet, Ann. l'Ecol. Norm Sup. 12 (1883) 47; J. H. Poincar, Les Methodes Nouvelles de la Mechanique Celeste, Vols. I, II, IV, (Paris, 1892, 1893, 1899).

[5] For review of Floquet methods for two-level systems, see, D.R. Dion and J.O. Hirschfelder, Adv. Chem. Phys. 35 (1976) 265.

[6] For reviews on generalized Floquet formalisms and methods, see, (a) S.I. Chu, Adv. At. Mol. Phys. 21 (1985) 197; (b) S.I. Chu, Adv. Chem. Phys. 73 (1989) 739; (c) S.I. Chu, Phys. Rep. (1996).

[7] S.H. Autler and C.H. Townes, Phys. Rev. 100 (1955) 703.

[8] Numerous methods for numerical solution of periodically time-dependent Schrödinger equations have been developed. See ref. 6(a) for a review. See, also, W.J. Meath, R.A. Thuraisingham, and M.A. Kmetic, Adv. Chem. Phys. 73 (1989) 307; P.P. Friedmann, Adv. Chem. Phys. 73 (1989) 197; R.E. Wyatt, Adv. Chem. Phys. 73 (1989) 231.

[9] S.I. Chu, J.V. Tietz, and K.K. Datta, J. Chem. Phys. 77 (1982) 2968.

[10] (a) S.I. Chu and W.P. Reinhardt, Phys. Rev. Lett. 39 (1977) 1195; (b) S.I. Chu, Chem. Phys. Lett. 54 (1978) 367; (c) A. Maquet, S.I. Chu, and W.P. Reinhardt, Phys. Rev. A27 (1983) 2946.

[11] S.I. Chu and J. Cooper, Phys. Rev. A 32 (1985) 2769.

[12] S.I. Chu, J. Chem. Phys. 75 (1981) 2215.

[13] T.S. Ho, S.I. Chu and J.V. Tietz, Chem. Phys. Lett. 96 (1983) 464; T.S. Ho and S.I. Chu, J. Phys. B 17 (1984) 2101; T.S. Ho and S.I. Chu, Phys. Rev. A 31 (1985) 659; T.S. Ho and S.I. Chu, Phys. Rev. A 32 (1985) 377; K. Wang, T.S. Ho, and S.I. Chu, J. Phys. B 18 (1985) 4539.

[14] J.V. Tietz and S.I. Chu, Chem. Phys. Lett. 101 (1983) 446; K. Wang and S.I. Chu, Phys. Rev. A 39 (1989) 1800.

[15] T.S. Ho and S.I. Chu, Chem. Phys. Lett. 122 (1985) 327; T.S. Ho, K. Wang, and S.I. Chu, Phys. Rev. A 33 (1986) 1798; K. Wang and S.I. Chu, J. Chem. Phys. 86 (1987) 3225.

[16] Several stationary approaches using generalized Floquet formalisms for (arbitrarily) time-dependent Hamiltonian systems have been developed. See, for example, (a) Nonadiabatic coupled dressed-states formalism: T.S. Ho and S.I. Chu, Chem. Phys. Lett. 141 (1987) 315; (b) Two-mode Floquet treatment of multiphoton excitation by (arbitrarily shaped) laser pulses: Y. Huang and S.I. Chu, Chem. Phys. Lett. 225 (1994) 46.

[17] E. Balslev and J.M. Combes, Commun. Math. Phys. 22 (1971) 280; J. Aguilar and J.M. Combes, Commun. Math. Phys. 22 (1971) 265; B. Simon, Ann. Math. 97 (1973) 247.

[18] See, for example, H.A. Yamani and W.P. Reinhardt, Phys. Rev. A11 (1975) 1144.

[19] G. Yao and S.I. Chu, Chem. Phys. Lett. 204 (1993) 381.

[20] J. Wang, S.I. Chu, and C. Laughlin, Phys. Rev. A 50 (1994) 3208.

[21] C.C. Marston and G.G. Balint-Kurti, J. Chem. Phys. 91 (1989) 3571.

[22] E. Layton and S.I. Chu, Chem. Phys. Lett. 186 (1991) 100.

[23] S.I. Chu, Chem. Phys. Lett. 167 (1990) 155.

[24] R.M. Potvliege and R. Shakeshaft, Adv. At. Mol. Opt. Phys. Sup. 1 (1992) 373.

[25] S.I. Chu, K. Wang and E.J. Layton, J. Opt. Soc. Am. B 7 (1990) 425.

[26] S.I. Chu, J. Chem. Phys. 94 (1991) 7901.

[27] D. Telnov and S.I. Chu, Phys. Rev. A 50 (1994) 4099.

[28] D. Telnov, J. Wang, and S.I., Chu, Phys. Rev. A 51 (1995) 4797.

[29] D. Telnov and S.I. Chu, J. Phys. B 28 (1995) 2407.

[30] For a recent review on stabilization of atoms in superintense laser fields, see, M. Gavrila, Adv. At. Mol. Phys. Sup. 1 (1992) 435.

[31] G. Yao and S.I. Chu, Phys. Rev. A 45 (1992) 6735.

[32] (a) G. Yao and S.I. Chu, Chem. Phys. Lett. 192 (1992) 413; (b) G. Yao and S.I. Chu, Phys. Rev. A 48 (1993) 485.

[33] J. Wang and S.I. Chu, Chem. Phys. Lett. 227 (1994) 663.

[34] D. Telnov, J. Wang and S.I. Chu, Phys. Rev. A 51 (1995) 4797.

[35] S.I. Chu, Z.C. Wu, and E. Layton, Chem. Phys. Lett. 157 (1989) 151; E. Layton, Y. Huang, and S.I. Chu, Phys. Rev. A 41 (1990) 42.

[36] T.S. Ho, S.I. Chu, and C. Laughlin, J. Chem. Phys. 81 (1984) 788; T.S. Ho, C. Laughlin, and S.I. Chu, Phys. Rev. A 32 (1985) 122.

[37] K. Wang and S.I. Chu, Chem. Phys. Lett. 153 (1988) 87; S.I. Chu, Comm. At. Mol. Phys. 25 (1990) 101.

[38] T.W.B. Kibble, Phys. Rev. Lett. 16 (1966) 1054.

[39] See, for example, P. Agostini, P. Breger, A. L'Huillier, H.G. Muller, and G. Petite, Phys. Rev. Lett. 63 (1989) 2208.

[40] (a) C.Y. Tang, P.G. Harris, A.H. Mohagheghi, J.C. Bryant, C.R. Quick, J.B. Donahue, R.A. Reed, S. Cohen, W.W. Smith, and J.E. Steward, Phys. Rev. A 39 (1989) 6068; (b) W.W. Smith, C.Y. Tang, C.R. Quick, H.C. Bryant, P.G. Harris, A.H. Mohagheghi, J.B. Donahue, R.A. Reeder, H. Sharifian, J.E. Steward, H. Toutounchi, S. Cohen, T.C. Altman, and D.C. Rislove, J. Opt. Soc. Am. B 8 (1991) 17.

[41] C.Y. Tang, H.C. Bryant, P.G. Harris, A.H. Mohagheghi, R.A. Reeder, H. Sharifian, H. Toutounchi, C.R. Quick, J.B. Donahue, S. Cohen, and W.W. Smith, Phys. Rev. Lett. 66 (1991) 3124.

[42] C. Laughlin and S.I. Chu, Phys. Rev. A 48 (1993) 4654.

[43] A.L. Stewart, J. Phys. B11 (1978) 3852; A.W. Wishart, J. Phys. B 12 (1979) 3511.

[44] For a review on ATI processes, see, R.R. Freeman and P.H. Bucksbaum, J. Phys. B 24 (1991) 325.

[45] N. Bloembergen and E. Yablonovitch, Phys. Today 31 (1978) 23.

[46] P.A. Schulz, A.S. Sudbo, D.J. Krajnovich, H.S. Kowk, Y.R. Shen and Y.T. Lee, Ann. Rev. Phys. Chem. 30 (1979) 311.

[47] A. Carrington and J. Buttenshaw, Mol. Phys. 44 (1981) 267.

[48] C. Laughlin, K.K. Datta and S.I. Chu, J. Chem. Phys. 85 (1986) 1403; S.I. Chu, C. Laughlin and K.K. Datta, Chem. Phys. Lett. 98 (1983) 476.

[49] C. Cornaggia, D. Normand, J. Morellec, G. Mainfray and C. Manus, Phys. Rev. A 34 (1986) 207.

[50] T.S. Luk and C.K. Rhodes, Phys. Rev. A 38 (1988) 6180.

[51] P.H. Bucksbaum. A. Zavriyev, H.G. Muller and D.W. Schumacher, Phys. Rev. Lett. 64 (1990) 1883.

[52] A. Giusti-Sugor, X. He, O. Atabek and F.M. Mies, Phys. Rev. Lett. 64 (1990) 515.

[53] A. Zavriyev, P.H. Bucksbaum, J. Squier and F. Saline, Phys. Rev. Lett. 70 (1993) 1077.

GLOBAL RECURSION POLYNOMIAL EXPANSIONS OF THE GREEN'S FUNCTION AND TIME EVOLUTION OPERATOR FOR THE SCHRÖDINGER EQUATION WITH ABSORBING BOUNDARY CONDITIONS

VLADIMIR A. MANDELSHTAM*

Abstract. We review and revise some recently developed iterative numerical techniques of solving the Time-Independent Wave Packet Schrödinger Equation $(E - \hat{H})\Psi = \chi$, which are especially suitable for calculations associated with a large many body scattering problem, requiring large basis sets (grids). The methods we consider take advantage of the particular way the wave equation depends on the energy E. Namely, when global polynomial expansions of the resolvent operator are used the energy-dependent solution can be generated simultaneously at many energies from essentially a single iterative procedure. A general problem in constructing a well behaved polynomial expansion of a function of operator argument, $f(\hat{H})$, is how to incorporate absorbing boundary conditions (ABC) which would eliminate reflection effects caused by an artificial truncation of an infinite grid where $\hat{H}$ is defined. It is shown that this problem can be solved naturally by modifying the auxiliary equations (recursion relations) used to generate the interpolating polynomials; this avoids using non-Hermitian operators. In particular Chebyshev and Newtonian interpolation schemes are considered. While the latter is more general, the former allows one to obtain analytically very simple global recursion polynomial expansions for the ABC Green's function and ABC time evolution operator.

1. Preliminaries. $\mathcal{L}^2$ methods for solving scattering problems are those in which the time-independent Schrödinger equation is solved by expanding its solutions entirely in a square-integrable basis. The most demanding computational step in such calculations is the solution of a system of linear equations associated with the inhomogeneous Schrödinger equation (see, e.g., [1-14]),

$$(1.1) \qquad (E - \hat{H})\psi = \chi.$$

Eq.1.1 arises in different contexts when the corresponding homogeneous system with certain asymptotic boundary conditions has to be solved.[1] The use of $\mathcal{L}^2$ methods for scattering problem stimulated great progress in developing efficient numerical techniques of solving Eq.1.1. We devote this article to the same issue.

Once the system Hamiltonian $\hat{H}$ is represented in an $\mathcal{L}^2$ basis set a proper choice of method of solving Eq.1.1 should be based on the properties of the $\hat{H}$ matrix and the inhomogeneity χ. When a large many dimensional system is considered and the basis size, N, gets too big, direct inversion of $E - \hat{H}$ is not feasible. We distinguish two basic competing strategies to solve the problem.

---

* Department of Chemistry, University of Southern California, Los Angeles, California 90089-0482.

[1] Going ahead we note that when χ in Eq.1.1 is energy independent we refer to this equation as the Time-Independent Wave Packet Schrödinger Equation as it first appeared in references 6-9.

The first strategy tries to exploit some physical properties of the system (such as an adiabatic separation of variables) in order to construct a physically more reasonable basis set having a much smaller size N (see, e.g., the recent paper [15] and references therein). Typically such basis reduction (adiabatic contraction) procedures allow one to reduce the size of the basis by one or two orders of magnitude (in some favorable cases this factor might be even larger) depending on the particular problem. However after the basis reduction is done the Hamiltonian matrix loses its convenient mathematical structure so that the linear algebra problem associated with Eq.1.1 can only be solved by standard methods based on the LU-matrix decomposition procedure. In such a case one again will face the problem of very unfavorable scaling of the numerical effort with respect to the basis size, namely, for the LU-matrix decomposition the storage scales as N^2 (the number of elements of the Hamiltonian matrix), and the CPU-time as N^3.

The other strategy is to exploit the mathematical properties of the Hamiltonian matrix represented in a primitive basis set rather than improve this basis set. In particular, using grid bases such as those based on the DVR (discrete variable representation) [16] is often advantageous since the DVR Hamiltonian matrix is very sparse and its nonzero elements can be produced on the fly. In such a case iterative methods are preferred since they require as an "elementary" step only a multiplication of a vector by the H-matrix. When the DVR is used the latter can be made extremely efficient. Iterative methods typically do not have enormous storage requirements since no transformations of large matrices need to be performed, and therefore only large vectors (not large matrices) have to be stored. Furthermore, the scaling of the CPU-time with the size of the basis is usually much more favorable than for the direct inversion methods.

1.1. All energy methods. Once the iterative methods have become our choice, the next key point to be considered is that the scattering results are often needed at many energies E, and when that is the case Eq.1.1 has to be solved many times. This could eventually lead to a very large total number of iterations. Hence it would be a significant advantage if all energies could be studied in the same calculation. This property was historically reserved for time dependent methods (see, e.g., [17]) which were able to obtain the scattering solution simultaneously at many energies from the time-to-energy Fourier transform of the time dependent solution:

$$(1.2) \qquad \Psi(E) = -i \int_0^\infty dt\, e^{iEt} \Psi(t),$$

where the latter would be propagated in time by a convenient iterative procedure using sufficiently short time steps Δt,

$$(1.3) \qquad \Psi(t + \Delta t) = e^{-i\hat{H}\Delta t} \Psi(t).$$

Notice that use of Eqs.1.2 and 1.3 assumes the initial condition $\Psi(0) = \chi$ to be energy independent.

The attractiveness of the above scheme is due both to its simplicity and to the possibility of eliminating so called reflection effects caused by truncation of the grid. The latter is usually done [18] by simply multiplying $\Psi(t + \Delta t)$ after each time step by a damping factor $e^{-\hat{\gamma}}$,

$$(1.4) \qquad \Psi(t + \Delta t) = e^{-\hat{\gamma}} e^{-i\hat{H}\Delta t} \Psi(t),$$

where for $\hat{\gamma}$ one can use any slowly rising in the asymptotic region positive function of the coordinate $\gamma(r)$, in which case $\gamma(r)$ is obviously related to having an effective negative imaginary potential (see also ref. [2]), i.e., when Δt is sufficiently small, $-iW(r) \approx -i\gamma(r)/\Delta t$. Such a potential $W(r)$ is sometimes called an absorbing potential or an optical potential.

Another way to look at such an absorbing potential [5] is to consider it to be a convergence factor for the Green's function expression,

$$(1.5) \qquad \hat{G}^+ = \frac{1}{E - \hat{H} + i\epsilon} \approx \frac{1}{E - \hat{H} + i\hat{W}},$$

where the constant, infinitesimal convergence factor is allowed to be a finite operator, i.e., $\epsilon \rightarrow \hat{W}$. A simple choice for the optical potential $\hat{W}$ is a function of coordinates $W(r)$ which vanishes in the interaction region, so that it does not perturb the correct dynamics in this region. It should also rise both rapidly and smoothly enough in the asymptotic region to move the real spectrum of $\hat{H}$ well down to the lower half of the complex plane so that the corresponding resolvent operator is analytic and well behaved on the real axis.

The main disadvantage of the procedure associated with Eqs.1.3 and 1.4 seems to be a roundoff error which is accumulated at each iteration making long-time propagation inaccurate. The roundoff error problem in time propagation procedure was resolved by Tal-Ezer and Kosloff [19] who introduced a global-in-time Chebyshev polynomial expansion of the time evolution operator:

$$(1.6) \qquad \hat{U}(t) = e^{-i\hat{H}t} = \sum_{n=0} a_n(t) T_n(\hat{H}_{\text{norm}}).$$

As opposed to the short-time propagation scheme this expansion is able to provide any desired accuracy for a very long propagation time since its convergence is exponential with respect to the number of terms included in the sum. Here the operator sequence $T_n(\hat{H}_{\text{norm}})$, $n = 1, 2, 3...$, satisfies the Chebyshev recursion relations:

$$T_0(\hat{H}_{\text{norm}}) = \hat{I}$$
$$T_1(\hat{H}_{\text{norm}}) = \hat{H}_{\text{norm}}$$

$$\cdots$$

$$(1.7) \qquad T_{n+1}(\hat{H}_{\text{norm}}) = 2\hat{H}_{\text{norm}} T_n(\hat{H}_{\text{norm}}) - T_{n-1}(\hat{H}_{\text{norm}}).$$

$\hat{I}$ defines the identity operator; and $\hat{H}_{norm}$ denotes a Hamiltonian that is scaled according to the formula

$$(1.8) \qquad \hat{H}_{norm} = \frac{\hat{H} - \bar{H}}{\Delta H},$$

where $\bar{H} = \frac{1}{2}(H_{max} + H_{min})$, $\Delta H = \frac{1}{2}(H_{max} - H_{min})$, and H_{max} and H_{min} are respectively upper and lower estimates of the maximum and minimum eigenvalues of the real symmetric Hamiltonian $\hat{H}$ (We can always assume that $\hat{H}$ is represented on a grid thereby making the spectral range ΔH finite). Thus, the real spectrum of the scaled Hamiltonian $\hat{H}_{norm}$ belongs to the interval $(-1, 1)$ and this assures stability of the Chebyshev polynomials.

The time dependent coefficients a_n are defined via the Bessel functions J_n as

$$(1.9) \qquad a_n(t) = (2 - \delta_{n0})e^{-i\bar{H}t}(-i)^n J_n(\Delta H t).$$

Recently it was further realized [20, 6] that the formal time-to-energy Fourier transform of the time propagator given by Eq.1.6,

$$(1.10) \qquad \hat{G}^+(E) = -i \int_0^\infty dt e^{iEt} \hat{U}(t),$$

immediately gives the formal global in energy Chebyshev polynomial expansion of the Green's function,

$$(1.11) \quad \hat{G}^+(E) = \frac{1}{E - \hat{H} + i0} = \frac{1}{i\Delta H \sin\varphi} \sum_{n=0} (2 - \delta_{n0})e^{-in\varphi} T_n(\hat{H}_{norm}),$$

thus avoiding the need to actually perform the Fourier transform of the time-dependent solution $\Psi(t)$ as in Eq.1.2.

An iterative procedure associated with recursion relations (1.7) that generate the polynomials $T_n(\hat{H}_{norm})$ is independent of energy, the energy dependence of the series in Eq. 1.11 being given by the energy dependent coefficients $e^{-in\varphi}$, where the phase $\varphi = \varphi(E)$ is

$$(1.12) \qquad \varphi = \arccos\left(\frac{E - \bar{H}}{\Delta H}\right).$$

Such an energy separable expansion is also called "global in energy".

In a series of papers [6-9] the authors explored the case where the r.h.s., χ, in Eq.1.1 is energy independent; this case allows one to take advantage of the energy separability of the Green's function expansion, Eq.1.11. In this case the solution $\Psi(E)$ of the Time-Independent Wave Packet Schrödinger Equation, Eq. 1.1, can be generated simultaneously at many energies from essentially a single iterative procedure. In particular, this makes it possible

to obtain the scattering matrix elements using a very convenient expression
based on the Kohn variational principle [9]:

$$(1.13) \qquad S_{lp}(E) = \frac{i\sqrt{k_l k_p}}{m(2\pi)^2 A(k_l) A^*(k_p)} \langle \chi_l | \hat{G}^+(E) | \chi_p \rangle,$$

where $A(k_l)$ is the Fourier component of the incoming (outgoing) energy
independent wavepacket χ_l that is placed outside the interaction region
and corresponds to the l-th channel state.

Utilizing similar ideas it was recently proposed to first produce (by a
polynomial expansion) a set of linear independent solutions of the wave
equation [12],

$$(1.14) \qquad \psi_p(E) = \frac{1}{E - \hat{H}} \chi_p^{ext}, p = 1, 2, .., N_o,$$

where N_o indicates the number of open channels at energy E. If the Hamiltonian $\hat{H}$ is a local operator (which is often the case) the solution $\psi_p(E)$
will satisfy the homogeneous Schrödinger equation in the region where χ_p^{ext}
vanishes. For the complete description of the scattering process at the energy E the $\psi_p(E)$'s have to satisfy the homogeneous Schrödinger equation
only inside the interaction region (see eqs. 2.1 and 2.2 of ref. [12]). This
can be easily achieved if the N_o linear independent, otherwise arbitrary,
wavepackets χ_p^{ext} are taken to be non-zero only in a negligibly small edge
space of the interaction region, where the corresponding solutions $\psi_p(E)$
are allowed to be arbitrary. In contrast to Eq.1.13 the inverse of $(E - \hat{H})$ in
Eq.1.14 may be, but is not necessarily a Green's operator satisfying some
particular scattering boundary conditions. This makes it possible to significantly reduce the grid (basis) size which could be essentially equal to the
size of the interaction region.

Notice that no matter how the Hermitian Hamiltonian is represented
in a basis the Chebyshev polynomial series for the time evolution operator
$e^{-it\hat{H}}$, Eq.1.6, converges exponentially since the Bessel functions $J_n(\Delta H t)$
decay exponentially for $n > \Delta H t$. The same, however, is not true for the
expansion of the Green's function, Eq.1.11, because it is not an analytic
function on the real energy axis. As it was noted in ref. [12] one can analytically continue the expansion by replacing φ with $\varphi - i\epsilon$. This would
certainly make the Chebyshev series to converge exponentially for any infinitesimally small ϵ since the coefficients $e^{-in(\varphi-i\epsilon)}$ decay exponentially
with n. At the same time such an analytic continuation does not lead to
rapid uniform convergence of the series nor does it make the expansion
physically miningful unless the Hamiltonian is represented on a very large
subspace of the configuration space. [2] The latter was done in ref. [6]. How-

---

[2] Due to obvious analogy to the time dependent approach such an unfavorable spatial
convergence of the causal Green's function is also referred to as "artificial reflection
effects".

ever having a basis (grid) expanded too far out in the configuration space does not seem to be practical for large many-dimensional systems where one looks for methods that allow basis sets to be as small as possible.

It is worthwhile to point out that even though the convergence of the series itself, as defined by Eq.1.6, cannot be spoiled by the reflection effects, this does not mean that the physical convergence for the evolution operator could easily be achieved if no care is taken of reflection effects. In other words for a finite grid representation of the Hamiltonian operator $\hat{H}$ after a long enough time the numerical solution $\Psi(t) = e^{-it\hat{H}}\chi$, being reflected by artificial boundaries, loses any physical meaning.

In the case of the short time propagation scheme, Eq.1.4, incorporation of ABC was quite natural, which effectively corresponded to adding to the system Hamiltonian a negative imaginary potential. When global polynomial expansions are involved generalization to the case of the non-Hermitian operator variable is not straightforward. In particular, a naive replacement of the Hermitian Hamiltonian by its non-Hermitian analogue,

$$(1.15) \qquad\qquad \hat{H} \to \hat{H} - i\hat{W},$$

in Eq.1.7 would make the Chebyshev polynomials exponentially unstable.

1.2. Other global polynomial expansions. Several different polynomial expansion (interpolation) methods allowing use of absorbing potentials were proposed recently. We would like to distinguish two different approaches. The first approach outlined in this subsection is the most straightforward as it tries to construct a generic well behaved polynomial expansion of an analytic function $f(\hat{H})$ [3] of an operator argument $\hat{H}$ that might be but not necessarily is Hermitian [7,8,21-23]. Most generally such an expansion would read

$$(1.16) \qquad\qquad f(\hat{H}) = \sum_{n=0} c_n P_n(\hat{H}).$$

The above expansion is iterative when the polynomials $\hat{P}_n \equiv P_n(\hat{H})$ satisfy a simple K-term linear recursion relation:

$$(1.17) \qquad\qquad \hat{P}_n = \Phi_n\{\hat{P}_{n-1}, \hat{P}_{n-2}...\hat{P}_{n-K}\},$$

Here the symbol Φ_n defines the recursion rule. Eq. 1.17 consequently yields the iterative expression for the wavefunction:

$$(1.18) \qquad\qquad \Psi = f(\hat{H})\chi = \sum_{n=0} c_n\xi_n,$$

---

[3] Even though the theory is formulated for any analytic function, for us the most important applications are the time evolution and resolvent operators.

where the recursion relation for the polynomials $\hat{P}_n$ implies the recursion relation for the vectors $\xi_n = \hat{P}_n \chi$:

$$(1.19) \qquad \xi_n = \Phi_n \{\xi_{n-1}, \xi_{n-2} \cdots \xi_{n-K}\}.$$

Apparently Chebyshev polynomials are not the only ones which could be used in such a context. In ref. [7] the Faber polynomials which are a generalization of the Chebyshev polynomials were shown to provide stable and uniformly converged expansions of both the ABC time evolution operator and the ABC Green's function. It has also become clear [8] that the main difficulty associated with the use of Faber polynomials is the need for an accurate and analytically simple estimation of the spectral domain of the non-Hermitian operator $\hat{H}$ in the lower half of the complex plane. While the size of the estimated spectral domain directly affects the convergence of the series, its complexity leads to many-term recursion relations.

As it was demonstrated in ref. [21] (see also the review [22]) even more generic expansions may be obtained using the Newtonian interpolation scheme in which an analytic function $f(z)$ is approximated as a polynomial of degree $M - 1$,

$$(1.20) \qquad f(z) \approx \mathcal{P}_{M-1}(z) \equiv \sum_{n=0}^{M-1} b_n \prod_{j=0}^{n-1} (z - x_j),$$

coinciding with the function f(z) on the set of the M sampling points x_j, i.e., $f(x_j) \equiv \mathcal{P}_{M-1}(x_j)$. The coefficients b_n can be obtained recursively by

$$b_0 = \mathcal{P}_0(x_0) = f(x_0)$$
$$\cdots$$
$$(1.21) \qquad b_n = \frac{f(x_n) - \mathcal{P}_{n-1}(x_n)}{\prod_{j=0}^{n-1}(x_n - x_j)}.$$

To construct the operator function $f(\hat{H})$ one can simply replace z in Eq.1.20 by $\hat{H}$, i.e.,

$$(1.22) \qquad f(\hat{H}) \approx \sum_{n=0}^{M-1} b_n \prod_{j=0}^{n-1} (\hat{H} - x_j).$$

For the price of having to determine the coefficients b_n numerically one now has a flexibility to choose the function $f(\hat{H})$.

As was shown later [23] both the choice and the order of the sampling points x_j are crucial in the algorithm. Note that in the non-Hermitian case the sampling points have to be taken on a curve in the lower half of the complex plane encircling the spectral domain, while for a Hermitian Hamiltonian the sampling points could be chosen on a real interval $[H_{\min}, H_{\max}]$.

While the Newtonian interpolation approach was used successfully for long time propagation [23], more work should be done to find out how efficient it is for the expansion of the resolvent operator $(E - \hat{H})^{-1}$. However some preliminary results already exist for this problem [24].

2. Absorbing boundary conditions for recursion relations. In this section we abandon the idea of using a non-Hermitian argument $\hat{H}$ and present an alternative way to introduce ABC in the polynomial expansion of an operator function $f(\hat{H})$ given by Eq.1.16. We assume that the expansion converges in the case of a Hermitian Hamiltonian $\hat{H}$ represented on an infinite grid. As such the starting expressions which are to be generalized could be the Chebyshev polynomial expansions, Eqs.1.6 and 1.11, as well as the Newtonian interpolation scheme given by Eq. 1.22. Our goal is to modify Eq.1.16 in such a way that the result in the physically important region would not be affected by a truncation of the grid sufficiently far outside this region. This is done exactly in the spirit of Eq.1.4 where ABC were used to eliminate the reflection effects, while the key difference with Eq.1.3 is that in place of the time-dependent wavepacket $\Psi(t) = \hat{U}(t)\chi$ we consider the n-dependent one $\xi_n = \hat{P}_n \chi$ or the corresponding polynomial solution $\hat{P}_n$ as the true victims of the artificial reflections. This analogy immediately gives a solution [12], which is to damp the "uninteresting" part of the wavepacket ξ_n in the edge (ABC) region after each iteration:

$$(2.1) \qquad \xi_n^\gamma = e^{-\hat{\gamma}} \Phi_n \{ \xi_{n-1}^\gamma, \xi_{n-2}^\gamma \cdots \xi_{n-K}^\gamma \}.$$

Consequently for the damped polynomials the ABC recursion relation should read:

$$(2.2) \qquad \hat{P}_n^\gamma = e^{-\hat{\gamma}} \Phi_n \{ \hat{P}_{n-1}^\gamma, \hat{P}_{n-2}^\gamma \cdots \hat{P}_{n-K}^\gamma \},$$

where the new polynomials $\hat{P}_n^\gamma \equiv P_n(\hat{H}, e^{-\hat{\gamma}})$ are now the polynomials in two generally noncommuting operator variables $\hat{H}$ and $e^{-\hat{\gamma}}$.

Clearly the damping factor $e^{-\hat{\gamma}}$ in the cases of both Eq.2.1 and Eq.1.4 serves the same role, namely it removes (absorbs) those parts of the propagated in n (time) wavepacket that escape from the physically relevant region assuring that this outgoing part would never be reflected back by the edges of the grid. This means that the damped and therefore localized wavepackets ξ_n^γ are expected to mimic the unlocalized "true" wavepackets ξ_n in the region where the damping potential γ is zero. That is for a matrix element between any two initial wavepackets χ and χ' located in this region we have

$$(2.3) \qquad \langle \chi' | P_n^\gamma(\hat{H}) | \chi \rangle \approx \langle \chi' | P_n(\hat{H}) | \chi \rangle.$$

Using the above important relation we can now replace Eq.1.16 by its ABC analogue:

$$(2.4) \qquad f(\hat{H}) \approx \sum_{n=0} c_n P_n^\gamma(\hat{H}),$$

where the basis set no longer has to cover the whole configuration space since the wavepackets ξ_n^γ have a finite support in space and therefore can be represented by a finite grid. The damping potential $\gamma(r)$ should be considered as a convergence factor whose optimal shape and amplitude have to be determined by numerical experimentation.

Operationally the approach associated with Eq.2.4 introduces no new complications when it is compared to Eq.1.16. In particular the problem of determining the spectral domain of a non-Hermitian operator does not occur; the stability of the polynomials $P_n^\gamma(\hat{H})$ can be easily achieved by scaling the Hermitian Hamiltonian using Eq.1.8.

It should be also mentioned that the effect of the same damping factor $e^{-\hat{\gamma}}$ can be different when used with different polynomial interpolation schemes and does not have to lead to the same convergence properties. In the next subsection we show how the approach can be applied to the Chebyshev polynomials. In particular we show that it provides an exact recursion polynomial expansion of the Green's function with an analytically known complex absorbing potential. This knowledge can be used advantageously in certain applications of Eq.2.4 (see refs. [30,31]).

2.1. Absorbing boundary conditions for the Chebyshev recursion relations. The Green's function expansion. Let us consider the Chebyshev recursion relations, Eq.1.7, modified with a damping factor [12,13]:

$$T_0^\gamma(\hat{H}_{\text{norm}}) = \hat{I}$$
$$T_1^\gamma(\hat{H}_{\text{norm}}) = e^{-\hat{\gamma}}\hat{H}_{\text{norm}}$$
$$\ldots$$

$$(2.5) \quad T_{n+1}^\gamma(\hat{H}_{\text{norm}}) = e^{-\hat{\gamma}}\left[2\hat{H}_{\text{norm}}T_n^\gamma(\hat{H}_{\text{norm}}) - e^{-\hat{\gamma}}T_{n-1}^\gamma(\hat{H}_{\text{norm}})\right] .$$

(Although the above modification differs from the prescription of Eq.2.2 it can be easily shown that both recursion relations with the damping factors lead to conceptually similar results, while Eq.2.5 is more convenient.)

Notice that in the trivial case when $\hat{\gamma}$ is equal to a constant factor ϵ, the number $e^{-\epsilon}$ commutes with the Hamiltonian, and hence we have $T_n^\epsilon(\hat{H}_{\text{norm}}) = e^{-n\epsilon}T_n(\hat{H}_{\text{norm}})$. However the interesting case in which a much more efficient absorbing effect can be achieved corresponds to the non-commuting $\hat{H}$ and $e^{-\hat{\gamma}}$. In such a case the relation given by Eq.2.3 should hold when the damping function $e^{-\gamma(r)}$ decay with r sufficiently slow. That is, for any two wave packets which do not overlap the ABC region we can write

$$(2.6) \qquad \langle\chi'|T_n^\gamma(\hat{H}_{\text{norm}})|\chi\rangle \approx \langle\chi'|T_n(\hat{H}_{\text{norm}})|\chi\rangle.$$

This leads to an important consequence based on the properties of the Chebyshev polynomials, namely

$$2\langle\chi'|T_n(\hat{H}_{\text{norm}})T_m(\hat{H}_{\text{norm}})|\chi\rangle = \langle\chi'|T_{n+m}(\hat{H}_{\text{norm}})|\chi\rangle\langle\chi'|T_{|n-m|}(\hat{H}_{\text{norm}})|\chi\rangle$$

(2.7) $\approx \langle \chi' | T^\gamma_{n+m}(\hat{H}_{\text{norm}}) | \chi \rangle + \langle \chi' | T^\gamma_{|n-m|}(\hat{H}_{\text{norm}}) | \chi \rangle.$

It is worthwhile to notice again that Eqs. 2.6 and 2.7 as well as Eq.2.3 can only be valid if the wavepackets χ and χ' are located in the region unperturbed by the damping potential $\gamma(r)$. In order to see how Eq.2.7 can be used in particular applications we refer the reader to refs. [25-27].

Let us now multiply Eq.2.5 by $e^{-in\varphi}$ and sum it over n as it appears in the r.h.s. of Eq.1.11. After some simple manipulations we have

$$e^{-i(\varphi - i\hat{\gamma})} \sum_{n=1} e^{-in\varphi} T^\gamma_n(\hat{H}_{\text{norm}}) + e^{i(\varphi - i\hat{\gamma})} \sum_{n=1} e^{-in\varphi} T^\gamma_n(\hat{H}_{\text{norm}}) -$$

$$2\hat{H}_{\text{norm}} \sum_{n=1} e^{-in\varphi} T^\gamma_n(\hat{H}_{\text{norm}}) + e^{-i(\varphi - i\hat{\gamma})} - \hat{H}_{\text{norm}} = 0$$

Gathering all terms together finally leads to the desired result

(2.8) $\hat{G}^+ \approx \dfrac{1}{E - (\hat{H} - i\hat{W})} \equiv \dfrac{1}{\Delta H} \dfrac{1}{\cos(\varphi - i\hat{\gamma}) - \hat{H}_{\text{norm}}}$

$$= \dfrac{1}{i\Delta H} \sum_{n=0} (2 - \delta_{n0}) e^{-in\varphi} T^\gamma_n(\hat{H}_{\text{norm}}) [\sin(\varphi - i\hat{\gamma})]^{-1} ,$$

where the relation between the effective complex absorbing potential $-i\hat{W}$ and dimensionless damping potential $\hat{\gamma}$ is

(2.9) $-i\hat{W} = \Delta H [\cos \varphi (1 - \cosh \hat{\gamma}) - i \sin \varphi \sinh \hat{\gamma}].$

Notice that the $\hat{\gamma} \to 0^+$ limit implies $Re\hat{W} \to 0^+$ which in turn provides the correct analytic continuation for the causal Green's function. Typically the absorbing potential is taken as a negative imaginary potential which is not the case here. However recent numerical experience [28,29] showed that as long as $Re\hat{W} \geq 0$ there is no reason to make it purely imaginary. In practice the potential $\hat{\gamma}$ is what is chosen by numerical experimentation but if a particular $\hat{H} - i\hat{W}$ is desired $\hat{\gamma}$ can be trivially obtained from Eq.2.9. If $\hat{\gamma}$ does not depend on energy (which makes the operators $T^\gamma_n(\hat{H}_{\text{norm}})$ energy independent and which is the most practical choice) the optical potential $-i\hat{W}$ will be a function of the energy E due to the phase $\varphi = \varphi(E)$.

Some properties of this expansion have been discussed elsewhere. As such for a model collinear $H + H_2$ reactive scattering system [12] it was demonstrated that the convergence of a series as in Eq.2.8 is rapid and uniform in energy. In refs. [30] and [31] the approach was successfully applied in the context of filter diagonalization to the calculation of bound states and resonances of the HCO and HO_2 radicals. The latter work has shown that the method is very efficient in the case of a complex system.

2.2. ABC time propagator. We note [13] that use of the energy-to-time Fourier transform of Eq.2.8 immediately gives a global in time, iterative formula for the ABC time evolution operator $\hat{U}(t)$, namely

$$(2.10) \qquad \hat{U}(t) = \frac{1}{2\pi i} \int\limits_{-\infty}^{+\infty} dE e^{-iEt} \hat{G}^+(E) \approx \sum_{n=0} a_n(t) T_n^\gamma(\hat{H} \text{ norm})$$

where the time dependent coefficients $a_n(t)$ are the same as in the Chebyshev polynomial expansion corresponding to the case of Eq.1.6. The evaluation of the integral for $\hat{U}(t)\psi(0)$ using Eq.2.10 is exact as long as the initial wavepacket $\psi(0)$ does not overlap the damping region, where $\gamma(r) \neq 0$, in which case

$$(2.11) \qquad\qquad \frac{1}{\sin(\varphi - i\hat{\gamma})}\psi(0) = \frac{1}{\sin\varphi}\psi(0).$$

Notice the important difference between the two expressions, equations 1.6 and 2.10. In the former case the operator $e^{-i\hat{H}t}$ is unitary by definition while the $\hat{U}(t)$ defined by Eq.2.10 is not. The numerical consequence of this difference is that for a given finite basis set representation of the time propagators and initial wavepacket $\psi(0)$, the solution $e^{-i\hat{H}t}\psi(0)$ will stay in the finite region of the configuration space forever. In contrast $\hat{U}(t)\psi(0)$ will vanish in the same finite region in the $t \to \infty$ limit in a way which correctly simulates the dynamics of the system represented by an infinite grid.

3. Conclusions. We have revised some recently developed iterative techniques of solving the Time-Independent Wave Packet Schrödinger Equation by expanding the corresponding resolvent operator in a polynomial series. The possible applications of these methods are much more general than the particular application emphasized in this paper, namely the wave equation for a scattering problem. The need for efficient representation of an analytic function $f(\hat{H})$ of an operator argument $\hat{H}$ arises in various contexts in different fields of computational physics.

As a starting point we first considered a formal generic polynomial interpolation of $f(\hat{H})$ with Hermitian Hamiltonian $\hat{H}$ assumed to act on an infinite grid. The condition that the argument of the function f is Hermitian makes the problem of its polynomial interpolation much easier. This is due to the fact that the spectrum of the Hermitian $\hat{H}$ is real which allows a simple scaling of $\hat{H}$ as in Eq.1.8. On the other hand the class of polynomials suitable for the expansion of $f(\hat{H})$ in the case of non-Hermitian $\hat{H}$ is narrower, and the implementation of corresponding methods seems to be less straightforward.

Existence of an interpolating formula of $f(\hat{H})$ for the Hamiltonian $\hat{H}$ formally defined on an infinite grid is not yet sufficient for its numerical implementation. Since in practice only finite grids can be used the problem of simulating the effect of an infinite grid must be also solved. In other words, the problem is how to eliminate artificial reflections from the edges after the grid is truncated. Instead of treating this problem by adding to the Hamiltonian $\hat{H}$ a complex absorbing potential we incorporated absorbing boundary conditions into the auxiliary equations (recursion relations) which generate the interpolating polynomials. This anzatz allowed us to obtain well behaved global polynomial expansions (if the original ones were global) for a generic analytic function $f(\hat{H})$, and in particular, when the Chebyshev recursion relations were modified exact analytically simple expansions were derived for the Green's function and time evolution operator. At the same time the Newtonian interpolation scheme, being more general since it can be easily utilized for any function f, does not provide analytic coefficients which have to be generated recursively.

All this makes the approach presented here very flexible and powerful when it is applied to a variety of large systems which cannot be treated by direct methods involving large matrix transformations because they would require too large a core memory and too long a computation time.

Acknowledgment. The main results presented in the article were obtained in collaboration with Howard Taylor [12,13]. I also wish to thank Ronnie Kosloff and Donald Truhlar for careful reading the manuscript and making many useful comments. This work was supported by DOE Grant number DE-FG03-94ER14458.

REFERENCES

[1] W.H. Miller and B.M.D.D. Jansen op de Haar, J. Chem. Phys. **86**, 6213 (1987).
[2] D. Neuhauser and M. Baer, J. Chem. Phys. **91**, 4651 (1989);
[3] D.E. Manolopoulos, R.E. Wyatt and D.C. Clary, J. Chem. Soc. Faraday Trans. **86**, 1641 (1990).
[4] G.J. Tawa, S.L. Mielke, D.G. Truhlar and D.W. Schwenke, J. Chem. Phys. **100**, 5751 (1994).
[5] T. Seideman and W.H. Miller, J. Chem. Phys. **97**, 2499 (1992).
[6] Y. Huang, W. Zhu, D.J. Kouri and D.K. Hoffman, Chem. Phys. Lett. **206**, 96 (1993).
[7] Y. Huang, D.J. Kouri and D.K. Hoffman, Chem. Phys. Lett. **225**, 37 (1994).
[8] Y. Huang, D.J. Kouri and D.K. Hoffman, J. Chem. Phys. **101**, 10493 (1994).
[9] D.J. Kouri, Y. Huang, W. Zhu and D.K. Hoffman, J. Chem. Phys. **100**, 3662 (1994).
[10] G.C. Groenenboom and D.T. Colbert J. Chem. Phys. **99**, 9681 (1993)
[11] S.M. Auerbach and C. Leforestier, Comp. Phys. Comm. **78**, 55 (1993).
[12] V.A. Mandelshtam and H.S. Taylor, J. Chem. Phys. **102**, 7390 (1995)
[13] V.A. Mandelshtam and H.S. Taylor, J. Chem.Phys. **103**, 2903 (1995).
[14] U. Peskin, A. Edlund and W.H. Miller, J. Chem. Phys. (1995), in press.
[15] J. Antikainen, R. Friesner and C. Leforestier, J. Chem. Phys., **102**, 1270 (1995).
[16] J.C. Light, I.P. Hamilton and J.V. Lill, J. Chem. Phys. **82**, 1400 (1985).
[17] D.J. Kouri, Y. Sun, R.C. Mowrey, J.Z.H. Zhang, D.G. Truhlar, K. Haug and D.W. Schwenke, in *Mathematical Frontiers in Computational Chemical Physics*, edited by D.G. Truhlar (Springer-Verlag, New York, 1988), p. 207;
Y. Sun, D.J. Kouri, D.W. Schwenke and D.G. Truhlar, Comp. Phys. Comm. **63**, 51 (1991).
[18] R. Kosloff and D. Kosloff, J. Comput. Phys. **63**, 363 (1986).
[19] H. Tal-Ezer and R. Kosloff, J. Chem. Phys. **81**, 3967 (1984).

[20] B. Hartke, R. Kosloff and S. Ruhman, Chem. Phys. Lett. **158**, 238 (1989); R. Kosloff, J. Phys. Chem. **92**, 2087 (1988).

[21] M. Berman, R. Kosloff and H. Tal-Ezer, J. Phys. A **25**, 1283 (1992).

[22] R. Kosloff, Annu. Rev. Phys. Chem. **45**, 145 (1994).

[23] G. Ashkenazi, R. Kosloff and H. Tal-Ezer, J. Chem. Phys. (1995), submitted.

[24] R. Kosloff, private communication.

[25] M.R. Wall and D. Neuhauser, J. Chem. Phys. **102**, 8011 (1995).

[26] R. Baer and R. Kosloff, J. Chem. Phys. **99**, 2534 (1995).

[27] G. Katz, R. Baer, R. Kosloff, Chem. Phys. Lett. **239**, 230 (1995).

[28] U.V. Riss and H.-D. Meyer, J. Phys. B: At. Mol. Opt. Phys. **26**, 4503 (1993).

[29] D. Macias, S. Brouard and J.G. Muga, Chem. Phys. Lett. **228**, 672 (1994).

[30] T.P. Grozdanov, V.A. Mandelshtam and H.S. Taylor, J. Chem. Phys. **103**, 7990 (1995).

[31] V.A. Mandelshtam, T.P. Grozdanov and H.S. Taylor, J. Chem. Phys. **103**, 10074 (1995).

IMA SUMMER PROGRAMS

1987   Robotics
1988   Signal Processing
1989   Robustness, Diagnostics, Computing and Graphics in Statistics
1990   Radar and Sonar (June 18 - June 29)
        New Directions in Time Series Analysis (July 2 - July 27)
1991   Semiconductors
1992   Environmental Studies: Mathematical, Computational, and
           Statistical Analysis
1993   Modeling, Mesh Generation, and Adaptive Numerical Methods
           for Partial Differential Equations
1994   Molecular Biology
1995   Large Scale Optimizations with Applications to Inverse Problems,
           Optimal Control and Design, and Molecular and Structural
           Optimization
1996   Emerging Applications of Number Theory
1997   Statistics in Health Sciences.

SPRINGER LECTURE NOTES FROM THE IMA:

The Mathematics and Physics of Disordered Media
     Editors: Barry Hughes and Barry Ninham
     (Lecture Notes in Math., Volume 1035, 1983)

Orienting Polymers
     Editor: J.L. Ericksen
     (Lecture Notes in Math., Volume 1063, 1984)

New Perspectives in Thermodynamics
     Editor: James Serrin
     (Springer-Verlag, 1986)

Models of Economic Dynamics
     Editor: Hugo Sonnenschein
     (Lecture Notes in Econ., Volume 264, 1986)

The IMA Volumes in Mathematics and its Applications

Current Volumes: